Modern Microwave Measurements and Techniques

The Artech House Microwave Library

E-Plane Integrated Circuits, P. Bhartia and P. Pramanick, eds.

Analysis, Design, and Applications of Fin Lines, Bharathi Bhat and Shiban K. Koul

Microwaves Made Simple: Principles and Applications, Stephen W. Cheung, Frederick H. Levien, *et al.*

Microwave Integrated Circuits, Jeffrey Frey and Kul Bhasin, eds.

Microstrip Lines and Slotlines, K.C. Gupta, R. Garg, and I.J. Bahl

Handbook of Microwave Integrated Circuits, Reinmut K. Hoffmann

Stripline Circuit Design, Harlan Howe, Jr.

Microwave Transition Design, Jamal S. Izadian and Shahin M. Izadian

Microwave Mixers, Stephen A. Maas

Nonlinear Microwave Circuits, Stephen A. Maas

Microwave Filters, Impedance Matching Networks, and Coupling Structures, G. L. Matthaei, L. Young and E.M.T. Jones

GaAs MESFET Circuit Design, Robert A. Soares, ed.

Filters with Helical and Folded Helical Resonators, Peter Vizmuller

Microwave Engineer's Handbook: 2 volume set, Theodore Saad, ed.

Handbook of Microwave Testing, Thomas S. Laverghetta

Modern Spectrum Analyzer Theory and Applications, Morris Engelson

Microwave Tubes, A.S. Gilmour, Jr.

Lossy Transmission Lines, Fred E. Gardiol

Introduction to Microwaves, Fred E. Gardiol

Solid State Microwave Devices, Thomas S. Laverghetta

High Power Microwave Sources, Victor Granatstien and Igor Alexeff, eds.

Handbook for the Mechanical Tolerancing of Waveguide Components, W.B.W. Alison

Microwave Transmission Line Couplers, J.A.G. Malherbe

Microstrip Antenna Design, K.C. Gupta and A. Benella, eds.

Materials Handbook for Hybrid Microelectronics, J.A. King, ed.

Microwave Transmission Line Filters, J.A.G. Malherbe

Gallium Arsenide Processing Techniques, Ralph Williams

Modern Microwave Measurements and Techniques

Thomas S. Laverghetta

Artech House

Laverghetta, Thomas S.
Modern microwave measurements and techniques.
Rev. ed. of: Microwave measurements and techniques.
Bibliography: p.
Includes index.
1. Microwave Measurements. 2. Microwave devices.
I. Title.
TK7876.L38 1988 621.38′3 88-8096
ISBN 0-89006-307-9

ARTECH HOUSE, INC.
685 Canton Street
Norwood, MA 02062

International Standard Book Number: 0-89006-307-9
Library of Congress Catalog Card Number: 88-8096

10 9 8 7 6 5 4 3 2 1

Dedication

To my midshipman son Tom, whom I love very much and of whom I am very proud.

Contents

Preface

The first edition of *Microwave Measurements and Techniques* was published in 1976 and was intended to provide an understandable text on microwaves. It has accomplished its purpose in many ways for many people. It is, however, time to bring the book into the 1980s and 1990s time frame. This is why this second edition is now presented. It is still in an understandable format, but has been updated to include information needed by microwave personnel to remain current in the microwave industry.

Chapter 1 is an introduction that describes microwaves and distinguishes them from other electronic areas. Frequency, wavelength, grounds, and general precautions are also covered.

Chapter 2 investigates microwave materials. Metals, dielectrics, ferrites, and the semiconductor materials used in microwaves are covered in an understandable way.

Chapter 3 covers transmission lines. Basic theory is covered as an introduction, followed by coaxial lines (flexible and semirigid), stripline, microstrip, suspended substrate, and waveguide. As a conclusion to the chapter, microwave connectors and adapters are included to illustrate the unique characteristics and requirements needed for high frequency operation.

Chapter 4 introduces microwave components. These are the devices encountered every day. Each of the components is described in an understandable way so that an individual can know why, and how, each of them performs/operates in a microwave system.

Chapter 5 presents microwave semiconductor devices. The chapter is broken into three sections: junctions (*pn* and Schottky), diodes (Schottky, *pin*, varactor, tunnel Gunn, IMPATT, TRAPATT), and transistors (bipolar, GaAs FETs, and HEMT). Each section is covered in detail and examples of each are included.

Chapter 6 enhances the transmission line theory using the Smith chart. The chart is introduced and completely constructed. A significant number of examples are included to aid in gaining an understanding of the chart.

In Chapter 7, the theory presented to this point is tied together to create a variety of test setups. The classification of tests are power, transmission, impedance, frequency, and noise. Also present are general test procedures to be followed to enhance your particular test setup.

Chapter 8 presents specialized equipment for making microwave measurements. This equipment is the *network analyzer*. The equipment and its operation are covered.

Chapter 9 covers microwave tubes. The devices covered are Traveling Wave Tubes (TWTs), Klystrons, Magnetrons, and Crossed-Field Amplifiers. It emphasizes that tubes are still an integral part of many microwave systems.

This new edition is designed to maintain the level of presentation of the original book while updating the material covered. Other areas are also added, such as HEMT (High Electron Mobility Transistor) and semiconductor materials (GaAs and InP) which provide the most comprehensive and understandable material to aid microwave personnel now and into the 21st century.

THOMAS S. LAVERGHETTA
AUBURN, INDIANA

Acknowledgements

When the first edition of this book was published, there were many people to thank for making the book a reality. There were the people who did artwork, typing, and supplied material for the book. My thanks went out to them at that time, and many times over the years.

Now that the second edition is complete, there are other people to thank for a job well done this time: The Hewlett-Packard Company and Wiltron who have supplied material to be used in the updated network analyzer chapter; my students over the years who have made suggestions and have tried out the Smith chart problems I used; Ron Emery, my department chairman, who offers a world of encouragement; and the fabulous staff at Artech House who have to be the easiest people in the world to work with and undoubtedly the most professional. This staff is so easy to work with, we put the first edition of this book together by mail and over the phone.

Finally, there is that group of people who not only encourage me, but set me straight when I start taking myself too seriously—my family. I know without their love and support I could accomplish nothing—THANK YOU.

Chapter 1
Introduction

What are microwaves? You probably have heard that question asked a thousand times, with answers ranging from "black magic" to "something used in ovens." But really, what are microwaves? If you consult Noah Webster you find the word "microwave" defined as "a very short electromagnetic wave."

The key word in this definition is *short,* which implies length; it can be shown readily that a length (a wavelength) is indeed involved. From Appendix D we can see that the wavelength (λ) of an electromagnetic wave is equal to the velocity (v) of that wave divided by its frequency (f). The velocity through a medium (a substance such as air) is constant throughout that medium. Therefore, the wavelength of an electromagnetic wave is inversely proportional to its frequency; that is, the higher the frequency, the *shorter* the wavelength. So what are microwaves? In simple terms, they are electromagnetic waves of a very high frequency.

Exactly where the microwave spectrum begins is a matter of some discussion. There are those who say it begins at 300 MHz (λ = 100 cm in air) and those who maintain it starts at 1.0 GHz (λ = 30 cm in air). The microwave spectrum should begin where conventional components (lumped capacitors, resistors, and coils) no longer exhibit their proper parameters and where it becomes practical and advantageous to build components in stripline and microstrip, as well as in coaxial, form. That point is 500 MHz. There are many arguments on this matter, but at least a few agree since many microwave receivers have been built with the lowest band being called C-Band (500 to 1,000 MHz).

As stated, the origin of the microwave spectrum should be where conventional components cease to display normal parameters. But why would such a change occur? One key reason is a phenomenon known as the "skin effect." At higher frequencies the energy of a wave tends to use only the outside skin of a conductor for the transferring of energy because larger and larger lines of flux are set up around a conductor, increasing its inductance. By once again consulting Appendix D, we can see that the inductive reactance (X_L) increases when the frequency does, thereby offering a larger resistance to the energy. Thus the high frequency component searches for the path of lowest resistance—on the outer skin of the conductor.

To understand further how the lines of flux force the energy to the outside edge of the conductor, consider this analogy. Imagine you have a cylinder filled with cotton, representing the conductor through which the energy has to flow. Now place a small aluminum ball in the center of the cylinder—the energy that has to propagate through the conductor. As the frequency of the energy increases, the cylinder spins faster. If we spin the cylinder slowly, simulating a low frequency system, the ball remains in or very near the center. As the cylinder spins faster and faster, and we approach our theoretical start of the microwave spectrum, the ball moves to the outside edge of the cylinder. In our analogy, the reason the ball moves to the outside edge is an old friend called centrifugal force. In an actual electromagnetic system we have a microwave centrifugal force—the inductance of the system. Because of this phenomenon, the energy travels only on the outside edge of the conductors; lumped components thus become very unpredictable and impractical at microwave frequencies. It has been observed that some capacitors actually look inductive at high frequencies. The capacitance set up in coils and wire-wound resistors when used at microwave frequencies completely cancel out the intended functions of each of these components. Special materials, usually of a ceramic base, are used in the production of capacitors and resistors for use at the higher frequencies. Coils are usually made of high impedance lines cut to a specific length.

Care must be taken not only in choosing the type of components; the individual must also observe other precautions when working at these higher frequencies. One very important consideration is lead length. Many components used in microwaves have no leads at all; when they do, those leads should be as short as possible. Ribbons leads on components are far superior to radial ones simply because of the skin effect.

Grounds are very important in microwaves; loops can be set up without realizing it until overall system performance is degraded. A ground loop is just that, a loop set up in a circuit as a result of two or more grounds with a different potential between each pair. Thus it is necessary to examine the placement of grounds, making sure that no potential difference is created. Many times a cold or bad solder joint at a ground point causes a loop to be formed. It also should be emphasized that ground planes play a very important role in microwaves. Often the return for energy is not focused in a single point, as it is in lower frequency systems, but in a plane the return must be uninterrupted and of low resistance. Any discontinuity would upset the wave and hinder performance, a factor which is particularly significant in stripline and microstrip circuitry.

One last point is the need of exercising care in component placement. Many circuits in microwaves rely on spacing of wavelengths, or fractions thereof, for their proper operation. If a component is randomly placed on a circuit board, there is a good chance that it will not operate properly. Always check before placing components on a board to ensure correct orientation.

Before discussing the information needed for one to work effectively in microwaves, a couple of remarks are necessary. One is that a non-engineer, or non-microwave engineer, has no need to derive equations. What *is* needed is a set of equations into which can be plugged variables to obtain answers. The ability to obtain meaningful information through the use of basic formulas is essential. The second point is that this information should be recorded, in ink, in a lab book. A well-kept lab book is an asset to experimenting engineers and technicians, a company's patent attorney, and, therefore, the company itself. Most importantly, an engineer's design is only as good as its practical realization; without good laboratory support, it will never see the light of day.

Chapter 2
Microwave Materials

Virtually every chapter in this book deals with microwave materials, either directly or indirectly. We will notice many terms being divided by ε_r (dielectric constant); metals will be referred to frequently throughout the text; the ferrite willbe an integral part of the list of microwave components presented; and the modern day semiconductor materials are a vital part of this solid-state world of today. Thus, it can be said that materials, and their role in microwaves, must be a basic part of any complete microwave text.

The four basic microwave materials to be covered are *metals, dielectrics, ferrites,* and *semiconductors*. Each of these materials is related to its particular applications in microwaves, and includes its advantages and disadvantages. The explanations are thorough enough to present the whole picture without needing to include a long series of formulas. The idea is to present a concise explanation of microwave materials in understandable, down-to-earth language.

2.1 METALS

Metals play a very important part in microwaves. They form the ground plane needed for an uninterrupted loop; they are the enclosures that keep circuits from "talking to one another"; and they create the very boundaries for RF energy when used in waveguides. In these and other applications, the metal used determines just how well the overall device or system operates.

Aluminum, copper, silver, and *gold* are four primary types of metals dealt with in the following discussions. Others metals are mentioned, some composed of one of the above four. The discussions also present additional metals termed *miscellaneous metals*. These metals are *indium, tin,* and *lead*. Although they are not primary metals for microwaves, each has its place in the overall microwave picture and are more prominent as combinations in solder than in a pure metallic form.

The question that arises at this point is "What makes a metal suitable and acceptable for microwave usage?" Two properties which are of prime importance when considering a metal for microwave usage are *good electrical conductivity* and *machinability*.

Good electrical conductivity is important because the metal used in microwave circuits must carry high frequency currents with as little loss as possible. This metal may be used as a conductor for a circuit (which is part of the ground plane). Regardless of the application, low loss (good conductivity) is necessary. Values of conductivity for the metals previously listed are shown below.

Material	*Conductivity (mho/meter)*
Silver	6.30×10^7
Copper	5.85×10^7
Gold	4.25×10^7
Aluminum	3.50×10^7
Indium	1.11×10^7
Tin	$.877 \times 10^7$
Lead	$.456 \times 10^7$

The metals listed are in order of conduction. That is, silver is the best conductor and lead is the worst of this group. This is not to say that lead is a poor conductor, only that it is poor compared to the rest of the metals listed. This, however, is not a problem since it is not used for metallization on substrates or for cases which must be a part of a critical ground plane; rather, it is used in solder with various other metals. Its conductivity by itself, therefore, is not a critical item.

Machinability of metals is important since very few microwave systems will fit into standard square or rectangular spaces. Most have a cutout or an overhang or a curved portion needed to fit a rounded area. For this reason the metals must be sheared, drilled, milled, or generally shaped to conform to the necessary space. In order to be cost effective and efficient, the metals used in microwave systems must be easily machinable.

With the importance of metals in microwaves introduced, we can now investigate the most widely used metals in more detail. We previously mentioned seven metals that find wide usage in microwaves. These will be covered in the following order:

- Aluminum
- Copper
- Silver
- Gold
- Miscellaneous Metals
    - Indium
    - Tin
    - Lead

Before discussing each individual metal, it might be interesting to look at a periodic table of the elements and see where each of these metals fall. Figure 2.1 shows a table with the seven metals we will cover. You can see that aluminum is

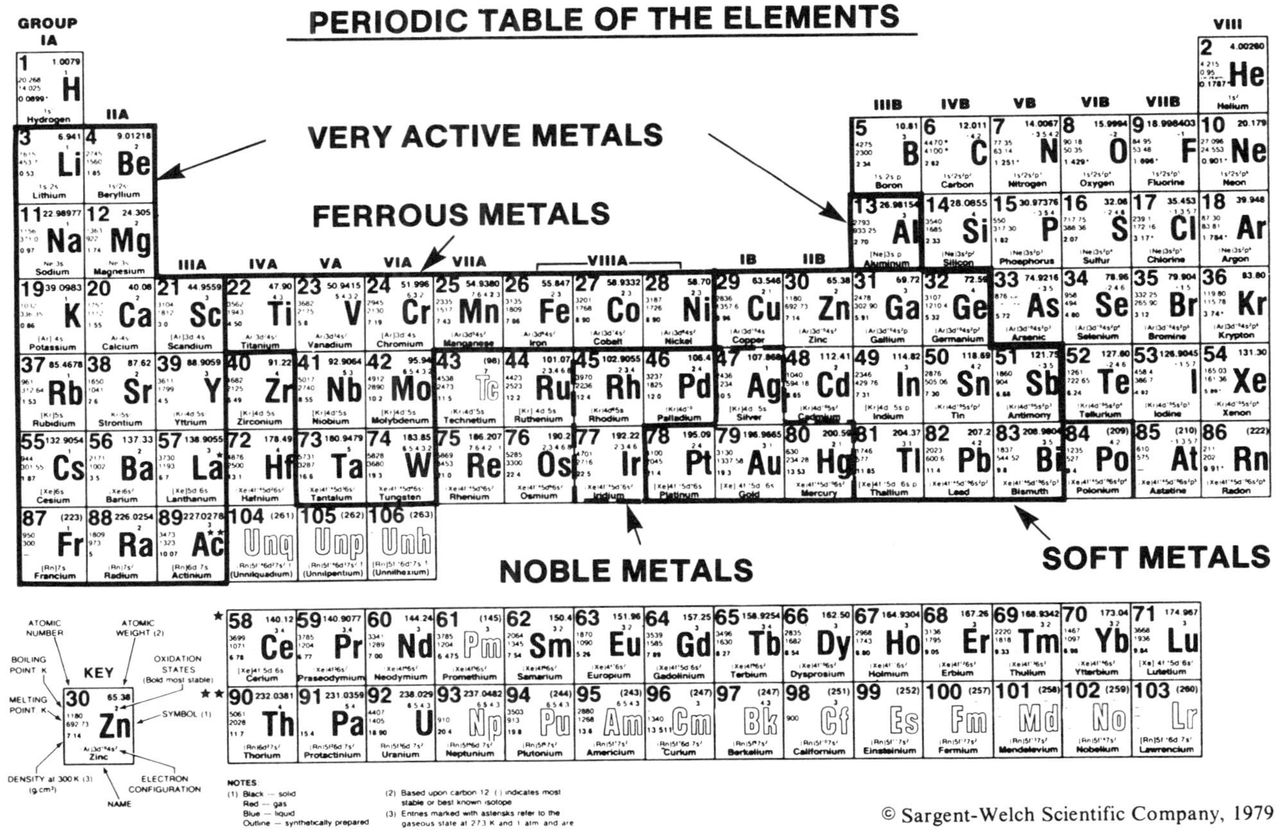

Fig. 2.1 Metal classifications
(Courtesy of Sargent-Welch Scientific Company).

by itself in the category classified as *very active metals*. The remaining six metals are in the category of *soft metals*. The metals silver (Ag) and gold (Au) are also listed in a separate group called the *noble metals*.

The area termed *ferrous metals* does not contain any metals which may be used in microwaves. (Although chromium (Cr) may be used as an adhesive, and nickel (Ni) as a plating material, these generally do not have the machinability or conductance properties of the soft metals.)

With the metals listed and their locations shown on the periodic table, it is now time to discuss each individual metal in the order listed previously.

2.1.1 Aluminum

Aluminum is usually first on the list of metals used in microwave applications: the majority of cases for microwave components are fabricated from aluminum. You see gold-plated aluminum cases, iridited aluminum cases, anodized aluminum cases, and plain painted aluminum cases. Many "caseless" components in microwaves have aluminum ground plane plates which enable the component to operate properly. The list goes on-and-on, but the message still comes through as to the importance of aluminum in microwaves. With a metal this important, it might be interesting to examine a little of its history.

In 1808 the existence of aluminum was predicted by the English scientist Sir Humphrey Davy. At that time it was considered to be the most abundant metal on the earth's crust but it was very difficult to extract from the earth. The first aluminum metal was produced in 1825 by the Danish scientist H.C. Oersted. At that time it sold for $160 a pound. This may be rather difficult to grasp when you think of how inexpensive and widely used aluminum is today.

In 1854, Deville, a French chemist, helped the economic situation of aluminum somewhat by using sodium as a reducing agent for the metal. This brought the price down to a mere $100 a pound. At this price, only a very few could afford anything made from this special metal. In the 1860s, Emperor Napolean III of France had his finest dinner spoons made of aluminum to impress his guests.

Charles M. Hall found a way to inexpensively refine the metal in 1886 which reduced the price to $8 a pound. Hall, who was a student at Oberlin College, filed his patent on the process just ahead of Paul Herould of France, who had simultaneously and independently developed the same process. Hall's process was bought in 1888, by the Pittsburgh Reduction Company, the first company to produce aluminum. The process the company purchased in 1888 is the same basic process used today in the current two-step process of aluminum refining.

Until World War II, the Aluminum Company of America (ALCOA), the successor to the Pittsburgh Reduction Company, was the only producer of aluminum in the United States. During the war, our aluminum production capacity was expanded greatly from the 1939 figure of about 164,000 tons. Although the majority of the plants built to handle this production were built and operated by ALCOA, their knowledge was made readily available to Reynolds Aluminum, the other prominent company.

The first cutback in aluminum production was made in December 1943, and by the end of 1944 the industry was operating at about one-half capacity. These cutbacks were needed simply because the supply of aluminum far exceeded the war needs. The huge industrial expansions that followed the war, however, more than made up for cutbacks made during the war. The price of aluminum has decreased significantly, even compared to the wartime prices. Following the war, in 1947, the price dipped to a record low of 15 cents per pound. The price, of course, did not hold at this level. Its present price of \$2 to \$3 a pound has been fairly constant for a number of years.

The only metal which has higher production quantities in the United States is steel. This is because of the wide application range and versatility of aluminum. Aluminum is a soft metal, although it is not listed as one in the periodic chart shown in Figure 2.1. You will recall that this group was called "very active metals." The metal is actually a soft metal which, almost contradictory to logic, becomes so brittle at high temperatures that it can easily be powdered. The melting point of the metal is 660°C (1220°F) and will boil at 2,057°C (3,734.6°F).

Aluminum is not the best conductor, but it has a very acceptable level of conductivity (3.50×10^7 mho/m as opposed to 6.3×10^7 mho/m for silver). Its softness enables it to be rolled into forms, cast, drawn, or stamped. It can, in its finished form, be milled, sheared, punched, or drilled. In short, it is a very machinable metal.

A general description of aluminum is given below:

Symbol	Al
Atomic weight (Referenced to carbon 12)	26.98
Specific gravity	2.7 (a little more than two and one-half times as heavy as water)
Color	Silvery-white with a bluish tinge
Properties	Soft, easily shaped, resists corrosion, nonmagnetic, good conductor of heat and electricity
Chief ore	Bauxite (named after the town of LesBaux in Southern France where it was first mined).

This is aluminum: a soft and machinable metal with an average conductivity that finds applications in microwaves as cases and ground plates. The important/main property which accounts for its wide usage is its light weight.

2.1.2 Copper

If the first metal you think of for use in microwaves is aluminum, the second would surely be copper. This is not a new metal. It was probably the first metal ever used by man. Objects from Egypt of hammered copper 8,000 years old have been found as have specimens of cast copper from Egypt and Babylonia that date from 4,000 B.C. The metal was given its name by the ancient Romans who called it *aes cyprum* (metal of Cyprus). The island of Cyprus was the chief source of copper for the Romans. Later, they called it *cuprum*.

When the first Europeans came to the new world, they found the Indians using copper for jewelry and decorations. The copper these Indians used was the from the Lake Superior area, deposited there by the action of the Ice Age . . . native copper broke off from exposed lodes in the Great Lakes region and scattered southward over an area of over 70,000 square miles.

The mining of copper and the development of the industry began in the 17th century. In 1664, the first copper mine was established in the United States in Lynn, Massachusetts. Copper pins were then produced from the native copper in 1666. In 1780 Paul Revere established a foundry after discovering the secret process used by the English by which copper could be made malleable enough to be hammered when it was hot. The government loaned him $10,000 to buy a site and establish a foundry in Canton, Massachusetts. In 1846 the first successful copper mine opened in the area of Lake Superior, Michigan.

Copper is one of the metals in the category of coinage metals because it is used extensively in the manufacture of coins. There is a very high percentage of copper in coins today, much higher than in the past. We may also call it a coinage metal because it costs so many coins to buy it. Its price has risen sharply due to its scarcity, but this factor should not influence anyone to change their copper applications to another metal. The world is in no danger of running out of copper. Also, many applications do not require a pure 100 percent copper. Such alloys as brass or bronze do an acceptable job in many cases.

Copper fulfills our two initial requirements for a metal to be used in microwaves. First, its conduction is excellent. Its conductivity (5.85×10^7 mho/m) is second only to silver (6.30×10^7 mho/m). Secondly, copper can be drilled, tapped, sheared, sawed, and milled. One thing to remember when machining copper, however, is that special lubricants may be necessary since copper is harder than metals such as aluminum.

Copper is also one of the easiest metals to solder. It can also be heliarced, welded, or brazed. This ease of connection makes it ideal for low-loss microwave circuits.

Copper, although it is not as widely used today, (mainly because of its cost), still has many applications in microwaves. If you looked in a waveguide catalog you would find extensive use being made of alloys of aluminum, magnesium, brass, and silver. These all make excellent conductors, are light-weight, and cost less than copper. The laminate market, however, still uses 100 percent copper because it has found nothing to match copper's excellent conductive properties for the price.

A general description of copper is given below:

Symbol	Cu
Atomic weight (compared to carbon 12)	63.54
Specific gravity	8.9 (nearly nine times as heavy as water)
Color	Reddish brown (pink when freshly made)
Properties	Soft and easily shaped, becoming hard when cold-worked; good conductor of electricity and heat; resists corrosion by atmosphere and sea water
Chief ore	Chalcopyrite (a compound of copper, iron, and sulfur)

The importance of copper to the microwave laminate and substrate market is immeasurable, as you will see from our discussions. Its versatility and properties should make it an important part of the microwave industry for many years to come.

2.1.3 Silver

A second coinage metal is silver. This metal is probably more recognizable than copper because most coins show the silver. Copper is usually only visible on the edges after extended usage wears the silver plating away.

Use of silver dates back to the earliest of times. Ornaments of silver found in the Near East date back to about 3,500 B.C. The Book of Genesis mentions silver as part of Abraham's wealth. The first recorded mining of silver in Europe, however, was no earlier than 500 or 600 B.C.

Despite the demand for silver throughout the world, very little was produced during the Middle Ages because Europe's supply of silver was limited and its mining techniques were not highly developed. This shortage diminished when, in the 1500s, silver mines were discovered in Mexico, Peru, and Bolivia. Silver was not discovered in the United States until the 1700s.

During the Civil War, silver dollars and many other silver coins disappeared from circulation because the demands of industry for the metal caused it to be more valuable than the worth of silver coins. After the war, in 1865, large silver discoveries in the Rocky Mountains decreased the price of silver.

Another silver shortage occurred in the early 1900s when large amounts were sent to India to help avoid a collapse of its currency system. A silver-purchase act passed in 1934 directed the Treasury to purchase silver, resulting in a large surplus.

This surplus did not last, however, because ever-increasing uses were found for the metal. In July 1965, the US Government changed the metallic content of dimes, quarters, and half-dollars. The result is the silver coin with the copper edges you see today.

The use of silver for microwave applications is almost entirely in plating. Waveeguide and tuned cavities are made of a light weight and inexpensive material (aluminum, *etc.*) and plated with a thin coating of silver to increase conduction of microwave energy. The coating is very thin since the microwave energy travels only on the "skin" of the waveguide or cavity due the phenomenon known as the "skin effect." It is much more economical and practical, therefore, to machine the cavity or fabricate the waveguide and plate it with the skin-depth thickness of silver. The price of silver is not quoted, in this text because it changes from day to day. If you want to know the going rate for the metal you will have to check your local business report for the current price.

A general description of silver is given below:

Symbol	Ag (from the Latin *argentium*)
Atomic weight (compared to carbon 12)	107.88
Specific gravity	10.49 (10 and one-half times as heavy as water)
Color	White
Properties	Soft and easily shaped; the best conductor of heat and electricity; resists corrosion by the atmosphere

2.1.4 Gold

Most of the statements made in the previous section on silver could be repeated for gold, starting by saying its use dates back many thousands of years. The use of gold predates modern civilization and the many ornaments found in Neolithic remains verify this statement.

A major use of gold in microwaves is for plating. Stripline and microstrip circuits are plated. Flanges and threads of microwave components are gold plated, as is the entire case or substrate carrier plate, to increase conduction and prevent corrosion of any base metals used.

Gold is an excellent metal for use in microwaves but is being replaced—and in many areas, has already been replaced—because of its price. For many years the price of gold was a nearly fixed figure and any industry which had applications for it could plan on a price and be assured of its accuracy. In recent years, however, the price of gold has varied from day to day (sometimes hour to hour) and has jumped in price to a point where it is impractical in many cases to use it. To save money many circuits are now using tin/lead plating, and connectors now use a stainless steel finish. These variations in plating have kept microwave components within a reasonable price range.

There are times when a gold plating is really the best metal to use for a particular application. When this is the case, the designer usually makes the case in sections and gold-plates only the base plate, for example, which will act as the ground plane for the circuit. The remainder of the case is iridited and the circuit usually operates very satisfactorily.

A general description of gold is given below:

Symbol	Au (Latin *aurium*)
Atomic weight (compared to carbon 12)	196.96
Specific gravity	19.32 (nearly 20 times the weight of water)
Color	Golden yellow when pure; impurities cause various shades of yellow
Properties	Very soft and easily shaped; extremely resistant to corrosion; excellent conductor of heat and electricity; reacts to very few chemicals

2.1.5 Miscellaneous Metals

Under this category of miscellaneous metals we will cover three metals that are in the "soft metal" group and whose primary use is in solders. The metals are indium, tin, and lead. Figure 2.2 shows the location of the three metals on the periodic table. You can see that they are all in a group and thus exhibit much the same properties. Each, however, has its unique features.

2.1.5.1 Indium

When you consider some of the metals we have discussed thus far, indium is a relative infant. It was discovered in 1863 by two German scientists, Reich and Richter. Indium is widely distributed in nature but occurs in commercially practical concentrations only in the ores of zinc, iron, lead, and copper. Because of this it did not have many applications and was considered a laboratory element. It was not until 1824 that Dr. William S. Murray decided to take indium out of the laboratory and put it to work commercially. In 1926, Dr. Murray received a patent for his indium electroplating process. In 1934, he founded the Indium Corporation of America in Utica, New York, the world's first commercial producer of indium metal.

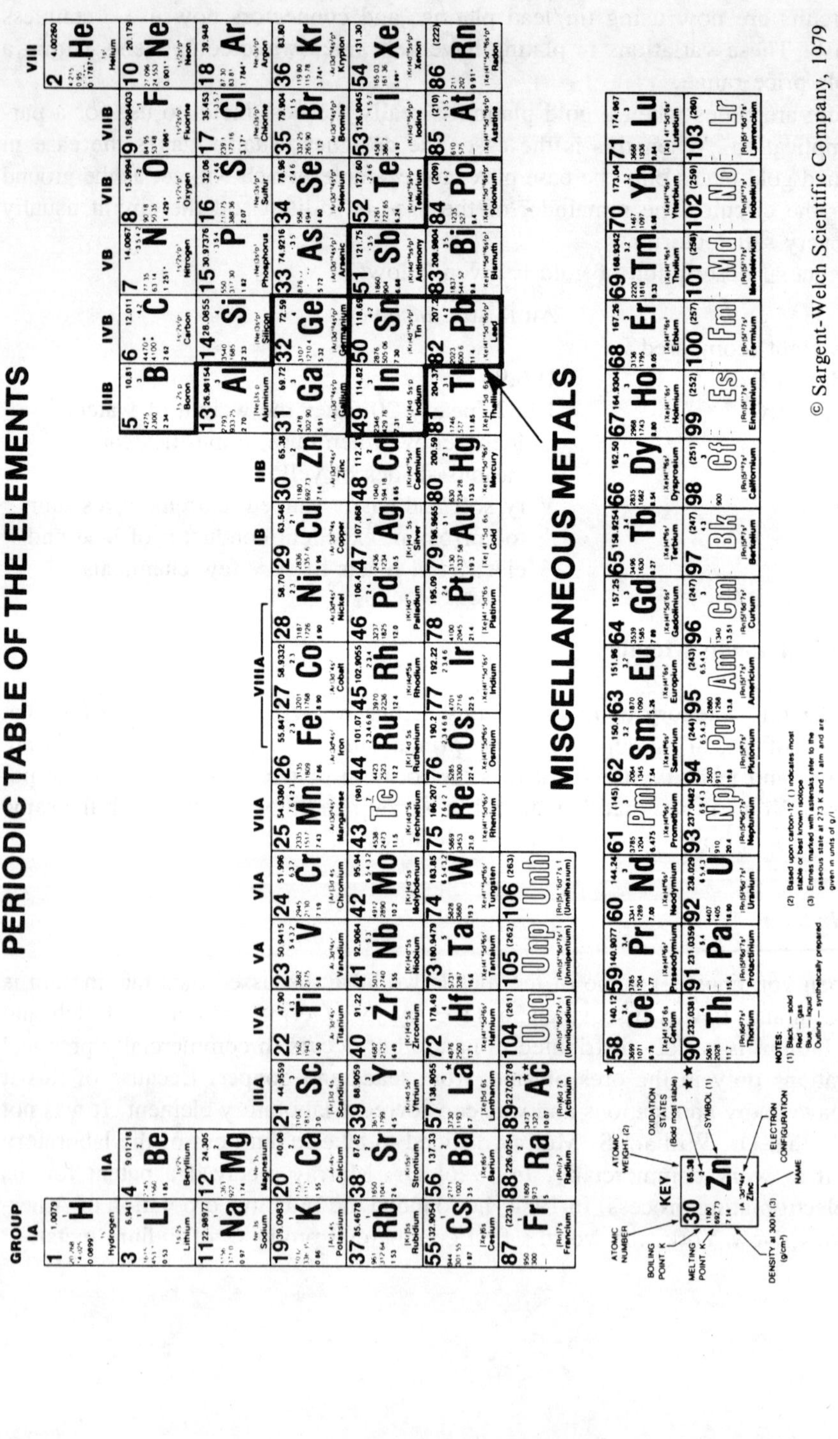

Fig. 2.2 Soft metals
(Courtesy of Sargent-Welch Scientific Company).

Indium has many properties that can be of great value for microwave application and in aiding the properties of other metals which could possibly be used in microwaves. Indium, for example, is so soft that you can scratch it with your fingernail. It will maintain this softness even at low temperatures. This is an excellent property if you are operating at low temperature and do not want your material to become brittle at these low temperatures.

Indium has no *memory*. That is, if you shape it one way, it will not return to its original shape but will stay in its new shape. One area where this property has found application is when using pill terminations in stripline circuits. The case in which the stripline is placed will have tolerances on its milled portions. For this reason, some of the cases could put extra pressure on the termination while others may just barely make ground contact. If you make the depression for the pill intentionally deep and put indium in it, the indium will fill the void, conform to the pill shape, make good electrical contact, and remain in that shape, even if the pill is removed from the case. There are other similar applications where indium could perform as well. Indium can be added to other metals to make them:

- Harder
- More fatigue reistant
- More ductile (hammered thin without breaking)
- Easier to melt
- Have higher heat conduction
- Have higher electrical conduction
- Easier to bond

These properties make indium a very useful metal. The improvements mentioned above when added to other metals make indium an excellent choice for use as a solder for microwave circuits. A primary area where indium solders are used is when soldering to gold or silver. An indium/lead (In/Pb) solder is an excellent choice over the conventional tin/lead (Sn/Pb) solder since tin alloys soldered against either gold or silver tend to scavenge the gold or silver. (Scavenging is the attack of molten solder on a metal surface, with the solder actually dissolving the base). The choice of using indium-based solder would be an excellent one if, for example, you were using alumina substrate with chrome/gold metallization. This would ensure that the solder joints would be solid, reliable, and exhibit high electrical conduction (something tin/lead solder probably would not do).

With all of the properties listed above, indium, and indium solders, merit strong consideration for use in your microwave systems. Be aware, however, that the metal also exhibits some properties that may be less than favorable. Research indicates that indium solder may develop long-term problems if all intermetallic interfaces are not considered. These problems seem more evident in space applications. With this in mind, be sure to investigate all of the conditions, both short-term and long-term, before choosing any solder or metal to be used for your application.

A general description of indium is given below:

Symbol	In
Atomic weight (compared to carbon 12)	114.82
Specific gravity	7.3
Color	Silvery white with brilliant metallic luster
Properties	Very soft material; stays soft at low temperatures; retains shape when shaped; good conductor of heat and electricity.

2.1.5.2 Tin

Most of the time tin is either a tin/lead solder or a tin/lead plating. About the only time you hear of tin used alone is in tin cans, and even these are not pure tin, but a thin tin plating on an inexpensive metal. Today many tin cans are not tin at all, and in the not-too-distant future, any reference to a pure tin product may be a thing of the past.

Tin is another of the metals that have been around for ages. Tin articles were known to exist at least as early as 1,400 B.C. in Egypt. Homer called tin *kassiteros*. because he said it came from Cassiterides in the Atlantic. (The Cassiterides, or Tin Islands, were the British Isles). The metal later got the same *stannum* which resulted in its present symbol, Sn.

The only important ore of tin is cassiterite, or tinston, SnO_2, which is found in veins and streams, called alluvial deposits, mainly in Southeast Asia. (Bolivia and Africa also mine a certain amount of the metal.) Although practically no tin is produced in the United States, the US ranks first in users of this silvery-white metal.

As previously stated, tin is primarily used in microwaves in solder and for plating applications. The metal is excellent for both of these tasks. Remember, however, the precaution in the previous section about using tin/lead solder on gold.

A general description of tin is given below:

Symbol	Sn
Atomic weight (compared to carbon 12)	118.7
Specific gravity	7.29 (about seven and one-half times as heavy as water)
Color	Silvery white with a bluish tinge
Properties	Soft and malleable; melts at a little over twice the boiling point of water; at very low temperature crumbles into a gray powder; conducts heat and electricity reasonably well.

2.1.5.3 Lead

Lead, like tin, is one of the metals that do not appear alone in microwave applications. You will recall how we referred to tin/lead solder and tin/lead plating. Similarly, indium/lead (In/Pb) solder is an excellent substitute solder when gold is involved. Regardless of the combination, lead always is used with another element.

Use of lead also has its origins back many centuries. A small statue in the British Museum dates back to the year 3,400 B.C. Lead is also mentioned in the bible. The ancient Greeks rarely used lead, but the Romans used large quantities of what they called *plumbum,* mainly for water pipes. Actually it is the word *plumbum* that resulted in our word *plumbing,* and provided the origin of the symbol *Pb.*

The microwave applications, as previously mentioned, are in solders and in plating. These uses are somewhat limited but provide excellent materials for joining or protective plating microwave materials and components.

A general description of lead is given below:

Symbol	Pb (from the Latin *plumbum*)
Atomic weight (compared to carbon 12)	207.21
Specific gravity	11.3 (more than 11 times as heavy as water)
Color	Bluish-gray
Properties	Soft and easily shaped; resists corrosion by sea water, air, and many chemicals; fairly low melting point; the worst conductor of electricity compared to all the metals discussed (although not a bad conductor)

2.1.5.4 Kovar and Invar

Two other "metals" should be mentioned here since they also find applications in microwaves where thermal considerations are important. These are kovar and invar, which are both alloys, (as opposed to pure metals), and will be listed here to be used as references later.

- *Kovar*: An iron/nickel/cobalt (Fe/Ni/Co) alloy with a coefficient of expansion similar to that of glass and silicon. Its thermal characteristics are similar to those of alumina substrates. It is used as a material for the mounting of alumina substrate to aluminum cases to compensate for differences in expansion, for headers, and in any glass-to-metal seals. (Notice that the three metals contained in this alloy are all in the ferrous metal group of the chart shown in Fig. 2.1). This metal can be difficult to machine.

- *Invar*: An alloy containing 63.8 percent iron, 36 percent nickel, and 0.2 percent carbon. It has a very low thermal coefficient of expansion and is used where thermal consideratitons are of prime importance and a minimum of machining is necessary. This material also can be difficult to machine.

2.2 DIELECTRICS

If you threw a baseball down an air-filled chamber to a target at the other end, the ball's travel would be fairly unhindered. If the chamber were filled with loosely packed feathers, the ball would be slowed somewhat; but if the chamber were filled with water, the ball would be hindered considerably, possibly not reaching the target at all. The velocity of the baseball depends upon the makeup of the medium through which it has to travel. This condition is also true of microwave energy. The velocity of this energy in air is the velocity of light. When something more dense than air is inserted, the velocity is decreased by a factor $\sqrt{\varepsilon}$—the dielectric constant due to the construction of the particular medium (or dielectric). Though this term has been mentioned, it has never been examined. Now, the questions are posed—"What is dielectric?" and "What is a dielectric constant?"

To help answer these questions, we relate the dielectric and its properties to a common example. Suppose that you have to drive from Chicago to San Francisco in less than one week. Before you start out, you would probably check to find out what sort of weather is expected along the way. If a heavy snow storm were due in Colorado, you would probably swing farther south in an attempt to avoid it. You would also check your route to find out what gas stations were along the way and about how far apart they were. Other items to be considered would be major road construction, restaurants, and motels. In effect, you would check the conditions to be encountered while traveling through your specified route (Chicago to San Francisco, in this case). The same precautions are necessary for the travel of microwave energy. These conditions (contained in the dielectric) must be known before an efficient system can be put into operation. Just as the planned route affects your trip, the dielectric determines the path of RF energy through its particular medium. With this information, a definition of *dielectric* is now possible.

A *dielectric* is a material that, due to its internal construction, alters the characteristics of an RF wave as it passes through the material. That is, the wave encounters different conditions along the way that vary its characteristics from what they are in free space.

To be completely realistic, the dielectric could be called an obstruction to the RF wave. It doesn't stop the wave, but slows it down by an amount determined by the material and construction of the dielectric.

It should not be too difficult to determine what is a dielectric constant. Unlike the proposed auto trip, the conditions in the dielectric do not vary drastically over the area of operation. The trip may include many different weather conditions, road conditions, gas station locations, and assorted other variables; but the dielectric material is of fairly consistent construction. In other words, the dielectric contruction, and thus its effect, is *constant*. The term *dielectric constant* indicates how close the material comes to the free space conditions and how constant these conditions hold through the material. For example, given a dielectric constant of 2.55, the characteristics of the wave would be changed by 1/2.55 or 0.63. That is, the velocity through the material is 63 percent of that of free space; the wavelength is 63 percent of that of free space; and the time delay is also 63 percent of the delay in free space. So, as mentioned before, the *dielectric constant* is a comparison of the material with free space.

When we hear the word *dissipation* we usually think of something being lost. When you think of heat dissipation, for example, you can visualize heat pouring from a radiator or furnace. When you hear on the news that the early morning fog will dissipate by 10 A.M., you picture it as being gone. The term *dissipation* in microwaves has a similar meaning. As we have previously said, dissipation is a loss. In microwaves, dissipation is the loss of energy in the form of heat. Thus, in the case of microwave laminates and substrates, we want the material to *store* energy within the structure rather than *dissipate* it. This storing of energy will result in a low loss path for the microwave energy. (When we speak of storing energy in the laminate or substrate, we mean that the energy remains in the circuit rather than escaping as heat and causing the material to exhibit a high loss).

The term *dissipation* is not generally the term you use when characterizing microwave laminates or substrates. *Dissipation factor* is the more recognized phrase which appears on all microwave material data sheets and it defines the loss characteristics of the material. This phrase (termed Tan δ) is actually a ratio. It is the ratio of energy dissipated to energy stored.

$$\tan \delta = \frac{\textit{Energy Dissipated}}{\textit{Energy Stored}}$$

(tan δ is also termed the tangent of the loss angle of the material. The loss angle is determined by the relationship between loss term and the conductivity term of a material).

If we consider the energy dissipated as being a loss, or resistance, and energy stored to be the measure of conductivity of the material, we have a relationship where:

$$\textit{Dissipation factor} = \frac{\textit{Loss (Resistance)}}{\textit{Conductivity}}$$

It should be obvious from our discussions that to have an efficient laminate or substrate you need a low value of dissipation factor. From the relationship above you can see that this is possible when the resistance loss is low and/or the conductivity is high. Some representative dissipation factors are shown below.

Material	*Dissipation Factor (Tan δ)*
Glass epoxy	.0180
PTFE* glass	.0018
Alumina	<.0001
Polyethylene	.0002
Mica	.0026
Silicon rubber	.0032
Ethyl alcohol	.0620

*PTFE = Polytetrafluorthylene

The dielectric thickness of a microwave laminate or substrate is of prime importance. You must be aware of exactly what thickness we are referring to in order to appreciate the importance of the dielectric thickness. When you look at a microwave laminate or substrate, you will see a piece of material with copper on both sides and some dark brown or white material in the center; gold completely surrounding a base material; or a white base material with gold on two sides. These are shown in Figure 2.3. Figure 2.3(a) shows the laminate or substrate with copper (or gold) on both sides but not on the edges. Figure 2.3(b) is a representation of a substrate with gold metallization on all six surfaces. (Actually, there is more than just gold on the substrate but this will be discussed in detail later in this chapter). From these drawings you could arrive at a thickness, but it would be the wrong thickness. The dimension we want is shown in Figure 2.4. You can see that it is the dimension of the dielectric *only,* not the total laminate or substrate. Any cladding or metallization dimension must be subtracted from the total dimension. Thus, when the data sheet says that laminates are available in .031 inch and .062 inch sizes, they are referring to only the dielectric material dimension with no cladding or metallization. This is the *unclad laminate dimension*. Be sure that when checking the dimension of a laminate or substrate you do not confuse the *total* dimension with the *dielectric thickness*. This confusion can cause serious problems with your calculations and thus your final results.

We mentioned *cladding* above when we talked about the dielectric thickness of a microwave laminate or substrate. We will now discuss this term and also expand on it by covering *metallization*. Generally speaking, when the term *cladding* is used it refers to microwave laminates with a copper cladding on one or both sides. *Metallization* is used when we refer to the covering on ceramic (alumina), sapphire, or quartz substrates. We will discuss laminate cladding first.

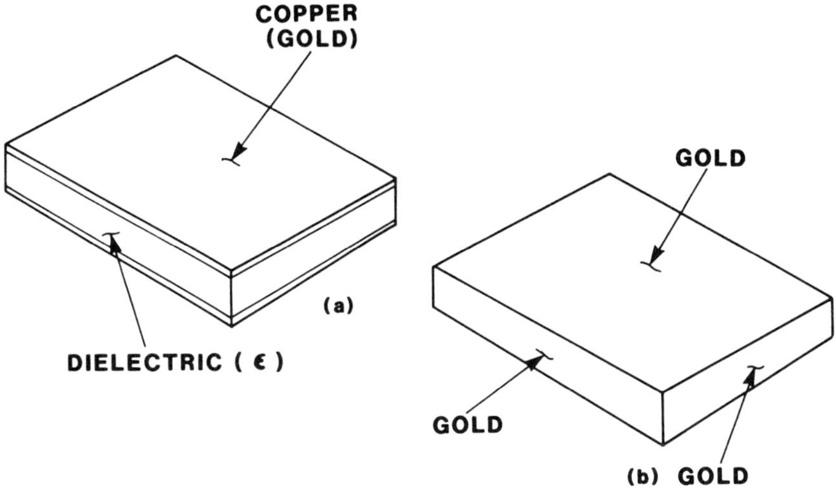

Fig. 2.3 Microwave laminates and substrates.

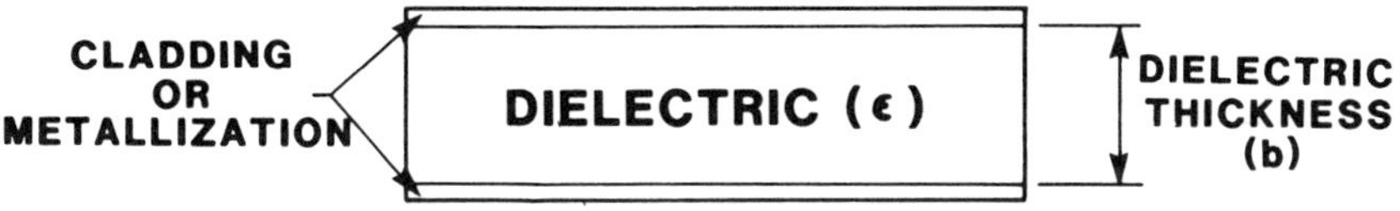

Fig. 2.4 Dielectric thickness dimension.

When you talk about cladding you must give two very important pieces of information with the term: the weight of the metal (copper), and how it is fabricated for the laminate (electrodeposited or rolled). The *weight* of the copper as specified on a laminate data sheet usually has standard values of 1/2 oz, 1 oz, and 2 oz. These numbers refer to the fact that one square foot of copper will weigh 1/2 oz, 1 oz, or 2 oz, depending on what weight you choose. Each of these weights, logically, has a definite thickness associated with it. If we have three different weights of copper, all one square foot in size, there must be a difference in their thickness to cause the difference in weights. The relationship between weight and thickness is shown below.

Weight (ounces)	*Thickness (inches)*
0.5	.0007
1.0	.0014
2.0	.0028

A 1/4 oz copper is also available from most manufacturers. This, however, is usually a special order. The 1/4 oz may be needed for very narrow lines or when narrow gaps between lines are required. Generally, however, the 1/2 oz copper will suffice and is a standard cladding. You can understand this if you refer to Figure 2.5. In Case I, we have a laminate with 2 oz copper at the left and either 1/2- or 1/4 oz copper at the right. (Since we have said that generally the 1/2 oz copper is used because it is a standard cladding as opposed to the special order 1/4 oz, we will make all of our comparisons with 2-oz and 1/2-oz from this point forward.) In this case we are required to etch a very narrow gap (<0.005 inch) between two transmission lines. You can see that the lines etched on the 2-oz copper are dramatically undercut. That is, they are narrower on the underside of the line than at the top. This is because of the time required to etch the thicker copper. The etchant does not etch the top portion because there is photoresist on that surface. The lower portion is thus subjected to the etchant for a long duration of time and over etches. The 1/2-oz copper portion is virtually undisturbed, as can be seen in the figure, because of the smaller volume of copper that needs to be etched to arrive at two transmission lines separated by a narrow gap.

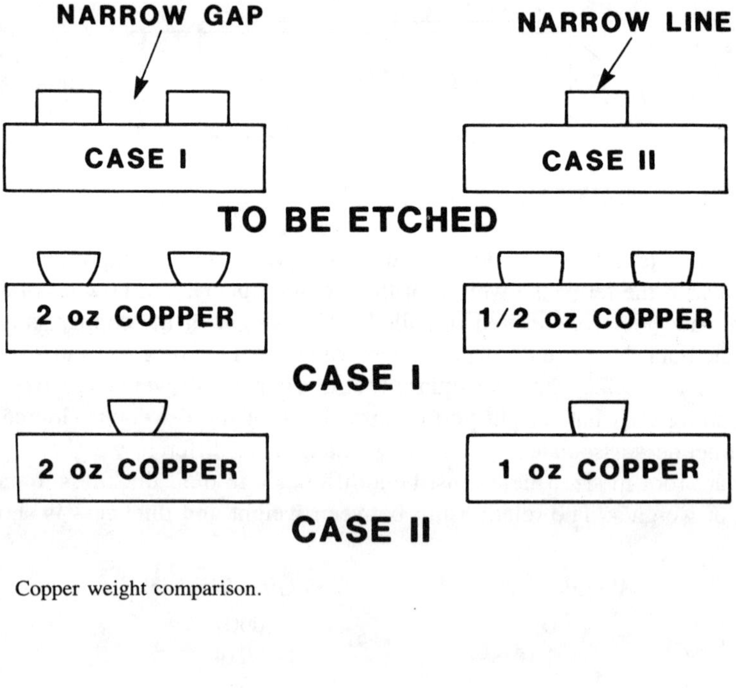

Fig. 2.5 Copper weight comparison.

In Case II we have only a single line to etch which is narrow (<0.010 in). As can be seen in the figure, the same conditions occur when 2-oz copper is used. Once again there is an undercutting which causes variations in dimensions of the line. You can see also that the 1/2-oz copper is virtually undisturbed when used for narrow lines. One condition must be watched in this case and in Case I discussed above: the severity of the undercutting. If the circuit is etched for too long a period of time, the undercutting will be so severe that the line lifts completely away from the laminate. This is shown in Figure 2.6. Care should be taken when etching narrow lines, whether on thick or thin copper.

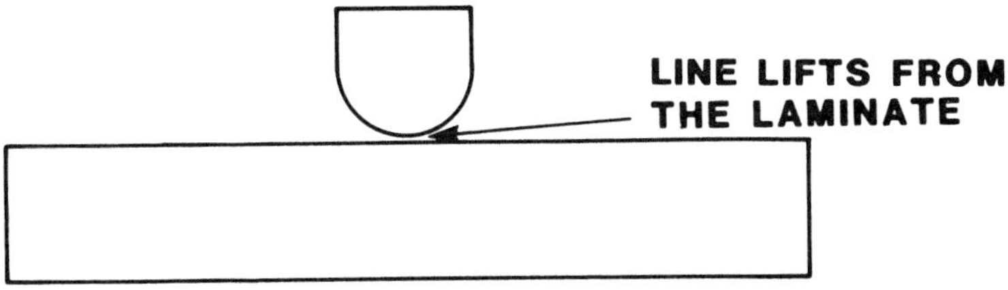

Fig. 2.6 Overetching of lines.

As a summary of our copper weights discussions for microwave laminates, consider the following advantages and disadvantages:

2-oz copper

- Good copper weight for high-power applications;
- Copper will undercut on narrow gaps and lines;
- Narrow lines will be inconsistent along a long length;
- Good for general purpose circuity;
- A standard weight.

1-oz. copper

- A very common weight for a variety of microstrip and stripline circuits;
- Less undercut on narrow gaps and lines than 2-oz;
- Can be used for medium power applications;
- A standard weight.

1/2-oz copper

- Used for low power applications;
- Excellent for narrow gap and line etching;
- Not used for high power circuits;
- A standard weight.

1/4-oz copper

- Should only be used for low power circuits;
- Excellent for narrow gaps and lines;
- A non-standard weight—usually available in special order only.

Now that we have looked at the different weights of copper, let us see how the copper is fabricated. Previously we mentioned two methods of copper fabrication—rolled and electrodeposited. The standard method used for most microwave laminates is electrodeposited (ED) copper. Rolled copper is available by special request and the data sheet usually has a statement saying the laminate can be supplied with rolled copper "for more critical electrical applications." We will cover the standard ED copper first and then the rolled.

If you were to look up the word *electrodeposition* in the dictionary, either a conventional or electronic dictionary, you would find the typical dictionary definition. It would say that electrodeposition is the process of depositing a substance on an electrode by electrolysis. Unless you are a chemist, this probably tells you no more than you knew when you first looked up the word. To clear the water a little you could look up *electrolysis*. At this point you would find that the dictionary says that electrolysis is the process of changing the chemical composition of a material by sending an electric current through it. If you put these two definitions together and do some thinking, you will get some idea of what ED copper is. Here is a basic definition for the term:

> *Electrodeposited copper* is a material produced by a chemical building process in which the individual copper particles are electrically joined to form the desired sheet thickness.

This can be likened to taking small pieces of wet clay and building one continuous sheet from the small pieces. If you had a certain number of these pieces of clay placed next to one another over some base surface, (as ED copper is plated on a rotating drum structure and then pulled off), you would have a structure that was a certain thickness and consistency. This might be likened, for example, to plating (or depositing) 1/2-oz copper (0.0007 inches). If you needed 1-oz copper (0.0014 inches) you would repeat the process until the desired 1.4 mils of thickness was achieved. Similarly, you would have to add more wet clay to your already existing structure in order to achieve the required thickness. This depositing process can be very precisely controlled by controlling the time and current used in the electrodeposition process.

As previously mentioned, ED copper is the standard copper found on microwave laminates. Rolled copper is obtained by special request to the manufacturers and is used for more critical applications. You will see why this statement is true when we complete our discussion of rolled copper and compare it with standard ED copper.

The easiest way to picture rolled copper is to compare it to putting something through an old style wringer washer. The copper block that is used to make a copper foil for laminates is compressed in much the same manner. Figure 2.7 shows this concept. The original thickness of copper is pushed into the pair of rollers which are being both rotated and forced together. (This vertical force depends on the final thickness required at the output.) As the copper is forced through the rollers, it is compressed to this previously determined thickness. Following this compression there are various processes that may be used on the copper to obtain the desired consistency or hardness. For electrical applications the copper should be relatively soft to increase conductivity in the material. This process, as previously mentioned, is used for more critical applications because the rolled copper is a much more evenly distributed and consistent copper than ED copper. This can be seen in Figure 2.8. Figure 2.8 shows cross-sections of ED copper and rolled copper. You can see how the rolled copper is in very even layers but the ED copper does not exhibit this obvious uniformity. This is not to say that ED copper is not a good conductor at microwave frequencies, but when critical applications are involved, it may be wiser for you to use a rolled copper laminate.

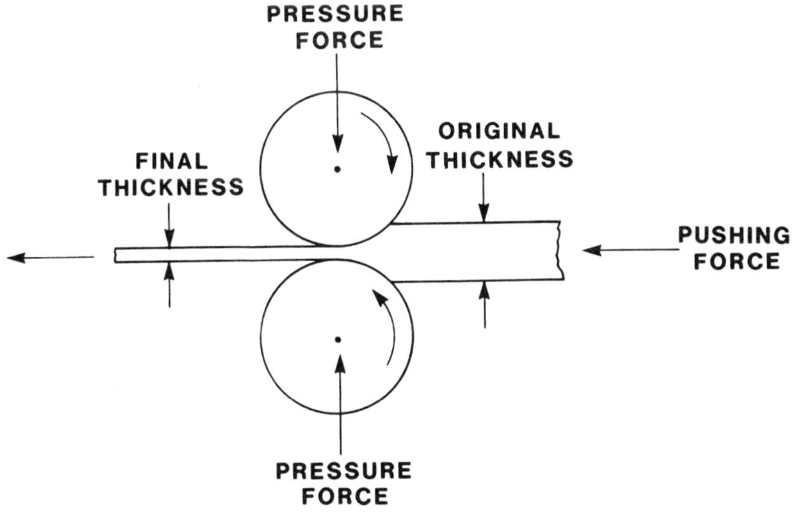

Fig. 2.7 Rolled copper diagram.

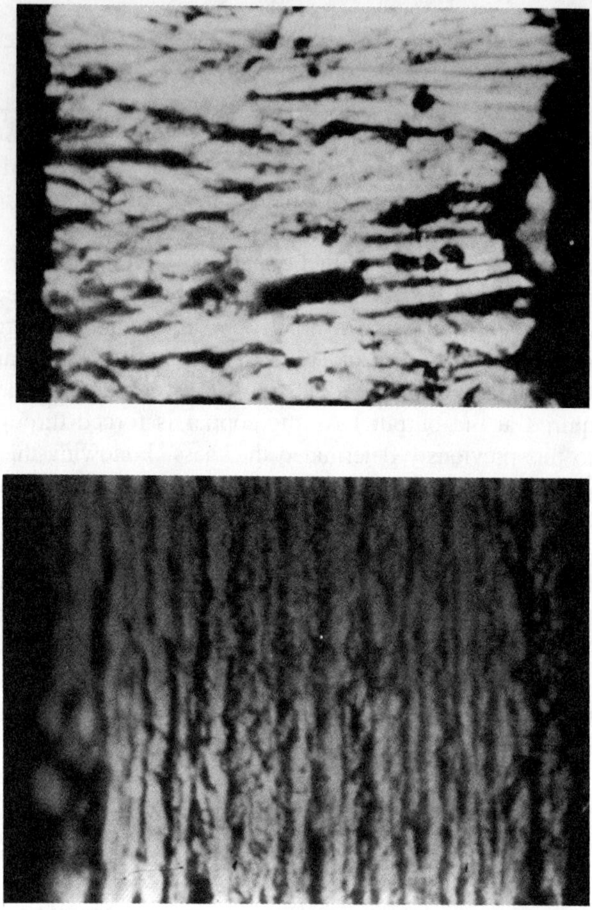

Fig. 2.8 ED and rolled copper
(Courtesy of the MICA Corporation).

Now that we have the weight and fabrication of the copper defined, the next step is to put the copper on a microwave laminate. All microwave laminates, whether using ED or rolled copper, are put together, as the name implies, by a lamination process. This basic process is familiar to all of us. If you have ever had your driver's license, social security card, or business card laminated, you will know what we are talking about. Consider the steps someone went through when they laminated one of your cards:

- The card is placed between two plastic laminating sheets;
- A press is heated to the lamination temperature;

- The card and laminating material is placed between the heater plates of the press and pressed together; and
- The card is now laminated and would probably be partially or totally destroyed if you attempted to separate the laminations.

This is the same basic process which is used to put the copper on a microwave laminate. One very important task must be performed before the copper can be laminated to the dielectric material, and that is the process of roughing up the surface of the copper so that it will "stick" to the dielectric. During the electrodeposition process a certain amount of roughness is encountered. Additional roughing is usually required for good adhesion. Since rolled copper usually is very smooth when completed, it requires more roughing. This surface roughing is usually accomplished by means of an acid etch similar to the roughing necessary when you iridite bare aluminum. The aluminum is subjected to an acid bath first to rough up the surface so that the iridite solution will "stick" to the surface of the aluminum. The same type of process is used for copper prior to the lamination process.

When combining copper with a dielectric material the same ingredients are used as with the business card—heat and pressure. With these two ingredients present and controlled, a highly reliable and efficient microwave laminate results.

Before we discuss specific types of laminates and substrates, let us spend a couple of minutes reviewing their history. The history of laminates and substrates for eventual microwave usage began in the 1950s. Early experiments used low-dielectric-constant plastics in a microstrip configuration. These first attempts were not very successful because the effects of radiation and intercircuit coupling were not taken into consideration.

Polystyrene was found to be a uniform low-loss material but was shown to crack or soften at temperatures above ambient. Attempts to harden and reinforce the material by adding glass fibers destroyed many of the desirable properties it originally had. As an alternative to the polystyrene, PTFE/glass cloth laminates were considered. This material was originally developed for other uses and although it had some rather unpredictable dielectric properties, it fulfilled many of the necessary requirements for a microwave laminate.

By the early 1960s PTFE/glass cloth laminates had their dielectric constant reduced from their original 2.65 to 2.75 to the more usable 2.55 of polystyrene. With this advance the PTFE/glass laminate replaced the styrene product and eliminated many problems. By reducing the glass content even further, the dielectric constant was reduced to 2.45 to create the first industry standard laminates, which became the GX type product in MIL-P-13949.

Dielectric constants were further reduced by the introduction of irradiated polyolefin ($\varepsilon = 2.32$) in the 1960s. This is a uniform and reproducible material that does not crack like the polystyrene. It does, however, have some rather serious mechanical instabilities, especially when temperature cycled.

PTFE/glass micro-fiber laminates were also introduced in the 1960s. The original material duplicated the dielectric constant of polyolefin (2.32) but had far superior dimensional stability and a much greater temperature range. A dielectric constant of 2.2 was achieved in the micro-fiber material by, once again, reducing the glass content. These two materials are found today characterized as type GR material in MIL-P-13949.

A popular dielectric material in the late 1960s was PPO (polyphenylene oxide). This material was touted as the ultimate in microwave laminates when it was first introduced. Although it had the ideal dielectric constant of 2.55, had very uniform and reproducible electrical properties, and was very stable both dimensionally and over temperature, we learned PPO had a large chemical problem. Not only would etching solutions make the edges of your boards very sticky and appear like a piece of pizza coming out of a pan if touched to some other surface, but it had a habit of developing fine cracks at unpredictable times after the material was attached to a case. These cracks (attributed to improper processing) usually occurred in the drilling process. The manufacturers of PPO eventually specified that the material should not be punched but drilled. Most people adhered to this recommendation but still ran into problems because in order to drill PPO, you needed a new drill bit, or one that had just been sharpened. If anything else was used the properties of the material would be such that the bit appeared to act as a punch and fine cracks would show up at a later date. Usually these cracks occurred after the circuit had been tested and was well on its way to being shipped. PPO finally disappeared from the microwave laminate scene.

The high purity (99.5 percent) alumina substrate material arrived on the microwave scene in the middle and late 1960s. This was when microstrip technology began its rapid rise in the microwave field.) Its high dielectric constant (9.9 to 10) and smooth surface finish resulting in low-loss circuits made it an ideal material for high-density microstrip circuitry. Although attempts were made in the 1960s to fabricate high dielectric laminates to compete with alumina substrates, most were very inconsistent, unreliable, or had fabrication difficulties. A glass-cloth-loaded PTFE material with a dielectric constant of six was the only success in a long line of attempts. This material is still available today.

The 1970s is when microwave laminates and substrates came of age. This was due partly to new technologies, which made more consistent and reliable materials available, and partly because the new systems being designed in industry put pressure on laminate manufacturers to produce these materials.

During the 1970s alumina substrates increased purity (99.6 and 99.7 percent) with finer particle size. Techniques for *sputtering* and *deposition* are refined so that thinner, more consistent metallization is possible. During this time, laminates were moving in two directions as regards their dielectric constant. They were going lower (2.17) for applications in the millimeter range, and going higher (10.2) to give the alumina market competition. The road to these new laminates was not smooth, however. The first high-dielectric material that was released worked rather well on the bench for breadboard or unstressed applications. (By *unstressed* we mean no extreme temperatures or humidity.) When subjected to these factors, there were great inconsistencies in the material; also, the copper lines would often lift off the material if a soldering iron was applied. All of these conditions, of course, were unsatisfactory and made the microwave industry wary of anyone promoting a high-dielectric microwave laminate. This situation persisted for a number of years but, as in all things, time (and extensive research) heals all wounds. By the late 1970s the material was reintroduced and three separate manufacturers were supplying it. This time it proved to be an excellent material—in many areas where it replaced alumina, as well as in many new areas of application.

As a result of the history of laminates and substrates briefly outlined above, the microwave industry has a broad cross-section of materials from which to choose. A listing of some of them is given below.

Material	*Dielectric Constant*
Woven PTFE/glass	2.55
Woven PTFE/glass	2.45
Woven PTFE/glass	2.33
Woven PTFE/glass	2.17
Nonwoven PTFE/glass	2.33
Nonwoven PTFE/glass	2.20
Ceramic filled PTFE/glass	6.0
Ceramic filled PTFE/glass	10.2
Alumina	9.9–10

2.2.1 PTFE/Fiberglas Laminates

PTFE (Teflon®) fiberglas laminates are probably the most widely used materials in microwaves today. That is a very broad statement and some may argue with it, but stop and think of all the microwave circuits you have seen recently. With the exception of some microstrip circuits on alumina substrates, the circuits were on woven PTFE, microfiber PTFE, or high-dielectric ceramic bonded PTFE (E-10, Di Clad 810, 6010). The microwave magazines present this same array of materials in their advertising. The Teflon® fiberglas laminate is the workhorse of the microwave industry.

In order to have a strong, consistent, and reliable workhorse, the materials are governed by military specifications. The one for microwave laminates is MIL-P-13949. As of this writing the specification is revision F, effective 10 March 1981. The specification has the following title: *PLASTIC SHEET, LAMINATED, METAL CLAD (FOR PRINTED WIRING BOARDS), GENERAL SPECIFICATION FOR.*

We are now ready to discuss specific laminates. There are two types of the lower dielectric laminate (2.45 and 2.55) in use today: woven and microfiber. The battle between these two has been raging since the microfiber material hit the market in the 1960s. This text will not recommend one over the other because, quite honestly, there is no one best material when speaking of microwave laminates. The material used depends on the particular application. We will, therefore, present both types, point out the pros and cons of each, where applicable, and let you make a decision from the information presented. We will also present discussions on ceramic filled PTFE material that has a dielectric constant of 10 and compete with alumina substrate in many areas.

2.2.1.1 Woven PTFE Fiberglas Material

Woven PTFE (Teflon®) will be covered first because it has seniority over the microfiber. It has been on the market since the mid-1950s. It was not the exact material then, however, that it is today. Its dielectric constant (ε) ranged from 2.65 to 2.75 as opposed to its present 2.45 to 2.55. You will recall from our previous history lesson how the original PTFE/glass (Polytetrafluorethylene/glass), though developed for other uses, showed some promise for microwave circuitry even though it proved to be a bit unruly at times. However, by the early 1960s the dielectric constant of the material had been stabilized and reduced to the now recognizable 2.45 and 2.55. The laminate was then ready to become a major part of the ever-growing microwave industry.

As the name implies, construction of this laminate is a woven pattern of fiberglas strands imbedded into the PTFE material. If you were to look at your shirt under a microscope you would see a pattern very similar to that of the woven microwave laminate. A drawing of such a woven pattern is shown in Fig. 2.9. The uniformity of this pattern is evident immediately. From this you can readily see how such parameters as dielectric constant and dissipation factor are also very uniform and consistent throughout the material. Figure 2.10(a) is an actual photograph of a piece of woven PTFE/glass laminate magnified 25 times. Once more, what stands out is the uniform construction as the fibers criss-cross through the PTFE material. Figure 2.10(b) is the same material magnified many times.

This uniform and consistent construction not only results in excellent consistency in parameters, but also makes machining operations possible. This type of construction dictates that drilling operations should be accomplished with care. Carbide tip drills should be used with a slow feed to the material and a high speed on the drill press. The high speed is necessary since the material is relatively soft and the hole is distorted if the drill speed is too slow. The same distortion results if the feed (both when penetrating the material and backing out of it) is too fast. Of at least equal importance with the speed and feed of a drill bit used on PTFE/glass material, is the bit itself. Solid carbide bits are the most useful and efficient, even

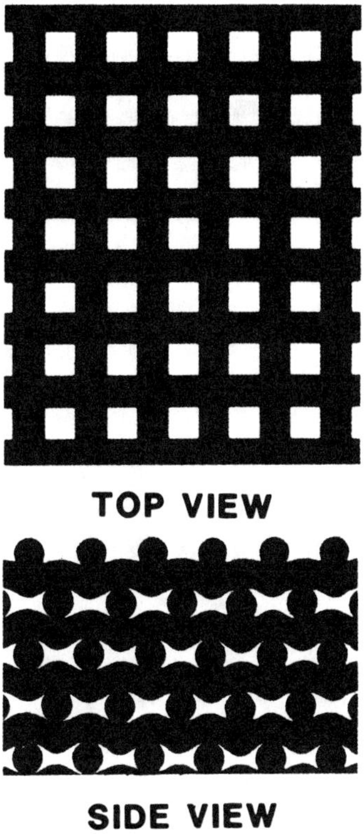

Fig. 2.9 Woven glass laminate construction.

though they need sharpening or replacement more often. Qualities a good drill bit should have are a 60° included angle, 30° clearance, and a hi-helix tape. The first two requirements are shown in Figure 2.11. The third one is a definition of twist on the drill. There are three types available on drills—slow, standard, and fast. The slow is a very shallow twist on the bit; the standard is a general twist; and the fast is the hi-helix type which is a fast- or sharp-pitch twist in the drill construction. This hi-helix design is best for PTFE/glass drilling because it provides the maximum *shearing action* and rapid removal of drill shavings as it goes through the laminate. The woven material can also be sheared, cut, or milled to various shapes to conform to a particular shape needed for specific applications. One thing should be remembered after cutting the material—since it is a woven fabric type of construction, there may be fabric edges left after machining. These edges should be removed with a knife or razor blade to ensure a smooth surface.

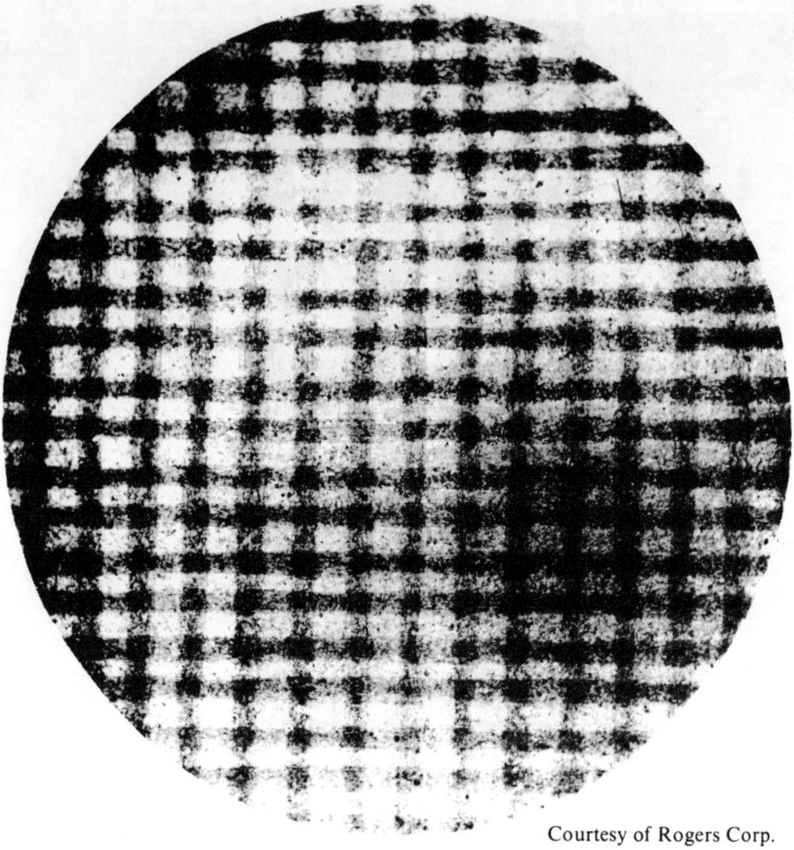

Courtesy of Rogers Corp.

Fig. 2.10(a) Woven PTFE (Actual).

The fabric edges can also be removed chemically with hydrofluoric acid. If an acid is used, however, the exposure time must be kept to a minimum and the laminate should be rinsed quickly and thoroughly to prevent a *wicking action*. This wicking action occurs because of the construction of the material. The fabrics that are woven together act in the way woven wicks used in kerosene lamps do.

The tools to be used when cutting or shaping this type of laminate are very important, just as the drill bits are in drilling operations. Helical two-flute or four-flute end-mills are excellent for cutting PTFE/glass material. Also, the mills should be solid carbide, just as with the drills, because of their long cutting life. Speed and feed rate are also important when using the end-mill, just as when drilling. Turning rates of 2,000 to 8,000 rpm and constant linear feed rates of four to ten in per minute will result in excellent cutting and a smooth finished material. (Speed and feed rates depend on the material thickness and the diameter of the end-mills.)

Courtesy of the MICA Corp.

Fig. 2.10(b) Woven PTFE (Magnified).

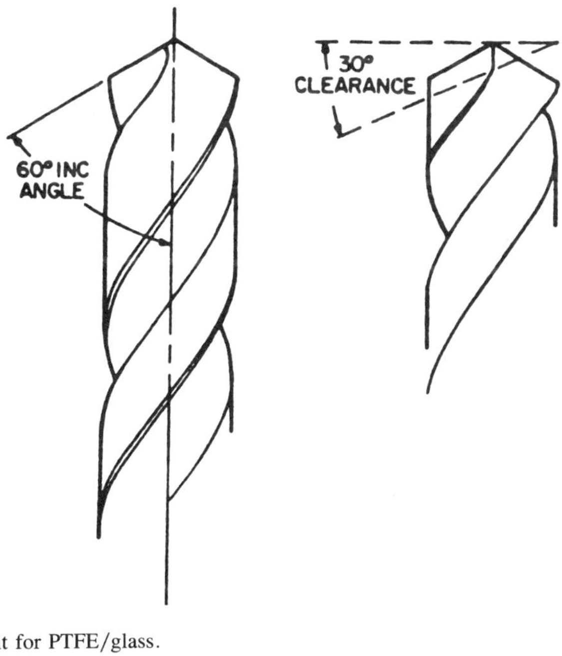

Fig. 2.11 Drill bit for PTFE/glass.

As previously mentioned, woven PTFE/glass material started out with a dielectric constant of 2.65 to 2.75. In the early 1960s it was reduced to 2.45 and 2.55. By varying the glass content of the material to an even lower percentage, dielectric constants of 2.33, 2.20, and 2.17 were put on the market in the latter part of the 1970s. This was a logical step to take when you consider that the dielectric constant of pure PTFE is in the area of 2.10. As you make the material closer to pure PTFE the dielectric constant will be lower. Thus when the 2.55 PTFE/glass material had the glass content lowered one point, a 2.33 or 2.20 dielectric constant resulted. Similarly, as the glass content is reduced even further the 2.17 dielectric constant resulted. This is about the lower limit with PTFE/glass laminates. Any further reduction in glass content will threaten the mechanical strength of the laminate. The 2.17 dielectric constant, however, is more than adequate for many applications in high microwave of the millimeter wave spectrum.

At the beginning of this chapter we covered four terms that characterized laminates. These were: *dielectric constant, dissipation factor, dielectric thickness,* and *cladding*. These terms, and others of value, are presented in Table 2.1 for four different woven PTFE/glass laminates. These laminates are those with dielectric constants of 2.17, 2.20, 2.30, and 2.50 and all commercially available. They can be ordered as standard products or specialized if you have requirements that differ from what is listed. These materials are available from three laminate manufacturers.

The cladding of each material is dependent on your particular requirements and on individual manufacturers. The manufacturer usually will have both available but will classify one as standard and one as a special order. Check the data sheet of each manufacturer for their cladding specifications.

One term in Table 2.1 has not been explained: *peel strength*. This is the amount of force required, per inch, to separate the copper cladding from the laminate. This should be a reasonable number, since the copper must adhere to the laminate fairly well to have the circuit operate. Notice that as the glass content is lowered in the laminates the peel strength decreases. This is because more PTFE is in the material and most materials do not adhere to predominately PTFE surfaces very well. (Think of your fried eggs sliding around a Teflon frying pan. Now picture these same eggs sticking in a variety of places where the pan has been scratched by metal utensils. This is analogous to the addition of glass to the laminate.)

Peel strength is also important when you are soldering to a copper-clad laminate. Since material with a low peel strength will have a tendency to have the copper lift away from the laminate when heat is applied, this is a good reason to have a material which has a high peel strength.

Laminates are available that intentionally have a very low peel strength. These materials are used for breadboard applications which can be hand cut and peeled with a knife. These materials have a peel strength in the order of four lbs/in and should only be used as a breadboard or for proving out a concept. *DO NOT USE THESE MATERIALS FOR FINISHED CIRCUITS*.

Table 2.1 Woven PTFE/Glass Laminates

Parameter	*Laminate No. 1*	*Laminate No. 2*	*Laminate No.3*	*Laminate No. 4*
Dielectric constant	2.50 ± 0.04	2.33 ±0.04	2.20 ± 0.02	2.17 ± 0.02
Dissipation factor	0.0018	0.0015	0.0012	0.0009
Thickness range	0.0035 in − 0.250 in	0.005 in − 0.125 in	0.005 in − 0.125 in	0.010 in − 0.120 in
Thickness tolerance	±0.0005 in to ± 0.006 in	± 0.0005 in to ± 0.005 in	± 0.005 in to ± 0.005 in	± 0.0006 in to ± 0.0005 in
Cladding	ED/Rolled	ED/Rolled	ED/Rolled	ED/Rolled
Peel Strength	12 lbs/in	8 lbs/in	8 lbs/in	8 lbs/in

Another point to notice in Table 2.1 is the relationship between the dielectric constant and the dissipation factor. You will note that as the dielectric constant (ε) increases, the dissipation factor also increases. This is due to the fact that the increased glass content in the material (to increase the dielectric constant) is considered to be an "impurity" added to the relatively low-loss pure PTFE material. This increases the loss in the material and, thus, a higher dissipation factor results. Increasing glass content no longer is feasible at the point where the losses become excessive at microwave frequencies. At this point the glass loading gives way to ceramic loading of the PTFE material. This combination gives results in a low-loss, high-dielectric (ε) microwave laminate. This material will be covered in detail in a later section of this chapter.

Since its release in the 1950s, the woven PTFE/glass laminate (the first "workhorse" laminate of microwave technology) has found many varied applications in microwaves. Today, as the microwave field expands to keep up with the fast-moving technologies, it still finds applications which make it an indispensable part of the huge and growing microwave industry.

2.2.1.2 Microfiber PTFE/Glass Laminates

As previously mentioned, there are two types of PTFE/glass laminates. The first, woven, was covered in the previous section. The second, microfiber, first appeared on the scene in the 1960s. When this laminate is copper-clad, it conforms to MIL-P-13949 types GP and GR. MIL-P-13949 defines GP as glass based PTFE resin and GR as glass based PTFE resin specifically for microwave applications. So the GP and GR designations readily define this non-woven PTFE material.

We have seen what the woven material looks like and how it resembles the weave of a shirt. What, then does microfiber PTFE/glass material look like? The official description of the material is a structure in which glass microfibers are individually encapsulated with PTFE and evenly dispersed throughout the material with random orientation in the X,Y plane of the laminate sheet.

Actually, the material resembles dark mashed potatoes that have become hard and smooth. It has a smooth consistent structure as compared to the woven material shown previously. This smoothness results in a laminate which has a dielectric constant which varies less than ± 1 percent and exhibits a low dissipation factor (0.0009 to 0.0012). The two microfiber materials available have dielectric constants of 2.20 and 2.33. Additional parameters are given in Table 2.2.

This material is softer than the woven type and certain precautions should be taken. Excessive clamping pressure should be avoided: if clamping is necessary, use more clamps with less pressure. Also, be sure the work area is clean; material resting on a steel chip can, from its own weight, cause the chip to become embedded in the material. Sawing should be done with skip tooth or coarse blade saws. A fine

Table 2.2 Microfiber PTFE/Glass Laminates

Parameter	*Laminate No. 1*	*Laminate No. 2*
Dielectric constant	2.20 ± 0.02	2.33 ± 0.02
Dissipation factor	0.0009	0.0012
Thickness range	0.010 in–0.250 in	0.010 in–0.250 in
Thickness tolerance	±3 percent (±0.003 in to ±0.0075 in)	±3 percent (±0.003 in to ±0.0075 in)
Cladding	ED/rolled	ED/rolled
Peel strength	15 lbs/in	15 lbs/in

tooth blade will tend to load up with material and will result in a rough edge. The material can be milled and drilled, just at its woven counterpart. Holes may become undersized because of heat generated during drilling. These should be checked at room temperatures for their final size.

The material also has many advantages. Because of its structure, etching and plating solutions do not wick as they have a tendency to do with woven material. This same structure results in cleaner edges and holes when cut, milled, or drilled. It is also very stable dimensionally as it can withstand temperatures up to 550°F. (287.7°C) without warping.

To sum up microfiber PTFE/glass laminates we can say the following:

- Excellent dimensional stability;
- Uniform dielectric constant and dissipation factor;
- Material is soft and susceptible to chips becoming embedded in it;
- Clamping should be kept to a minimum;
- No wicking occurs; and
- Good processing ability.

With both woven and microfiber laminates covered in detail, this would be a good time to put to rest the argument about which material is best. In Table 2.3 we will present four materials, two woven and two microfiber. The 2.33 dielectric constants will be compared and the 2.20 dielectric constant will be compared. In this way we will not be comparing apples with oranges, which would result in useless conclusions.

You can see from this table that there are very few differences in the materials. *The argument that the microfiber material had less loss than the woven is only slightly validated by this chart.* Actually the dielectric constant of both materials is lowered by reducing the glass content. When this is done, the dissipation factor (loss) is reduced. This reduction will be essentially the same in both materials, as the table shows.

Table 2.3 Comparison of Material Characteristics for Microfiber and Woven Laminates

	Microfiber Laminate No. 1	*Woven Laminate No. 1*
Dielectric constant	2.33 ± 0.04	2.33 ± 0.04
Dissipation factor	0.0012	0.0015
Thickness range	0.010 in–0.250	0.005 in–0.125 in
Peel strength	15 lbs/in	8 lbs/in
Water absorption	0.020 percent	0.024 percent
Surface resistance (megohms)	3×10^8	10^4
Cladding	ED/rolled	ED/rolled
	Microfiber Laminate No. 2	*Woven Laminate No. 2*
Dielectric constant	2.20 ± 0.02	2.20 ± 0.02
Dissipation factor	0.0009	0.0012
Thickness range	0.010 in–0.125 in	0.010 in–0.120 in
Peel strength	15 lbs/in	8 lbs/in
Water absorption	0.020 percent	0.020 percent
Surface resistance (megohms)	3×10^8	10^4
Cladding	ED/rolled	ED/rolled

Similarly, the argument that microfiber material is too soft can also be put to rest by discussing the warp of the material. This is one test for hardness. Microfiber material can withstand up to 550°F (287°C) without warping, rather conclusive evidence that the microfiber hardness should not be questioned.

In conclusion, we can say that all of the arguments for one material over another are based on personal preference. The ultimate answer to the arguments about woven and microfiber PTFE/glass laminates is that the best material is the one that does the job for you. There is no one *best* laminate.

2.2.1.3 High-Dielectric PTFE

When we discussed PTFE/glass laminates in the previous section we talked about varying the glass content to adjust the dielectric constant. A decrease in glass content meant a lower-dielectric constant and dissipation factor. Conversely, an increase in glass content resulted in a higher-dielectric constant but also increased the dissipation factor. There was a limit where the dielectric constant was higher than standard material and the dissipation factor was marginal (about $\varepsilon=2.7$). Unfortunately, this dielectric constant is not high enough for lower-frequency applications where the quarter-wavelengths become too long in standard PTFE/glass where $\varepsilon=2.55$.

Alumina substrates (covered in the next section), with their dielectric constant in the area of 10, are excellent for these applications. Unfortunately, they are difficult to machine, drill, and form to nonstandard sizes.

To fulfill the need discussed above, the PTFE laminates are *ceramic-filled* rather than glass-filled. The result is a PTFE board material with a high-dielectric constant, good loss characteristics, and flexible enough to be machined and drilled. This material is referred to as: *high dielectric Teflon®*, High K-material, or *E-10*. Although it is one of the most revolutionary materials developed in a long time, it was not always this accepted. When the material first came on the market in the 1970s it presented many problems: two prominent ones were water absorption and peel strength. Typical figures for the original ceramic-loaded material were 0.7 to 1 percent water absorption and a peel strength of less than 8 lbs/in.

Water absorption is a measure of the percentage of moisture absorbed by the material when it is soaked in water for 24 hours at a temperature of 23°C. The original numbers can be more fully appreciated when they are compared to woven and microfiber PTFE/glass.

Woven .01%
Microfiber .02%

You can see just how high a 0.7 to 1.0 percent absorption really is when compared to these materials. You do not want a material to absorb water since it will disrupt such parameters as dielectric constant and dissipation factor. The latest data sheets for high-dielectric ceramic-loaded material show that materials with dielectric constants of 10.2 and 6.0 have a water absorption of <.1 percent (typically 0.05 percent). This material has come a long way since its introduction in the 1970s.

The second problem, peel strength, has also been improved. Typical peel strengths for the most widely used materials used today were given in Table 2.4.

Table 2.4 Peel Strength of Various Laminates

Laminate	*Peel Strength (lb/in)*
Woven (ε = 2.50)	12
Woven (ε = 2.33)	8
Woven (ε = 2.20)	8
Woven (ε = 21.7)	8
Microfiber (ε = 2.33)	15
Microfiber (ε = 2.20)	15
Ceramic-loaded (ε = 6.0)	16
Ceramic-loaded (ε = 10.2)	16 (10 min)

The last item, ceramic-loaded ($\varepsilon = 10.2$) shows conclusively that the peel strength problem has been resolved. Even the 10 lbs/in minimum figure is much better than the original value of less than 8 lbs/in. This improvement, and a general overall material improvement, has made the ceramic-loaded PTFE laminate a viable competitor for the well-established alumina substrate in the microwave area.

Typical specifications for the high-dielectric material available today are shown in Table 2.5. You will note that the specifications are given for both $\varepsilon = 10.2$ and $\varepsilon = 6.0$. This is because there are applications where a dielectric constant of 10.2 is too high and one of 2.50 is too low. This was why the 6.0 material was created. Thus we present both materials as high dielectric laminates.

Table 2.5 Typical Specifications for High Dielectric Material

Parameter	*Laminate No. 1.*	*Laminate No. 2*
Dielectric constant	10.2 ± 0.25	6.0 ± 0.2
Dissipation factor	0.002	0.002
Peel strength	10 lbs/in min	16 lbs/in
Water absorption	< 0.1 percent	0.05 percent
Thickness	0.025 in and 0.080 in	0.020 in, 0.030 in, 0.050 in, and 0.062 in

Both of the materials listed above are available with metal backing attached to them. Typical thicknesses of type 6061 roll mill grade aluminum are 0.0625 in, 0.125 in, and 0.1875 in. This is an excellent way of using this material because of the inherent softness of the material without a backing. This softness will cause a warping after a normal etching process. With an aluminum (or copper) backing this warping does not occur. This backing of high-dielectric material is shown in Figure 2.12. This arrangement provides excellent dimensional and mechanical stability and also guarantees a uniform electrical ground plane for your circuit.

The main point we mentioned earlier concerning the replacement of alumina substrates with a high-dielectric PTFE laminate was the machinability, or lack of it, of alumina. The laminates can be milled, drilled, or sheared in a manner similar to lower-dielectric PTFE materials. (Remember that this still is a PTFE-base material, and proper ventilation is required.) It may also be sheared or cut with a razor blade or artists' cutting knife if only rough cuts are needed.

When connecting other circuits to those on high-dielectric material, many conventional methods can be used. Everyday hand soldering methods are very acceptable and are frequently used. Thermocompression bonding has been done to some samples using a 300°C (572°F) hot stage. When solid copper cladding was used, the bonds were good. However, when small pads or etched lines were used, they tended to lift from the substrate. This is because the thermocompression bonding hot stage

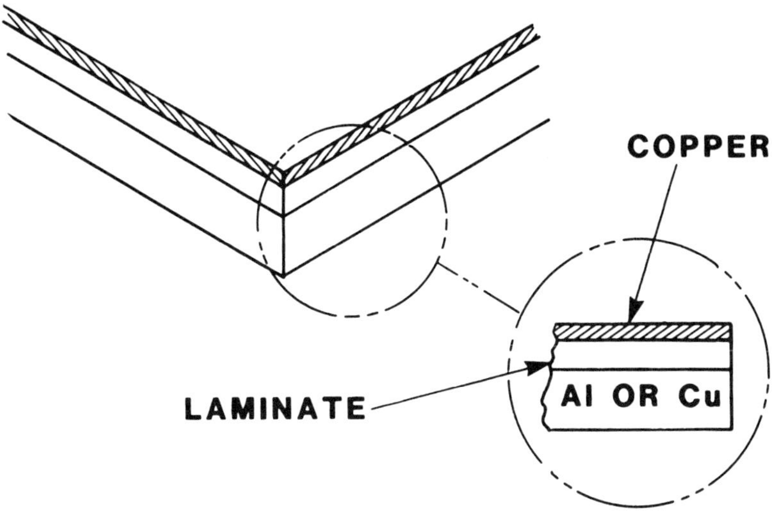

Fig. 2.12 Metal clad high-dielectric laminates.

softens the dielectric to a high degree. For this reason pulse or ultrasonic bonding should be used on these materials.

High-dielectric PTFE laminates are finding more and more applications throughout the microwave industry. It should be remembered, however, that they are not the answer to every microwave application. Just as woven or microfiber laminates on alumina substrates have their specific places, so do high-dielectric laminates. It is up to you to find where they fit and use them appropriately.

Before moving on from soft substrates to the world of hard (or inorganic) materials, we should briefly look at the term dielectric *anisotropy*. This is a difference in dielectric constant in a microwave laminate in the X, Y plane of the material compared to that in the Z direction. This difference comes about because fill materials (ceramic, glass, *et cetera.*) are added to a pure material (PTFE) to obtain dimensional stability in the laminate. The laminate is no longer completely consistent in all directions when these fill materials are added. The difference in consistency is measured as anisotropy and is the anisotropy ratio, $\varepsilon_{xy}/\varepsilon_z$. Figure 2.13 shows this ratio as a function of dielectric constant for both woven and microfiber PTFE/glass material. You can see that the ratio of ε_{xy} to ε_z is low (approaching 1.00) for low-dielectric constants and increases with dielectric constant increase. This is logical as lower-dielectrics have less fill material (glass) and are thus more consistent, since the laminate is almost a pure PTFE.

The figure appears to have a large distance between PTFE and microfiber but actually the materials are rather close. The ratios are as listed in Table 2.6.

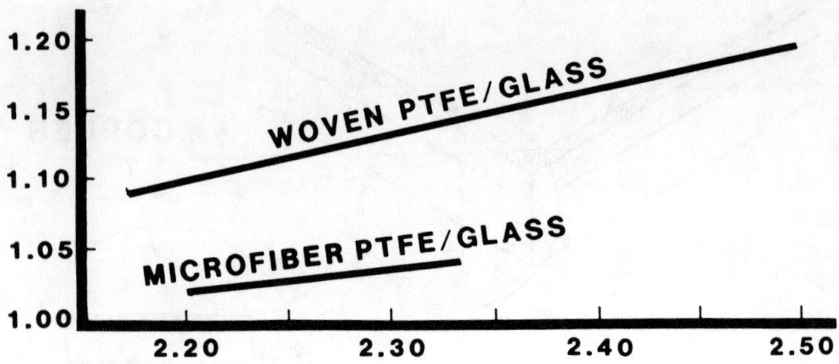

Fig. 2.13 Anisotropy *versus* dielectric constant.

Table 2.6 Anisotropy Ratios of Microfiber and Woven Material

Material		$\varepsilon_{xy}/\varepsilon_z$
Microfiber	$\varepsilon = 2.20$	1.025
Microfiber	$\varepsilon = 2.33$	1.040
Woven	$\varepsilon = 2.17$	1.090
Woven	$\varepsilon = 2.45$	1.160

We should briefly explain why dielectric anisotropy should be a concern to the microwave designer. The major area of concern is in microstrip and stripline circuits, where it produces additional fringing capacitance. This additional component has the most prominent effect on resonator elements, in narrow lines, and in couplers using edge coupling techniques. In all of these cases the effects of fringing capacitance must be known and taken into account. If you have a material with a high anisotropic ratio and you do not know it, your circuits will not perform as expected. When choosing a material to be used in your microwave circuit, do not forget the term *anisotropic*.

2.2.1.4 Alumina Substrates

We have referred to alumina substrates on various occasions throughout this chapter. Now it is time to explain them and where they fit into the microwave industry.

Until the 1970s, when high-dielectric PTFE material was introduced, alumina substrates were the only reasonably priced materials on which microstrip circuits could be built. These substrates were referred to as the familiar alumina, aluminum oxide, or Al_2O_3. Their popularity began around the mid-1960s when their purity went from the previous low of 96 percent to a much more acceptable 99.5 percent. Since then it has gone to 99.7 percent and, in some cases, to 99.9 percent. The most widely used alumina substrates today use a purity of 99.6 or 99.7 percent. This increase in purity resulted in the "as-fired" surface finish dropping from a previous 8 to 10 microinch thickness to approximately four microinches. The term *as fired* means that this is the substrate after its final firing without any additional machining or polishing. You might call these *rough* substrates.

The two items that are of importance in an alumina substrate and make it useful for microwave application are dielectric constant and dissipation factor. The dielectric constant of the substrate is affected by two factors: the preferred orientation of the alumina crystallites and the density of the material. Fine-grained alumina ceramic has a more random crystalline orientation than one with coarse-grained properties. This random orientation is difficult to regulate and depends very highly on an accurate process control during manufacturing. This type structure, however, is still the preferred structure for the higher dielectric constant and lower dissipation factor. This is because the fine grain structure results in a higher density in the material. The adhesion of circuit-forming films depends on the density of the grain boundary areas and the fine-grain sizes produce a higher density which reduces *microwave scatter* caused by a coarse grain structure. This scatter is similar to a dissipation and increases the loss. It is preferred, therefore, to have alumina substrates with a fine grain structure for the optimum dielectric constant and dissipation factor. This guarantees consistent substrate properties.

Other than the dielectric constant and dissipation factor, the flatness of a substrate is another important factor when choosing the right material for your application. Flatness is the term used to define the overall *camber* of a substrate. (Camber is likened to the warping of a material as shown in Figure 2.14.) This is usually expressed in the allowable inch out of flat per inch of substrate (0.001 in/in). The term *camber* is important since flat substrates (or as near flat as possible) are necessary to ensure proper operation of vacuum fixture during exposure, a good fit in holders during processing, and a good mask contact during the printing on the metallization.

Besides *flatness,* other terms used to describe the final finish of a substrate are *as-fired* and *polished*. We have previously described an as-fired substrate as the condition of a substrate after its final firing without any additional machining or polishing. You will recall that we called these *rough* substrates. These substrates, as might be expected, can have definite limits regarding flatness and dimension control. Most of the time these limits make the as-fired substrate unsuitable for microwave application. When this occurs you must specify *polished* substrates. These substrates offer excellent performance but are much higher in cost. These substrates are a fired material that is buffed smooth by mechanical grinding or lapping.

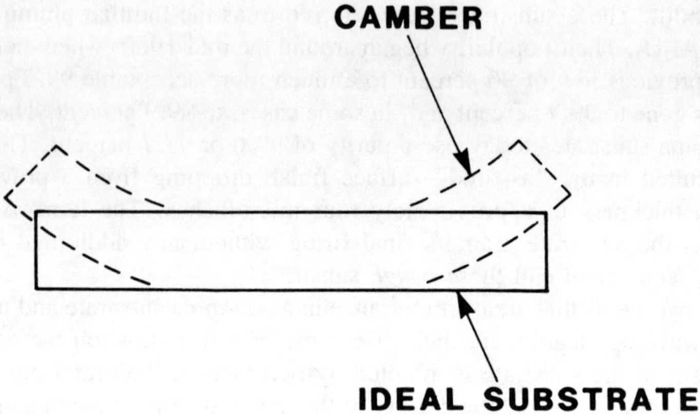

Fig. 2.14 Substrate camber.

With the primary terms presented, we will close our discussions on alumina with a typical data sheet:

Material	Al_2O_3 (99.6 percent)
Dielectric constant	9.90 ± 2 percent
Dissipation factor	0.0001
Metallization	Chrome-gold
Surface finish	3 microinches (as-fired)
Camber	0.001 in per in
Thickness	0.025 in ± 0.00005 in
Outside dimensions	0.5 in × 0.5 in ± 0.002 in

All of the terms in the data sheet above have been covered. This example should show you that alumina is an excellent substrate (electrically) for microwave applications. The only drawback in the material is its brittleness (it is rather like working with a dinner plate) and its difficulty to machine. Other than these factors, the substrate is an excellent choice for microwave usage.

2.2.1.5 Sapphire, Quartz, and Beryllia Substrates

These substrates probably should be put under a category of exotic. None of them are everyday materials that would be the first choice of a microwave designer.

Sapphire (also referred to as α Alumina) is a material with a dielectric constant of approximately 11.0. This is about as close as we can pin this material down, since it is not isotropic and dielectric constants ranging from 9.3 to 11.7 have been measured on it. The difficulties in the fabrication of alumina (cutting, drilling, *et cetera*) are also encountered with sapphire substrates. We primarily use such a material when a high thermal conductivity is desired. This would probably be the only real application of such a material for microwaves.

Quartz (SiO_2) is another brittle material that has the same fabrication difficulties as alumina and sapphire. Its dielectric constant is 3.78 which is somewhat higher than the woven PTFE/glass material. This may be useful for higher-frequency applications where smaller wavelengths need a low-dielectric constant to keep the dimensions reasonable. The dissipation factor is relatively impressive .0005. Quartz would be used only in very special and rather loosely specified applications.

The last of the materials in this section is beryllia (BeO). It has a dielectric constant of 6.9 and a dissipation factor of 0.0002. This dielectric constant is placed in the category of high-dielectric PTFE laminates ($\varepsilon = 6.0$) and in the alumina category of dissipation factor. Beryllia substrates require polishing since they have a coarse-grain structure resulting in an inherently rough surface and a poor resistance to chipping. These substrates are normally used only where an extremely high thermal conductivity is needed because the material is extremely expensive and has a high toxic hazard rating. Beryllia would normally be one of the last materials considered for typical microwave applications.

The dielectrics covered in this section are by no means the only ones available. Other types are cross-linked polyolefin, cross-linked polystyrene, glass reinforced polystyrene, polypheneline oxide (PPO), quartz-loaded teflon, and silicon based dielectric.

By means of special processes, the dielectric constant can be precisely controlled for very close tolerance application. Dielectric thicknesses can also be specified to close tolerances. Many manufacturers guarantee dielectrics at 0.030 in ± 0.001 in, and hold the thinner material (0.005 in) to a 3 to 4 percent tolerance.

One last parameter should be mentioned here—the power handling capability of dielectrics. Table 2.7 lists six dielectrics with their maximum operating temperature and their maximum average power.

Table 2.7 Maximum Operating Temperatures and Average Powers of Six Dielectrics

Material	*Maximum Operating Temperature (°C)*	*Maximum Average Power (W)*
Woven Teflon® fiberglass	265	560
Microfiber Teflon® fiberglass	260	620
Polyolefin	100	455
PPO	194	324
Polystyrene	100	175
Glass polystyrene	100	185

(0.125 Ground Plane Spacing; 50Ω line; 25°C at 2.5 GHz)

2.3 FERRITES

Ferrites usually suggest phase shifters, circulators, isolators, or components of this nature. They should also automatically warn you to keep your watch a good distance away to avoid damage. The devices mentioned above all contain ferrites. Since some are not adequately shielded, the associated magnetic field can do damage to a watch. However, most devices that are properly engineered present no danger from magnetic fields.

The operation of ferrite devices is probably quite familiar, but what exactly causes the ferrite material to react as it does when it is subjected to a magnetic field?

Ferrites of the spinal structure are of interest in the microwave region because the materials exhibit good dielectric and magnetic properties. Ferrites or ferrimagnetic materials are ceramic members of the ferromagnetic class of materials.

Since the structure of ferrite materials determines their microwave characteristics, the physical properties of these materials is covered. The ferrites with the simplest physical structure belong to a chemical group with the formula:

$$(MO)\ (Fe_2O_3)$$

Where (MO) is a metallic oxide of a divalent metal such as iron (Fe), manganese (Mn), cobalt (Co), nickel (Ni), magnesium (Mg), zinc (Zn), cadmium (Cd), *et cetera*. Ferrites are prepared by thoroughly mixing the iron oxide (Fe_2O_3) with the metallic oxide (Mo). The mixture is then fixed at a high temperature which is less than the melting point of either oxide. In the firing process, the oxides sinter together into a spinal crystal structure. A spinal structure is cubic—all the sides are equal in area and are at right angles. Because the oxygen atoms are much larger than the metallic atoms, the oxygen determines the size of the crystal. The structure tends to be as small as possible (in crystal language, it is close-packed). In Figure 2.15(A), a close-packed arrangement of one layer is shown. If another layer is placed on top, and the whole structure is made close-packed, it takes on the form of Figure 2.15(B). The points marked A and B in Figure 2.15(B) occur between the two layers of oxygen atoms. The dotted circles represent the top layer of atoms; the solid circles, the bottom layer. Metal atoms occupy all of the A and B sites. In this simple structure, the iron atoms are in the B sites and the M atoms are in the A sites. There are twice as many B sites as A.

The A sites are called *tetrahedral sites* because there are three oxygen atoms above the metal ion and one oxygen atom below. The B sites are called *octahedral sites* because they occur at the corners of a regular eight-sided figure. The spinal crystal structure consists of layers of oxygen atoms with its smaller metal atoms between them.

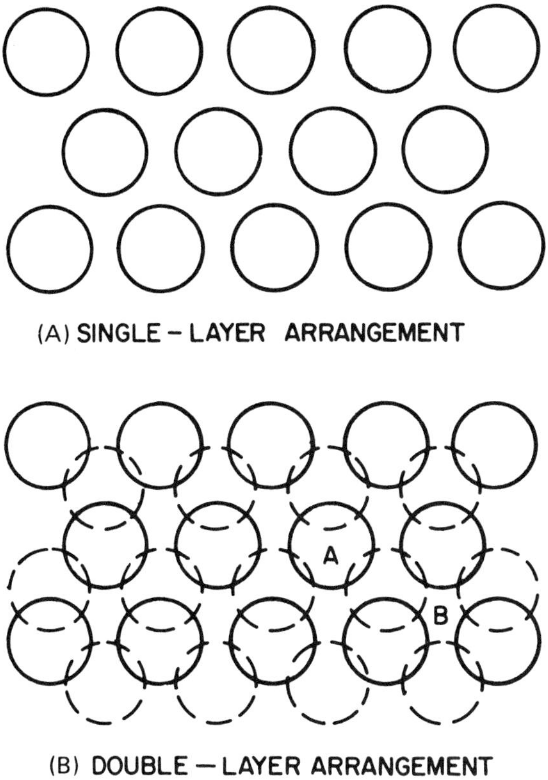

Fig. 2.15 Oxygen and metallic atoms.

The metal atoms are expected to contribute to the magnetization of the ferrite, but the sum of all the spins in the ferrite add up to a value considerably more than what is measured, meaning that some spins are up and others are down. In ferrite materials the iron atoms are normally distributed so that half of them are in each of the B and A sites, while all of the M atoms are in the remaining B sites. All of the metallic atoms in the A and B sites average themselves to form a system with a minimum amount of energy.

In the type of system described above the iron atoms have a tendency to reverse their spins in the A and B sites, leaving only the metal ions to contribute to the magnetization of the ferrite. However, materials can be mixed in ferrites to vary the saturation magnetization over a wide range. For example, by combining nickel and zinc oxides, the value of saturation magnetization can be varied by a factor of 50, depending upon the ratio of zinc to nickel used.

By now, the term *ferrite* has been adequately defined. As a double check, a dictionary of electronic terms defines a ferrite as "a powdered, compressed, and sintered magnetic material having high resistivity, consisting chiefly of ferric oxide combined with one or more metals."

Needed at this point is a discussion on how these ferrites perform in the components to be used in an application. The main reason that most of the devices work as they do is a phenomenon known as *gyromagnetic resonance* (also called *ferromagnetic resonance*).

Ferromagnetic resonance is the principle of operation of a great many ferrite devices. The electrons spin about their own axis but also precess about the stationary magnetic field (H_i) at a frequency determined by H_i and a constant called a *gyromagnetic ratio*. This action is shown in Figure 2.16. If microwave energy of this same frequency and polarization is incident upon the spins, the latter interact with it and couple energy. The coupled energy is used to widen the angle ϕ between the axis of the electron and the internal magnetic field. To detect resonance, the loss in dB of the incident microwave energy is measured after it passes through the ferrite. A plot of microwave energy loss *versus* the magnetic field is shown in Figure 2.17. Notice that the resonance is not a sharp effect, as some devices do exhibit. There is a definite spreading out of the loss region. The points A and A′ are 3 dB higher than the minimum point and the distance between them is termed the *linewidth* (ΔH). Normally, linewidth is measured in oersteds; the smaller the linewidth, the better the ferrite. However, a broad linewidth ferrite material has advantages for some applications.

Ferrites are used in circulators, isolators, phase shifters, and gyrators, to mention a few devices. A gyrator is defined as a two-port circuit element that exhibits 180° of differential, or nonreciprocal, phase shift; that is, it exhibits 180° of phase shift in one direction only. It does not have a predicted behavior in the reverse direction. Wherever a material is needed that provides directional information, as in directional couplers and mixers at lower frequencies, the material which fills the bill is a *ferrite*.

2.4 SOLID-STATE MATERIALS

To appreciate fully and understand how, and why, microwave solid-state devices function as they do, it is necessary to investigate the materials used to fabricate them. An understanding of why germanium was used for early devices; why silicon was, and still is, the primary material for microwave devices; what makes gallium arsenide acceptable for higher-frequency applications; and what the compound indium phosphide holds for the next generation of material for microwave transistors and diodes—all of these statements are part of the history of microwave solid-state devices and all of them involve a specific material. By knowing and recognizing each of these materials, an individual can make a much more intelligent choice for a particular application.

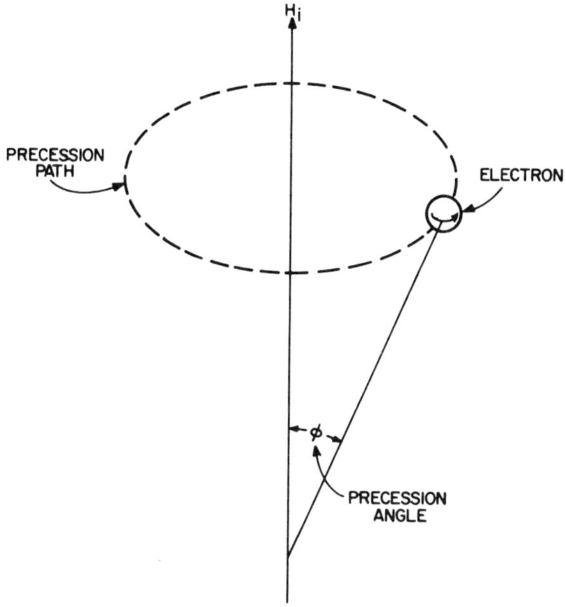

Fig. 2.16 Gyromagnetic action.

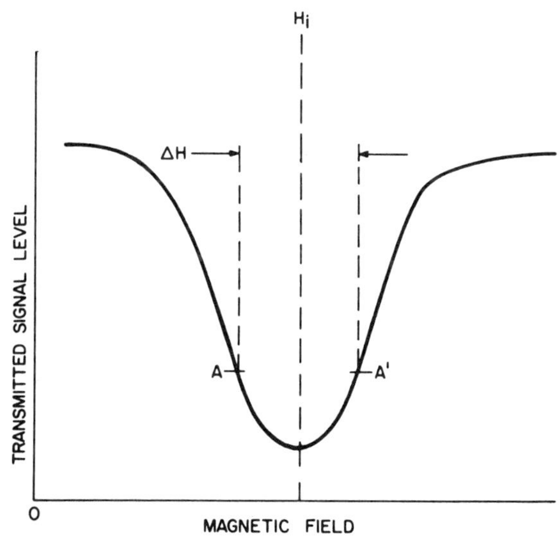

Fig. 2.17 Ferrite resonance.

The materials to be covered in this section have a variety of names. They are germanium, silicon, gallium arsenide, and indium phosphide. Each has distinct qualities which make them desirable for certain applications. Each has advantages and disadvantages. The one property they all have in common, however, is that they all are *semiconductors*. A dictionary defines semiconductor as follows:

> "Material whose electrical resistivity is lower than that of insulators and higher than that of conductors."

What this definition means is that a semiconductor is not a conductor, not an insulator, but somewhere in between the two. To realize where semiconductors fall between conductors and insulators, consider that the typical conductor has a resistivity of $10^{-6}\Omega$cm; and a typical semiconductor has a resistivity of $10'\Omega$cm. So it can be seen that the semiconductor actually falls between conductors and insulators. That is, a true *semi*-conductor.

What is it that makes these materials semiconductors, and how do they lend themselves to microwave applications? To help answer these questions refer to Figure 2.18. This is a periodic table of the elements. Notice how the materials are all grouped together being a part of either group three, four, or five. The initial materials mentioned, germanium and silicon, are by themselves in group four. The compounds referred to initially, gallium arsenide and indium phosphide, consist of one element from group three (gallium and indium) and one from group five (arsenic and phosphorus).

The base materials, those in column four, are the materials which usually have impurities added to them. This is a process called *doping*. If, for example, arsenic atoms are added to germanium, there would still be covalent bonding between atoms, but there would be an excess electron as a result of the combination. This is shown in Figure 2.19. This is because arsenic has five electrons in its outermost orbit as opposed to the four which are present in germanium. This excess electron is termed a free electron. This condition decreases the resistivity of the overall combination and is termed a *donor* composition and is said to form *n*-type germanium. The *n* in this term is a designation for *negative,* which is the charge of the electron which is the primary current carrier in this case.

Similarly, if an atom from the third column was added to germanium, for example, there would be too few electrons available to make a covalent bond (Figure 2.20). In this case there would be a void, or *hole,* present. This construction constitutes a condition which is termed an *acceptor* atom. This means that the element formed (a germanium and boron combination is this case) will accept an electron if one is available. This condition is also termed a *p*-type material with the *p* representing a positive charge, which is the charge of the vacated *hole*. Once again, the resistivity of the germanium would be decreased, since the hole is an electrical carrier which can be considered comparable to a free electron when considering current flow.

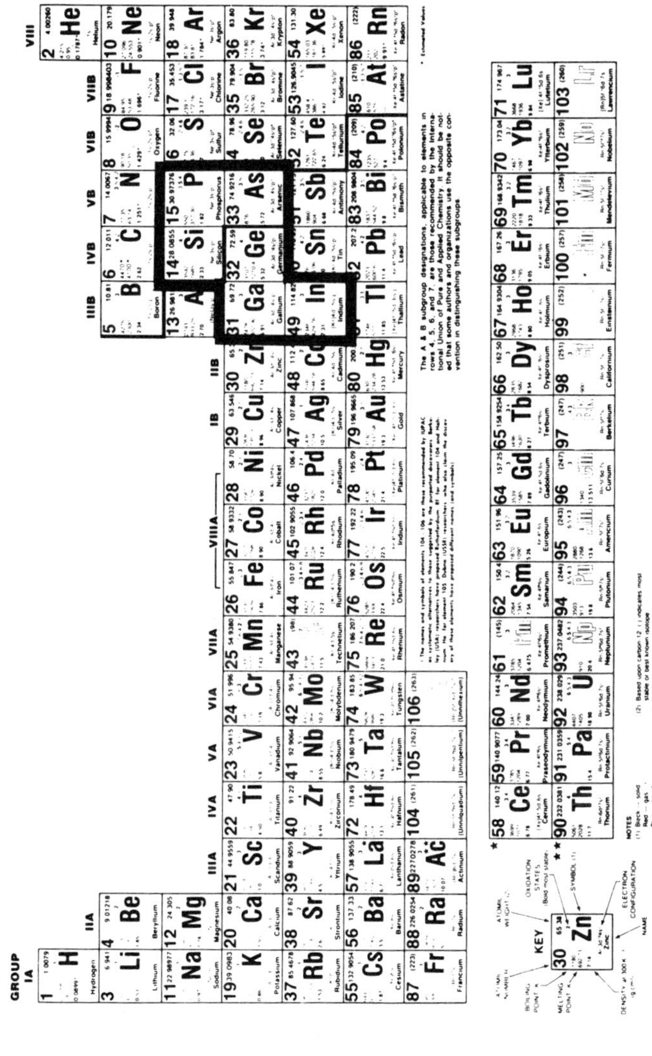

Fig. 2.18 Periodic table of the elements
(Courtesy of Sargent-Welch Scientific Company).

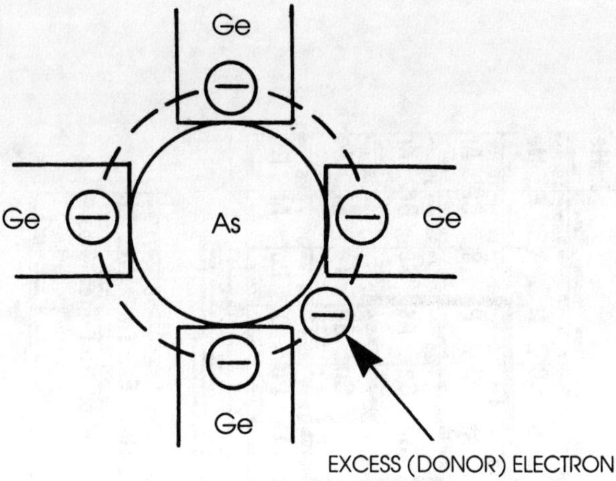

Fig. 2.19 Donor atoms.

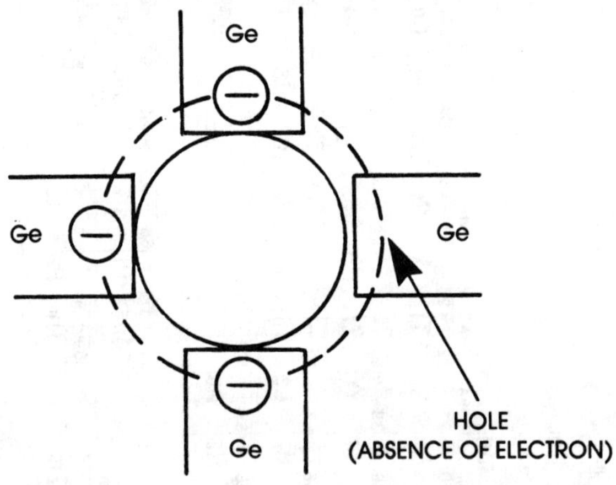

Fig. 2.20 Acceptor atoms.

Figure 2.21 shows this same relationship as gallium is combined with arsenic and indium with phosphorus. In each case an element from group three (gallium and indium) is combined with one from group five (arsenic and phosphorus). It can be seen that there is an excess of two electrons. This makes these materials excellent majority carrier materials; that is, materials which have electrons as their source of current flow. There is no *hole movement* in such structures. These are called *unipolar* devices and will be covered extensively in later chapters.

The discussion above has shown the properties of the materials presented with regards to being semiconductors. In order to understand how each lends itself to microwave applications, it is necessary to look at each one individually because each material has unique properties which can be used to the highest degree for microwave applications.

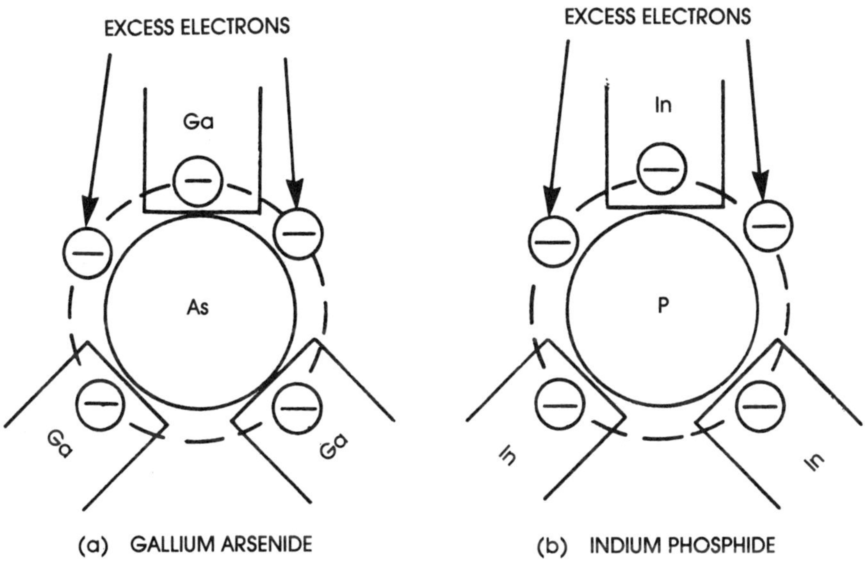

Fig. 2.21 Gallium Arsenide and Indium Phosphide.

An important part of semiconductor operation is the energy band within a particular element. Figure 2.22 shows a diagram of the two upper bands of energy within a material. There are other bands below the valence bond band, but for considering electrical properties of a material, the two upper bands are the only ones of interest. These are the ones on which we will concentrate.

The *conduction band,* in Figure 2.22, is a band of energy in which the level of energies is high enough to allow electrons at this level to move freely under the influence of any external field or force. That is to say, the electrons will sustain a current rather easily if a voltage is applied.

The *valence band* is a band of energies in which the energy is of the same level as that of the valence electrons. This means that the electrons which occupy this level are not free to move around as are those in the conduction band. These electrons, however, may be raised to the conduction band if sufficient energy is applied to the material. To reach the conduction band they must cross the gap betwen the *valence band* and the conduction band. This gap is called the *forbidden energy band*. It is the energy difference across this region that determines whether a material is a conductor, an insulator, or a semiconductor (such as we need for our application). Let us now look at specific types of microwave semiconductor material.

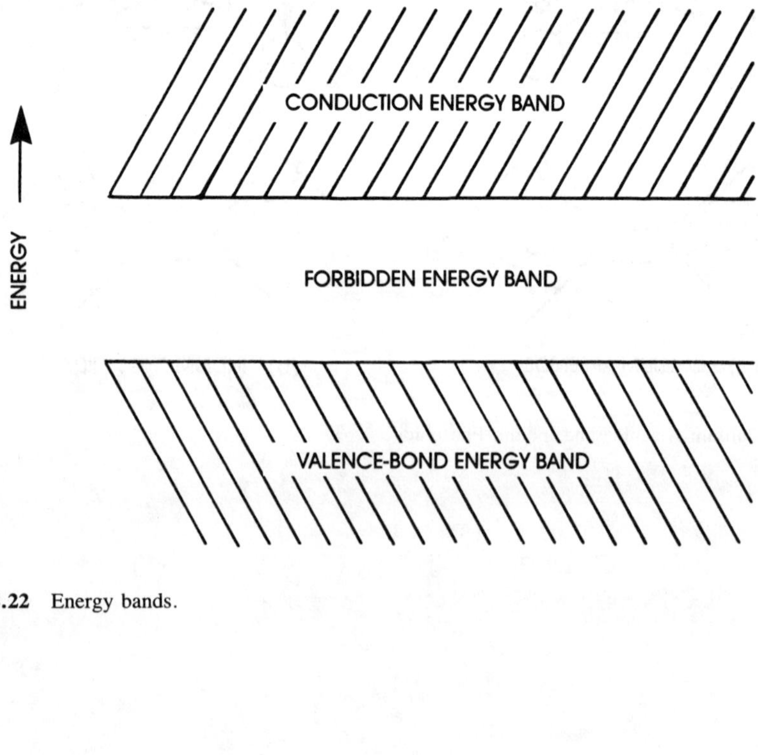

Fig. 2.22 Energy bands.

2.4.1 Silicon (Si)

Silicon follows only oxygen as the second most abundant element in the earth's crust. It occurs as silica in sand and quartz, and as metal silicates in most rocks. The use of silicon in diodes and transistors has become extensive and has, in many cases, completely replaced germanium for these applications.

The silicon we are concerned with for microwave applications is *pure silicon*. The methods of preparation of very high purity silicon (greater than 99 percent) may vary from manufacturer to manufacturer. Also, most of the processes and their details are proprietary and are not disclosed. There are, however, some general steps taken by all manufacturers which can be discussed. The first step is the conversion of metallurgical silicon (that silicon which is 96 to 99 percent pure, made by the reduction of silicon with coke at 1,770 K in an electric furnace either to silicon tetrachloride or, more commonly, trichloroethylene. The silicon halide is purified by fractional distillation and then reduced back to silicon by heating with hydrogen at 1,300 K (1,027° C or 1,880° F), either in a quartz tube or over a heated rod of pure silicon on which deposition of the reduced silicon takes place, thereby increasing the diameter as the process continues. Trichloroethylene is preferred to silicon tetrachloride at this stage because it is more rapidly reduced and it is easier to remove phosphorus and boron compounds by distillation.

The *pure* silicon obtained is polycrystalline and $>$ 99.0 percent. Most semiconductor devices require single-crystal, ultra-pure silicon with impurity levels of less than one part in 10^9. These levels of impurity are determined by activation analysis, mass spectrometry, or are empirically determined by measurement of resistivity and other electrical tests. The silicon obtained from reduction of the halide is further purified by the zone refining process in which narrow bands (or zones) are melted in a vertical silicon rod and made to travel repeatedly in the same direction along the length. The molten zone retains its integrity by surface tension. Impurities preferentially dissolve in the molten zone and are carried with it to the end where they are concentrated, leaving the rest of the rod in the ultra-pure form. Extremely small controlled additions of the required additives to give *p*- or *n*-type characteristics (as described earlier) are then added to the silicon. This doping may be of acceptor atoms (boron or gallium) to produce *p*-type material, or of donor atoms (arsenic or antimony) to produce *n*-type material.

After doping, a single crystal of silicon is grown from a small crystal seed by very slowly pulling the growing crystal from the melt. The doped silicon in the single crystal form is then sliced into very thin discs. This is the point where most people perceive the semiconductor process as beginning. It can be seen, however, that the actual process has begun many steps back. After the discs are sliced, subsequent stages of actual device manufacture include the construction of *pn* junctions (usually by vapor deposition of a layer of silicon with dope of the opposite type), division of the disc into tiny chips, selective etching, attachment of connecting leads, and final encapsulation. At this point you have a working silicon-based semiconductor.

Probably the greatest advantage of silicon over germanium semiconductors is its gap energy or forbidden band energy. In the case of silicon, it is 1.1 V as opposed to 0.7 V in germanium. This comparison is shown in the Table 2.8.

Table 2.8 A Comparison of Germanium and Silicon Semiconductors

Semiconductor	E_g *(electron volts)*	U_e	U_h
Ge	0.7	3,900	1,900
Si	1.1	1,500	500

A large energy gap (or forbidden band) is useful in semiconductor devices since a large gap gives a small level of thermal conduction. This is one of the disadvantages of germanium. It has the narrowest energy band of all the semiconductor materials. This gap is important when considering materials for semiconductor devices, since you only want conduction through the material because of bias voltages applied to it. In this way you can control the device to meet your particular needs. If you have a large level of conduction through the device due to the temperature the device is subject to, you will never know how that device will act, since you cannot possibly have complete control over it. Thus, a wider energy gap is very beneficial for semiconductor material.

The last two columns of Table 2.8 are indications of charge mobility. They are the electron mobility, U_e, and the hole mobility, U_h, within the material. Units of this mobility are cm^2/V-s. Of more importance to the operation of semiconductor devices is their ratio. Since germanium and silicon devices are generally characterized as *bipolar* devices, that is they rely on both electron and hole movement (or mobility) for their operation, characterizing the relationship of these two charges is a necessity. It can be seen that in germanium there is approximately a 2:1 ratio of electron-to-hole mobility. This is good since the electrons are designated as the primary carriers. Silicon, on the other hand, has a 3:1 ratio of electron-to-hole mobility. This tells you that its primary carrier conduction is better than that of germanium.

Another advantage of silicon over germanium is its higher avalanche or breakdown voltage. This higher voltage capability is due to a large extent on the carrier density and energy gap available within silicon. As a comparison, consider the fact that germanium diodes will exhibit a breakdown voltage in the order of 150 to 200 V. The same diode constructed in silicon will have a breakdown voltage of 1,000 V.

A major disadvantage that is always mentioned when speaking of silicon is its upper frequency limit of operation. This is due to the inherent characteristic of a relatively low value of the diffusion constant for charge carriers. This, as mentioned, is a characteristic of silicon and must be worked around to obtain the other desirable features of silicon devices.

Silicon is used in microwaves for diodes (Schottky, *pin,* IMPATT, and TRAPATT) and in bipolar transistors. The only limitation encountered is when using the bipolar transistor. Best performance as far as both power and noise performance is generally below 4 GHz. Devices designed and used in this range will generally perform well. When this limitation is kept in mind, and the proper design procedures used, silicon devices can be used very effectively for a variety of microwave applications.

2.4.2 Gallium Arsenide (GaAs)

Gallium arsenide, as the name implies, is a compound. It consists of a group three element (gallium) and a group five element (arsenic). Notice that we are now talking about a material that has two elements precisely joined, as opposed to the manufacturing of a pure element such as silicon. The first element, gallium, is a silvery metal with a melting point just slightly less than the surface temperature of the human body (29.8°C or 85.6°F). It is one of only four metals that can liquify at, or near, room temperature. The others are mercury (−38.87°C or −37.96°F), cesium (28.5°C or 83.3°F), and rubidium (38.5°C or 101.3°F). Gallium was discovered by Paul E. Lecoqde Boisbaudran in 1875 through the use of the spectroscope and electrolysis. The name *gallium* is derived from the Latin *Gallia* which means Gaul, the ancient name of France.

The second element in gallium arsenide is arsenic. This element is a dark gray semimetallic solid which can be traced back to around the year 1250 when it was discovered by Albertus Magnus. From its earliest beginnings it has been known as a lethal poison. If you mention the word *arsenic,* you will probably see the same association made by those who hear the term. Its name is derived from the Greek word *arsenikon,* which means *yellow opiment.* This element is presently classified as a suspected carcinogen by the Environmental Protection Agency. That is, it is suspected of causing cancer under certain circumstances.

If you think of these two elements by themselves, you would have a difficult time ever believing that when combined, they would form a compound that would allow the field of microwaves to make one of its largest advancements. This combination has ruled the field from the very time it first appeared in the early 1970s. Although it sounds like a very lethal combination, the material, GaAs, a grayish and brittle semiconductor, is relatively benign chemically and is safe to handle under normal conditions. Gallium arsenide is characterized as a direct-band-gap semiconductor with an isotropic minimum at the center of the Brillouin zone. This zone is named after Louis Brillouin for his contributions to the understanding of wave motion in periodic media. The zone is when the wave number, k, is equal to the range between $-\pi/a$ to $+\pi/a$. Figure 2.23 shows a comparison of energy levels for this zone for silicon (Figure 2.23(a)) and gallium arsenide (Figure 2.23(b)). It can be seen that there is a definite difference in the area of the forbidden zone. The silicon zone is much smaller. Also, the electron energy (mobility) is much greater in the gallium arsenide. This makes the material an excellent choice for high-frequency operation. Table 2.9 presents these points vividly.

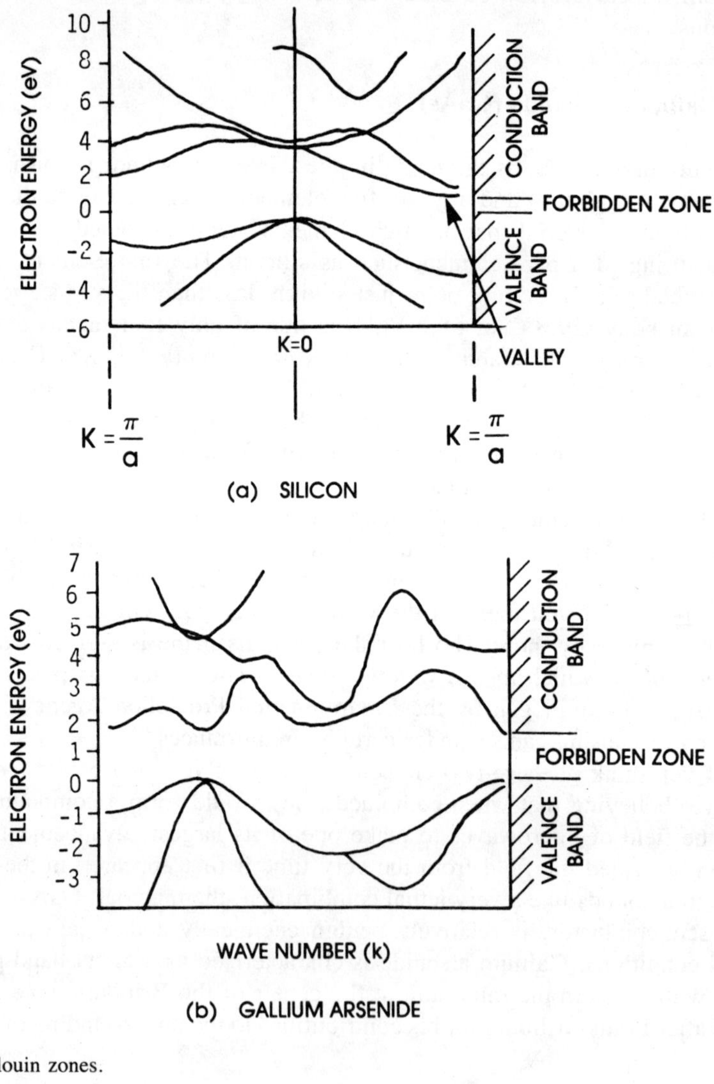

Fig. 2.23 Brillouin zones.

Table 2.9 Gap Energy of Various Semiconductors

Semiconductor	E_g *(electron volts)*	U_e	U_h
Ge	0.7	3,900	1,900
Si	1.1	1,500	500
GaAs	1.43	5,000	400

It can be seen that GaAs has a higher gap energy (1.43 V as opposed to 1.1 for silicon and 0.7 for germanium). This allows the devices to operate at a higher temperature than either of its two predecessors.

Another point to be emphasized is the higher mobility ($cm^2/V\text{-}s$) of the electron, in particular within the GaAs material. It can be seen that the ratio of electron mobility to hole mobility is 12.5:1 as opposed to 2 or 3:1 for germanium and silicon, respectively. This is necessary for high-frequency operation since many GaAs devices are *unipolar* devices as opposed to *bipolar,* which the germanium and silicon devices are. The term unipolar means that the primary means of conduction is by electrons. Thus, the ratio of electrons to hole mobility must be high. It can be seen that this is true for the GaAs material.

The combination of a large energy gap and high electron mobility make gallium arsenide an ideal material for high-frequency, high-temperature, and radiation resistant devices. In contrast, however, to the growth of bulk crystals from a monatomic and low vapor pressure material such as silicon, the synthesis of gallium arsenide is greatly complicated by the previously discussed volatile and toxic element, arsenic. By using special handling procedures and special growth techniques, very high quality single crystals can be grown and processed into high performance devices.

During the course of crystal growth, gallium and arsenic are first compounded in an exothermic (liberation of heat) reaction. Single crystal GaAs is then grown from the molten material which results. Substrate materials are then sliced and polished from the single crystals. The major commercial GaAs crystal growth methods fall into two categories: The first is the horizontal Bridgeman technique and the second is the vertical liquid encapsulated Czochralski (LEC) technique.

Figure 2.24 shows the horizontal Bridgeman technique. In this setup a GaAs charge is formed separately or compounded *in situ* in a quartz crucible or PBN boat with a seed crystal. Sufficient arsenic to maintain adequate vapor pressure during growth is also placed in the sealed quartz tube. The crystal is grown by moving the entire furnace so that the tube moves through a temperature gradient from the hotter to cooler section of the furnace. Careful observation and control of the seed-melt interface is required for initiation of single-crystal growth. Variations of this system have included use of four independent hot zones and a sodium heat pipe for precise control of the arsenic vapor pressure, and elimination of the viewing window and replacing it with thermal monitoring devices.

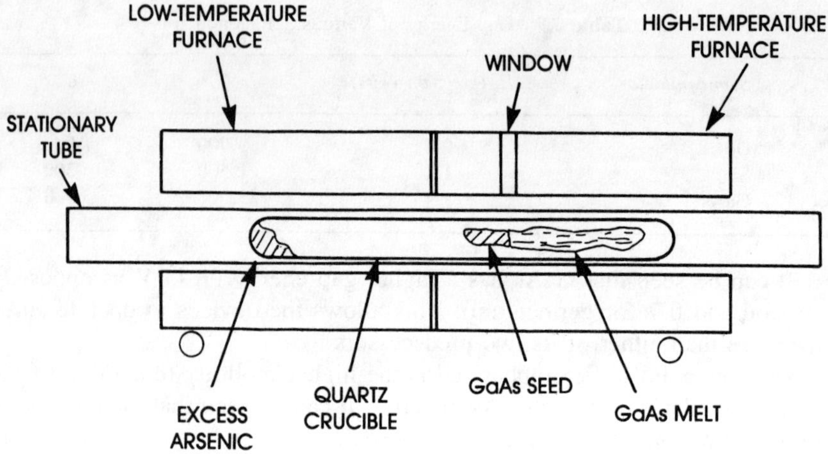

Fig. 2.24 Bridgeman horizontal technique.

The vertical Czochralski technique is referred to as the LEC (Liquid Encapsulated Czochralski) growth technique. This process is illustrated in Figure 2.25. In this technique the vaporization of arsenic (As) from molten GaAs is inhibited by placing a layer of nonreactive boric oxide (B_2O_3) on the melt surface. An inert gas pressure, which is higher than its arsenic partial pressure, is then maintained on top of B_2O_3 encapsulating layer, and growth proceeds as would take place in a standard Czochralski technique. This standard process is as follows:

> A seed crystal is dipped into a melt whose temperature is lowered until a small amount of crystalline material is solidified. The crystal is then slowly withdrawn from the melt at a typical rate of 1 to 10 mm per hour. The temperature of the melt is carefully adjusted to produce a crystal of the desired diameter.

Other methods that may be incorporated for growing GaAs are *horizontal gradient freeze, zone melting, Liquid-Encapsulated Kyropoulus* (LEK), *Magnetic LEC* (MLEC), and a variety of techniques tailored to individual manufacturers. The best method to use is the one which will provide the purest and most consistent GaAs crystals. Generally, either the LEC or MLEC method is used.

2.4.3 Indium Phosphide (InP)

The latest compound to make its presence known on the microwave scene is indium phosphide (InP). This material has made solid state devices operating efficiently up into the hundred gigaHertz region very practical and attainable. The compound, obviously, is made up of two elements: indium and phosphorus.

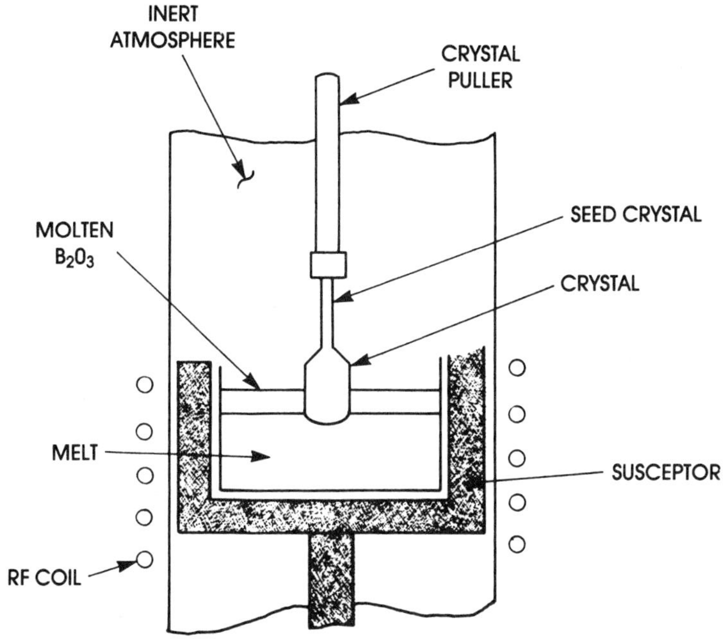

Fig. 2.25 LEC technique.

Indium, which was already covered in section 2.1.5.1, received its name because of its indigo blue spectrum, is a soft, silvery, malleable metal. This element has become very important for a variety of microwave applications, not the least of which is forming the compound InP for solid-state applications.

Phosphorous is a nonmetallic element which appears in three varieties:

- A yellow which is very reactive, soft, and poisonous;
- A red which is less reactive and non-poisonous; and
- A black phosphorous.

The element was discovered in 1669 by Hennig Brad by heating white sand with evaporated urine. Its name is derived from the Greek *phosphorus* which means "light-bearing." It is also classified as the eleventh most common element in the earth's crust.

Once again, as with GaAs, these two elements seem like an unlikely combination to put together, but their combination results in a material which shows excellent performance in the upper microwave and the millimeter bands. If we complete our table of gap voltage and mobility parameter by adding indium phosphide to it, we can see some interesting comparisons (see Table 2.10).

Table 2.10 Gap Voltage and Mobility Parameters of Semiconductors

Semiconductor	E_g *(electron volts)*	U_e	U_h
Ge	0.7	3,900	1,900
Si	1.1	1,500	500
GaAs	1.43	5,000	400
InP	1.25	4,000	100

One area to note is that the gap voltage is not as high as GaAs, but not as low as silicon. This indicates that indium phosphide will not show as good a performance in temperature as GaAs. The large advantage, however, can be seen in the mobility numbers (U_e and U_h). There is a ratio of 40:1 of electron to hold mobility. This accounts for the excellent higher frequency operation.

When we compare InP to GaAs, there is only one area where GaAs is better for millimeter and higher microwave application. That area is mobility, as shown in Table 2.4. All other areas are dominated by InP. The shorter time constant (0.75 ps as compared to 1.5 ps for GaAs), related to the electron energy transfer effect, allows the InP GUNN device to operate effectively in its fundamental mode through 100 GHz. These devices will be discussed further in Chapter 5 of this text. The higher operating velocities (*saturated velocity* and *effective transit velocity*) in InP provide dimensional advantages in the actual device lengths. These velocities are shown in Table 2.11:

Table 2.11 Higher Operating Velocities in GaAs and InP

Velocity	*GaAs*	*InP*
Saturated Velocity (cm/s at 500 K)	5×10^6	6×10^6
Effective Transit Velocity (cm/s)	0.7×10^7	1.2×10^7

The advantages of indium phosphide (InP) are numerous and will be expanded upon as applications are presented throughout the text. At this point it is sufficient to say that for application from 20 GHz to 150 GHz, there is no better material to use than indium phosphide.

2.5 SUMMARY

We stated in the beginning of this chapter that every chapter in this book deals with microwave materials—directly or indirectly. It should be obvious by now just how true this statement is.

By covering such topics as *metals, dielectrics, ferrites,* and *semiconductor materials,* we have covered every material that is used in microwaves. The chassis and cases for circuits are made of metal, as well as the metallization on laminates and substrates; the circuits are etched on dielectrics or microwave laminated (substrates); circulators, isolators, and phase shifters use ferrites; and the microwave diode and transistor rely on semiconductor materials. You can see just how important materials are to microwaves.

Chapter 3
Transmission Lines

A general definition for the term *Transmission Line* is *a device that transfers energy from one point to another*. This definition does not take into consideration the activity of this energy between the two points. Perhaps a more accurate definition would be *a device that transfers energy from one point to another with a minimum loss,* that is, a component which provides an efficient means of transferring energy. This meaning is accurate but still general and specialized. An efficient transfer of energy does not necessarily mean a very low loss, as in the case of pulses being transmitted. We could say that the *ideal* transmission line definition depends, to some extent, on the particular application.

A transmission line can take on many different forms. It may be a two-wire, a coaxial line, stripline, microstrip, or a waveguide (rectangular or circular). All of these types are efficient means of transferring energy from one point to another, or, more specifically, efficient means of transferring *microwave* energy. This qualification helps to explain why we choose only those lines listed above.

This chapter covers basic transmission line theory, coaxial lines, stripline, microstrip, and waveguides as they apply to the microwave spectrum. The discussions also cover such familiar terms as VSWR, characteristic impedance, and skin effect.

3.1 BASIC TRANSMISSION LINE THEORY

To illustrate the basic theory of transmission lines, the two-wire line is employed. The two-wire form is used because it exhibits certain properties that are common to all types of transmission lines.

The two-wire transmission line consists of two parallel conductors properly supported and insulated from each other. When a source voltage is connected to the pair, a current flow exists as a result of the flow of charges into the line, and, in turn, a magnetic field, proportional to this current, surrounds the conductors. The associated flux linkages per unit current (I) are called *inductance L*.

If a given length of this transmission line were to be repeatedly divided and subdivided, infinitesimal sections of the basic elements—resistance R, conductance G, inductance L, and capacitance C—ultimately would be formed. The circuit that would result is shown in Figure 3.1. When the source is applied to the line, it *sees* this first section of line which is made up of the basic circuit parameters. The series resistance and inductance are shown in one line only, but actually are formed in both lines and can be distributed equally or unequally. If the basic transmission line parameters are spread evenly along the entire length of the line, the constants are said to be distributed. The capacitance of the line is spread out along it, thereby causing an effect unlike that of a capacitance centered at one point.

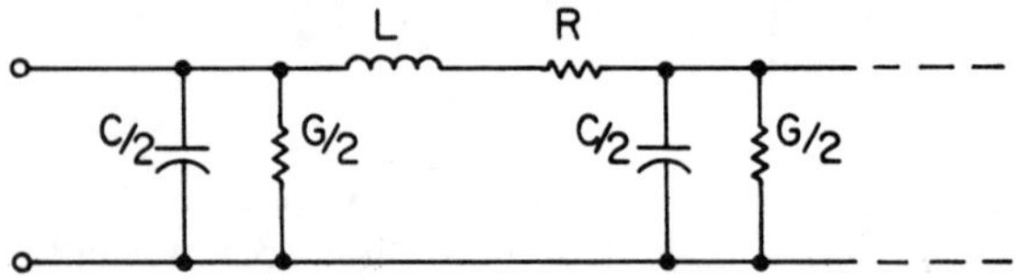

Fig. 3.1 Equivalent circuit of a transmission line.

Therefore, there is an inductance per unit length of line when a unit current is flowing. The charge on the conductor is proportional to the potential difference (voltage). Therefore, the line has shunt capacitance (C), previously defined as the charge per unit potential difference. If the dielectric medium between the conductors is not perfect, a conductive element must be assumed between the lines to account for the resulting loss in current—the conductance (G) per unit length of line. In addition to the above parameters, there is also a series resistance (R) associated with the conductors, since the perfect conductor cannot exist in practice. This resistance depends upon the resistivity of the material, the length and cross section of the conductor, and the distribution of currents in the cross section.

In most microwave transmission lines the losses are extremely small, if indeed there are any at all. If the loss parameters R and G are zero, then the equivalent circuit of Figure 3.1 is reduced to series sections of inductance and shunt sections of capacitance.

Reflections are caused by discontinuities in the transmission line structure, which may be regarded as changes in the characteristic impedance of the transmission line or guiding structure. The circuit in Figure 3.2 with a charging resistor (R_g) equal to the characteristic impedance (Z_o) simplifies the initial discussion of the behavior of the open-circuit termination. When the switch is closed, the line appears as shown in the equivalent circuit of Figure 3.2(B) because the battery sees Z_o in series with R_g. Waves of voltage and current, known as designated incident waves, start from the source and move toward the open termination; they are labeled E_i (incident voltage wave) and I_i (incident current wave).

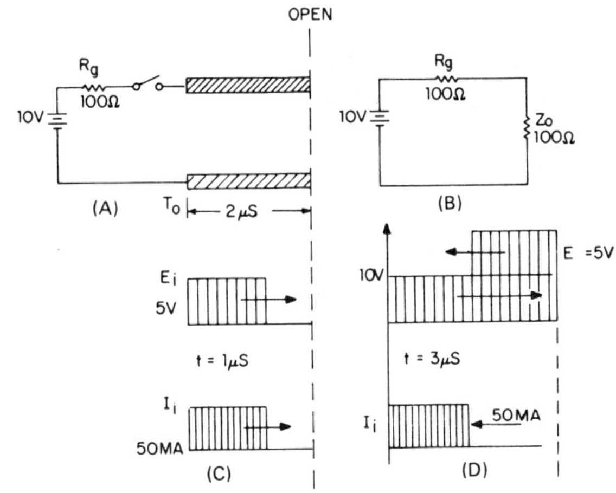

Fig. 3.2 Traveling waves on a transmission line terminated in an open circuit.

Assume that the velocity of propagation is such that a voltage wave and current wave can travel the length of the line in 2 μs, as indicated in Figure 3.2. At t_o, 5 V are dropped across R_g; a 5 V wave, accompanied by a 50 mA current, starts down the transmission line. These wavefronts travel together down the line at a velocity equal to $1/\sqrt{LC}$, for a maximum of 2 μs, and charge each element of line capacitance to 5 V. The collapsing magnetic field, associated with the inductance element at the end of the line, tries to maintain the original current flow into the capacitance. This additional current flow charges the end capacitance to 10 V ($2E_i$), and the current at the end of the line decreases to zero. This action continues from section to section back toward the battery, as shown in Figure 3.2(D); a correspondence is created with a reflected wave of voltage (E_r) equal to the incident wave (E_i) and a reflected wave of current (I_r) equal to the incident current (I_i). The reflected waves travel at the same velocity ($v = 1/\sqrt{LC}$) as the incident waves, and the ratio of E_r to I_r is equal to Z_o. The entire line is charged to 10 V when the reflected wave arrives at the battery after a total travel time of 4 μs. The steady-state condition is reached when a voltage of 10 V exists between the conductors and the current stops flowing. *Note*—At an open-circuit termination, E_r is both equal in magnitude to, and in phase with, E_i, and I_r is both equal to, and 180° out of phase with I_i.

Another form of reflection occurs whenever the transmission line is terminated in a short circuit and a zero voltage is thereby produced at the end of the line. The line behavior at this point can be described in terms of the inductance and capacitance of the line as considered for the open circuit. When the 5 V, 50 mA incident waves reach the short circuit, a reflected voltage of 5 V with reversed polarity starts back toward the battery. The line capacitance at the wavefront is discharged by a net current of 50 mA, and the voltage between the conductors drops to zero. This 50 mA of current adds to the 50 mA incident current, producing a total line current flow of 100 mA at the wavefront. When the reflected wave reaches the battery, the voltage between the conductors becomes zero, while the current flow remains 100 mA. Consequently, the steady state is reached, since the 100 mA of current equals the current drawn by the 100-ohm resistor (R_g) across a 10 V battery. At a short circuit E_r is equal to E_i but has opposite polarity, while I_r is equal to and in phase with I_i.

The two cases discussed above can be put to valuable use in microwaves. The properties of a transmission line repeat every half-wavelength and are opposite every quarter-wavelength, thus making the open-end or shorted transmission line very useful. For example, one frequency may be acted upon while others are left moderately untouched by inserting an open-ended line in the circuit. This line would be a quarter-wavelength at the unwanted frequency. Since one end is open, a short circuit would occur a quarter-wavelength removed, and the frequency would be at an RF short to ground. But what sort of impedance is there between the quarter- and half-wavelength of the line? The answer can be found in Figure 3.3 and 3.4. Figure 3.3 is a chart of the impedance characteristics of an open-ended line; Figure 3.4 is a similar tabulation for a shorted line. As viewed in these figures, there are areas where the line is capacitive (X_C) and other points where it is inductive (X_L). This phenomenon becomes more understandable in Chapter Six through a presentation of the Smith chart, whereby the effect is seen first hand. But, for the present, simply recall from studies of ac theory that, in an inductance, the voltage leads the current; in a capacitance, the current leads the voltage. In the voltage-current drawings in Figures 3.3 and 3.4, wherever an X_L is indicated, the voltage is, in fact, leading the current. Similarly, wherever an X_C is indicated, the current is ahead of the voltage. Wherever a value of R is indicated, keep in mind that either a short circuit (zero ohms) or an open circuit (an infinite resistance) is present.

Thus far we have dealt with transmission lines that are terminated in either a short- or an open-circuit. However, lines are also terminated in resistances. If the transmission line is terminated in a resistance equal to Z_o, it behaves as if it had an infinite length; since there are no reflections, the line is said to be *matched*. Also, the lack of reflections results in a maximum transfer of power to the load and maximum efficiency for the overall transmission system.

If the line is terminated in any resistance other than Z_o, reflections occur, the amplitude and polarity of which depend on the particular value of load and characteristic impedance. The amount of reflections is expressed as a ratio of reflected voltage (E_r) to incident voltage (E_i) and is called the *reflection coefficient,* (ρ)—the percentage of power, or voltage, reflected from the mismatched load. The expression for reflection coefficient is:

$$\rho = \frac{E_r}{E_i} = \frac{R_L - Z_o}{R_L + Z_o}$$

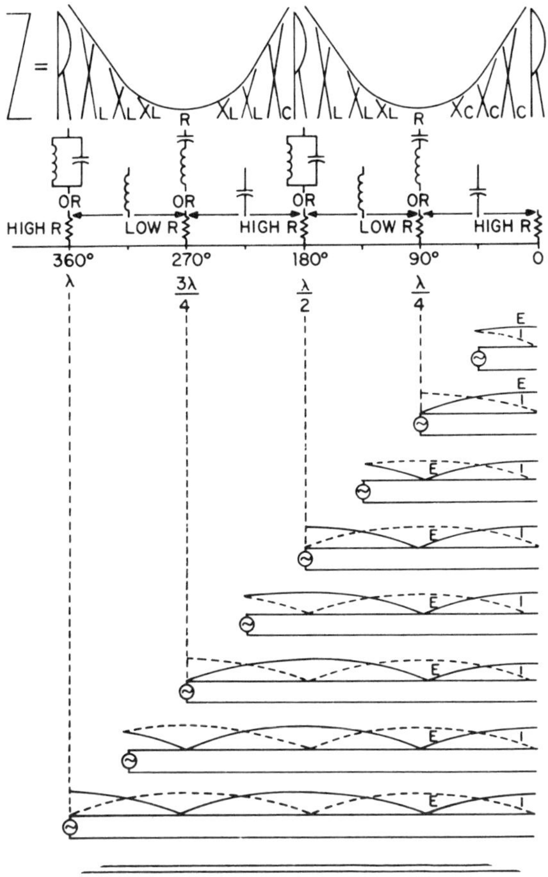

Fig. 3.3 Impedance characteristics of an open-end line.

The reflection coefficient can be calculated from this equation if the load is either a *complex impedance* or a *pure resistance;* in the latter case, it could be higher or lower than Z_o. If a transmission line is terminated in a pure resistance (Z_L) greater than the characteristic impedance, the reflected voltage wave (E_r) is in phase with the incident voltage wave (E_i) and the reflected current wave (I_r) is 180° out of phase with the incident current wave (I_i).

If a transmission line is terminated where Z_L is less than Z_o, E_r is 180° out of phase with E_i, and I_r is in phase with I_i.

It should be noted that if the source impedance is not equal to its characteristic, reflections occur from the *source* end of the line back toward the load. For this reason

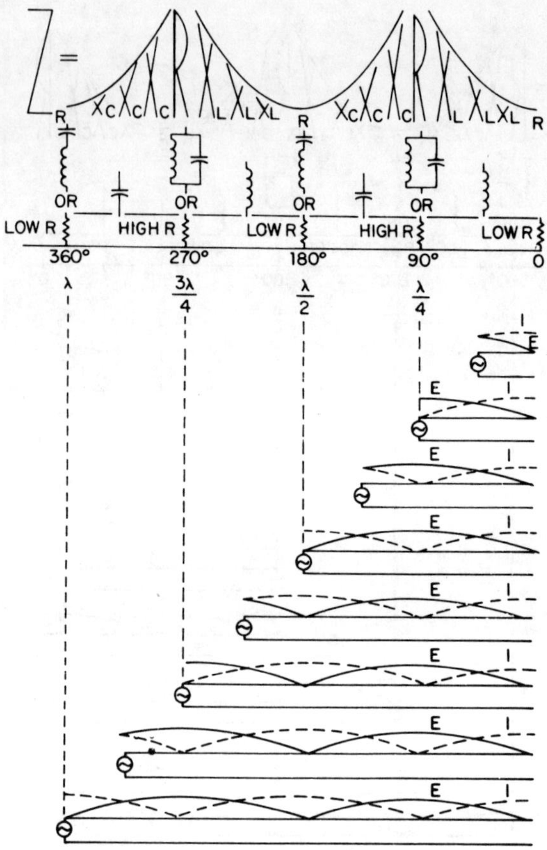

Fig. 3.4 Impedance characteristics of a shorted line.

many of the test setups previously discussed employ attenuators at the output of the source, thereby cutting down any reflections and improving the system match.

The discussions above apply to transmission lines at all frequencies. Now let us examine the transmission lines specifically at microwave frequencies. All microwave phenomena ultimately can be expressed in terms of frequency, wavelength, and power, as they apply to energy in the form of a wave movement. One method by which the wave propagates along a transmission line is in the *transverse electromagnetic mode* (TEM). In a TEM wave, both the electric and magnetic fields are transverse (at right angles) to the direction of travel; for lossless lines, there are no electric or magnetic components parallel to the wave motion. TEM waves exist in two-conductor transmission lines such as two-wire, stripline, microstrip, and coaxial line.

TEM waves are also called *traveling waves*. Figure 3.5 shows a sinusoidal traveling wave of voltage on a transmission line. This line has a physical length (l) measured in meters, and an electrical length measured in wavelengths (λ). The wavelength is defined as the distance between successive points of the same electrical phase in a wave, and is contingent upon the frequency (f), the velocity of propagation (v), and the nature of the medium through which the wave travels. In free space, the velocity is 300,000,000 meters/s (186,000 miles/s). This figure is more easily recognizable when put into powers of ten — 3×10^8 meters/s or 3×10^{10} cm/s.

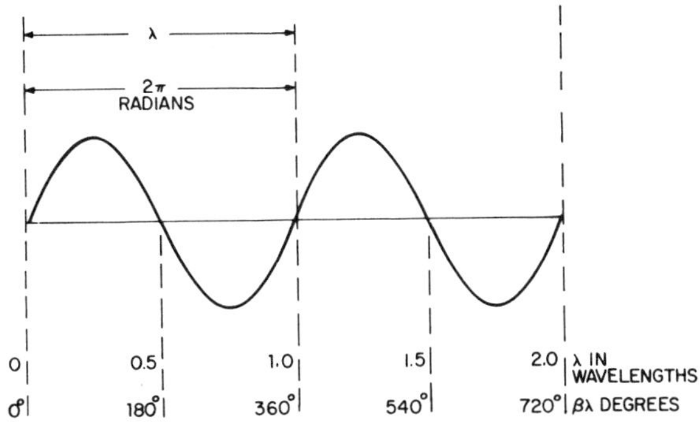

Fig. 3.5 Traveling wave on a transmission line.

If the medium has a dielectric constant other than one, (the dielectric constant of air), the wavelength is reduced by the factor $1/\sqrt{\varepsilon}$. The wavelength, λ now becomes:

$$\lambda = \frac{1}{\sqrt{\varepsilon}} \frac{v}{f}$$

A wavelength alternately is defined as the distance in which the phase changes by 2π radians (1 radian $= 180°/\pi = 57.3°$), as in Figure 3.5.

The term *voltage standing wave ratio* (VSWR) has arisen frequently in previous chapters. The VSWR is defined as the ratio of the maximum and minimum voltages on a transmission line; it can be determined in many ways:

$$\text{VSWR} = \frac{E_{\text{max}}}{E_{\text{min}}} = \frac{E_i + E_r}{E_i - E_r} = \frac{I_{\text{max}}}{I_{\text{min}}} = \frac{1 + \rho}{1 - \rho}$$

where:

E_i = Incident voltage
E_r = Reflected voltage
ρ = Reflection coefficient (magnitude of Γ)

Similarly, given the VSWR, the reflection coefficient can be ascertained:

$$\rho = \frac{\text{VSWR} - 1}{\text{VSWR} + 1}$$

If the transmission line is terminated in a short or open circuit, the reflected voltage (E_r) is equal to the incident voltage (E_i).

From the above equation, the reflection coefficient is 1.0, and the VSWR is infinite. If a matched termination is connected to the line, the reflected wave is zero, the reflection coefficient is zero, and the VSWR is 1.0. Therefore, the best VSWR is 1.0, while the worst is infinite (designating either an open or short circuit).

The most convenient method of calculating VSWR begins by measuring the return loss in dB, and working backwards as follows:

1. Find the return loss figure, which, in dB is 20 log ρ.
2. By working the above figures backwards a value for ρ (reflection coefficient) can be determined.
3. Given the value of ρ, VSWR = $1 + \rho/1 - \rho$.

By using standard, precomputed tables, a VSWR may be obtained easily, just by knowing the return loss figure.

A true transmission line is one which is used for a frequency high enough to cause the line to become more than a quarter-wavelength long. In this case interest is focused on a property previously mentioned—*characteristic impedance*. Both the source and the load should be matched to the impedance; an ideal way is by using adjustable transformers at the source and load ends of the line. If the transformers are perfectly adjustable and they can be set so that a maximum power first is drawn from the source through one into the line, then made available out of the other transformer into the load, perfect matching can be obtained. But if the line is broken at any point, the measured impedance becomes the characteristic of the line and is the same in either direction. This result, of course, is expected since maximum transfer of power takes place when $R_L = Z_o$. Perfectly adjustable transformers obviously are not available, but matching procedures exist which adequately match both the source and the load to the line; adjustable transformers illustrate just how a maximum transfer of power could be obtained.

The lossless transmission line has certain relationships and properties that vary depending on the type of termination found at the end of the line. The properties of terminated transmission lines are open, short, or matched, with resistive greater than Z_o, or less than Z_o. Pure reactances are shown below.

Open-Circuit Termination

1. The incident and reflected voltages are in phase at the open circuit and at one-half wavelength intervals thereafter.
2. The angle of the reflection coefficient is zero at the open circuit and at intervals of one-half wavelength thereafter.
3. The reflected current is equal in amplitude to, and 180° out of phase, with the incident current at the open circuit, and at one-half wavelength intervals from the load.
4. The magnitude of the reflection coefficient is 1.0.
5. The VSWR is infinite.
6. The first voltage minimum along the line is located one-quarter of a wavelength from the open-circuit termination.
7. The first current minimum is located one-half wavelength from the open-circuit termination.

Short-Circuit Termination

1. The incident and reflected currents are equal in amplitude and in phase at the short circuit and at one-half wavelength intervals along the line thereafter.
2. The VSWR is infinite.
3. The reflection coefficient is 1.0, and the angle of the reflection coefficient is 180°.
4. The first voltage minimum is located one-half wavelength along the line from the short circuit.
5. The first current minimum is located one-quarter wavelength from the short.
6. The input impedance of the line is a function of the line length.

Matched Termination

1. The reflected wave is zero.
2. There are no standing waves.
3. The VSWR is 1.0:1.
4. The reflection coefficient is zero.
5. The input impedance of the line is independent of the length of the line.

Pure Resistance Termination Greater Than Z_o

1. The incident and reflected waves of voltage are in phase at the load and at one-half wavelength intervals thereafter.
2. A voltage maximum exists at the load and at one-half wavelength intervals thereafter.
3. The angle of the reflection coefficient is zero at the load and at one-half wavelength intervals thereafter.
4. The amplitude of the reflected wave, the magnitude of the reflection coefficient, and the VSWR depend upon the values of Z_o and R_L.
5. The reflected current is 180° out of phase with the incident current at the load and at one-half wavelength intervals thereafter.
6. The wavelength locations of voltage and current maxima and minima follow the same pattern as for the open circuit except for amplitude variations.

Pure Resistance Termination Less Than Z_o

1. The incident and reflected currents are in phase at the load and at one-half wavelength intervals thereafter.
2. The incident and reflected voltages are 180° out of phase at the load and at one-half wavelength intervals thereafter.
3. A voltage minimum is located at the termination.

Pure Reactance Termination

1. The incident and reflected voltages are out of phase except at E_{max} and E_{min} points where they are either in phase or 180° out of phase, respectively.
2. The VSWR is infinite.
3. The reflection coefficient is 1.0.

Transmision line resistance differs greatly from transmission line impedance. This resistance—the type that causes attenuation down the line—can be attributed to many factors such as:

1. Conductor losses (*skin effect*)
2. Dielectric losses
3. Hysteresis losses
4. Mismatch losses
5. Losses due to radiation

The first three factors—conductor, dielectric, and hysteresis losses—are absorptive losses by nature; they dissipate energy. Mismatch loss and losses due to radiation reflect and guide energy away from the transmission line, respectively.

Conductor Losses (skin effect)

Conductor losses are caused by the series resistance of the conducting medium. Existence of the skin effect is known as a result of lower-frequency techniques. As frequency increases, skin effect becomes more critical; that is, at higher frequencies in the transmission line, the current is restricted to travel in the surface layer of the conductor. The penetration of the current flow is defined by the skin depth (δ)—the thickness of the layer in which the current density drops to $1/\varepsilon$ the value on the surface. Skin depth can be calculated from the following formula:

$$\delta = \frac{1}{2\pi} \sqrt{\frac{\rho}{f\mu_r}} \text{ (cm)}$$

where ρ is the specific resistivity of the conductor in terms of Ω-cm, μ_r is the relative permeability of the conductor (considered only when the material is ferromagnetic and has a relative permeability different from nonferrous material), and f is the operating frequency in terms of gigahertz. Table 3.1 gives the specific resistivity of a few metals. Since skin depth at microwave frequencies is a very small fraction of the conductor thickness (less than a thousandth of an inch), the conductor surface is usually plated. Plating of a couple of thousandths of an inch with a high conductive metal completely negates the effect of the base material since all the current flows in the plating.

Table 3.1 Specific Resistivity of a Few Metals

Metal	*Specific Resistivity (ohm-cm) × 10^{-6}*
Aluminum	2.83
Brass	6.4–8.4
Copper	1.724
Gold	2.44
Lead	22.0
Palladium	11.0
Phosphor bronze	10.5
Platinum	10.0
Rhodium	5.0
Silver	1.629
Zinc	6.1

Caution should be exercised when calcualting skin depth and effective losses for plated metals. Solid metals and plated surfaces have different specific resistivities since plating is as dense as solid metal and is porous in nature. If high conductivity is desired, plated surfaces are burnished in order to pack the plating more densely. It is good practice to plate to a thickness of at least three to five times that of skin depth. The surface finish of the plated material also comes into consideration. Because current penetration into the surface is so shallow, the currents must follow all surface imperfections, consequently increasing both the path of the current and the actual resistance of the conductor. When very low losses are desired, microwave transmission line components are usually highly polished to eliminate all machining marks; even microscopic scratches crossing the current flow can increase appreciably the equivalent resistance of the conductor. However, if the machining operation is planned in such a way as to keep the marks in line with the current flow, their effect on the resistance is negligible.

Dielectric Losses

The dielectric losses of a transmission line can be compared to the dielectric effect in a capacitor. The dielectric material in a capacitor increases the effective capacitance between the conductors, but not without some losses being involved.

Loss Tangent

The losses in a dielectric are designated by a loss tangent. Since the loss tangent of commonly used dielectric material is very small, it is approximately equal to the power factor of a capacitor. Consequently, the dielectric losses of a transmissison line, although small in nature, can be controlled by a careful choice of material.

Hysteresis Losses

For most practical purposes, hysteresis losses are included in the skin-effect formula. If material with permeability differing from nonmagnetic materials is used, plating is usually applied, and the effect of hysteresis losses becomes either negligible or is alleviated entirely. Hysteresis is discussed further in Chapter Two under *YIG Filters*.

Mismatch Losses and Losses Due to Radiation

Mismatch loss occurs when a discontinuity appears on a transmission line or when its termination is not equal to the characteristic impedance. Consequently, some of the power available at that point is reflected, rather than transmitted or propagated.

As a result, not all the energy is delivered into the load. The losses from the available and dissipated signals result from a mismatch. Discontinuities or openings on transmission lines are more serious at higher frequencies than at lower ones; the microwave energy—the electromagnetic wave—can radiate more easily out of the line and cause losses. For all practical purposes, this type of loss is not intentional, and cracks should be sought out and repaired.

At high microwave frequencies, some cables should be double- and triple-shielded to alleviate losses due to radiation.

3.2 COAXIAL CABLES

Each piece of an automobile engine is a marvel to behold. However, unless the separate parts are put together they are utterly useless. Without the nuts, bolts, belts, and hoses to connect the individual pieces, all the craftsmanship exemplified by them is wasted—they will rust away never having run a mile. The same points are applicable to the microwave components to be discussed in the next chapter. Each component is an amazing piece of circuitry designed for a specific purpose. But unless there is some means of putting them together to form an operative system, they might just as well be rusting automobile parts. That means of incorporation is through the use of cables. Indeed, the nuts and bolts of a microwave system are the cables, their associated connectors, and any necessary adapters.

The coaxial line is composed of two concentric conductors separated by an insulating material. The coaxial line illustrated in Figure 3.6 can be constructed in both rigid and flexible forms. The dielectric in the rigid coaxial line is usually air, and the center conductor is concentrically located within the outer conductor by means of dielectric insulating supports called beads. In flexible cable, the center conductor is surrounded throughout its length by a flexible dielectric material, such as polyethylene.

Just as there are certain terms that characterize the nuts and bolts in a car—such as a 10–24 × $1^1/_4$ in Hex head—so also are there definite specifications for cables to help an individual decide which type is suitable for a particular application. The list of terms is fairly extensive; each item, though, plays a role in helping the system perform optimally. There is much more to specifying cables than simply saying that you need a 50 Ω flexible coax. The terms covered are:

1. Characteristic Impedance
2. VSWR
3. Capacitance
4. CW power rating
5. Maximum operating voltage
6. Attenuation
7. Velocity of propagation
8. Electrical length stability

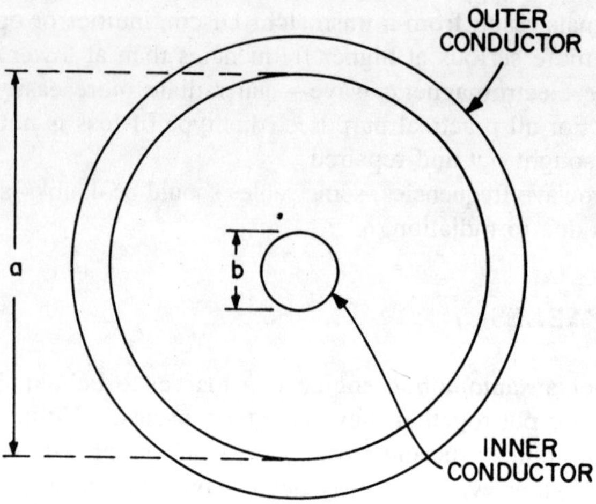

Fig. 3.6 Coaxial cable.

9. Pulse response
10. Shielding
11. Cut-Off frequency
12. Cable noise
13. Flexibility
14. Operating temperature range
15. Environmental resistance

The list is extensive, but no unnecessary terms are included. To prove this last statement, we study each and how it applies to a microwave cable.

Characteristic Impedance (Z_o)—Characteristic impedance is the value of resistance, present all along the cable, that would be obtained by an RF measurement at either end. Very simply, a cable with a characteristic impedance of 50 Ω, for example, is equivalent to a 50 Ω resistor that is long enough to connect two system components. This impedance value is determined by both the ratio of the inner diameter of the outer conductor (D) to the outer diameter of the inner conductor (d) and the dielectric constant (ε) of the insulator material between the conductors. These specifications are illustrated in Figure 3.7. Expressed mathematically:

$$Z_o = \frac{138}{\sqrt{\varepsilon}} \log_{10} \frac{D}{d}$$

The most common values of Z_o are 50, 75, and 93 Ω.

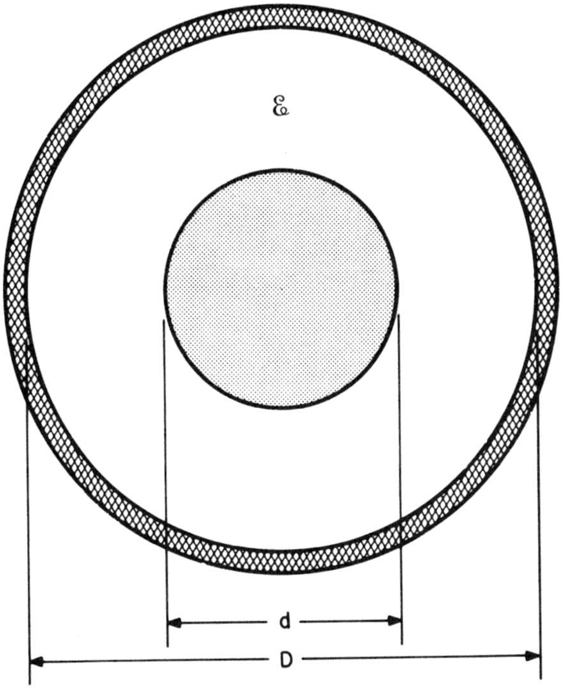

Fig. 3.7 RF cable diagram.

VSWR—VSWR is a measure of the reflections (which may be due to cable length, breaks in the dielectric or shield, or even discontinuities in construction) generated within the cable. Any condition that could upset the impedance and cause reflection produces a standing wave in the cable. A good VSWR figure on a data sheet ensures that a cable is of both good construction and a constant characteristic impedance.

Distributed Capacitance and Inductance—Figure 3.8 shows a schematic representation of a section of RF cable. The component L represents the inductance created by a current flowing through a wire; it is present in any conductor and can be figured by the expression

$$L = .1404 \log D/d \; (\mu H/\text{ft})$$

The component G is the amount of conduction through the dielectric material of the cable due to leakage; under most conditions, this figure is very small. The component C is the natural capacitance set up between the two conductors of the cable, which act as plates of a capacitor. The total cable capacitance is

$$\text{C} = \frac{7.354\varepsilon}{\log_{10} D/d} pF/\text{ft}$$

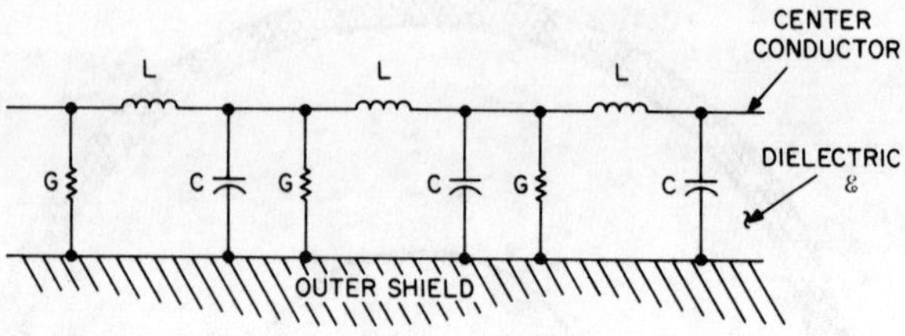

Fig. 3.8 Schematic representation of an RF cable.

The nominal value for a 50 Ω cable is around 25 to 30 pF/ft. Values for higher impedance cables are lower because the D/d ratio is larger; that is, the two conductors are farther apart and, thus, there is less capacitance between them.

CW Power Rating—This measurement depends on such things as frequency, temperature, altitude, and VSWR; it indicates the amount of power the cable can handle and still perform in the system as originally intended. Basically, this rating evidences how well the cable removes the heat generated by high RF power; its value decreases as the frequency of operation increases. High operating temperatures and high altitude also reduce the rating; both conditions decrease the ability of a cable to dissipate heat. A high VSWR reduces the power rating because there is a hot spot at each point of reflection of power. Care in choosing a cable, taking into consideration proper temperature, altitude, frequency, and system impedances, ensures that it can handle the level of power generated by your system.

Maximum Operating Voltage—A cable cannot operate continuously if excessive voltage causes an arc-over to exist. Noise is generated and dielectric damage, or even eventual breakdown, can occur. The maximum operating voltage must be less than the arc-over level. (This restriction should not be confused with the dielectric strength determination of the cable, which involves a test voltage applied for one minute during manufacture.) Maximum operating ac (rms) voltage or peak levels are given on cable data sheets; the maximum dc voltage is considered conservatively to be three times the ac. This term goes hand-in-hand with that of power handling discussed above; if either is exceeded, performance is degraded and there is a real possibility of cable damage.

Attenuation—Attenuation is of prime importance in a microwave cable, especially if your system needs long runs of cables between circuits that may be sensitive to an input level. Very simply, it is the amount by which the RF energy is decreased (or lessened) as it progresses down a length of cable. Usually, attenuation is given in dB/100 ft. Expressed mathematically it is

$$\alpha = 4.34 \frac{R_T}{Z_o} + 2.78\sqrt{\varepsilon}\,(\mathrm{f}) \tan d \text{ dB/100 ft.}$$

when:

$$R_T = \frac{1}{D} + \frac{1}{d}\sqrt{f}$$

[f = frequency (MHz)]

Tan d is the dissipation factor given in data sheets. The actual attenuation can usually be approximated by 4.34 R_T/Z_o; the other terms are small enough to have little effect on the overall value. Basically, attenuation is a function of the physical construction (D and d) and frequency of operation.

Velocity of Propagation—The velocity of propagation of a cable is determined primarily by the dielectric constant of the insulating material between the conductors; it is explained as a percentage of the velocity of light in free space and calculated as $1/\sqrt{\varepsilon} \times 100$. Spiral or star-shaped dielectric areas are frequently used to approach the velocity factor of 100 percent; a dielectric support is utilized nonetheless.

Electrical Length Stability—Applications such as antenna feed systems require many cable assemblies that are trimmed to specific electrical lengths. The changes in length with temperature, flexing, tension, and other environmental factors are critical. Therefore, these influences (collectively called the electrical length stability) must be known for such applications.

Pulse Response—There are many microwave applications where there is a pulse response instead of a CW signal. In these cases, special parameters (not critical to a CW application) must be considered—impedance and reflection, rise time, amplitude, overshoot, and pulse echoes. The impedance term refers to impedance variation; fluctuation of ±5 percent are common and tolerable. Greater changes increase reflection and detract from the performance of the cable in a pulse application. The rise time and amplitude are functions of the length of the cable. As the cable grows longer, the rise time increases and the amplitude decreases. The increased capacitance in the longer cable causes exponential (gradual) growth in a pulse and, thus, a longer rise time—it takes more time to attain a specified level. The decrease in amplitude is due to the increased attenuation of the longer cables. An overshoot is when the leading edge of a pulse rises and extends above the desired level, then settles back down. This condition usually arises because of cable construction; techniques using different configurations for the dielectric eliminate the problem. When a narrow pulse is placed on a cable, a small burst of energy may emerge after the initial one arrives. That is, two pulses of different levels are present when there should only be one. This pulse echo is caused by finite, periodic reflections within the cable; normally, it is of such a small value that it can be neglected.

Shielding—The shielding of a cable depends entirely on the construction of its outer conductor. The most common are single braid (brass, tinned, or silver-plated round copper wires), double braid (two single braids with no insulator between them), strip braid (flat strips of copper rather than round wires), and solid sheath (aluminum or copper tubing, used in semirigid cables). This shielding accomplishes two functions—it keeps outside interference from upsetting the RF energy in the cable and it prevents the internal energy from interfering with outside circuitry.

Cut-Off Frequency—This frequency is where modes of energy other than the normal ones can be generated. The standard mode of RF energy transmission is termed the TEM—the transverse electromagnetic mode. The cut-off frequency, f_c, is a function of the diameter of the cable conductor and the velocity of propagation.

Expressed mathematically it is

$$f_c = \frac{1.18 \times 10^4 v}{\pi\left(\frac{D + d}{2}\right)} \text{ MHz}$$

where

$$v = \frac{1}{\sqrt{\varepsilon}}$$

Modes higher than the TEM are generated only at impedance discontinuities; in many cases the cable can be operated above cut-off without a substantial increase in either VSWR or insertion loss. However, it is recommended that, in general usage, the cut-off frequency not be exceeded.

Cable Noise—A phenomenon noted is that when flexed a cable generates both acoustical and electrical noise. The acoustical noise is a function of mechanical motion within the cable; the generation of electrical disturbances is attributed to an electrostatic effect. These noise voltages can be minimized by preventing motion between dielectrics and conductors or by using semiconductor layers to dissipate electrostatic charges between the two. Low noise construction of cables must take into account their life expectancy and the environmental conditions to which they are subjected.

Flexibility—Coaxial cables with a stranded center and braided outer conductors are intended for use in applications where the cable must flex repeatedly while in service. Standard braid construction can withstand over 1,000 flexes through 180° if bent over a radius equal to 20 times the cable's outside diameter. Flexible cables may be stored, and are normally shipped, on reels with a hub radius equal to 10 times their outside diameter. If a flexible cable is to be installed in a fixed, bent configuration, the minimum recommended bend radius is five times the outer diameter.

Coaxial cables with a tubular aluminum outer conductor (semirigid cables), cannot withstand more than ten 180° bends over a radius equal to 20 times the outside cable diameter. Semirigid cables are normally shipped on reels with a hub diameter equal to this figure.

Semirigid cables may be bent for installation. The minimum recommended bend radius is 10 times the outside diameter—a bend of only five times may cause a cable to exhibit mechanical and electrical degradation.

Operating Temperature Range—The operating temperature range of a cable is determined primarily by that of the dielectric and jacketing material.

The low end of the temperature range for dielectric is −250° C; most, though, can withstand a minimum of only −65° C. Jacket material is rated as low as −100° C; the majority of types are rated at −60° C. For dielectrics, the high end maximum is +250° C; most materials are rated only to +80° C. Some jacket material goes to +200° C though most is rated only to +125° C.

One additional point—only silver plated conductors are suitable for long term use at temperatures over 80° C.

Environmental Resistance—The length of the life of a cable depends on many factors. The materials used, ultraviolet (sunlight) exposure, humidity, galvanic action, and salt-water corrosive vapors are prime causes of cable failure; each must be considered in its appropriate place, depending on a particular application.

With the above list of terms and explanations as a background for our study of cables, let us proceed to the various flexible and semirigid cables. For each we investigate properties and characteristics. Once again, we also provide sample data sheets and discussions of the listed specifications.

3.2.1 Flexible Cable

It should be clear that, at microwave frequencies, cables are more than just a means by which to bridge gaps between components—they are components that take great skill to design.

Flexible cable is the most widely used for testing and in systems—it is the black cable you have seen many times on benches, connecting generators to components, or transferring a dc voltage to an X-Y recorder. Flexible *coax* is available in a large number of varieties; in a catalog, cables range from RG-4/U to RG-389/U (RG is the military designation for cable). Although there are many different types, the basic construction of each is the same, and is shown in Figure 3.9. The overall cable diameters vary from less than $^1/_4$ inch to over one inch, determined primarily by power requirements. Some cables use a stranded inner conductor while others employ solid wire; sometimes the outer conductor is a single braid, other times it is double. The outer jacket plays no part in the electrical performance of the cable—it simply holds everything together and supplies a waterproof covering.

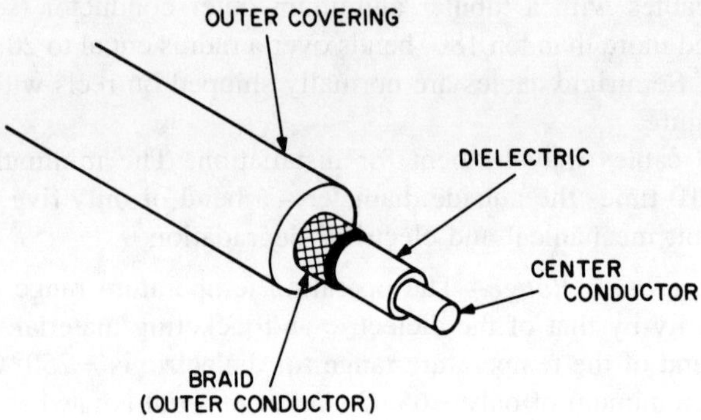

Fig. 3.9 Basic cable configuration.

Though there are a great many different types of flexible cables, only a few of them are widely used. These few are the ones on which we concentrate—RG-8, -9, -58, -59, -62, -196, -214, and -223. (Other cable types are listed in Appendix P.) Figure 3.10 shows the electrical characteristics and dimensions of each. Clearly, there are many specifications to consider when deciding on a cable for your particular needs.

To illustrate the various flexible cable parameters, and how you would go about choosing the proper type, we use a theoretical example: We need a cable for a 50Ω system operating at 1.0 GHz in a pulse operation. The pulse rise time is of a medium duration; its leading edge must be preserved as much as is possible.

The cable run is to be from a 50 W transmitter to an antenna 25 feet away. Existing cables must be used, and there is a maximum allowable space of $^1/_2$ in diameter.

From Figure 3.10 we should be able to determine an appropriate cable; the process may take a while, though, in order to point out the specific properties of each cable along the way. In Table 3.2 we list all nine cables and see how each fits our set of requirements.

We can eliminate two cable types immediately—the RG-59A and -62A—because each has a characteristic impedance greater than 50Ω. (The Z_o for the first is 75Ω; for the second, 93Ω. These cables have applications in 75Ω and 93Ω systems, respectively.)

The next specification, that of pulse response, can be met by choosing a low value of cable capacitance per foot. Obviously, it would be a toss-up if this prerequisite were the only one to be satisfied. Since values range from 28.5 pF to 30.8 pF, (not too broad a spectrum) any of them would be suitable.

RG#	CAP(pF/FT)	WEIGHT lb/FT	MAX OPER VOLTAGE (RMS)	Z_o	ATTEN (DB/100FT) FREQ (GHZ)					POWER (WATTS) FREQ (GHZ)				
					.4	1.0	3.0	5.0	10.0	.4	1.0	3.0	5.0	10.0
8A	29.5	.106	5,000	52Ω	4.6	9	19	28	47	350	190	95	65	37
9A	30.0	.140	4,000	51Ω	4.6	9	19	28	47	350	190	95	65	37
58A	28.5	.029	1,900	52Ω	11	20	41	--	--	75	44	22	15	--
59A	21.0	.032	2,300	75Ω	6.7	11.5	25.5	41	--	135	77	40	27	15
62A	13.5	.038	750	93Ω	5.2	8.5	18.4	29.5	--	138	70	40	27	15
174	30.8	.008	1,500	50Ω	17.5	31	64.3	97	185	25	16	--	--	--
196	29.4	.006	1,000	50Ω	27.5	45	78	115	172	123	78	41	28	14
214	30.8	.126	5,000	50Ω	4.6	9	19	28	47	350	190	95	65	37
223	30.8	.034	1,900	50Ω	8.8	16.5	36	51	85	90	53	28	20	10

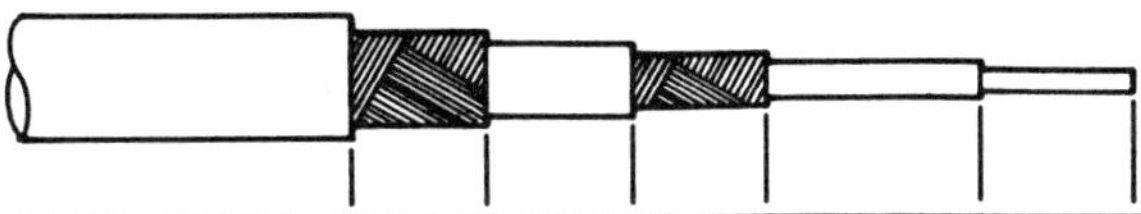

MILITARY NO.	OUTER JACKET	OUTER BRAID	INNER JACKET	INNER BRAID	DIELECTRIC	CENTER COLUMN	IMPEDANCE (OHMS)
RG-8A/U	.405	.340			.285	.089	50
RG-9A/U	.420	.360		.348	.280	.085	51
RG-58A/U	.195	.150			.116	.031	50
RG-59A/U	.242	.191			.146	.083	75
RG-62A/U	.242	.191			.146	.025	93
RG-174/U	.100				.060	.019	50
RG-196/U	.080	.054			.034	.012	50
RG-214	.425	.360			.285	.087	50
RG-223/U	.216	.146			.113	.037	50

Fig. 3.10 Cable data.

Table 3.2 Cable Types

RG#	$Z_o(\Omega)$	*Capacitance (pF/ft)*	*Power (W) at 1 GHz*	*Attenuation (dB/25 ft) at 1 GHz*	*Outside Diameter (inches)*
8A	52	29.5	190	2.25	.405
9A	51	30.0	190	2.25	.420
58A	52	28.5	44	5.0	.195
59A	75	21.0	77	2.875	.242
62A	93	13.5	70	2.125	.242
174	50	30.8	16	7.75	.100
196	50	29.4	78	11.25	.080
214	50	30.8	190	2.25	.425
223	50	30.8	53	4.125	.216

The number of available cables dwindles when the 50 W of power is taken into consideration. Suddenly, only four have this type of power handling capability—8A, 9A, 196, and 214. The RG-223 cable is rated at 53 W, but this specification is applicable only at a lower temperature and under controlled conditions. In a real situation, wrapped inside other cables, the 53-W capability would be questionable.

The additional qualification of low loss over the 25 foot run cuts down the choices even further. The RG-196 quickly falls by the wayside since 11.25 dB of loss over 25 ft would be intolerable.

We are left with three possible choices with outer diameters that fit our requirements. We need some additional information to make our final decision. One point is that the RG-214 is a recommended substitute for the RG-9A, because the former is a double shielded cable. That is, instead of having a single layer of metallic braid covering the dielectric and shielding the center conductor, the RG-214 has a double thickness which improves the cable's shielding ability and makes it more rugged. the RG-8A cable also is found to be a single shield cable; it is recommended for only up to 1 GHz. The RG-214, on the other hand, can be used up to 10 GHz with good results.

Our choice is quite clear at this point. The RG-214 has a characteristic impedance of 50 Ω and adequate cable capacitance/ft figure; it can handle over 50 W of power easily (it is rated at 190 W at 1 GHz) and has 2.25 dB of attenuation in 25 ft; and its outside diameter is .425 in. All of these properties indicate that the cable will perform well in our theoretical example.

This hypothetical situation makes it clear that there is more to choosing a cable than calling for a piece of 50 Ω flexible coax. The characteristic impedance is of great importance, but your specific application usually requires much more than that single qualification. As we have seen, capacitance/ft, attenuation/25 ft, power handling capability, and size are all influential; dielectric material, weight/ft, operating voltage, and operating temperature range also may need to be investigated.

One point should be emphasized about flexible cables—the term flexible does not mean that they can be flexed in any direction an infinite number of times. Instead, a flexible cable has a center conductor (made of some type of metal), a dielectric, and an outer conductor (a metal braid) that can bend a certain number of times at a specified angle (usually 180°) and minimum radius and then return to their original shape with no degradation in operating performance. However, any metal eventually breaks or distorts if it is bent (or flexed) enough times—discretion must be used regarding flexible cables in your system or test setup. Such components last much longer if they are flat on a bench or in a system instead of kinked up in a tight space.

3.2.2 Semirigid Cable

The center conductor in a flexible cable can be solid or stranded; the outer conductor may be either a single or double braid. But in semirigid cable both the center and outer conductors are solid, from which the name semirigid is derived. Semirigid cable can bend, but generally it cannot return to either its original shape or characteristics. It is neither rigid nor flexible—it is semirigid. The cable consists of a core (insulated center conductor) and an outer conductor of seamless tubing. Compared to the flexible variety, semirigid cable offers greater shielding effectiveness and, therefore, reduced attenuation and an increased power rating. Space savings also are realized, since a given insertion loss can be met by a semirigid cable of a smaller diameter than a comparable flexible one. Semirigid construction is shown in Figure 3.11.

Semirigid cable can be manufactured from a number of materials, some of which are listed below, grouped according to the particular portion they compose.

Inner Conductor—Silver plated copper, silver plated copper-clad steel, silver plated beryllium-copper, silver plated aluminum, and resistance wire.

Dielectric—Polyethylene, polytetrafluoroethylene (TFE Teflon®), polytetrafluorethylene-hexafluoropropylene (FEP Teflon®), and kapton.

Outer Conductor—Copper, aluminum, stainless steel, and special copper alloys.

Selecting a particular type of semirigid cable involves the consideration of electrical, mechanical, and environmental conditions. Aside from characteristic impedance, electrical characteristics are attenuation, power rating, higher-mode cut-off frequency, dielectric strength, and corona extinction voltage.

The relationship of a cable diameter, attenuation, and power rating is demonstrated early in the chapter; also, each term is defined.

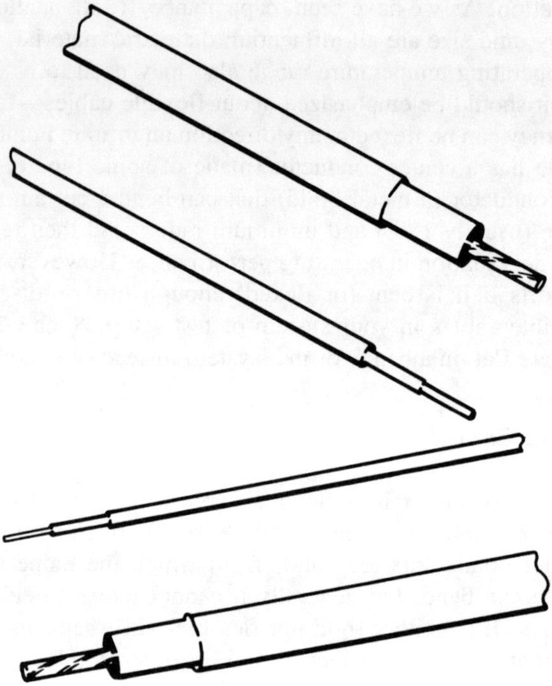

Fig. 3.11 Semirigid cable.

This interconnection is such that the diameter of a cable is chosen so that, for a given space requirement, a cable run has minimum loss and carries maximum power. However, the initial decision also must take into account the fact that a coaxial line (semirigid cable in this case) operated at increasingly high frequencies reaches a point where higher-order modes (other than the desired TEM) are supported. The term that describes the frequency where the higher modes are supported is the *90 percent higher mode cut-off frequency*. Care must be taken that the highest frequency at which your system has to operate does not exceed this figure. The mathematical expression below reveals the dependency of the higher mode cut-off frequency on the dimensions of the cable conductors.

$$90\%\, f_c = \frac{6.75}{\sqrt{\varepsilon}(d + D)}$$

Where

d = the inner conductor outside diameter (inches)
D = the outer conductor inside diameter (inches)

There are times when the selection of a cable involves consideration of a maximum voltage rating. While it would appear to be safe to operate a cable up to and including the *dielectric strength test potential* (a voltage applied during manufacturing to determine the strength of the cable dielectric), it is better to avoid use above the *corona extinction potential* (voltage where an arc-over, or corona, appears and causes noise, electrical damage, or even cable breakdown). There is good evidence that the corona discharge can cause permanent damage to polytetrafluorethylene, a very common semirigid cable dielectric.

To meet mechanical or environmental requirements (and thus improve overall electrical performance) when selecting a cable, the following factors should be kept in mind:

1. Inner conductors of copper-clad steel are stiffer and have a higher tensile strength than those of copper. Hence, copper-clad steel wire may be preferable if a cable is to be terminated with an SMA-type connector, the design of which incorporates the cable inner conductor as the connector pin.
2. Copper wire cannot be used in small-diameter cores with relatively heavy insulation; its tensile strength is insufficient to withstand the stresses of the insulation process. Copper-clad steel wire, though, has magnetic properties and hence should not be employed (or else adequate precautions should be taken) near components whose proper operation is disrupted by magnetic interference. Alternatively, beryllium-copper wire is a good choice for a non-magnetic conductor of high tensile strength and stiffness.
3. Tubular inner conductors reduce the weight of large-diameter cables and enable low VSWR performance over a broad frequency spectrum by utilizing connectors whose pins can be flush-mounted by screwing into the cable inner conductor. This latter attribute eliminates the need for a compensation of the discontinuity caused by a contact barrel.
4. When specifying cables, dielectric materials should receive great consideration. As compared to copper, their temperature coefficients of expansion differ by an order of magnitude—construction of a semirigid cable whose dielectric does not grow upon heating is impossible. Paste-extruded PTFE is reported to have a small temperature coefficient of expansion, and so would be preferable; but manufacturing limits exist.

FEP dielectric has superior resistance to nuclear radiation effects, but it also exhibits both greater variations in dielectric constant and loss tangent with operating frequency. These problems may be objectionable in broadband applications.

Polyethylene dielectric, because of its low melting point, is not suitable for use in system designs involving soldering to the cable outer conductor.

Having covered the basic construction and terminology of the semi-rigid cable, let us see what cables are available for microwave system use.

Types (or sizes) available are 0.020, 0.034, 0.047, 0.056, 0.070, 0.085, 0.141, 0.215, 0.250, and 0.325. (These figures refer to outer conductor outside diameters, expressed in inches.) Of the ten cables listed, three are widely used in microwave—0.085, 0.141, and 0.250. Table 3.3 shows the characteristics of this trio.

Two important characteristics are not in the Table 3.3—cable attenuation and power rating. Because of their great importance, they are listed below in Table 3.4 as a function of frequency:

The cables all have a solid dielectric with a constant of 2.07, resulting in a velocity of propagation of approximately 69.5 percent. The ideal velocity, of course, is 100 percent with an air dielectric—an impossiblity since the center conductor would have no support. We can, however, come close to this ideal by using a combination of air and dielectric material. If the dielectric is in the shape of a corkscrew (spiral), a much higher volume of air is present in the cable. Although this method accomplishes the initial purposes of lowering the dielectric constant and raising the velocity of propagation, it creates the problem of periodicity effects because the density of dielectric inside the cable repeats itself at regular intervals. Periodic VSWR changes also result; thus, any lengths of this sort of cable must be designed very carefully into your particular system. For short runs, though, there is little noticeable effect.

There are other means by which to achieve the same initial goal. Cables are available that provide inner conductor support within the outer conductor by using longitudinal splines of paste-extruded polytetrafluorethylene. These cables are called air-articulated and are shown in Figure 3.12. Such construction results in a lightweight component with minimum attenuation, free of any periodicity effect, and having a dielectric constant of 1.3 and a velocity of propagation of 87 percent. For equal diameters, air-articulated cables have a lower attenuation figure and a higher maximum operating frequency than do those having solid dielectrics.

As is the case with most components, special types of cables are available upon request—those specifically designed to be used at very low temperatures (cryogenic cables), ones with resistance wire as a center conductor to result in a lossy cable, and those that can take excessive vibration or flexing without serious changes occurring in their characteristics.

Since the construction of semirigid cable is different from that of flexible cable, its handling is also different. When working with the former cable, certain procedures should be followed to ensure maintenance of the intended performance. A complete section on cutting and bending of semirigid cable can be found in Appendix Q.

Table 3.3 Characteristics of Three Cables Widely Used in Microwave

	Part No.		
	SRC-085	*SRC-141*	*SRC-250*
Inner conductor OD	.0201 ± .001 in	.0359 ± .0005 in	.0641 ± .001 in
Dielectric OD	.066 ± .001 in	.1175 ± .001 in	.210 ± .002 in
Outer conductor OD	.085 ± .001 in	.141 ± .001 in	.250 ± .002 in
Safe bend radius	.125 in	.250 in	.375 in
Weight	1.31 lbs/100 ft	3.25 lbs/100 ft	10 lbs/100 ft
Maximum operating temperature	125° C	150° C	175° C
Impedance tolerance	±1Ω	±1Ω	±1Ω
Dielectric constant	2.07	2.07	2.07
Dielectric strength rating	5 kV (rms) 60 Hz	5 kV (rms) 60 Hz	7.5 kV (rms) 60 Hz
Corona extinction voltage	1.5 kV (rms) 60 Hz	1.9 kV (rms) 60 Hz	3.15 kV (rms) 60 Hz
90% higher mode cut-off frequency	44.9 GHz	30.4 GHz	17.2 GHz

Reference—Precision Tube Co., Inc.; "Precision Coaxitube" Catalog 3725M

Table 3.4 Cable Attenuation and Power Rating

	.085		.141		.250	
Frequency (GHz)	*Attenuation (dB/100 ft)*	*Power (kW)*	*Attenuation (dB/100 ft)*	*Power (kW)*	*Attenuation (dB/100 ft)*	*Power (kW)*
.50	11.6	.370	7.2	1.000	4.3	2.10
1.0	18.7	.222	11.6	.600	7.3	1.20
3.0	34.0	.115	21.5	.310	14.0	.60
5.0	46.0	.082	28.5	.230	18.9	.44
10.0	73.2	.048	44.5	.160	29.0	.28
12.0	84.0	.040	51.0	.140	33.0	.25
18.0	115.0	.028	68.0	.100	—	—

Reference—Precision Tube Co., Inc.; "Precision Coaxitube" Catalog 3725M

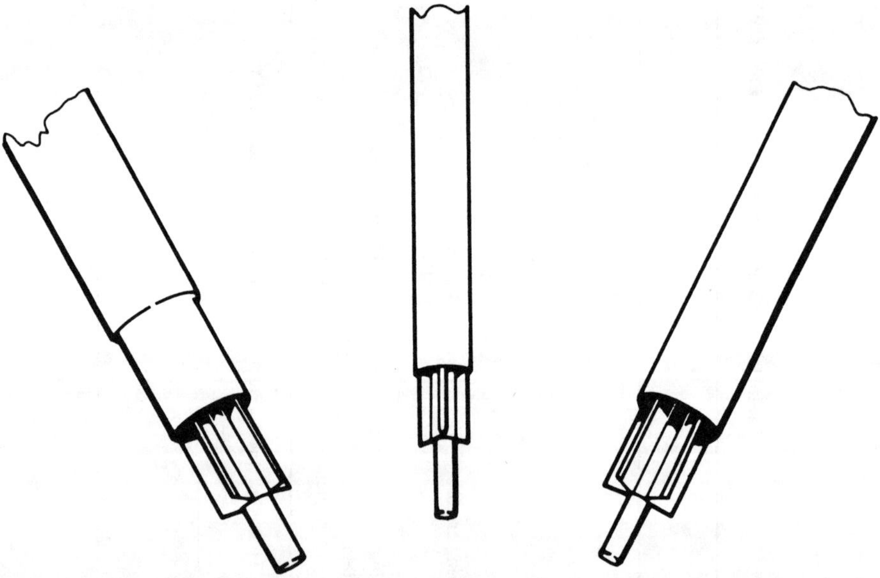

Fig. 3.12 Air-articulated cable.

3.3 STRIPLINE

The terms *tri-plate, sandwich package, strip-transmission line,* and *stripline* arise repeatedly in microwaves; the seemingly insignificant packages are vital to microwave systems.

A cutaway view of a section of stripline is shown in Figure 3.13. It depicts a narrow, flat strip of copper sandwiched between two outer layers of dielectric. The outer surfaces of the dielectric sheets are faced with copper foil. In the most common single-registration form of stripline, the circuit is photo-etched on one side of a double copper-clad laminate, then laid against the dielectric side of a single-clad board. The sandwich produced is held together by means of either eyelets through the board, screws, heavy clamping plates, or a resin bond formed under heat and pressure. Connections to the stripline are made either at the edge of the package or through the side; either special stripline end launch connectors or metal tabs are used.

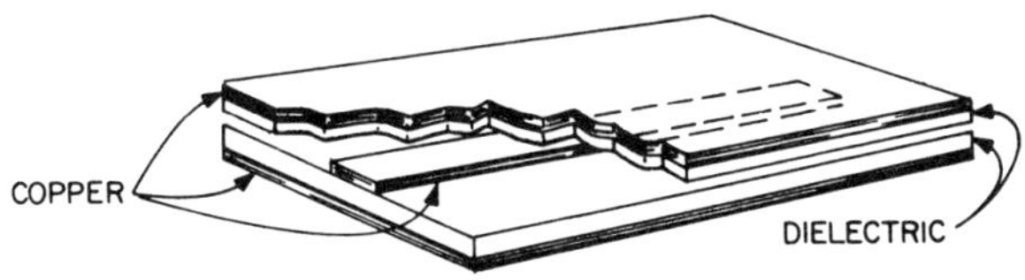

Fig. 3.13 Cutaway view of stripline.

It is helpful to recognize that stripline may be interpreted as a flattened form of coaxial cable. This relationship is shown in Figure 3.14. The arrows indicate the electric field between inner and outer conductors at an instant when the inner conductor is positive. When the lines are far apart, the field is weak. There is no field outside a coaxial cable because the outer conductor is continuous.

Proceeding from step (C) to step (D) in Figure 3.14, the outer conductor is interrupted and it becomes two plates—but the field extends only a short distance on each side of the center strip. If the plates extend well beyond the strip, the field at their edge is essentially zero. Therefore, properly designed stripline, like coaxial cable, is largely self-shielding, although some leakage may occur through the exposed edges of the dielectric. To counteract this potential problem, a conductive tape or strap is folded over the edge of the package to make contact with, and to electrically bond together the two ground planes, as shown in Figure 3.14(D). If clamping plates are used, they can be designed to make contact with each other over the edge of the laminate.

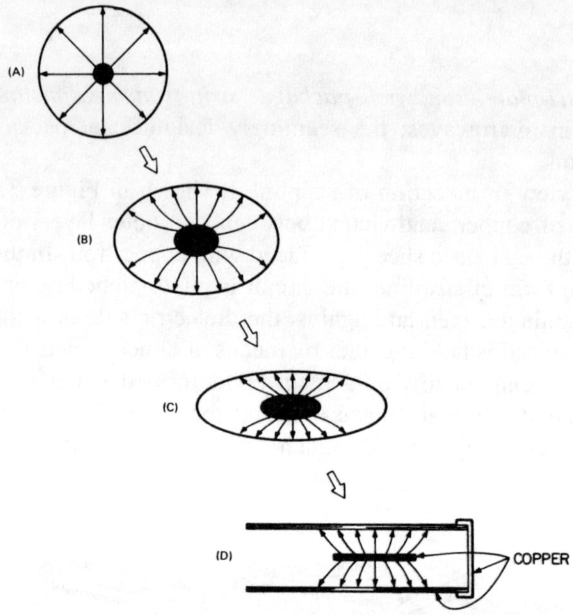

Fig. 3.14 Evolution of stripline from coaxial cable.

When the dielectric is a solid (as opposed to being air), as is usually the case, the speed at which the wave travels along the transmission line (velocity of propagation) is reduced, as is the wavelength. The actual stripline wavelength (λ) is equal to the free space wavelength (λ_o) divided by the square root of the dielectric constant ($\sqrt{\varepsilon}$):

$$\lambda = \frac{\lambda_o}{\sqrt{\varepsilon}}$$

To emphasize the importance of the dielectric constant to the physical size of stripline, Table 3.5 below shows five frequencies and their wavelengths in air and in two types of dielectrics.

It is evident that by increasing the dielectric constant, the required size of stripline components may be reduced. Previously, the only way to achieve this diminution was through the use of aluminum ceramics as dielectric substrates. However, today *Teflon*® fiberglas products are available with dielectric constants of 4, 6, 8, and up to 10.3. From the table above the advantages of these products are obvious—they enjoy the features of fiberglas, the dielectric constant of a ceramic, and are not fragile.

Table 3.5 Frequency Wavelengths in Air and in Two Types of Dielectrics

	Wavelength					
	λ_o *(air)*		$\lambda_1(\varepsilon = 2.55)$		$\lambda_2(\varepsilon = 9.0)$	
Frequency (GHz)	*cm*	*in*	*cm*	*in*	*cm*	*in*
.50	60	23.60	37.7	14.80	20	7.87
1.00	30	11.80	18.8	7.42	10	3.93
3.00	10	3.93	6.29	2.47	3.33	1.31
5.00	6	2.36	3.77	1.48	2.00	0.79
12.00	2.5	0.98	1.57	0.62	0.83	0.33

Because the dielectric constant controls the wavelengths of signals in a stripline circuit, it is a critical property in all applications; however, the thickness of the dielectric is often of equal importance. The *characteristic impedance* (Z_o)—a fundamental design parameter for all stripline circuits—depends on the dielectric constant, the width and thickness of the conductors, and the thickness of the dielectric layers.

Figure 3.15 shows the dependence of the product $Z_o\varepsilon$ on stripline dimensions. This curve, one of the most widely used in stripline design today, illustrates that, for a 50Ω line between dielectric layers with an $\varepsilon = 2.6$ the value of $\sqrt{\varepsilon}Z_o$ should be 80Ω and the required ratio of width to ground plane spacing (w/b) is about 0.7 for 1 oz copper (which has a thickness of 0.0014 in). If a 50Ω line with a ground plane spacing of 1/8 inch were needed, the width should be 0.087 in.

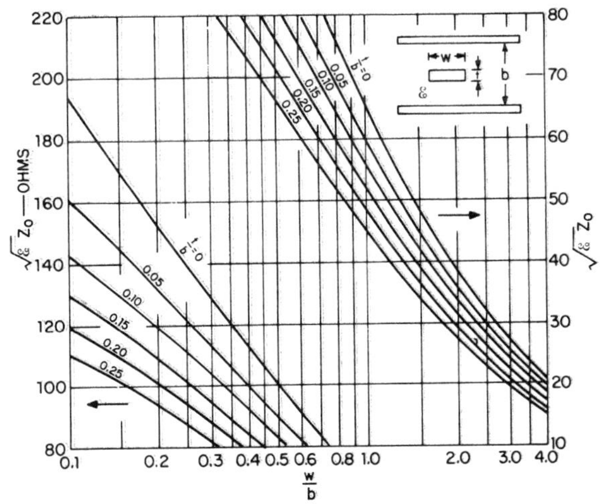

Fig. 3.15 Z_o *versus* w/b for various values of t/b (Cohn, 1955).

Figure 3.15 also illustrates that characteristic impedance is very sensitive to change in dielectric thickness. Because a perfectly uniform and reproducible material for dielectric layers does not exist, most applications are characterized by variations of 3 to 5 percent in thickness. Dielectrics of 0.030 ± 0.001 in are not uncommon; laminate manufacturers guarantee such tolerance.

Another property of dielectrics for stripline use which demands consideration is that of dielectric loss; however, there is some confusion as to its significance and magnitude. The following arbitrary classifications may help to clarify the issue—a dissipation factor below 0.0005 might be termed very low loss; between 0.0005 and 0.002, low loss; and from 0.002 to 0.005, medium loss. Except in the most critical and highest frequency circuits, the loss in signal strength (attenuation) due to dielectric loss is generally tolerable for materials with dissipation factors up to 0.002.

Very low loss circuitry is needed where high Q or very low signal attenuation is needed. High Q circuits—those with a highly selective frequency response—are necessary in filters or devices which control frequency.

Receiving or other circuits for low level signals also need low loss stripline circuits to prevent deterioration of signal strength.

One of the greatest advances in stripline circuitry today comes from the thin material now available which makes possible the production of dielectrics with a thickness of 3, 5, and 10 mils (with a close tolerance). The single component that has benefited the most is the quadrature hybrid coupler. With material of a set thickness, a 3 dB coupler can be obtained and can be repeated almost interminably. Once the line widths and lengths have been determined, the couplers can be mass produced. However, great care should be taken when working with this thin material—the dielectric is only 0.005 in thick or less; some of the lines get narrower than 0.025 in and are made of 1 oz or $^1/_2$ oz (0.0014 in and 0.0007 in) copper. Any unnecessary flexing or bending of these lines can produce hairline cracks which cannot be seen unless they are magnified or checked with an ohmmeter.

The basic elements of stripline circuitry have thus been introduced. For your own interests, a catalog or data sheet pertaining to copper-clad laminates should be consulted in order to match a circuit to a particular application. Items such as dielectric constant, dielectric thickness, copper thickness, and dissipation factor are all important parameters to be considered.

The use of stripline—admittedly a great advance in the microwave field—is feasible only within a certain range of frequencies. As seen from the Table 3.5 above, at frequencies below 1.0 GHz, the size becomes too large and cumbersome. Similarly, above 18.0 GHz, the wavelength (and therefore the stripline) become too small, especially as most components rely on a quarter-wavelength dimension for proper orientation. (A wavelength at 18 GHz = 0.436 in; $\lambda/4$ = 0.109 in.) Therefore, as a general rule of thumb, it can be said that the world of stripline exists between 1.0 GHz and 18.0 GHz.

3.4 MICROSTRIP

The many different types of transmission lines available today vary widely in their characteristics and parameters; however, they all have only one type of dielectric. The dielectric may be air, PTFE, or any other appropriate material, but only one variety is used in a given line. Conversely, microstrip lines have two types of dielectrics—air and a solid.

A microstrip transmission line consists of a metalized strip and a ground plane separated by a solid dielectric, as shown in Figure 3.16. This configuration has become a popular alternative to the classic shielded stripline, especially for miniature microwave circuits. However, there are many applications where stripline still is preferred over its successor.

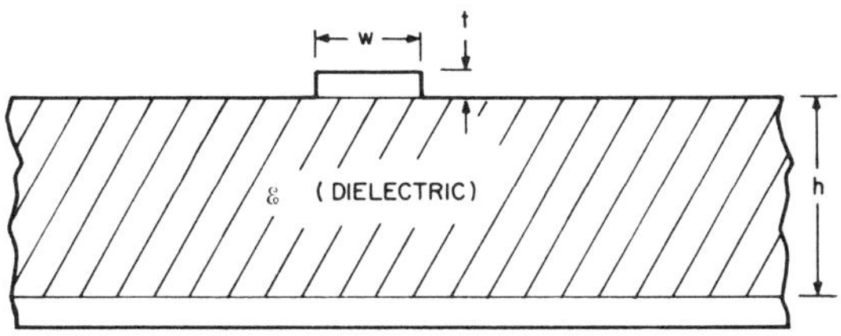

Fig. 3.16 Microstrip transmission line.

Microstrip is usually chosen wherever active devices (transistors) are used. Because the transistors can be placed on top of the board in many instances, there is no need for any cutting of the ground planes. This convenience is also true for any circuit whose components must be mounted to a board. Any resistors, capacitors, or diodes that are to be attached can be soldered directly onto the microstrip board.

The dielectric constant and thickness of the board material are of equal importance in microstrip as in stripline. Length, of course, varies by the same constant, $1/\sqrt{\varepsilon}$, as with stripline or any other transmission line. The dielectric thickness, once again, determines the width of the microstrip line for various impedance values. Figure 3.17 is a plot of strip width in inches versus impedance for various thicknesses of board material. Notice that a 50Ω line for a common $^{1}/_{16}$ inch Teflon® fiberglas board is 0.160 in. Also see that if the thickness of the board is cut in half (to $^{1}/_{32}$ in), the width of a 50Ω line does not decrease proportionally. A 50Ω line is not 0.086 in, hence the width of a microstrip line is *not* a linear function.

Figure 3.18 is an enlarged photo of a microstrip amplifier. Notice the one-sided open construction and the variations in length and width to create the appropriate values of capactiance and inductance. Notice also the accessibility of the components, as they are mounted to the top of the board. Circuits of this type should

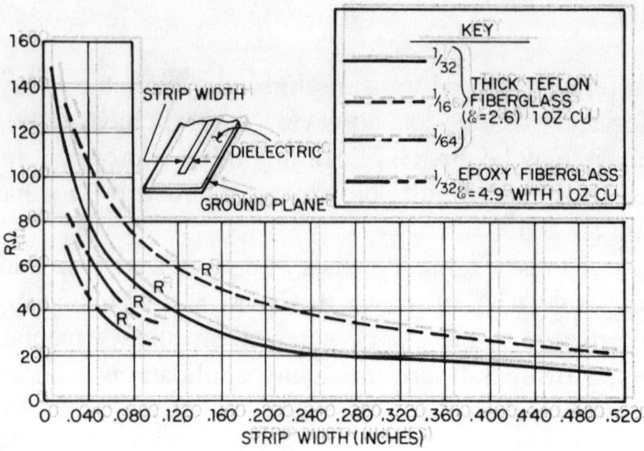

Fig. 3.17 Microstrip line impedance *versus* strip width (*courtesy of Microwave Engineers' Handbook*).

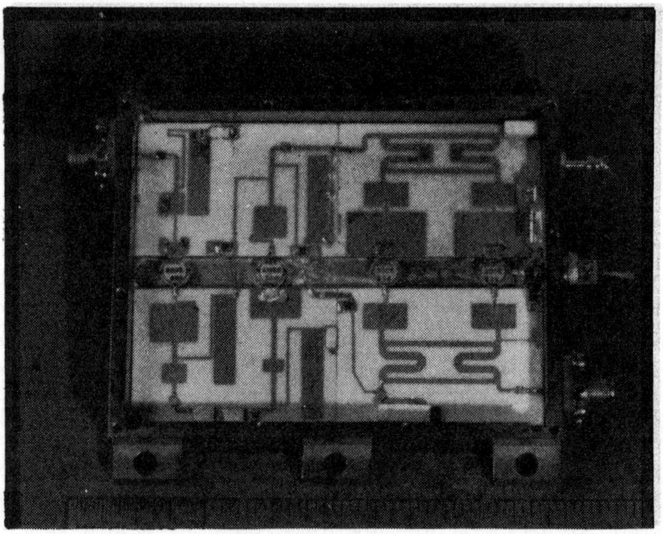

Fig. 3.18 Microstrip amplifier.

represent microstrip construction. Whenever active devices are found on a board, or many components are to be mounted to that board, microstrips should be employed.

Although the rule stated above applies 90 percent of the time, microstrip is sometimes advisable for the simple reason of ease of construction. For example, suppose you have a string of amplifiers and wish, for some reason, to monitor the output of each stage. The easiest means is by placing a directional coupler at the output of each stage. If all of the amplifiers are of a microstrip construction, there is no need to put a stripline coupler in between each microstrip stage. Thus, for ease of construction (everything printed on one board and no sandwich packages), microstrip couplers should be used as monitors. Indeed, only the conventional printed circuit board is easier to assemble.

If there is one word which describes microstrip, it is *simplicity*. It is possible to build a microstrip circuit, which can perform highly complex operations, with the ease of using a carving knife. People have looked at a microstrip amplifier and asked, "Where are all the parts?"

When building microstrip, one precaution should always be taken—be sure that any grounds present in the circuit are established on the underside of the board. One method often used is soldering copper straps to the top and bottom edges of the board along its entire length. If the ground is located at some point other than along the outside edge, a long narrow slot should be cut into the board through which a copper strap is run and soldered to the board at its top and bottom. *Do not* use a piece of brass wire as a jumper through a microstrip board. Remember that in microwaves, grounds are not points, but planes.

Thus, the *microstrip transmission line* has been introduced and categorized.

3.5 WAVEGUIDE

In actuality, every type of transmission line is in a sense, a waveguide, since it is designed primarily to guide or conduct energy (an electromagnetic *wave*) from one point to another. Whether the waves are thought of as electromagnetic field quantities or simple sinusoidal distributions of current and voltage, the meaning is the same. A transmission line is essentially like a track which steers the energy wave along a certain direction. Common usage, however, dictates that the term *waveguide* be restricted to indicate what might more completely be called a hollow-pipe waveguide, or at least a waveguide in which two distinct conductors are not present.

Waveguides are used at microwave frequencies for two reasons: they are often easier to fabricate than are coaxial lines; secondly, they can often be made to have less attenuation. Coaxial line requires that a center conductor be supported somehow in the center of a cylinder; a waveguide requires no center conductor and its dielectric is air. Losses in a coaxial line over any distance (usually greater than a couple of feet), along with cable leakage at the higher frequencies, make the waveguide structure even more desirable.

One way of commonly deriving rectangular waveguide is from a two-wire transmission line, as illustrated in Figure 3.19. The transmission line is supported by two quarter-wavelength sections which, since the input impedance of each is theoretically infinite, have no effect on the transmission of power. If the number of stubs is increased to infinity, the rectangular waveguide is formed as shown. The *a* dimension of the waveguide cannot be less than one-half wavelength; in fact, it must be slightly greater in order to completely accommodate the transmission line function

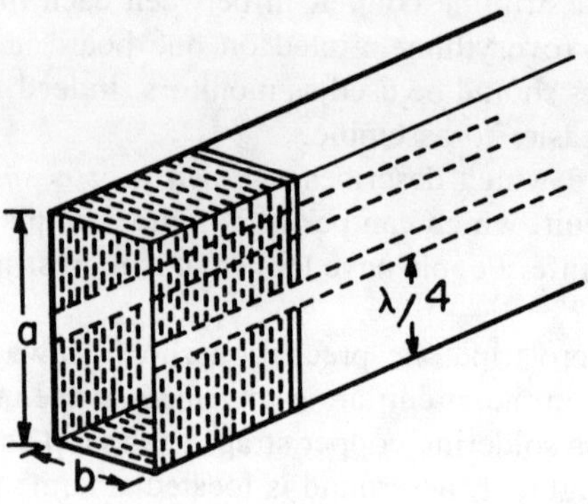

Fig. 3.19 Waveguide derived from quarter-wave sections.

while preserving the insulating properties of the quarter-wave sections. Any frequency which makes the *a* dimension less than one-half wavelength causes the circuit to become an inductive shunt which allows no propagation. The frequency at which the *a* dimension is one-half wavelength (free-space wavelength) is called the cut-off frequency and is designated f_c. The free-space wavelength associated with this cut-off frequency is the cut-off wavelength, designated λ_c. ($\lambda_c = 2a$.) At dimensions greater than $2a$ some attenuation is experienced. Except for complications that may arise in attempting to avoid higher order modes when $\lambda_c < a$, all shorter wavelengths are passed readily. Thus, a waveguide is very similar to a high-pass filter, in this respect.

As mentioned before, microwave energy in a transmission line is, in effect, a wave. This concept is especially appropriate to the propagation of RF energy in a waveguide.

Waves of this type are of great importance and, more than a mere combination of two plane transverse waves, they take on an identity of their own. They exist in either of two forms, designated as TE or TM modes.

In the TE (transverse-electric) mode, the electric field is transverse to the direction of propagation. In the TM (transverse-magnetic) mode, the magnetic field is transverse to the direction of propagation.

In addition, subscripts are used to describe the electric and magnetic field configurations. The general symbol is TE_{mn} or TM_{mn}, where the subscript m indicates the number of half-wave variations of the electric field intensity along the *a* (wide) dimension of the guide. The second subscript n indicates the number of half-wave variations of the electric field in the *b* (narrow) dimension of the guide. The TE_{10} mode has the longest operating wavelength and is designated as the dominant mode, in which the lowest frequency that can be propagated in a waveguide is found. Therefore, knowing the mode of operation and the dimension of the waveguide, the cut-off wavelength can be ascertained from the equation below:

$$\lambda_c = \frac{2}{\sqrt{(m/a)^2 + (n/b)^2}}$$

where a and b are measured in centimeters.

Rectangular waveguides are usually chosen so that only the dominant mode exists over a certain frequency range. The above consideration determines the *a* dimension. The *b* dimension is important because of the following considerations:

1. The attenuation loss is greater as the *b* dimension is made smaller.
2. The *b* dimension determines the voltage breakdown characteristics and therefore determines the maximum power capacity.

The power capacity depends on the maximum electric field intensity, *a*, and *b*. Therefore, *a* and *b* should be made as large as possible for higher power capacity. In practice, the dimension *b* is usually chosen to be about one-half of *a*, forming a ratio of 2:1. Standard waveguide dimensions are shown in a complete table in Appendix O.

The *characteristic wave impedance* of a waveguide is analogous to the characteristic impedance of two-wire and coaxial lines. The wave impedance represents solely the ratio of the electric and the magnetic fields at a certain point in the waveguide. The actual wave impedance is of little use and is seldom determined. There is no advantage in extending calculations beyond the determination of the per-unit impedance since the actual measurement in a waveguide results in a normalized value. For reference purposes, the wave impedance equation for TE and TM modes are:

$$Z_{TE} = \frac{\eta}{\sqrt{1 - (\lambda/\lambda_c)^2}} = \frac{\eta}{\sqrt{1 - (f_c/f)^2}}$$

$$Z_{TM} = \eta\sqrt{1 - (\lambda/\lambda_c)^2} = \eta\sqrt{1 - (f_c/f)^2}$$

where η is the intrinsic impedance of the transmission line medium (377Ω for free space).

Notice that the *waveguide impedance* depends on both the frequency for which the waveguide was built and on the medium used (usually free space).

As is the case with ordinary wired circuits, waveguide impedance usually implies more than just the ratio of two numbers representing the peak, average, or rms values of the sinusoidally varying fields. The impedance may, as a special case, be pure resistance, or it may be reactive.

Waveguide is not always found in the familiar rectangular form—sometimes it is circular. This variety is used in many special applications in microwave techniques. The circular guide has the advantage of greater power-handling capacity and lower attenuation for a given cutoff wavelength, but it is also larger and heavier.

Also, the polarization of the transmitted wave can be altered by minor irregularities of the wall surface of the circular guide, whereas the rectangular cross section definitely fixes the polarization.

A section of waveguide efficiently allows the transferral of energy from one point to another. However, unless this energy can be coupled from one piece of waveguide to another, or from waveguide to a coaxial line, the waveguide is completely useless to us.

Several methods which can be used to effect this coupling are shown in Figure 3.20.

Magnetic coupling into a waveguide by means of a loop is shown at (A). The loop may be placed at any position as long as it links with the magnetic lines of force. The degree of coupling can be altered by rotating the loop relative to the direction of the magnetic line. The probe can sometimes be positioned where the magnetic field is weaker.

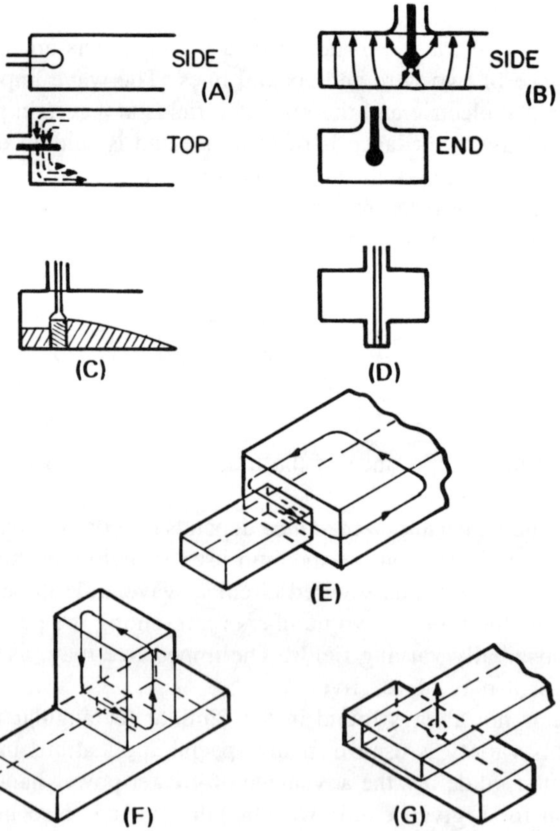

Fig. 3.20 Examples of coupling between transmission lines.

The coaxial line may be coupled to the waveguide by extending the center conductor of the coaxial line into the wavelength parallel to the electric field at, or near, a point where the electric field has its maximum value. The conductor is located in the center of the guide, as shown in (B). The outer conductor terminates at the wall of the guide. The waveguide is usually closed at one end in order that the waves reflected from this point can reinforce those from the probe, which is located one-quarter wavelength from the closed end.

The coupling probe can be extended across the waveguide as shown in (D). An adjustable short circuit can be connected to the center conductor beyond the junction so that the transition can be adjusted for best match over an appreciable frequency band.

The tapered transition at (C) is used when the waveguide impedance must be lowered appreciably. The taper can be either a continuous or step type, as indicated by the dotted lines. This type transition is therefore useful for broadband applications.

As shown at (G), electric coupling is effected through the round hole when components of the electric field normal to the common surface are present. Magnetic coupling exists if the magnetic field has a component parallel to the magnetic field at the adjoining surface. These conditions are sometimes determined by field distributions set up by certain types of restricted enclosures connected to the waveguide, such as resonant cavities.

Magnetic coupling occurs through narrow slots. Electric coupling is usually small, but magnetic coupling is somewhat more pronounced.

There are many variations and combinations of the different types of coupling structures. Figure 3.21 shows some waveguide components where coupling is necessary for proper operation.

Before investigating the applications of waveguides in microwaves, a review of some of the important points covered thus far, along with a comparison of waveguides with coaxial or two-conductor lines, is in order.

1. The characteristic impedance and phase velocity are essentially independent of frequency in the TEM mode of propagation. Lossless transmission lines carrying TEM waves are nondispersive.
2. Propagation occurs in a waveguide only above a critical frequency called the cutoff frequency. Thus, a waveguide carrying a TE or TM wave may be considered to be a high-pass filter as far as the particular mode is concerned. At frequencies far above the cutoff point, the propagation characteristics approach those of free space; for example, the waveguide wavelength approaches the free unbounded wavelength in the medium contained inside the waveguide.
3. In the waveguide, the phase and group velocities and the wave impedance of TE and TM waves are functions of frequency. Therefore, the waveguide is a dispersive medium, in that it tends to dissipate waves of different frequencies which may be applied simultaneously at a common point. This dispersion phe-

nomenon causes distortion in waves which have modulation components extending over a wide frequency range.

4. TEM waves cannot exist in closed waveguides without inner conductors, but TE and TM waves may exist as higher order modes on two-conductor or coaxial lines if the operating frequency is sufficiently high.

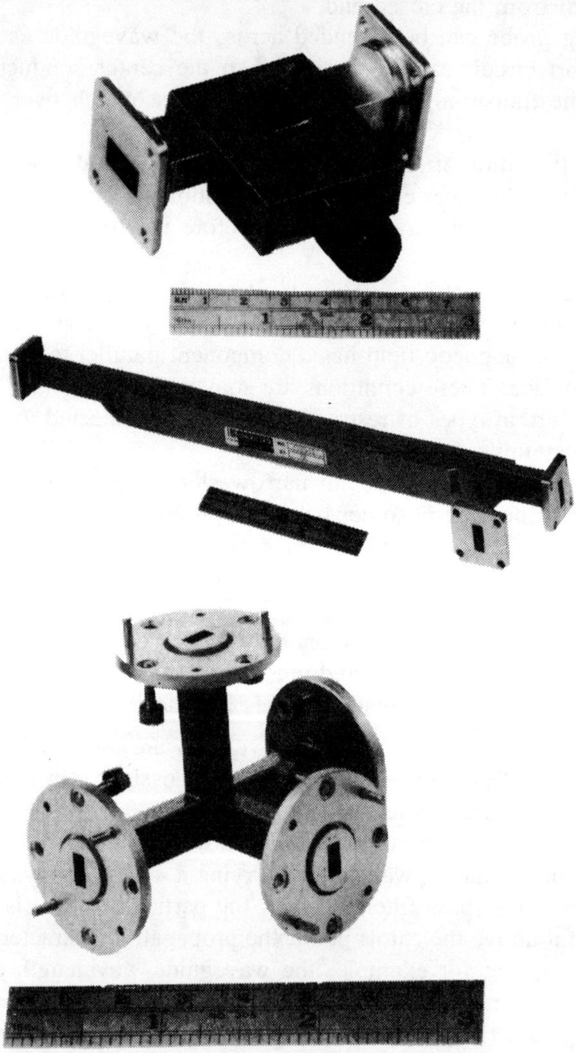

Fig. 3.21 Waveguide components.

3.5.1 Impedance-Changing

Impedance-changing devices are introduced into the waveguide near the sources or reflected waves in order to eliminate standing waves. If diaphragms of good conductivity and of a thickness considered small in relation to wavelengths are extended into the waveguide, the necessary susceptance can be introduced to reduce the standing waves to nearly zero. Such diaphragms therefore play an important part in microwave techniques. The element illustrated in Figure 3.22(A) adds inductive susceptance across the waveguide and is called an inductive window, or inductive iris. The capacitive window or iris is shown at (B) and can be explained by the presence of the electric field in the vicinity of the diaphragm. The inductive iris has an advantage over the capacitive iris in high-power waveguide installations. If the combined effects of the inductive and capacitive windows are employed, the composite type illustrated in (C) and (D) is obtained, known as the resonant window or iris. Shown at (E), the circular iris, providing inductive susceptance, is usually preferred in the circular waveguide.

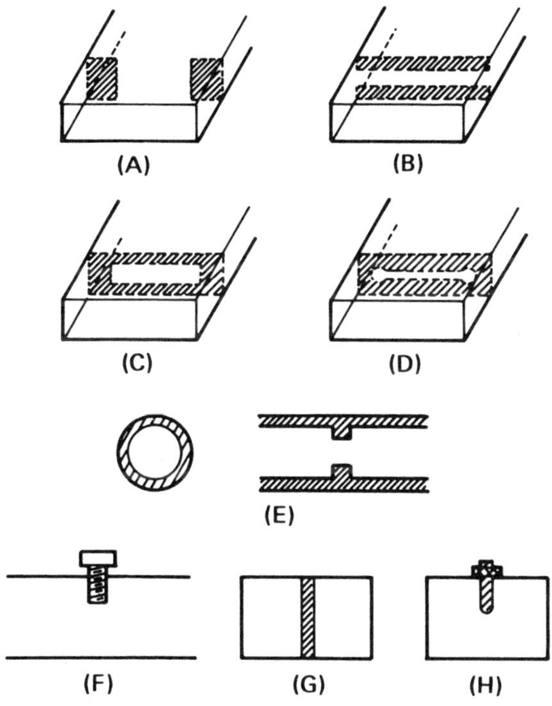

Fig. 3.22 Waveguide impedance-changing devices (waveguide discontinuities).

The tuning screw shown at (F) is capacitive, but if it is long enough and is inserted into the guide, it becomes series resonant when the depth of penetration equals one-quarter of a wavelength and inductive at any greater depth. When the screw is inserted across the guide, the inductive post in (G) is obtained. If the diameter of the post or screw is decreased, or if it is moved away from the center of the waveguide, the susceptance of the equivalent inductor or capacitor decreases. A considerable range of susceptance can be obtained by variation of the probe insertion and distance along the line, as shown in (H). The illustrations in Figure 3.22 are only representative of the many possible configurations and combinations of these elements.

3.5.2 Waveguide Short

During the discussion of coaxial test systems, the capacitance of a short was introduced. One of the most useful waveguide elements is a piece of waveguide closed at one end (a short circuit). Either a fixed closed-end or an adjustable short circuit which allows the reference plane of the short to be varied may be employed. The latter is more convenient and finds many more applications in microwave measurements.

3.5.3 Junctions and Tees

We spoke earlier about coupling between waveguides and between waveguides and other types of transmission lines. At this point, waveguide *junctions* or *tees* must be considered. These elements play an important part in waveguide circuitry and techniques. The two most common forms of junction are the E-plane and the H-plane tees. The former is also called a *series tee,* because the axis of the side arm is parallel to the E-field of the main transmission line; the latter is referred to as a *shunt tee,* because the axis of the side arm is parallel to the planes of the magnetic field of the main transmission line.

A combination of both tees forms a hybrid wave guide junction called a *magic tee,* shown in Figure 3.23. The term is sometimes reserved for the hybrid junction in which matching structures have been introduced to improve the match of the E- and H-arms to the junction.

The characteristics of the hybrid in which the E- and H-arms are symmetrically placed are such that energy applied to either arm divides equally between arms one and two, and no energy emerges from the opposite arm. If the fields entering arms one and two are equal in amplitude and of the same polarity, the net field in the E-arm is zero, and the total energy emerges from the H-arm.

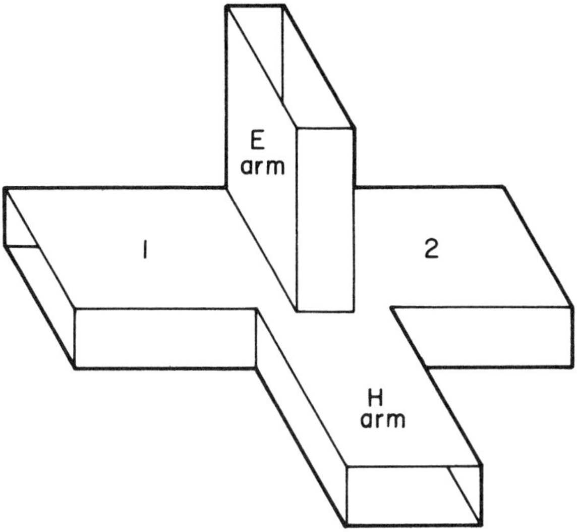

Fig. 3.23 The magic tee.

At the junction, each arm is effectively terminated by the two other arms of equal impedance. Therefore, discontinuities are inevitable unless special precautions such as compensated jucntions are observed. One manner in which the magic tee can be matched is by the use of tuning rods, whereby one rod is placed normal to the E-field in the series arm, and the other is placed normal to the E-field in the shunt arm. The precise location and configuration of the various matching structures are usually determined empirically.

The magic tee can also be used as a phase shifter when the series and shunt arms are terminated with adjustable short circuits. Furthermore, the device has been used with microwave receivers where crystal detectors are placed in arms one and two. The signal frequency enters the H-arm, while the local oscillator signal is fed into the E-arm. The tee provides isolation between sources of the two signals mixed in the crystals. The magic tee also is employed in microwave phase-measuring systems and various microwave bridge circuits.

3.5.4 Attenuators

A very useful waveguide component is the waveguide attenuator. It consists of an attenuator plate supported within the guide either by metal rods or by moving the plate in and out of the guide at a fixed location. In the latter method, a resistance card is inserted through a longitudinal slot in the top (center) of the waveguide, creating what is known as a flap attenuator. In order to reduce leakage, absorbent material is placed next to the slot.

Cross sections of the two types of waveguide attenuators are shown in Figure 3.24. A drive mechanism is attached to the struts in order to move the vane in and out of the electric field to obtain the variable attenuation characteristic. Also, a dial is attached to provide accurate calibration of the attenuator. Waveguide attenuators can be matched by various means; the resistive transformer, shown in the diagram, is only one such method. A match can also be obtained simply by tapering the vane at each end—a one-half wavelength taper provides a good match. In addition to the frequency sensitivity of the variable glassvane attenuator, considerable phase shift is encountered since the moving vane is a dielectric material. (A much improved version of a waveguide attenuator is the *rotary vane attenuator,* by virtue of its precision and direct reading feature. It has a calibrated range of 50 dB, accurate within two percent of the decibel reading at any frequency in the waveguide band. The phase-shift variation of the attenuator is less than one degree between 0- and 40-dB variations.)

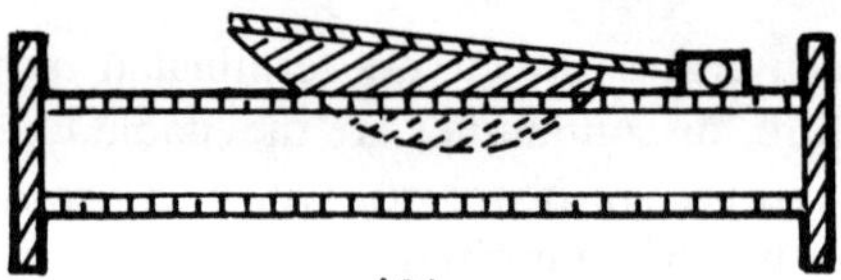

(A)

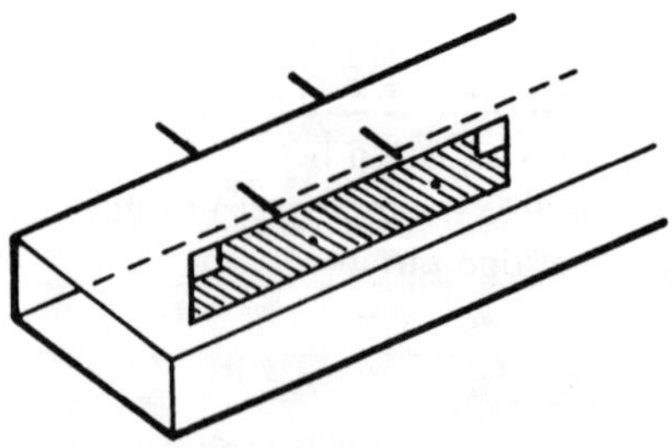

(B)

Fig. 3.24 Waveguide attenuators.

3.5.5 Directional Couplers

A useful form of hybrid known as a directional coupler is constructed by placing an auxiliary section of uniform waveguide along the narrow or wide dimension of the main guide with appropriately located apertures connecting the two.

The Bethe hole coupler consists of a single coupling aperture in the wide dimension of the guide. The electric field which leaks into the auxiliary guide is actually a combination of two fields: the fringing electric field through the hole, and the field across the hole caused by the flow of charge through it. This charge flow is caused by the transverse component of the magnetic field of the wave. The directions of the component electric fields in the auxiliary guide are such that the fields tend to cancel in the forward direction from the aperture and reinforce in the reverse. The magnetic coupling is a function of the angle between the guide axis, whereas the electric coupling is essentially independent of the angle. Therefore, the maximum directional property can be obtained by an optimum setting of the angle between the auxiliary guide and the main guide, as shown in Figure 3.25(A). The coupling and directivity are sensitive to changes in frequency.

The two-hole directional coupler is illustrated in the cross-section diagram in Figure 3.25(B). The two auxiliary holes are placed one-quarter wavelength apart. The signal which travels back to the first hole from the second hole is 180° out of phase. The two signals tend to cancel in the reverse direction; in the forward direction, they are in phase and reinforce each other. The coupling and directional properties are impaired by off-frequency operation when the distance bewteen the holes is no longer one-quarter wavelength.

The coupling factor is defined as the ratio, expressed in decibels, of the power entering the main line input to the power output of the auxiliary guide:

$$C = 10 \log \frac{P_i}{P_o}$$

From the definition and illusutration of Figure 3.25(C), the directivity is given by:

$$D = 10 \log \frac{P_{oi}}{P_{or}}$$

Directivity is defined as the ratio, in dB, of the powers out of the auxiliary guide when a given amount of power is alternately applied in the forward and reverse directions to the main guide.

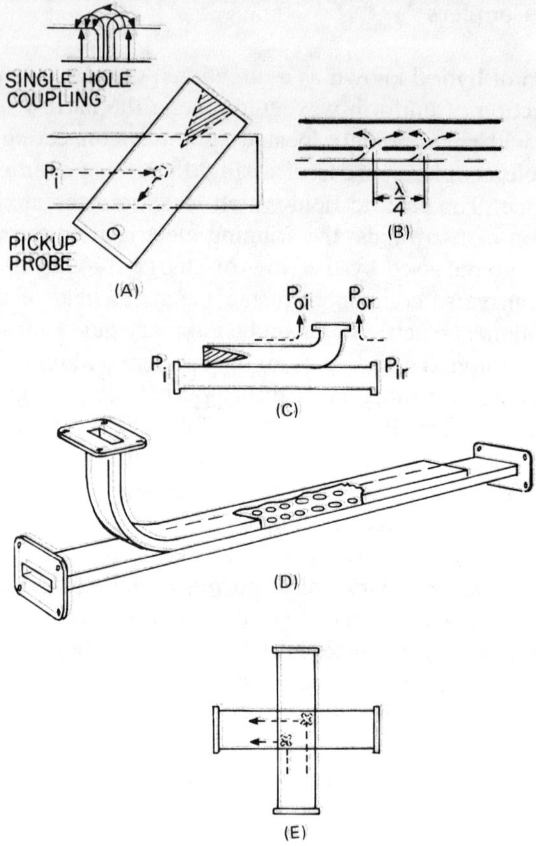

Fig. 3.25 Waveguide couplers.

The leakage signal due to the intrinsic directivity of the coupling holes, the reflected signals from the internal termination, and any discontinuities at the output flange of the coupler determine the total directivity signal.

Multihole couplers operate on the same basic principle as does the two-hole coupler. The coupling array, providing high directivity, is illustrated in Figure 3.25(D). The coupling *versus* frequency variations are ±0.5 dB over the entire waveguide frequency range. Coupling is obtained through a series of graduated holes accurately machined along the broad or narrow face of the waveguide. In couplers designed for high power operation, the holes usually are drilled in the narrow dimension of the waveguide.

In applications where a multihole coupler is not required, the less expensive and more compact crossguide coupler, shown in Figure 3.25(E), is used.

3.5.6 Terminations

Terminations are special devices designed to absorb incident energy with very little reflection. They are constructed by mounting power-absorbing material in the space near the end of a closed section of waveguide. Two low power terminations are shown in Figure 3.26. Figure 3.26(A) shows a tapered resistance strip termination; (B) illustrates a movable polyiron termination. At higher power, radiating fins are added to the waveguide casing to aid in removal of heat, as seen in Figure 3.27.

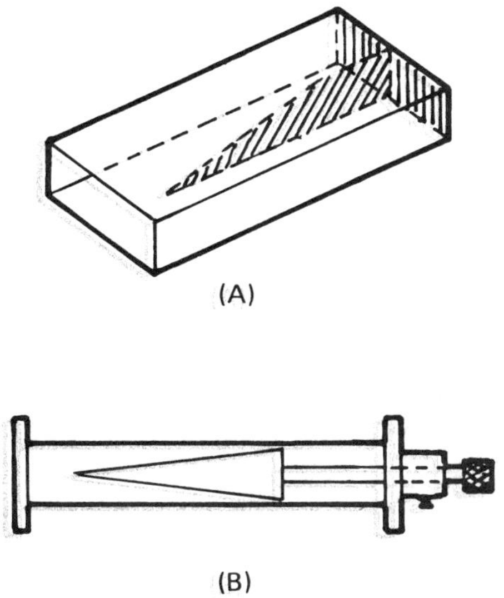

Fig. 3.26 Low power waveguide termination.

Fig. 3.27 High power termination.

3.5.7 Detector Mounts

The usefulness of the microwave-detecting elements is intimately related to the design of the mounting structure. The latter should be arranged to transform the RF impedance of the element to equal the impedance of the reflections. For maximum utility, the structure should maintain the above condition (low VSWR) over as broad a frequency range as possible. The importance of obtaining a match is indicated by the fact that approximately one percent of the input power is reflected from a VSWR of 1:2.

If the detecting element is connected directly across the open end of the waveguide, both a conductance component (associated with radiation into open space) and a susceptance component (associated with reflections from the open end of the waveguide) arise. The radiation is eliminated and the open circuit is obtained at the detecting element by attaching a closed quarter-wave section beyond it. Figure 3.28 shows a series of waveguide detector mounts that are commercially available today.

Fig. 3.28 Detector mounts.

One very important parameter in a detector mount is *efficiency*—a means of determining the extent to which the available power is absorbed by the detecting element and the fraction of power which is dissipated by stray losses in the mount. Its value is determined by comparing the power values obtained when known amounts of dc and RF power are alternately applied to the mount. An accurately calibrated dc bridge is used to measure the dc power entering the detecting element. If all the power entering the mount were used in heating the bolometric element, the efficiency would be 100 percent. Efficiency is usually given as

$$\eta = \frac{\text{power absorbed by the bolometer}}{\text{power delivered to the bolometer mount}}$$

The efficiency of commercial bolometer mounts may range from 95 to 98 percent.

3.6 CONNECTORS

To allow the cables previously discussed to operate, something is needed at each end of a cable to permit it to be attached to a component—that something is a *connector*.

The reason for using connectors is either to join two components or modules, or to gain access to a certain point in a circuit. A connector's primary qualification is to provide a connect/disconnect capability for your circuit (or system) without disturbing a measurement in progress or the overall system's performance. In other words, it must have a very low VSWR.

The coaxial connector can be considered as a separate microwave component; as such, there are certain parameters (or defining terms) which help to specify each type. Besides the characteristic impedance, the major areas of concern are upper frequency limit, CW power capability, size and weight, environmental conditions (vibration, moisture, temperature), cost, field replaceability, and contact characteristics. If the connector is to be used for measurements in a lab, attention also should be given to repeatability, life (connect/disconnect cycles), VSWR, and mechanical strengths (to support a weight extending horizontally).

With all these terms in mind, let us investigate some of the more important ones.

The first is that of characteristic impedance. Most connectors are 50Ω. But why is that figure chosen? Figure 3.29 answers that question. Minimum attenuation is achieved (for air dielectric) at an impedance of 77Ω; maximum power capability is attained at 30Ω. Therefore, 50Ω is considered to be an appropriate compromise to maximize both aspects. A 75Ω system though, may be used when attenuation in long lines becomes a problem, as in CATV or data communications. Coaxial cables

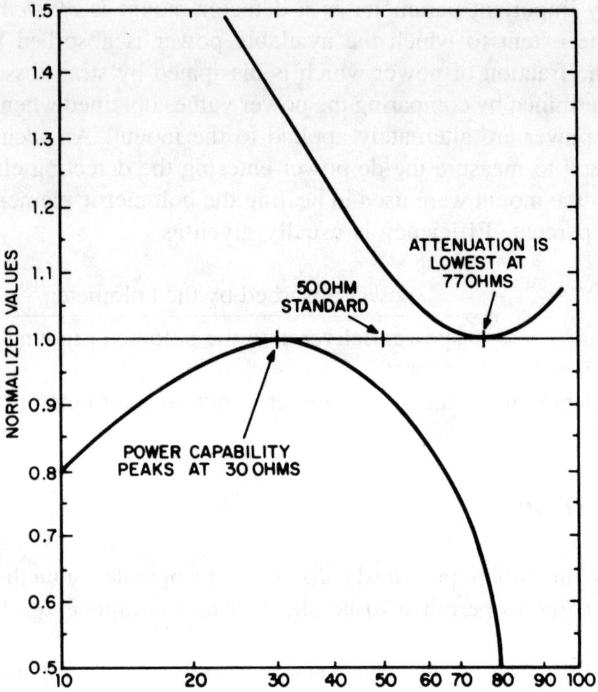

Fig. 3.29 Attenuation and power capability.

used in such applications, whether 50Ω or 75Ω, are usually "specials" supplied by the company that makes the particular communications equipment.

The power handling capability (or maximum average power that may be transmitted), of the connector unlike with other components, is not a function of the connector itself. Rather, it depends on the power handling capability of the cable used. The peak power is determined by the voltage breakdown point specified for each individual type of connector.

The frequency range of a coaxial connector-cable assembly is from dc to the point where a higher-order mode can be supported. The cut-off wavelength is given as:

$$\lambda_c = \pi \frac{D + d}{2} \sqrt{\varepsilon}$$

when

λ_c = cut-off wavelength;
D = inside diameter of the outer conductor;
d = outside diameter of the inner conductor;
ε = dielectric constant.

The cut-off frequency (or upper frequency limit) is

$$f_c = \frac{v}{\lambda_c}$$

where

v = velocity of propagation.

When we specify a connector, we indicate mechanical properties as well as electrical ones. The details, however, depend on the type of connector. There are sixteen readily available varieties that are covered by an assortment of standards—3.5 mm, 7 mm, 14 mm, SMA, SMB, SMC, C, SC, BNC, TNC, N, LT, HN, QDS, QC, and TPS. Of this group, we show seven of those most widely used in order to illustrate their dimensions. Figure 3.30 through 3.36 show the SMA, N, TNC, BNC, SC, 7 mm, and 14 mm, respectively. A special version of the SMA connector is the K connector. This connector is rated to operate above 40 GHz.

Obviously, one of the most important mechanical properties of connectors is how they go together; that is, what sort of interface is there between the male and female of a specific type. There are two basic methods of joining with the outer conductor of a connector—butt-mating and wiping contacts—which are illustrated in Figure 3.37. From this figure it is easy to see how each gets its name. With butt-mating (Figure 3.37(A)) contacts, such as those on an SMA connector, good electrical continuity depends on a proper tightening of the connectors when joined. Manual tightening (fingers) provides approximately 1 to 3 in-lb or 1.11 to 3.5 cm-kg; this amount of pressure applied to the coupling nut of an SMA connector should be adequate for typical laboratory electrical requirements. However, for use in equipment that will be subjected to shock and vibration, tightening torque should be increased to 7 to 10 in-lb (8 to 11 cm-kg).

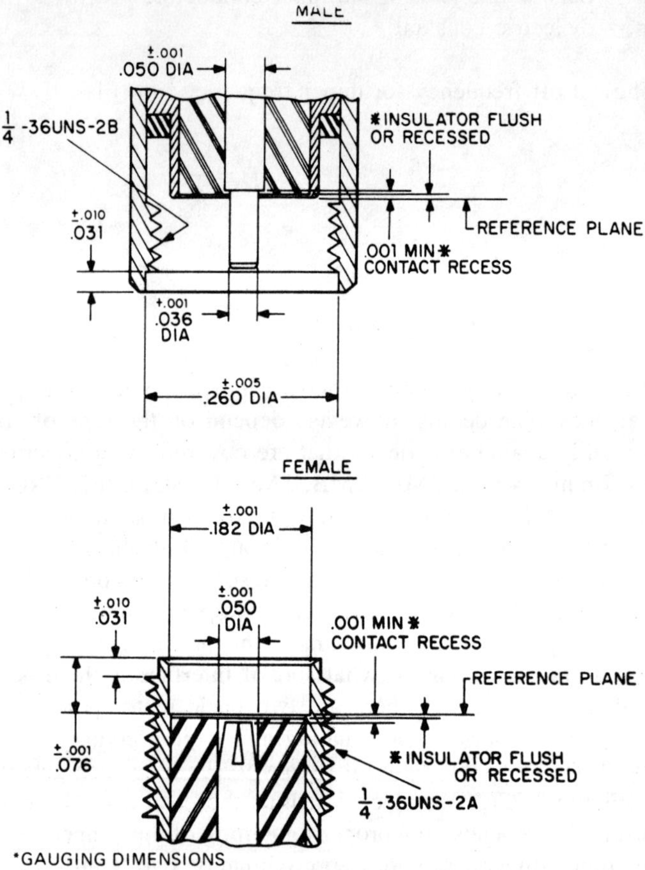

Fig. 3.30 Dimensions for type SMA connectors.

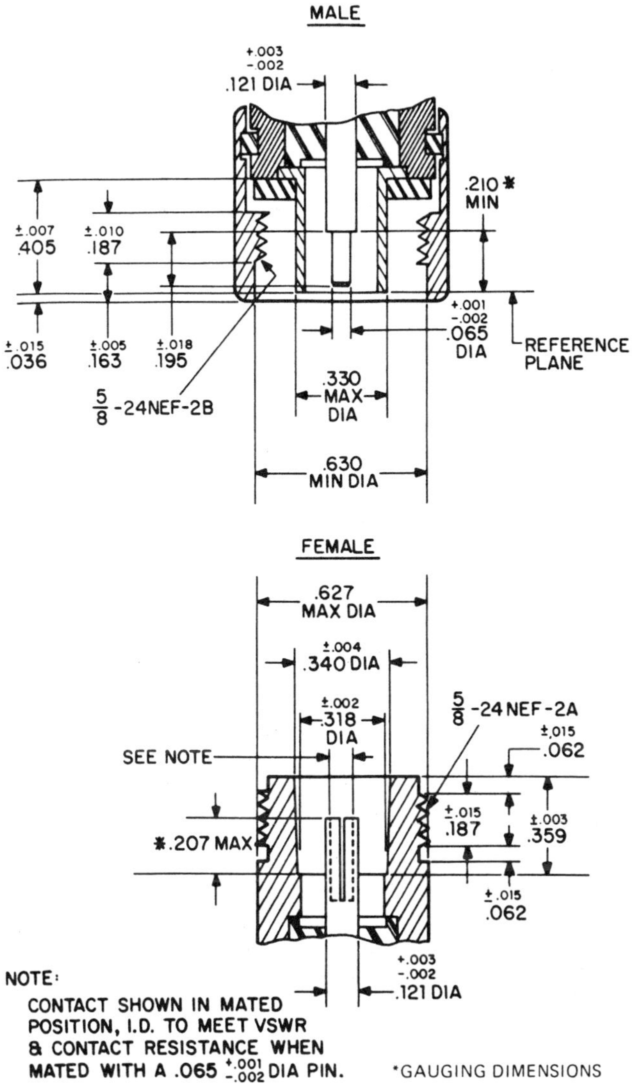

Fig. 3.31 Dimensions for type N connectors.

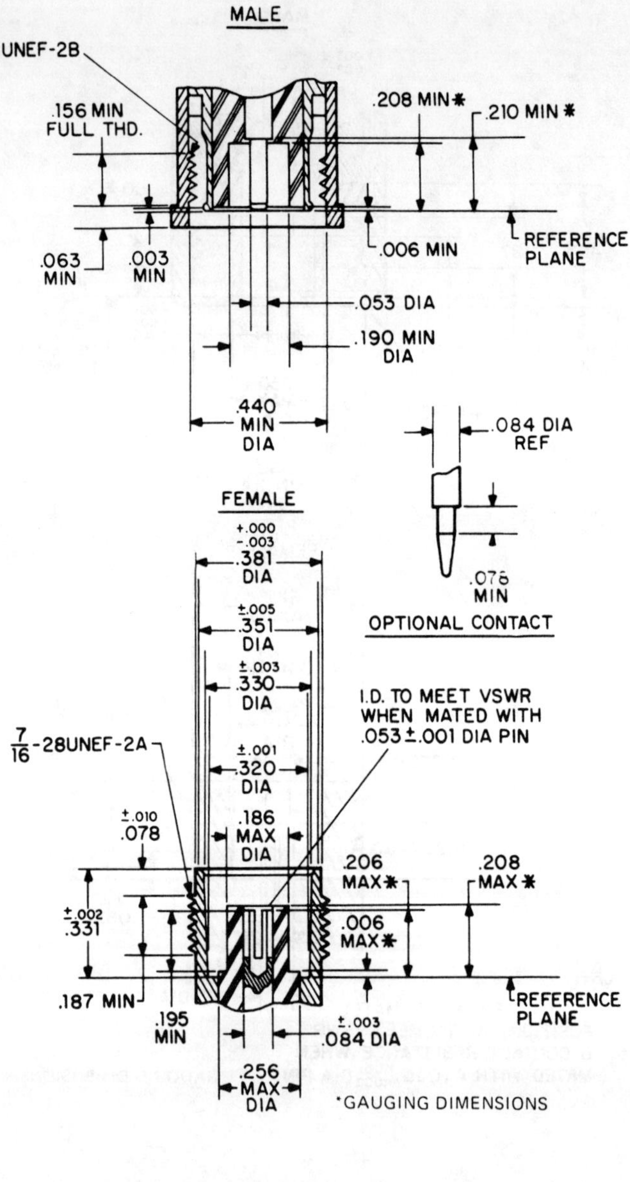

Fig. 3.32 Dimensions for type TNC connectors.

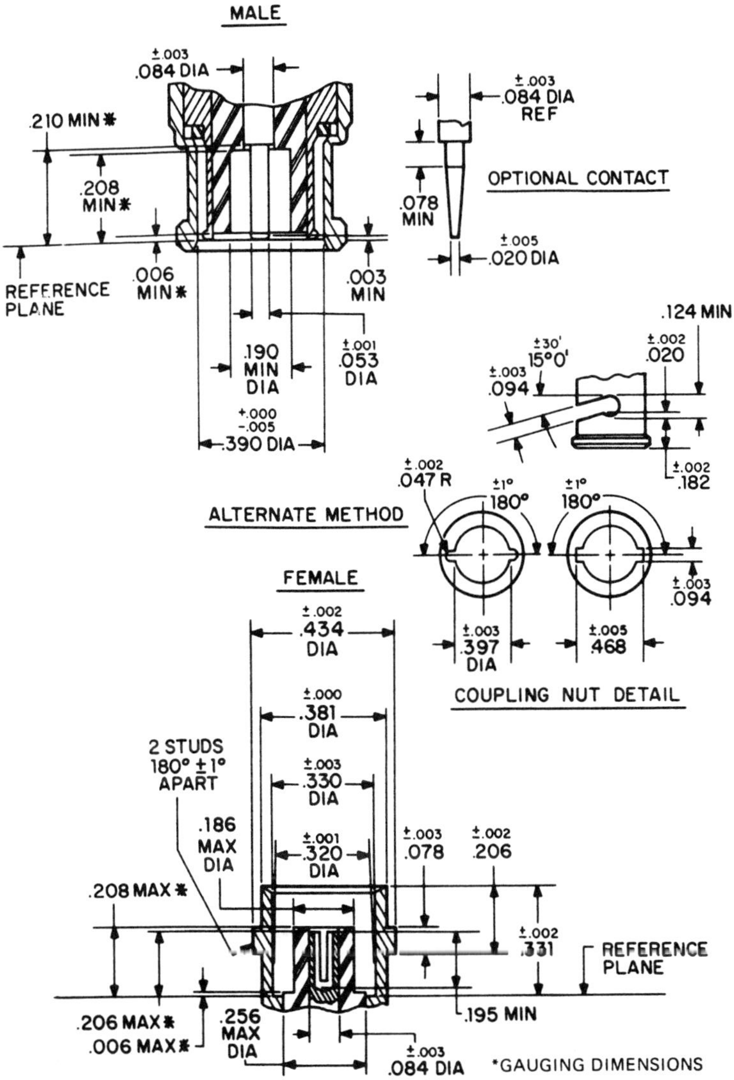

Fig. 3.33 Dimensions for type BNC connectors.

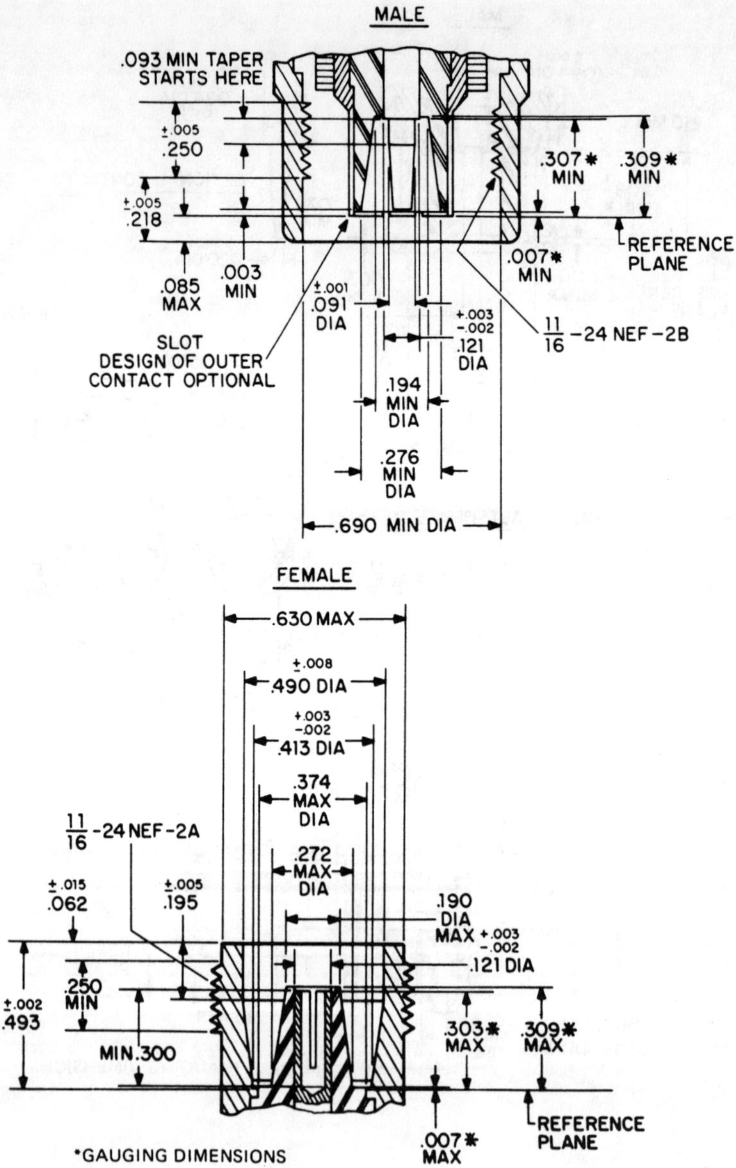

Fig. 3.34 Dimensions for type SC connectors.

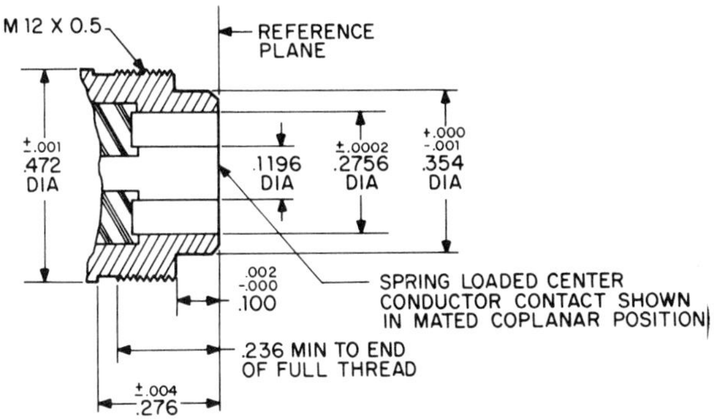

Fig. 3.35 Dimensions for type 7 mm precision connectors.

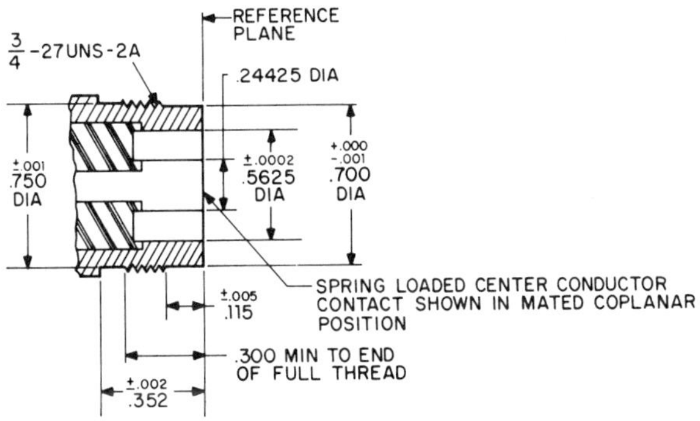

Fig. 3.36 Dimensions for type 14 mm precision connectors.

With wiping contacts (Figure 3.37(B)) such as those on the type N, BNC, TNC, and SC connectors, the mating conductors overlap. The inside conductor generally is slotted and the outside one tapers slightly inward, increasing the tension as they are tightened together. This "wiping" action helps to clean the contacts as the connector is used.

The two remaining types of connectors illustrated are the 7 mm and 14 mm types. They are called precision connectors and have butting inner *and* outer conductors instead of pin and socket contacts. The size referred to in their designation is the inside diameter of the outer conductor. The cut-off frequency of the precision connector is inversely proportional to that dimension. For example, the theoretical frequency limit for a 14 mm connector is about 9 GHz; when using the very popular 7 mm connector, the boundary is increased substantially to 18 GHz. (As a point of interest, the 3.5 mm connector mentioned earlier operates up to 26.5 GHz; work is being done on a 1.75 mm connector that could extend the range even farther.)

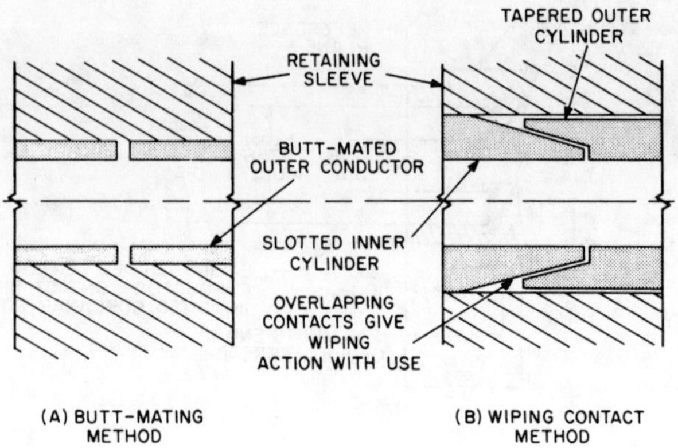

Fig. 3.37 Butt-mating and wiping contacts methods of interface.

One very important point should be made concerning precision connectors. As can be seen from Figure 3.38, there is no protruding pin to line them up as they mate; the joining involves the outside threaded sections of the connector body. To properly seat the connectors, run the threaded section all the way counter-clockwise on one and completely clockwise on its mate. Join the two connectors and turn the loose outer section clockwise until the two are mated snugly. This method ensures a good electrical contact and eliminates any potential for damage to an expensive connector by cross-threading or forcing.

Fig. 3.38 Precision connectors.

Three basic methods are used to attach connectors to the outer conductor of a cable—soldering, clamping, and crimping. For many years, clamping the braid of a flexible cable between the connector body and a threaded retainer has been the most popular method of assembly. However, crimping is becoming increasingly popular. The braid is positioned between the connector body on the inside and a crimp sleeve cylinder on the outside; a crimping tool then secures the connection. This method of assembly offers several distinct advantages over the clamping—cost is reduced, more uniform assemblies are produced because the process is simple and straightforward, and retention strength is greater (the braided outer conductor of the coax normally breaks first).

As is with everything, though, the crimping procedure has a couple of disadvantages—the assembly is not reusable and special tools are required.

Generally, electrical performance is not compromised when crimp-retention methods are used for the outside coax connector. However, attaching to the center conductor can be considerably more critical, and care should be taken to follow the manufacturer's specifications.

Solder connection of the outer conductor is the method used with semirigid cable. A low temperature solder or a liquid flux is employed to ensure good mechanical and electrical connections. The center conductor is almost always soldered; the only exception is in the case of certain SMA types in which the center pin of the connector is the actual cable center that has been filed to a point.

Cable size is the primary factor influencing microwave connector choice; both should have the same general diameter to prevent any abrupt impedance transitions.

The six categories of microwave connectors are ultraminiature, microminiature, subminiature, miniature, medium, and large. Table 3.6 lists the categories, their dimensional ranges, and an example of each.

Table 3.6 Microwave Connectors

Type	*Diameter (inches/mm)*	*Example*
Ultraminiature	$^5/_{32}$ (3.96)	3.5 mm
Microminiature	$^1/_4$ (6.35)	SMB, SMC
Subminiature	$^5/_{16}$ to $^7/_{16}$ (7.9 to 11.1)	SMA
Miniature	$^9/_{16}$ to $^{21}/_{32}$ (14.3 to 16.6)	BNC, TNC
Medium	$^3/_4$ to 1.0 (19 to 25)	N, SC
Large	Over 1-$^1/_4$ (over 31.75)	QC, LT

Figure 3.39 shows actual SMA, BNC, TNC, and N connectors and their relative sizes. When deciding upon an appropriate connector to use, there are many items to consider.

Sometimes, existing hardware determines the types. Listed below are six additional specifications (and a brief explanation of each) which should be checked.

Cable Size—For optimum electrical performance, the connector should approximate the cable in size, thereby minimizing the number of changes in diameter in the transition between cable and connector.

Impedance—Most connectors are designed with a nominal impedance of 50Ω to be used with cables having the same value.

Frequency Range—The available operating frequency range decreases as the connector size increases. For example, the standard SMA series connectors are usable to 18 GHz while the N series is limited to 11 GHz. With specially designed connectors, a frequency range may be extended past the recommended limits of a series.

Voltage Rating—The voltage and power handling capabilities of connectors are dependent on many factors, the most important of which are size, interface, and cable entry designs. A higher voltage rating necessitates an increase of the creepage path between center and outer conductors. This alteration is accomplished by telescoping the interface dielectrics and counterboring that in the cable entry portion. The latter action has a limiting effect on the operational frequency range because of the resulting impedance mismatch at the connector cable entry.

Connector Coupling Method—Connector series also differ in the type of coupling used to engage mating connectors; the most popular varieties are threaded and bayonet. The bayonet type can quickly connect or disconnect a circuit. However, if this ability is not needed in a given application, the threaded type is preferable. It generally exhibits better electrical performance and can withstand high vibration levels without producing excessive noise.

Contact Captivation—Contacts may be either captivated or noncaptivated. Captivated contacts exhibit better mechanical stability by preventing axial movement due to temperature changes or extreme cable flexing. Connectors using noncaptivated contacts can be designed for better electrical performance, shorter length, and lower manufacturing cost.

Fig. 3.39 Coaxial connectors.

In some cases, all these considerations are applicable; other times, only a few are appropriate. Yet each should receive great attention in order to achieve the best possible performance from your system. It is very discouraging to design a microwave component, only to have its performance degraded by an improper connector or cable assembly. In the long run, care exercised throughout the selection and utilization processes pays off handsomely—obviously, proper assembling of cables and connectors is of prime importance; instructions to this end are shown in Appendix L.

3.7 CABLE TESTING

Having discussed microwave cables (both flexible and semirigid) and connectors, it is only logical to deal with testing them to ensure that we did a good job combining the two. RF cable testing is an extensive process and is not treated here; it is covered in its entirety in Chapter Seven. For now, we concentrate on dc checks of our cable assemblies.

There are two basic tests that should be made on a completed cable assembly—one for continuity, and another which we refer to as a "ground check," as illustrated in Figure 3.40. The only test instrument needed is a multimeter placed on the RX1 scale.

Before beginning either test you must "zero the meter," which is done by touching the two test leads together. While the leads are in contact, adjust the zero control for a full-scale reading—you are now ready to check your cable. This procedure may seem trivial and unnecessary but, at microwave frequencies, it does not take much resistance to cause an intolerable insertion loss. The few seconds it takes to accurately zero your meter is more than offset by the improvement to be gained when making measurements.

To check for cable continuity (or simply that of the center conductor), place one meter lead on the center conductor of one connector and the other lead on the other center conductor. You should receive a true full scale, or zero, reading on the meter. If so much as $^1/_2$ Ω registers on the meter, you should first recheck your meter zero and then check for bad solder connections or other flaws in your cable assembly. This procedure is shown in Figure 3.40(A).

The "ground check" may be made in either of two ways—with one connector open or with it terminated. What you are checking for is a short between the center and outer conductors. Once again, zero the meter as described above. Place one lead on the center conductor of one connector and the other lead on the outer conductor—there should be no reading at all on the meter. To check that this is truly an open circuit, set the meter to the highest resistance scale (usually RX1MEG), zero it again, and repeat the procedure outlined above. Once again you should get no reading on the meter. This setup is shown in Figure 3.40(B).

To check for grounded connections with one end terminated, set the meter on the RX1 or RX10 scale (whichever gives the best display of 50Ω) and zero it. Place a 50Ω termination on one end of the cable assembly and test the other as outlined above. If your assembly is put together properly, you should read 50Ω on the meter; if there were a short, the reading would be zero. Figure 3.40(C) illustrates the setup for this check.

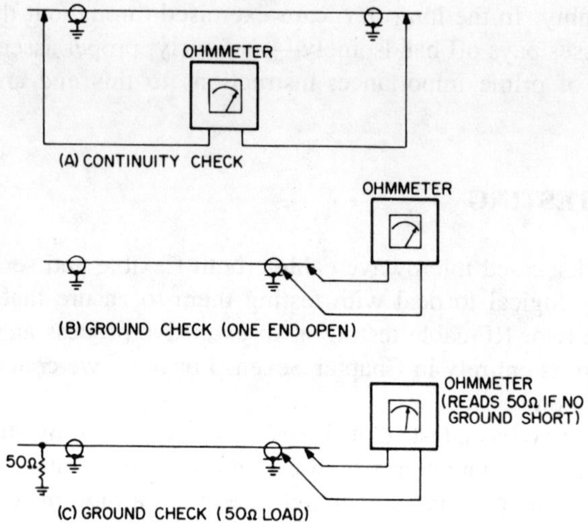

Fig. 3.40 Dc cable tests.

The tests described above are very basic and take very little time. However, the few minutes you spend immediately after the systems are assembled could save hours of trouble-shooting later on. Take a few minutes now and prevent a lot of headaches in the future.

3.8 ADAPTERS

Throughout this text we stress selecting the proper connectors for your components so that you would not need adapters. However, there are times when there is no choice but to use adapters. For these times, you need to know what is available, how they should be used, and what some of their properties are.

A basic concept to grasp is that an adapter is *not* just a mechanical device to join two unlike connector types—it can be a very complex item. To understand this fact, remember that one of the most important specifications for a cable connector is an approximate equality in size so as to minimize mismatch. Also, imagine the amount of work expended in the design of some of the adapters available in a variety of sizes and therefore, impedances. An adapter is much more than two connectors put back-to-back.

Some of the adapters available are shown in Figure 3.41. The important specification for adapters is that of VSWR; usually they are described as having a fixed VSWR plus a correction figure (a constant times the frequency of use in GHz). For example, the type N to SMA adapter shown in Figure 3.41 is rated to operate from dc to 12.4 GHz with a VSWR of 1.05 + 0.005f (GHz). If we were operating at 1.0 GHz, the actual VSWR would be 1.05 + 0.005 (1) or 1.055. Both the constant

Fig. 3.41 Coaxial adapters.

VSWR and the correction figure vary with the specific adapter considered; VSWR has been noted to range from 1.015 to 1.09 and the factor, from 0.002f (GHz) to 0.009f (GHz).

The VSWR shown above may seem small, but they result in a loss through the adapter that could be 0.1 or 0.2 dB. If a number of adapters are used, it does not take long to accumulate 1.0 dB or 1.5 dB of loss. Therefore, careful consideration must be given to any system or test setup that must use adapters. If adapters are an absolute necessity, use a minimum number and go directly from one series to another—do not go N-SMA, SMA-TNC, and TNC-BNC if you want a type N-BNC connection. Though this example may sound exaggerated, arrangements such as these occur in many a test setup simply because someone does not want to take the time to find a single, proper adapter. Where adpaters are concerned, discretion and common sense should be used.

3.9 SUMMARY

The preceding chapter has shown microwave transmission lines all the way from basic theory to coaxial cables (flexible and semirigid), stripline, microstrip, and waveguide. It also has shown the importance of knowing the different connectors and adapters that can be used for microwave circuits.

Each transmission line described has its own place in microwaves. This ranges from 1.0 to 4.0 GHz for coaxial cables; 1.0 to 12 GHz for stripline and microstrip; and 8.0 to 90 GHz for waveguide. Each transmission line is designed to provide the user with specific characteristics.

Chapter 4
Microwave Components

If you have ever played chess, you know that the success of your game depends on your knowledge of each individual piece and its abilities. Such is also the case concerning microwaves—your success in the microwave field depends very heavily on your knowledge of the many individual pieces, or components, and what each can do. These microwave components can range all the way from a simple two-way power divider to a complex mixer.

Each individual component has a specific purpose. Its function may be to couple, mix, filter, isolate, divide, add, act on a system's phase, convert RF to dc, attenuate, or any combination of the above. Knowing when and where to use each of these components is an essential part of your contribution to the microwave systems that you employ.

Components to be covered in this chapter are directional couplers, detectors, mixers, quadrature hybrids, power dividers, filters, circulators and isolators, attenuators, and antennas. You must be familiar with, and understand, each of these components if you are to be the type of individual who is truly versed in microwaves at a useful level.

To understand fully each of these microwave components, it is necessary to answer three basic questions for each of them:

1. What does it look like?
2. How does it work?
3. What does a data sheet or catalog say about that component?

4.1 DIRECTIONAL COUPLERS

One of the most widely used components in microwaves is the directional coupler. It may be employed by itself to monitor, sample, or attenuate; or it may be used as part of another component, such as a mixer. You have heard the term "directional coupler" many times—but what exactly is it? To explain we break the actual term up into two individual terms—"directional" and "coupler."

To clarify our definition refer to Figure 4.1, which is one way of schematically showing a directional coupler. Other methods are illustrated later.

Let us take the word "coupler" first in our definition. The basic definition of "couple," from which the word "coupler" is derived, is "to bring two electrical circuits into such close proximity as to permit mutual influence." In other words, two circuits are side-by-side so that they have an effect on one another. How much of an effect depends on just how close they are. The "circuits" could be two pieces of waveguide, two coaxial lines, or two copper lines in a stripline coupler. One of the circuits is the main line energy of the coupler; the other circuit is the coupled line.

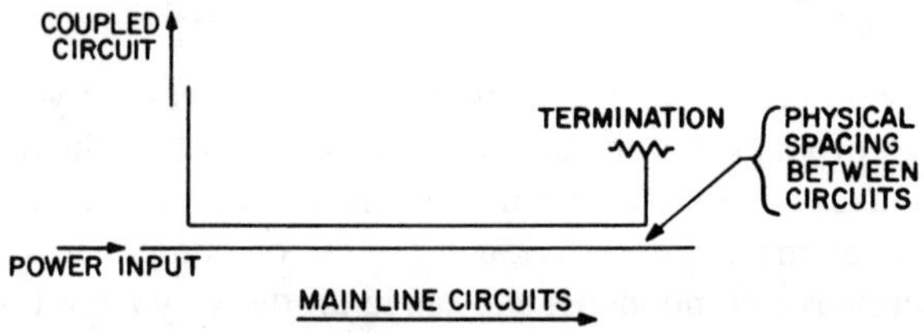

Fig. 4.1 Basic directional coupler.

As Figure 4.1 shows, the coupling, or transfer of energy, is accomplished between the two parallel circuits of the coupler. In a coaxial or stripline coupler, that transmittal is a function of the spacing between circuits; in a waveguide coupler the actual physical spacing is set by the dimensions of the waveguide. These dimensions are a function of the frequency of operation; the coupling is determined by a series of coupling holes in the waveguide. Figure 4.2 illustrates: (A) a coaxial coupler, (B) a stripline coupler, and (C) a waveguide coupler. It can be seen that there is a physical gap between the circuits in the coaxial and stripline couplers; the waveguide coupler has a series of holes to provide coupling from one circuit to another. These holes may also be slots or irises, depending on the coupling needed and the mode of operation used.

It should be clear by now to what the term "coupler" refers. A coupler is a device for transferring energy from one circuit to another without utilizing any direct connection.

Thus far, the definition would apply to any two circuits placed close enough together to act on one another. What makes the component we are dealing with unique is its outstanding directionality.

Figure 4.3 helps in our explanation of the term "directional." If energy is applied to port one in (A), most of the energy appears at port two, while a fraction appears at port three. The actual amount at port three depends on the coupling value; standard couplers are of 6, 10, 20, and 30 dB. A 3 dB coupler, called a quadrature hybrid, is also available, but it is treated separately later.

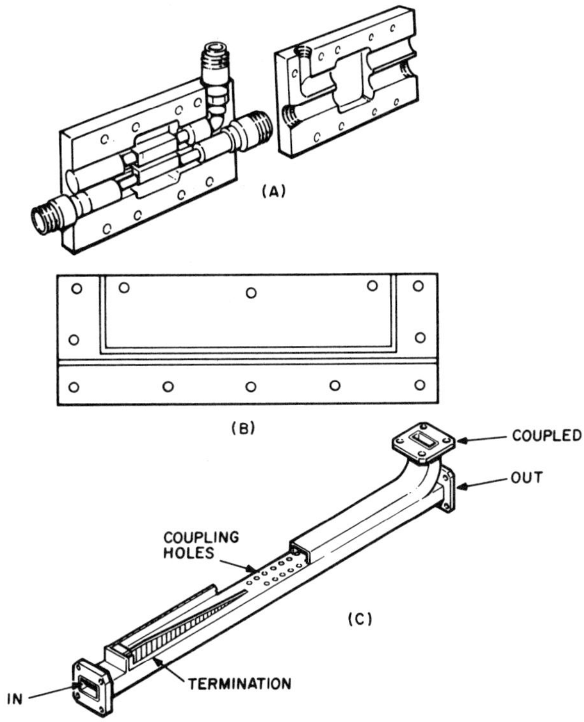

Fig. 4.2 Directional couplers.

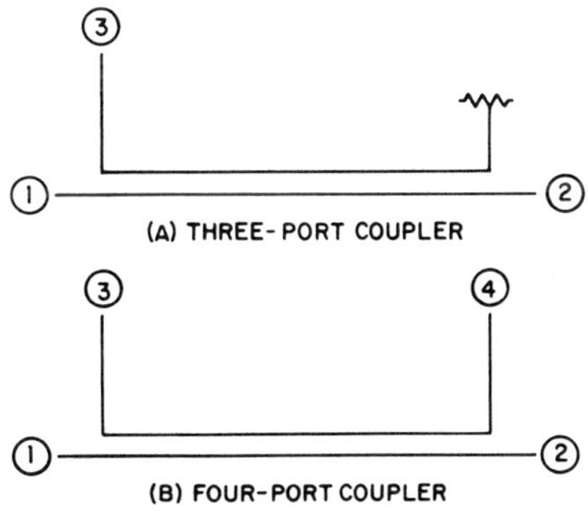

Fig. 4.3 Directional (3- and 4-port) couplers.

If the coupler were turned so that the power were applied at port two, there would be very little energy at port three because it no longer would be the coupled port. Rather, it would be the isolated port. The same is true of Figure 4.3(B). If power is applied at port one, the coupled port is three, the straight-through port is two (small loss), and the isolated port is four.

The actual difference between the power available at port four, and that at port three, is called the *directivity* of the coupler. Another way of phrasing this definition is to say that the difference between the desired and undesired couplings is a measure of the directivity of a coupler.

The essence of this discussion is that a coupler only works as designed when power is applied to the proper port; that is, it only couples properly in one direction. Referring to Figure 4.3(B), if we apply power to port one and expect the coupling to take place as designed, our coupled output indicator should be at port three. Port four only exhibits the coupling value in a reverse direction, that is, with power applied at port two. This property is the outstanding attribute of a directional component and can be very useful since it isolates the coupled circuit from any mismatches present in the main line.

Now that the terms of the component have been explained, we can put together all of the information needed to define a directional coupler:

> A directional coupler is a component which allows two microwave circuits to be combined into one integrated system in one direction, while being completely isolated from each other in the opposite direction.

One method to expand the basic directional coupler is through the construction of a dual-directional coupler. Just as the name implies, this procedure involves two couplers placed back-to-back. Figure 4.4 is a diagram of a dual-directional coupler. The one great advantage of this type of coupler is that it allows you to monitor both forward and reflected power at the same time. It greatly aids in obtaining data from a device which indicates the input match and, thus, its VSWR.

Figure 4.5 is a picture of various directional couplers as they would appear on a bench or in a catalog. The type of coupler that should be used depends on many factors. It may be necessary for it to have a large value of coupling, a very high directivity (as in accurate VSWR measurements), or a low input VSWR of its own. Any of these specifications may be of great importance when choosing the appropriate coupler for your job.

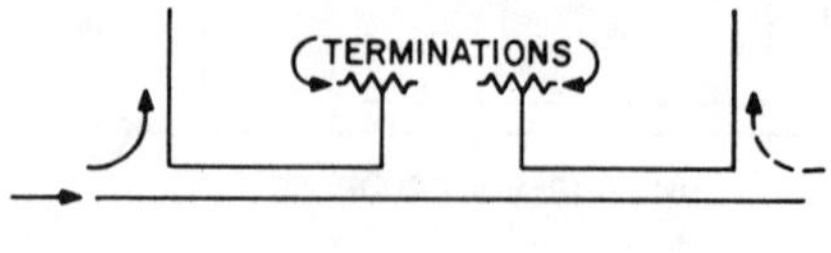

Fig. 4.4 Dual-directional coupler.

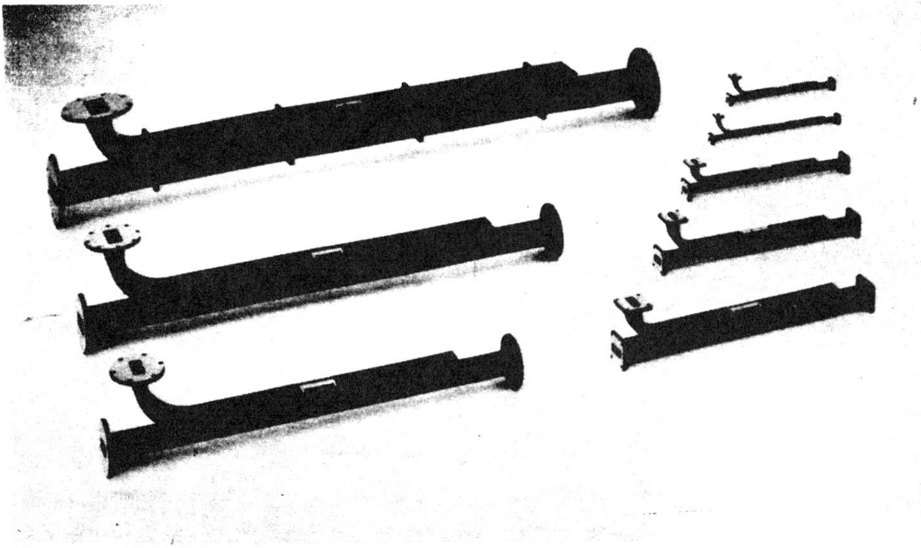

Figure 2.5/Directional Couplers

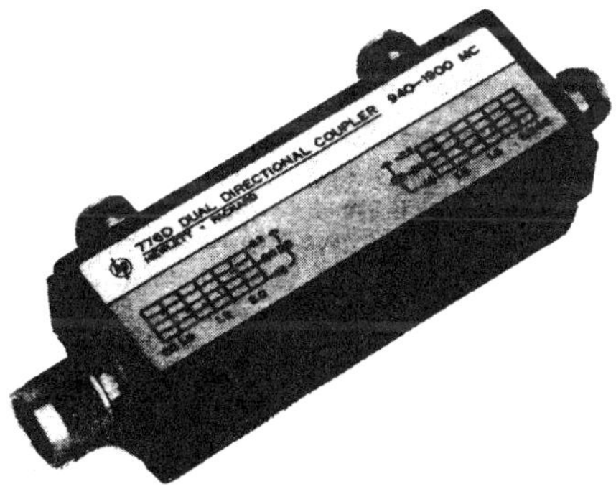

Fig. 4.5 Various directional couplers.

Below are sample specifications for a directional coupler as they could appear on a data sheet or in a catalog. We investigate each term and define those that have not already been covered.

Model number	001–20
Frequency	1.0–2.0 GHz
Coupling	20 dB
Coupling deviation	±1.0 dB
Insertion loss	0.2 dB
Directivity	25 dB
Impedance	50 Ω
Input VSWR	1.2:1 (maximum)
Power	200 W (CW)
	10 kW (peak)
Connectors	Type N (female)
Weight	1 lb (0.5 kg)
Size	4.093 × 2.125 × 0.875

Additional information may include price, reverse power rating (if a three-port coupler), coupled line VSWR, or the different values of any of the above parameters, if the coupler is designed to be used over more than one frequency band and some of the specifications vary with frequency. It should be noted that very seldom do all of the above specifications appear for all directional couplers. It is up to the individual manufacturer to decide which specifications he wants to publish and which ones someone must query. Also, the specifications vary if you are talking about a three-port coupler, a four-port coupler, or a dual-directional coupler. Thus, as pointed out before, the listing above is only a sample of what might be on a data sheet or in a catalog. Let us now discuss each of the terms listed.

To enable you to have ready access to each of these items, they are arranged as a glossary of terms in alphabetical order. You choose the term and the explanation you need.

Connectors—This portion of coupler specifications is very important since it tells you if you need adapters to fit the coupler into your system, or you need to special order your desired connector from the manufacturer. Remember that additional adapters mean higher overall losses in a system. Connectors may be type N, BNC, SMA, APC-7, or any special variety required; the various types were explained in Chapter Three.

Coupled Line VSWR—This specification indicates what sort of match the coupled port (port three in Figure 4.6) offers to any external load. A low VSWR would mean a good match and would cause little or no problems to any external circuit.

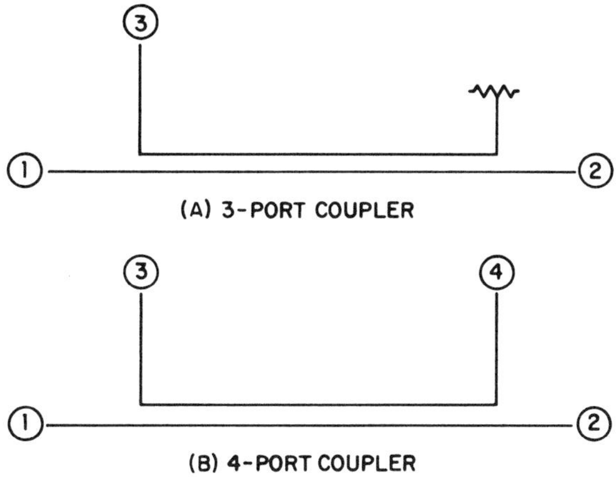

Fig. 4.6 Three-port and four-port directional couplers.

Coupling—The ratio of the power available at the coupled port to the power at the input port. (Port three to port one in Figure 4.6). It is the amount of attenuation of the input power as a result of the coupling structure. This term is expressed in dB and has standard values of 6, 10, 20, and 30.

Coupling Deviation—This measurement indicates how flat the coupling is over a specified band of operation; it is expressed as a (+) or (−) dB value. If you rely very heavily on a specified value of coupling over a wide range of frequencies, the term should be low.

Directivity—This figure indicates how accurately you are able to measure parameters in your system. As mentioned before, it is the difference between the desired and undesired couplings. If a coupler has low directivity, the forward and reverse powers in the coupler interfere with one another and cause great inaccuracies to occur. (Ports one and four in Figure 4.6(B) are the undesired coupling; one and three are the desired combination.

Frequency—This term indicates the range of frequencies for which all the other specifications are true. Above and below this range, none of the specifications apply. Most couplers are designed to operate over an octave band (1 to 2 GHz, 2 to 4 GHz, *et cetera*).

Impedance—Impedance is an RF resistance reading which ensures that any component with a similar impedance value connected to the coupler will operate properly. The most common impedance is 50 Ω, which is considered a standard in most systems. Such a value allows the coupler to have reasonable size conductors and lines and to adapt to conventional connectors with relative ease.

Input VSWR—This specification is identical to the coupled line VSWR in that it is an indication of what sort of match is offered by the input port of the coupler; it is usually of low-value, 1.15:1 for example. (In Figure 4.6(A) it is the match at port one; in (B) it is either port one or two).

Insertion Loss—The insertion loss indicates the power lost in the main line of the coupler primarily through dissipation. In coaxial couplers it is simply a loss which occurs in a conductor as energy is passed through it; in a stripline coupler, the copper losses of the conductors; and in a waveguide, a loss within the guide with the only contributing factor usually the length of the guide. In Figures 4.6(A) and (B), the insertion loss is that from either ports one and two, or two and one. In lower value couplers (3 dB or 6 dB, for example) the coupled energy due to the coupled port also is included in the total through-line insertion loss, thereby increasing the overall value.

Model Number—The model number is simply a way of referring to, and identifying, the coupler. In most cases the last two numerals indicate the coupling value.

Power Rating—This specification refers to the amount of both CW and peak power that the coupler is capable of handling. Any power level greater than this value may cause arcing or a deterioration of performance. This term is one of the most important to take careful note if you will be operating at a high power level.

Price—This item is one that, more often than not, does not appear with coupler specifications. If it does, it is either a price for an individual component or a 2 to 9 quantity price. Usually, a manufacturer's representative or the manufacturer himself has to be contacted if larger quantity prices are desired.

Reverse Power Rating—This term applies only to a three-port coupler; it is usually a lower value than that of the forward power rating because the fourth port is internally terminated and so, generally, cannot handle any amount of power. This specification is also quoted as a CW and peak power number; it is the power handling capability of the coupler if power is applied in a reverse direction (At port two, instead of port one in Figure 4.6(A)).

Size—The actual physical size of the coupler. The size and location of mounting holes is also given to aid in laying out any external mounting diagrams. This term is of a secondary nature except when you are required to put the coupler in a restricted space.

Weight—Weight is self-explanatory term which only comes into play when an overall weight restriction is placed on a system, such as with airborne equipment.

To conclude the discussion on directional couplers, we show how the properties described in the preceding paragraphs can be put into practical application. So that the individual may recognize a directional coupler in a practical application, Figure 4.7 shows three symbols that are employed.

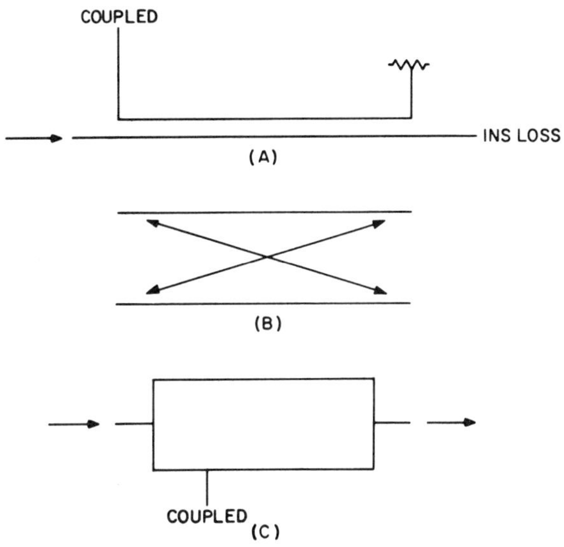

Fig. 4.7 Directional coupler symbols.

The most obvious use of a directional coupler is to monitor an output power. Its directional properties ensure that only the forward power is coupled; its property of circuit isolation leaves the main line power undisturbed. Figure 4.8(A) shows the orientation of a directional coupler when used as a power monitor.

A scheme similar to the monitoring technique described above is shown in Figure 4.8(B). The big difference in such a setup is that a diode detector is placed at the coupled port to provide a dc voltage proportional to any main line power variations. That voltage operates leveling circuitry which corrects the RF level and eliminates differences in the output power over a certain specified frequency band. This technique is used widely where swept frequency measurements are taken.

Figure 4.8(C) is an application which takes advantage of the directional couplers' ability to handle higher CW power; it is an actual example of a system used to check a 100 W power amplifier. The only components available were two-watt attenuators, five-watt terminations, and the directional coupler. It can be seen that a more adequate setup was created by utilizing the coupler's 200 W capability and directivity of 25 to 30 dB. None of the components used had its power rating exceeded; the directional coupler was doing the majority of the work. This particular application is discussed more thoroughly in the test setup chapter later in this book, but, for now, study the setup and see how many properties of the directional coupler you can see being put to use.

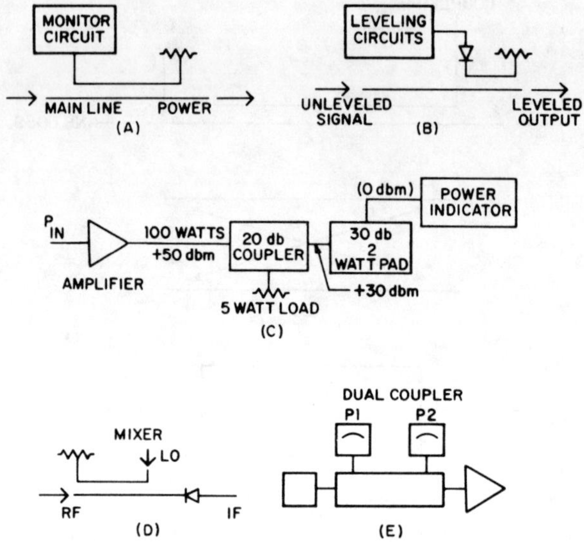

Fig. 4.8 Applications of directional couplers.

Figure 4.8(D) is an example of what we were referring to when we stated that the coupler could be part of another component. In this case it is a very important part of a single diode mixer. The circuit is of a very basic mixer, but it does show that by using a coupler, the RF and LO signals have a high degree of mutual isolation.

The dual-directional coupler, as stated before, is very useful in VSWR measurements (Figure 4.8(E)). It is employed to measure the return loss which then is converted to VSWR. Power meter number one reads the forward power to the device under test; meter number two measures any reflected power. By comparing these two figures, a return loss is determined and converted to the input VSWR of the particular device under test.

By now you can probably realize why we call the directional coupler one of the most widely used components in microwaves. Its property of being directional, its isolation, its low input and coupled line VSWR, and its adaptability to being three- or four-port or dual-directional make it a vital part of the growing knowledge necessary for a microwave engineer or technician to stay informed.

4.2 DETECTORS

The key to a reliable, trouble-free system is simplicity. One component whose operation is based on simplicity is the microwave detector. This concept of a simple and reliable component is the reason why the detector is used so extensively in microwave systems.

The basic construction of the microwave detector is illustrated in Figure 4.9. The heart of the detector is, of course, the diode itself. Today, Schottky diodes are used most often because of their excellent RF characteristics, which include high sensitivity. The diode is matched to the driving circuit (usually 50 Ω) so that maximum power transfer is obtained. In this manner maximum efficiency of the diode is ensured since any reflections due to mismatches are eliminated and, thus, all of the input power reaches the diode. The dc return, besides acting as a ground for the diode, also has the second function of acting as an RF choke so that no RF is shunted to ground. A low-pass filter is placed after the diode to eliminate all of the high frequency ripple caused by the detection process and allow only dc to be present at the output.

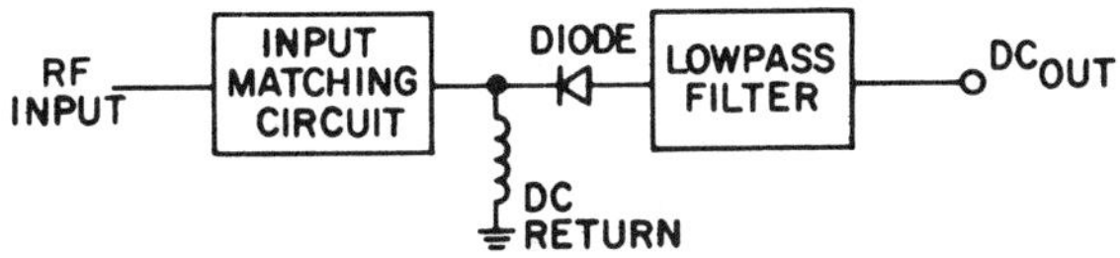

Fig. 4.9 Basic microwave detector.

Figure 4.10 shows actual detectors that are available; note the variations in size and shape. The one right for you depends on your particular application. A simple diode symbol is used to schematically show that a detector is in a system.

A typical set of specifications for a microwave detector is given below. You can look at the specifications, then we define the important terms.

Model number	101
Frequency	0.01–12.4 GHz
Frequency response	±0.5 dB
Sensitivity	0.4 mV/μW
Impedance	50 Ω
Maximum input	100 mW
Polarity	Negative (see options)
VSWR	1.3:1 (typical)
Connectors	Input—N (male) Output—BNC (female)
Weight	0.25 lb
Dimensions	2.75 in long 0.65 in dia.

Options: 101-1: Positive output, 50 Ω
101-2: Negative output, 75 Ω
101-3: Positive output, 75 Ω

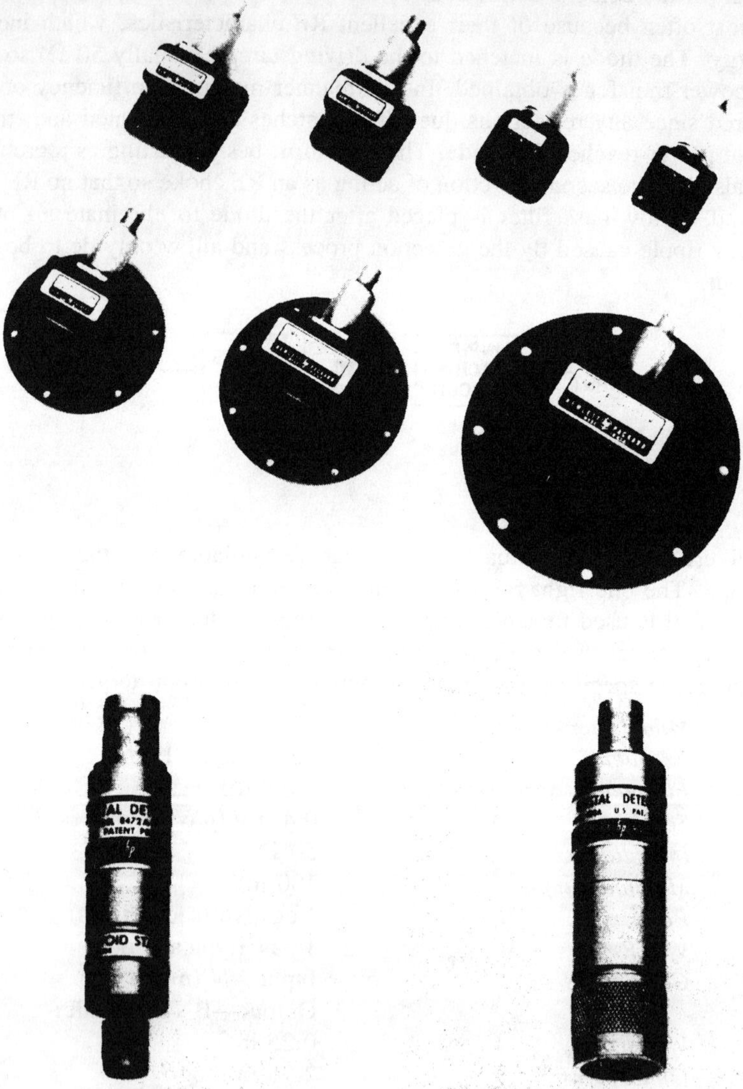

Fig. 4.10 Microwave detectors.

Frequency Response—Frequency response is a term used to indicate the RF performance of the detector; it is sometimes given in decibels per octave or in one figure for overall performance. Basically it is a measure of the variation in sensitivity of the detector, expressed in dB.

Maximum Input Power—This term refers to one of the most important considerations when using a detector—the highest level of RF power that can be applied to its input. It is always good practice to stay below this figure by a reasonable margin. The reason that there is a maximum level is that there is a diode directly in line in the circuit; wherever there is a semi-conductor, there is a certain potential power level that could destroy the detector.

Polarity—Polarity is either positive or negative. It is important to know the polarity of the output of your detector since it will have to be joined with some external circuit that requires a certain polarity to operate properly.

Sensitivity—Figure 4.11 is a typical sensitivity curve for a microwave detector. It can be seen that one factor affecting sensitivity is the RF power applied to the input of the detector; another is the dc output voltage. Therefore, the detector sensitivity is the dc voltage produced at the output for a specific power input, usually expressed in millivolts per microwatt CW. Sensitivity is simply how much power you need to produce a certain voltage.

It is important to note, before leaving detector specifications, that you must always consider the type of connectors on the detector. It cannot be emphasized strongly enough that when the devices utilize the proper connectors, the electrical and mechanical operation of your system is improved.

The microwave detector finds applications wherever there is a need to control or convert an RF signal to a dc voltage for display purposes.

Figure 4.12 shows three such instances. Figure 4.12(A) is one of the most common applications, that of monitoring an RF line. Power is coupled off the main line; the detector provides a dc voltage proportional to this power. The monitoring device, which has been previously calibrated, is deflected to give an instant and accurate reading of the main line power.

Figure 4.12(B) is a scheme that is very useful when a leveled power is needed, such as in swept measurements. Once again, a coupler and detector work together to sample the main line RF; the detector converts the RF to dc to activate the leveling circuits. This dc voltage causes either the RF power out of the source to be attenuated, or the attenuation to be removed in order to raise the power level. The result of this instantaneous pattern of action and reaction of coupler and detector is that the RF power remains level at the system output. Automatic leveling control (ALC) in microwave systems often employs just such a process.

Figure 4.12(C) is a method of displaying RF energy on an oscilloscope. It is impractical to attempt to view signals at frequencies above 100 MHz directly on a conventional oscilloscope. Storage scopes are available with responses as high as

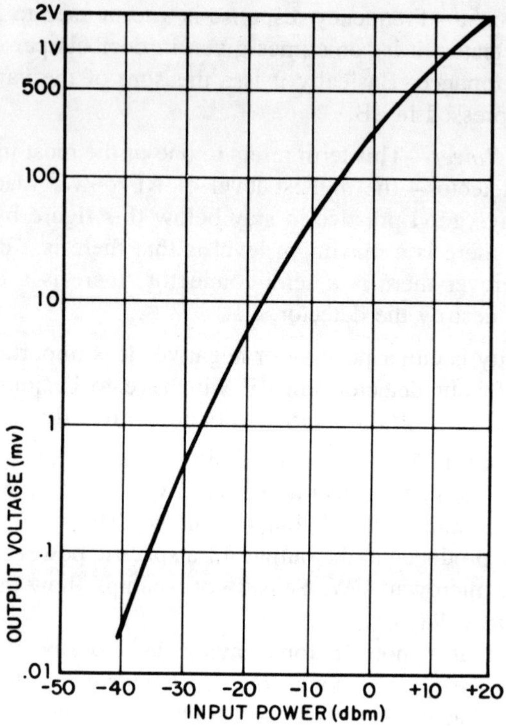

Fig. 4.11 Detector sensitivity curve.

1.7 GHz, but what if you would like to look at the output of a sweeper that is ranging from 2 to 4 GHz? If you had a network analyzer you would have no problem; otherwise Figure 4.12(C) is the answer. A simple combination of an attenuator and a detector do the job very nicely. The attenuator is of great importance in the setup, remembering that you should keep below a maximum power level. The attenuator ensures that you do, and it also keeps you in the linear range of the sensitivity curve of the detector (Figure 4.11), giving you much better results.

You can see that the idea of simplicity can produce a component which fits into many applications and gives months of reliable operation if its specifications are not exceeded. The detector seems to bear out two old sayings—"Good things come in small packages" and "The simpler the better."

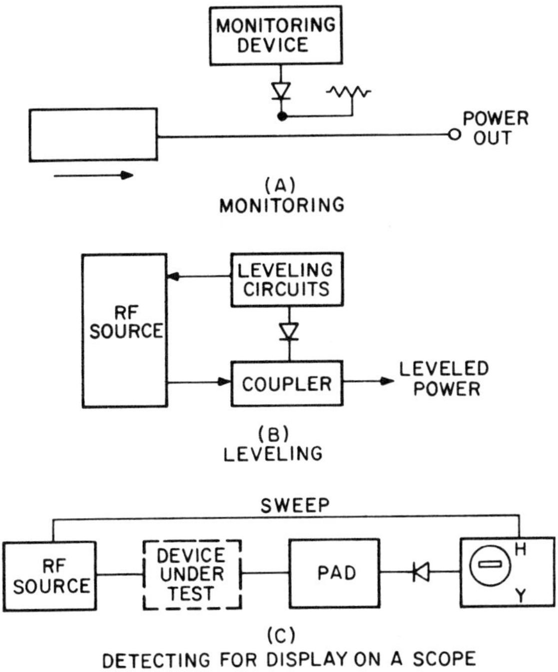

Fig. 4.12 Applications of microwave detectors.

4.3 MIXERS

There are many times when the available frequency is not the one you want for a system—it may be either too high or too low. You may be able to multiply or divide the frequency but what happens if you need a frequency that is not a harmonic or sub-harmonic of the one you have? A very versatile component called a mixer is the solution to the problem. By using a mixer and carefully selecting your RF and LO (signal and local oscillator frequencies), you can derive any frequency. That is not to say, however, that the only function of the mixer is to produce new frequencies. It is a very useful tool and has many applications; all of these are covered later in this section.

Figure 4.13(A) is a schematic representation of a mixer. You can always recognize the symbol on a schematic by its distinctive X inside a circle.

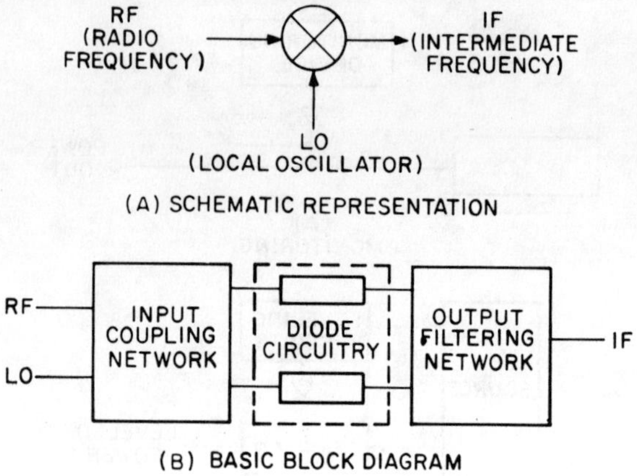

Fig. 4.13 Microwave mixer circuits.

Figure 4.13(B) is a basic block diagram of the three fundamental parts of a mixer—an input coupling network, the diode circuitry, and an output filtering network. The input coupling network is the circuitry which allows the RF and LO signals to be combined properly before being applied to the diode portion. This network could be a directional coupler, a transformer, or a quadrature hybrid; its prime feature should be that, under operating conditions, it combines the two signals while providing isolation between them. The diode circuitry consists of the diode(s), any matching circuitry used, and a dc return for the diodes; this area is where the actual mixing process occurs. When the RF and LO signals are applied to the diode circuitry, many combinations of the two are generated. The signals are at different levels but the ones which are most prominent are each of the originals and their sum and difference. In other words, if the RF signal is called f_1, and the LO signal called f_2, the output would have f_1, f_2, $f_1 + f_2$, $f_1 - f_2$, and every other combination of fundamentals and harmonics, ($2f_1$, $2f_2$, $2(f_1 + f_2)$, *et cetera*).

These combinations of frequencies are the reason why the final block is needed in a mixer—the output filtering network. Usually, the upper or lower sideband (that is, $f_1 + f_2$ or $f_1 - f_2$) is the desired output; everything else must be eliminated to produce the desired and necessary clean signal. The filtering usually is accomplished by a low-pass network which might be as simple as a shunt capacitor, or as complex as a large number of low-pass filter poles; it could be a series of trap circuits which would eliminate only specific unwanted signals. For certain applications, the filtering could be performed by a bandpass or highpass filter. The type of filtering and device used depends, of course, upon the individual situation.

Figure 4.14 shows some of the available mixers; notice their variations in size and the different methods of connecting them to a circuit.

To aid in your understanding of the various types of mixers we will concentrate on three basic configurations—single diode, balanced, and image rejection.

Fig. 4.14 Microwave mixers.

4.3.1 Single Diode Mixer

Figure 4.15 shows a single diode mixer; it clearly fulfills the three building block requirements for a mixer as previously discussed. It has a coupling network (a directional coupler), a diode circuit, and an output filter network (a shunt capacitor). It provides isolation by applying the RF signal (f_1) to the main line port and the LO signal (f_2) to the opposite coupling, or the isolated port; in this way the two signals are combined at point A, but are kept separate by means of the coupler.

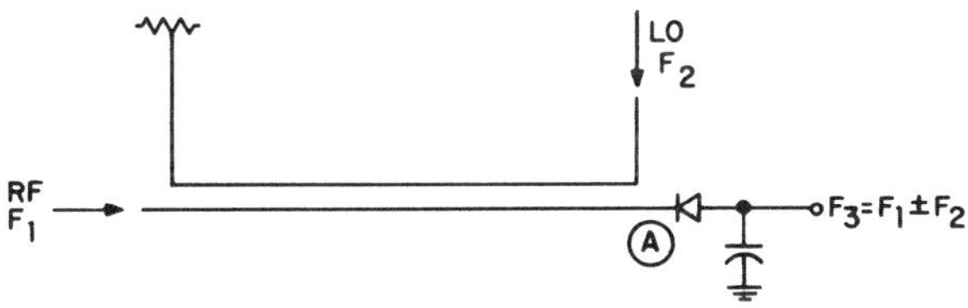

Fig. 4.15 Single diode mixer.

The advantages of the single diode mixer are both its simplicity of construction and design and, the fact that, since it needs only one diode to work, it is low in cost.

The disadvantages, however, far outweigh the advantages in today's microwave system market. Since only one diode is used, this mixer is a very narrow band device and the diode characteristics determine mixer operation. There are no provisions for eliminating noise in the mixer, which is a primary concern of the sophisticated low-noise systems now in use; a single diode is unable to cancel the noise terms which are prevalent in a non-linear circuit. And finally, conversion loss is high because of inefficiencies in the coupling network. Our diagram of the single diode mixer used the directional coupler, which produces optimum results; other possible coupling networks are a capacitive probe, an inductive loop on the main line, or even a straight T connection. All of these setups, however, including the coupler, result in a higher conversion loss than is tolerable when used in a single diode mixer.

It is apparent that there must be something better to use than the single diode mixer; the simplicity of the device does not make it the best choice. However, the single diode mixer can still be found in the front ends of many radar systems today, working very well for their particular system and giving adequate performance.

4.3.2 Balanced Mixer

Figure 4.16 is a balanced mixer. Notice that this diagram fulfills the requirements set down for a mixer, but certain differences make it much more useful than the single diode version.

One difference that can be noted is that the input network is termed a *balanced* coupling network, either a 90° or 180° hybrid. Whichever type is used, the efficiency of the coupling network is vastly greater than that of any single diode mixer. The increased efficiency of the hybrids means that virtually all of the power reaches the diodes, and there is very little loss due to mismatches. By eliminating the variations which result from a mismatch condition, the diodes have the power applied to them necessary for efficient operation. This increased efficiency results in a conversion loss through the mixer which is of a reasonably low value. Also, this type of coupling network usually operates over an octave band, another great improvement.

The second difference is that there are two diode circuits instead of only one. This point is significant because the second circuit improves the overall performance figures, such as noise level, a great deal. The lowering of the noise level can be attributed to the fact that, since there are two diode circuits working within the mixer, the noise voltages generated are cancelled to a great extent. It is generally true that the noise figure of such a mixer is equal to, or slightly higher than, the conversion loss.

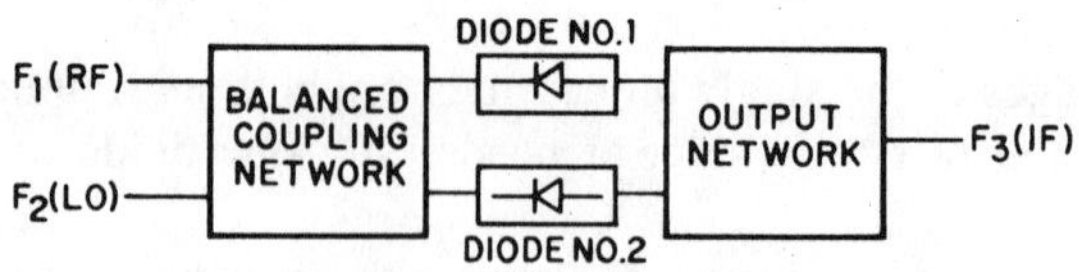

Fig. 4.16 Balanced mixer.

We have been referring to the mixing apparatus as diode circuits rather than just diodes—an intentional action since each side may contain more than one diode. A configuration which contains four matched diodes called a *quad,* is now available. A quad makes up a balanced configuration with two matched sections in each side, having the effect of lowering conversion loss and noise.

The advantages of balanced mixers already should be quite apparent—they have very good noise characteristics and low conversion losses; they operate over a wide band of frequencies (both LO and RF); and, when proper methods are employed, they have an acceptable input VSWR on the RF and LO ports.

The only disadvantage is that the balanced mixer is a more complex circuit to fabricate—multiple diodes and sophisticated input network are needed. However, the improved performance of the balanced mixer more than outweighs this aspect of design.

The next logical step is to do something to ease the restrictions for output filtering; that is, find a mixer that would give the desired sideband ($f_1 + f_2$ or $f_1 - f_2$) with a minimum of filtering. The component to accomplish this task is the image-rejection mixer.

4.3.3 Image-Rejection Mixer

Figure 4.17 shows a block diagram of an image-rejection mixer. In a basic mixer, the local oscillator may be above or below the RF signal in frequency; for the sake of discussion, let us suppose an LO below the RF. When the mixer is operating there is an IF signal equal to the difference between the two frequencies. This operation is a proper one. However, there is also a signal separated from the RF by the same IF amount, but it is above the RF signal and is the image frequency generated by the mixer. Obviously there are occasions where this signal would create problems; thus, a means of rejecting it is needed. The image-rejection mixer provides the solution.

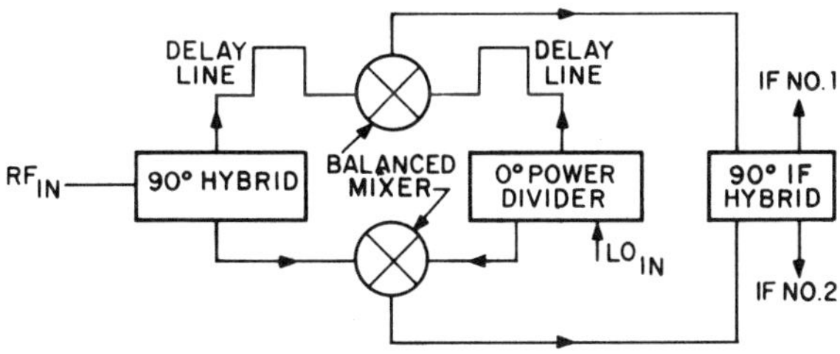

Fig. 4.17 Image-rejection mixer.

It can be seen readily that this mixer is much more complex than any circuit we have discussed thus far. Yet we still can identify the three basic building blocks necessary for a mixer.

The input coupling network consists of a combination of a 90° hybrid for the RF, a 0° power divider for the LO, and the input coupling networks of the two very important balanced mixers. The diode circuitry is that contained in the mixers; it is identical to that previously discussed—two diodes in each mixer. The output filtering consists of the 90° IF hybrid, a quadrature hybrid designed to operate at the IF frequency. One output port (No. 1) produces the upper image frequency; the other (No. 2), creates the lower. The quadrature characteristic (90° phase relationship) of the hybrid results in only one frequency being present at each port when the opposite port is terminated, thus providing rejection of the opposite image frequency. The low frequency design of the hybrid also provides rejection of the original frequencies (f_1 and f_2).

One great advantage of the image-rejection mixer is that it does as its name indicates—it rejects the image frequency that is generated in the mixer. When a clean IF signal at the desired frequency and phase is wanted, this ability is of great importance.

The only disadvantage of the image-rejection mixer may be that it is a complex circuit which takes a certain amount of skill to fabricate.

Listed below is a typical data sheet for a mixer. A balanced mixer is chosen because it has been explained already and is an important part of the image-rejection mixer.

Model number	MIXA1-2
Frequency (RF & LO)	1–2 GHz
Frequency (IF)	dc-500 MHz
Conversion loss	7.0 dB (maximum)
Isolation (RF/LO)	25 dB (minimum)
Noise figure	7.5 dB
VSWR: LO	2.5:1 (maximum)
RF	2.0:1 (maximum)
LO Power	+10 dBm
Connectors	SMA

To explain the mixer operation more fully we now define and discuss the most important terms from the above list of specifications.

Conversion Loss—Conversion loss is the level of the IF output as compared to the level of the RF input. Usually expressed in dB, it can be considered as a measure of the efficiency of the mixer; that is, it indicates how efficiently the two input frequencies are combined to form the IF output.

Isolation—It may be any one of three combinations of measurements referred to by this term—RF/LO, RF/IF, or LO/IF. In each case, isolation is a measure of the undesired frequency at an auxiliary port. An example would be as follows—the RF/LO isolation figure, in dB, is the amount of RF power measured at the LO port compared to that at the RF port. (The other two isolation figures are found the same way.)

LO Power—LO power is the amount, usually expressed in dBm, necessary to provide proper operation of the mixer. This power level is used to bias the diodes in a non-linear region so that mixing action occurs.

Noise Figure—Every network, whether a mixer or other component, has a certain amount of internal noise associated with its operation. A comparison of this level of noise to that of the desired signal in the network is called the signal-to-noise ratio. There are input and output signal-to-noise ratios for each network under consideration. The noise figure measures the reduction of a signal-to-noise ratio by the network under consideration (for example, the balanced mixer). It is a term expressed in dB which compares the input and output signal-to-noise ratios. Expressed mathematically it is:

$$\mathrm{NF} = 10 \log \frac{S_i/N_i}{S_o/N_o}$$

where:

NF = noise figure in dB
S_i/N_i = input signal-to-noise ratio
S_o/N_o = output signal-to-noise ratio

At times, you may hear noise figure being referred to as a "figure or merit" of a device. In a broad sense, such an evaluation is appropriate. The noise figure of a device is an indication of how its internal noise is handled in the process of trying to produce as nearly a noise-free output as possible. Effective execution of such a task could be described as a measure of the merit of a device.

VSWR—VSWR is an important term which indicates the match, or return loss, of each port under actual operating conditions. If measurements are made under abnormal circumstances, the figures obtained are useless. Only when the diodes in the mixer are being driven properly are true readings obtained.

The mixer finds many applications in today's microwave systems. The first is in a receiver front end; Figure 4.18 shows a typical instance of how the mixer is

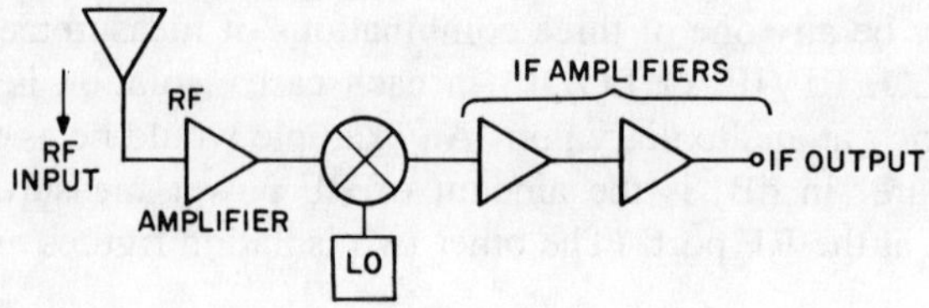

Fig. 4.18 Mixer application.

incorporated. It is used to convert the high frequency input signal to a lower frequency in order to obtain the transmitted information; it is much easier to operate on this lower frequency than on the input signal as it appears at the antenna.

The same concept is used in microwave network analyzers. The input GHz signals are converted down to a set frequency in the kHz range for easy operation and greater accuracy. In applications of this type, it is very important to have low noise mixers with a high degree of isolation and minimum conversion loss.

Other applications are as an up-converter in a transmitter (to increase the frequency for transmission), as a balanced modulator (which incorporates a 180° hybrid and reversed diodes in a balanced mixer), as a single sideband modulator, and in many instances where signal processing must be accomplished.

One application in testing is to use a mixer to down-convert a signal for noise figure measurements. Some noise figure meters have set frequencies as inputs; therefore, it is sometimes necessary to convert an IF signal to a lower frequency.

The key to understanding mixers in use in today's microwave systems is having a knowledge of the balanced mixer. From this basic circuit you can build more complex mixing schemes and are equipped to use the mixer as a valuable tool in needed test setups.

4.4 QUADRATURE HYBRIDS

A term you probably have heard many times is *quadrature hybrid;* but has anyone ever been able to give you a good definition of it? You may have heard one much like that I received when I first asked:

> A quadrature hybrid is a four-port, 3 dB coupler capable of dividing an input signal into two mutually isolated quadrature phased outputs while maintaining isolation of the fourth port from the input.

If I got many definitions like that one, I would probably seriously consider becoming a carpenter and leaving the field of microwaves to someone else.

Fortunately, there are easier and more understandable ways by which to define microwave components and their functions. The key to comprehending what a quadrature hybrid is, lies with knowing what the separate terms *quadrature* and *hybrid* mean. For reference purposes, Figure 4.19 shows two schematic representations of the device.

Quadrature is the easier to define. For any terms, whether signals, objects, or graphical points, to be in quadrature, they must be separated by 90°. This quality is one of the most important of the quadrature hybrid component—that of having output signals in quadrature, electrically separated 90° from one another. In Figure 4.19 ports two and three would be in quadrature with an input at either port one or four.

Before defining *hybrid* first we need to provide a meaning for *conjugate pair*. A network has a conjugate pair if an input applied to one of the branches has no effect on the other; that is, two branches form a conjugate pair if they are isolated from one another. In Figure 4.19 ports one and four are a conjugate pair, as are ports two and three.

A *hybrid,* or hybrid junction, is a four terminal network with two conjugate pairs of terminals. Power flowing into any terminal of the hybrid does not appear at its conjugate, but divides equally between the other two terminals. For example, if power is applied to port one in Figure 4.19, ports two and three would divide the power equally between them and port four would be isolated from the input. So a hybrid is a directional coupler which utilizes the concept of conjugate pairs to provide equal level, but mutually isolated signals at two outputs.

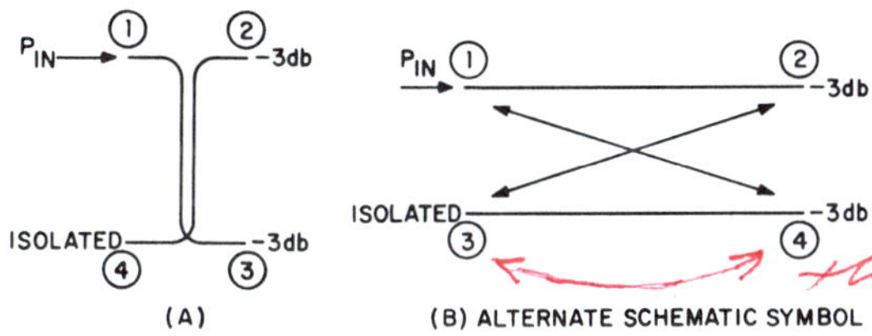

Fig. 4.19 Quadrature hybrid symbols.

From these individual meanings we can put together a basic definition of the entire term as follows:

> A quadrature hybrid is a directional coupler whose two outputs are equal in amplitude and separated from one another by a constant 90° phase.

Figure 4.20 pictures some of the quadrature hybrids that are available. There is a great variety, and your choice depends on your frequency range and circuit application.

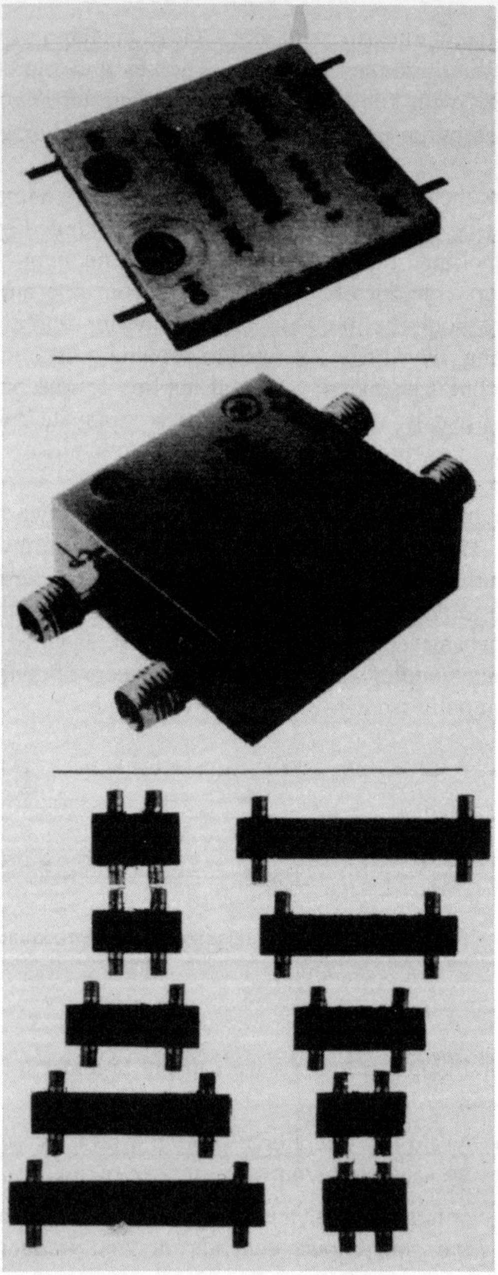

Fig. 4.20 Quadrature hybrids.

Shown below is a typical data sheet for a quadrature hybrid. Following an investigation of it, we define some of the most important terms.

Model number	QHI-2
Frequency	1–2 GHz
Isolation	25 dB (minimum)
VSWR	1.2:1 (maximum)
Insertion loss	0.25 dB
Amplitude balance	±0.5 dB
Phase balance	±1.5°
Power	100 W (CW), 1 kW (peak)
Impedance	50 Ω
Connectors	SMA

The following four specifications are generally considered most important, but your emphasis would, of course, depend on the particular application.

Amplitude Balance—This figure is a comparison of the power levels of the two output ports on the hybrid—ideally there should be no difference in amplitude between the two outputs, but in the real world there naturally is some variation. It is necessary for this figure to be as low as possible so that both circuits driven have the same level applied to them.

Insertion Loss—Figure 4.21 illustrates the insertion loss of a quadrature hybrid—basically the difference between the maximum height of the coupled curve (1–2) and the "straight-thru" curve (1–3). This internal loss of the device causes its output response to deviate from an ideal 3 dB figure across the desired band.

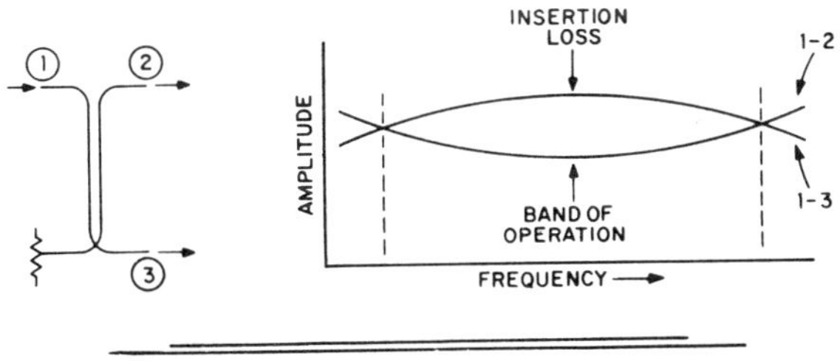

Fig. 4.21 Insertion loss of a quadrature hybrid.

Isolation—This term tells you how good a conjugate pair is formed by the terminals. The isolation between terminals is very important to consider if the hybrid is going to be used to drive two active devices, such as transistors or diodes. If the figure is not high enough, any imbalance in one device affects both of them and may eventually cause one or both to be destroyed.

Phase Balance—The quadrature hybrid is designed to have a 90° separation between the output terminals. But, because of variations within the hybrid, the phase of the output varies with frequency. The phase balance figure on a data sheet is a measure of how well the two output terminals track over the band of operation.

Other items to consider when choosing a hybrid are VSWR, power handling capability, and connectors. The VSWR is important since any mismatch on the output hinders operation of the overall circuit, a factor which is particularly significant if two transistor stages are being driven.

Power handling capability should be considered if there is any reason to believe that input power could exceed a conventional rating.

Finally, you should make note of the type of connectors on the hybrid. A careful check will eliminate the need for adapters which would cause extra expense and create a possible power loss.

The list of applications for a quadrature hybrid is almost endless; Figure 4.22 shows a few. Since they provide isolation between the RF and LO while applying equal signals to each of the diodes, quadrature hybrids are an integral part of a balanced mixer. They are ideal for a matched detector also, because of the equal level applied to the detector diodes. They are also appropriate as a means of dc biasing an RF circuit—they supply a continuous dc path while providing isolation for RF.

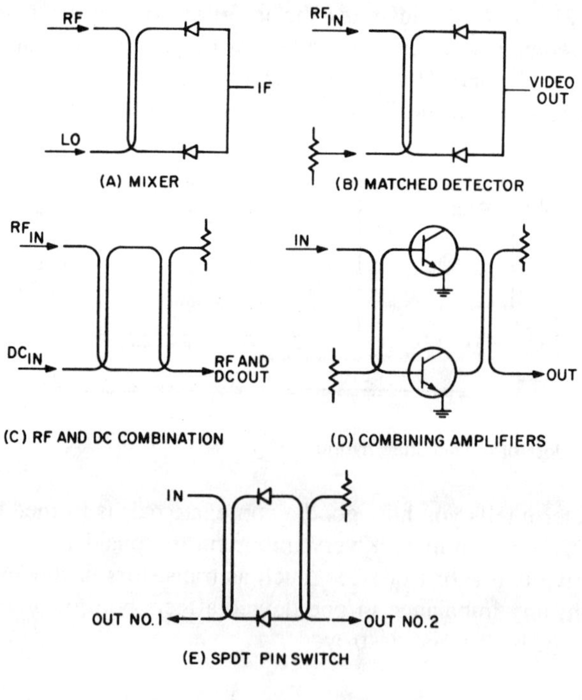

Fig. 4.22 Applications of quadrature hybrids.

Perhaps the two most common applications are shown in Figures 4.22(D) and (E), these being the combining of amplifiers and the very versatile *pin* switch. In both cases the ability of the hybrid to provide isolated signals of equal amplitude to a device is of great value. The quadrature effect (90° separation) allows the split of signals at the input and their recombination at the output.

Other areas in which quadrature hybrids are used include two-way power dividers, time delay circuits, and modulators, and as variations on the simple *pin* switch shown in Figure 4.22(E).

Obviously, the quadrature hybrid is a component with many applications. When a device is needed to provide two output ports of equal amplitude and which are isolated from one another, the hybrid satisfies that requirement. Its quadrature characteristics are also of great value when division and recombination are to be accomplished with active devices.

One very important reason that the quadrature hybrid has been so widely used of late is its adaptation to stripline, made possible by the new thinner dielectric material which allows the component to be produced rapidly, at a relatively low cost, and with exceptional repeatability.

4.5 POWER DIVIDERS

There are times when a single source of RF power is available and there may be two, three, or four circuits which need power. In this situation, the source power must be divided to supply each of the circuits.

In previous discussions it has been noted that the quadrature hybrid has applications as a power divider. In many cases, though, the hybrid does not achieve the proper results. One point is the hybrid's 90° difference in phase between its outputs; most of the time, power is required to be in phase. Another drawback is that a hybrid can only be constructed with a binary number (2^n) of output ports (that is two, four, eight, sixteen) (Figure 4.23).

Though your application may consist of an even number of ports and not have an in-phase requirement, in the majority of cases an in-phase condition must be met. To accomplish this end, and to obtain the proper number of output ports, a separate component is needed—the microwave power divider, sometimes called a resistive power divider. Figure 4.24 shows some schematic representations of this device.

The simplest method of dividing power is to tie a number of lines together and drive them with a source; this procedure divides the power, but the impedance is nowhere near a matched 50 Ω and there is no isolation between the output ports. A change in any of the output circuits is felt immediately at all of the remaining ports. Thus, special precautions must be taken during design to ensure that matched conditions occur within the divider and that there is good isolation between ports—the two primary requirements for a good power divider.

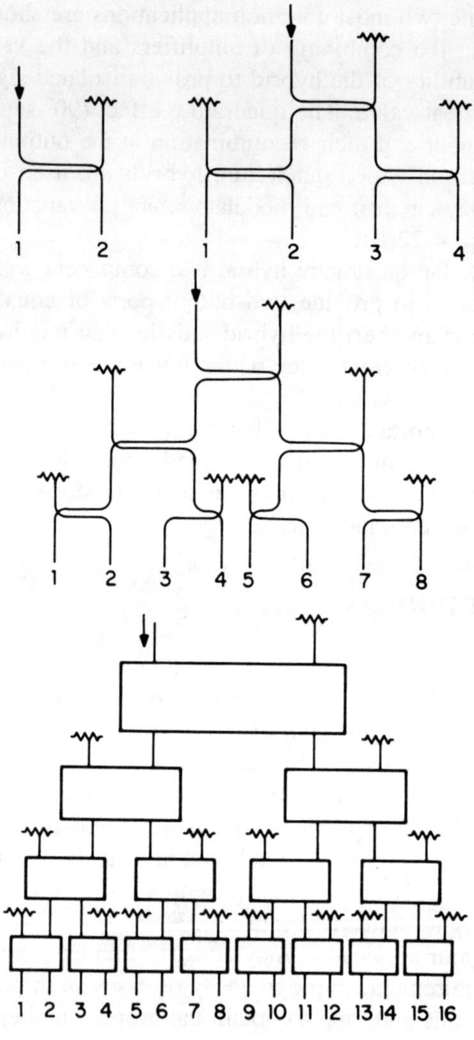

Fig. 4.23 Quadrature hybrid binary division.

Shown below is a chart of in-phase power dividers. N is the number of output ports; the loss is the best figure that can be obtained if there are no other losses in the divider. This loss is due simply to the reduction in power caused by division. That is, a two to one divider has 3 dB loss because half the power comes out either port, a four to one divider has 6 dB loss because one-fourth the power comes out each port, and so forth.

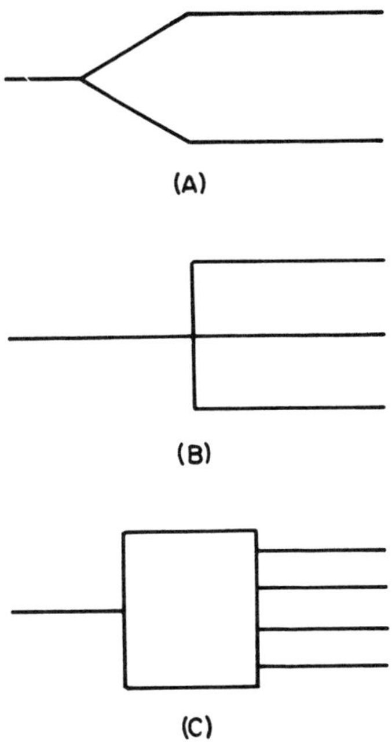

Fig. 4.24 Power divider schematics.

N	*Loss (dB)*
2	3.00
3	4.78
4	6.00
5	7.00
6	7.80
7	8.40
8	9.00
9	9.45
10	10.00

These are the values you can expect if you have any of these dividers in your circuit. There is also the additional insertion loss of the device.

The data sheet can tell us a great deal about a power divider and aid in the decision of whether or not it is suitable for our use. Following is a typical set of specifications—we examine these specifications and provide additional explanations of some of the more important terms at the end.

	two-way	*four-way*
Model number	PD-2	PD-4
Frequency	1–2 GHz	1–2 GHz
Isolation	20 dB (minimum)	20 dB (minimum)
Insertion loss	0.3 dB	0.5 dB
Amplitude balance	±0.2 dB	±0.5 dB
Phase balance	±1.5°	±3.0°
VSWR	1.3:1 (maximum)	1.4:1 (maximum)
Power	5 W	1 W
Connectors	SMA	SMA

Important terms to consider are *amplitude balance, insertion loss, isolation,* and *phase balance*. Let us examine and explain each of these.

Amplitude balance—How closely the output ports track with respect to one another is indicated by the amplitude balance. For example, if we have a four-way power divider and port one is at 6.05 dB, the amplitude balance is the value of ports two, three, and four with respect individually to port one.

Insertion loss—This figure is the internal loss of the power divider, which causes the output ports to vary from the theoretical values listed in the in-phase power divider chart above. If a three-way divider is being used and the ports are at 4.93 dB instead of the ideal 4.78, the insertion loss of the device is 0.15 dB.

Isolation—Isolation is the amount of attenuation from one output port to another. It is of prime importance since each port of the divider must be independent of all others for proper operation of the many circuits that the divider must drive.

Phase balance—This specification is how closely the output ports track one another with respect to phase; the dividers are designed to have in-phase outputs. Such a power divider can have its output phase balance destroyed merely by the addition of unequal lengths of line to its output connector. Depending on your particular application, this change could be beneficial or detrimental.

Other factors to consider are the VSWR (which should be within reason at the output); the power handling capability (which is important if high-power circuits are being used); and the connector type (which makes the power divider compatible with the rest of your system without the use of adapters).

Applications of power dividers include their use as signal dividers in a test setup where the reference signal and the measured signal must track; in mixer circuits to supply one of the input signals in-phase to a pair of balanced mixers; and where multiple antennas are to be fed in phase by a single source.

The power divider has many advantages and applications in microwave systems, but we have been speaking solely in terms of these devices as dividers. The power divider also is very useful when working in the opposite direction, that is, as a combiner. If we applied a signal at the input of a four-way divider, placed another four-way power divider directly at the four ports and summed their separate outputs back together, we would have the same level of signal at the output as we did at the input (minus a small insertion loss for each divider). What we have done is to place two power dividers back-to-back, thereby dividing a signal in one and adding it back in the other. So when we call the component a *divider,* it should also be remembered that what we divide in one direction we can re-combine in the opposite.

4.6 FILTERS

When someone mentions a filter, many things come to mind. You could think of an oil filter for your car, a furnace filter in your home, various types of filters at work in a factory, or maybe even a filter on a swimming pool. Regardless of what type of filter comes to mind, they all have one thing in common—eliminating something unwanted in order to keep a system working. The microwave filter is no different in that respect; its job is to eliminate a certain portion of a frequency band in order to keep the overall system working properly. This unwanted portion may be on either side of a band of frequencies, or it may be above or below a specified frequency. Whichever area is designated as unwanted in the system, the microwave filter removes it.

Generally speaking, we could call every component we have discussed thus far in this chapter a filter of some sort. Each component has a term on its data sheet that is called *frequency*. Each time this specification appears it is indicated over what range of frequencies the component operates. Implicit is that, on either side of this band of frequencies, the component attenuates the signals to some extent. That is, it acts as a filter to eliminate any frequencies not in its designated operating band.

In this section we discuss five basic types of filters—*bandpass, low-pass, high-pass, tunable,* and *YIG*. For each we provide a general description, explain the characteristics of the filter, show a typical data sheet and explain its terms, and, finally, list applications.

4.6.1 Bandpass Filter

The *bandpass filter* is designed to do exactly what the name implies, pass a specific band of frequencies with as small a loss as possible.

Figure 4.25 shows a bandpass filter response. An examination of the response shows that there is an area called the "passband" where there is a minimum loss, on either side of which the response falls off quickly to form what are known as the "skirts" of the filter.

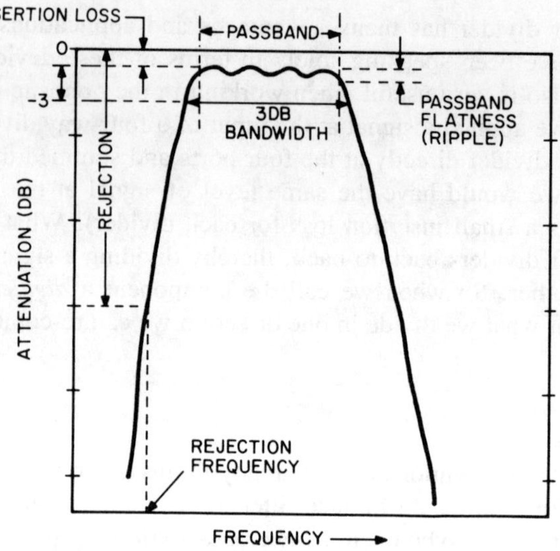

Fig. 4.25 Bandpass filter response.

The steepness of these skirts depends on how close the rejection frequency is to the edge of the passband—the closer, the steeper. This rejection frequency is very often termed the 60 dB bandwidth figure; that is, when the skirts are 60 dB below the insertion loss, then this figure is the bandwidth of the filter response at that point.

Shown below is a typical bandpass filter data sheet. There are many important terms on this sheet; they are explained after this listing.

Model number	BPF-XN1
Passband	1.5–3.0 GHz
Insertion loss	0.75 dB
VSWR	1.5:1 (maximum)
Rejection	60 dB (0–1.3 GHz, 3.2–10.1 GHz)
Impedance	50 Ω
Power	25 W
Connectors	SMA

The important terms to be explained are *insertion loss*, *passband*, and *rejection*.

Insertion loss—This loss is that internal to the filter in the desired band of frequencies; it includes any ripple in the filter response. Every filter has a certain amount of ripple, or variation in the response, due to the fact that all filters are made up of a combination of resonant sections that add together to form the complete response.

Each section has a response of its own; when added together, small variations in amplitude are evident. This insertion loss figure is valid only when the generator and load are matched. Any variation in either of these impedances and the insertion also varies.

Passband—Passband is the desired band of frequencies in which the only loss is the insertion loss of the filter. It is the actual band of frequencies which the design allows to pass through with very little effect on system performance.

Rejection—In most cases rejection is an attenuation figure of 60 dB. There are frequencies on the response curve that are undesired within the overall system and which are curtailed greatly. The term rejection indicates the amount of attenuation of specific frequencies within the undesired range.

The bandpass filter finds many applications in a microwave system. It serves well in reducing the noise and harmonic content of a system by limiting the band of frequencies seen by the system. Many times, bandpass filters are placed at the front end of a receiver to decrease the out-of-band noise that can enter; they can be used at the output of a transmitter to reduce, or eliminate, harmonics and spurious outputs generated within the system.

A very important use of the bandpass filter is to ensure a *clean* transmission. After a circuit, which produces a comb of frequencies as a result of a multiplication process, has been designed, a filter is used at the multiplier output. A similar application in a test setup would be to filter the output of a high power generator. Otherwise, low frequency components—which could cause oscillations and destruction of microwave transistors—might be produced.

4.6.2 Low-Pass Filter

The *low-pass filter* is designed to pass all frequencies below a specified cut-off; any signal above this frequency is attenuated. Figure 4.26 shows the response of a low-pass filter with a portion, which could be called the passband, ranging from *zero* to the cut-off frequency. The term zero is used very loosely. It does not mean that if we had a low-pass filter with a cut-off frequency of 8.0 GHz there would be an insertion loss of 0.5 dB from dc to 8.0 GHz. The zero referred to is some low frequency limit where the insertion loss figure does apply. Commerical low-pass filters are not specified as to their low-end characteristics. However, they pass dc simply because of their series inductance tendencies.

The rejection frequency shown in Figure 4.26 is usually that where the response is attenuated 60 dB. This figure varies from manufacturer to manufacturer and from filter to filter; it may be 40, 50, 60, or even 100 dB.

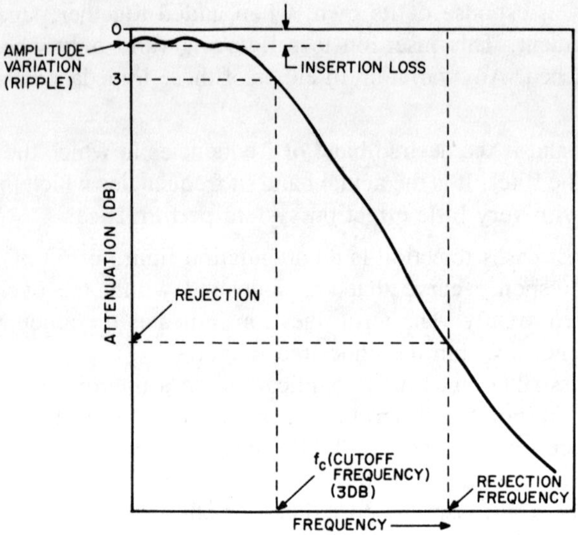

Fig. 4.26 Low-pass filter response.

Shown below is a typical data sheet for a low-pass filter. Some manufacturers give a range of values for terms such as *cut-off frequency, impedance, insertion loss,* and *stopband rejection*. What follows is what could be found if you had a specific application.

Model number	LPFXN-1
Cut-off frequency	2.0 GHz
Insertion loss	0.4 dB
VSWR	1.5:1 (maximum)
Rejection	50 dB (4.0–12.0 GHz)
Impedance	50 Ω
Power	10 W
Connectors	SMA

Only one term from the above data sheet need be clarified—the cut-off frequency of the low-pass filter. All other terms are defined in the bandpass section and apply to the low-pass.

Cut-off frequency—This frequency is 3 dB below the insertion loss of the filter; it denotes the point at which the filter is considered to be rejecting unwanted signals. Below this frequency the filter is said to be in its passband and exhibits a low loss.

From the beginning of a system design, great care should be taken to note such specifications as power handling capability and connectors for the low-pass filter, in order to save time and money later.

Although the low-pass filter can be used in a variety of applications, two are very popular.

The first is at the output of a mixer circuit. When mixing occurs, the result is a lower frequency (IF) which is used for various operations later on in the system. This IF signal should be free of any RF components which could cause problems in the IF circuitry. Therefore the low-pass filter is used to provide a low loss for the IF and to reject the high RF signals that are present at the output of the mixer. In the process, a good match is provided for the mixer at the IF frequency, which makes interconnection with following circuitry much easier.

The second application is that of *pest control*. One of the biggest problems where active devices are employed, particularly in high power systems, is removal of harmonics and spurious outputs. The low-pass filter is ideal for this situation since it rejects the higher frequencies which compose the harmonics and majority of spurious signals. Any signal of frequency other than that desired can have an adverse effect on any following circuitry; most times the best way to refer to these signals is as we did in the beginning—*pests*.

These two applications of the low-pass filter are by no means the only ones. Another important job is to remove harmonics and spurious signals at the output of a test generator. The low-pass often can be used when a bandpass filter is not available for some of the applications listed earlier.

When the portion of the frequency spectrum to be eliminated is the high frequency signals, the low-pass filter is the device to use.

4.6.3 High-Pass Filter

The *high-pass filter* is designed to do just the opposite of what the low-pass does; that is, it passes all frequencies *above* a specified cut-off frequency. Once again, it should be noted that the filter does not pass every frequency from the cut-off to white light with a minimum loss; as higher and higher frequencies are reached, the attenuation increases. Figure 4.27 shows the response of a high-pass filter; it also has a passband, from the cut-off frequency on up (though, as mentioned above, there is an upper limit).

The rejection frequency in Figure 4.27 is where the response is usually down 50 to 60 dB. The specific value varies from filter to filter, but it is usually in the neighborhood mentioned.

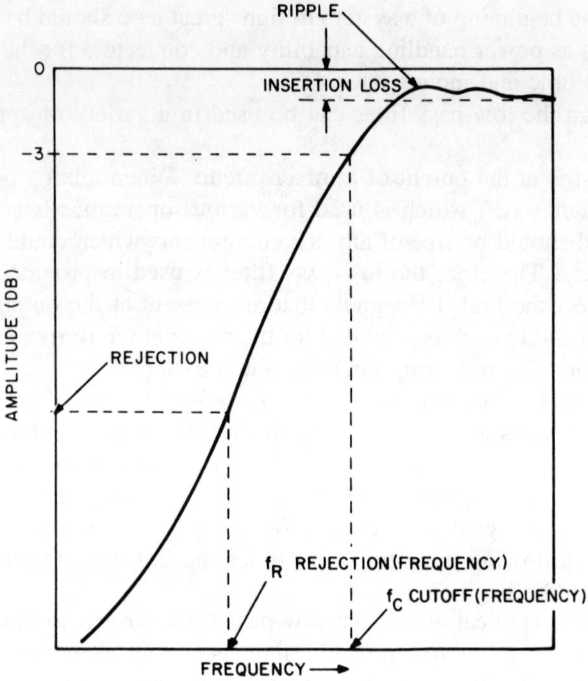

Fig. 4.27 High-pass filter response.

A typical data sheet for a high-pass filter follows. Once again, it should be mentioned that some manufacturers give a range of available cut-off frequencies, impedances, and rejection figures; this list is only a sample.

Model number	HPFXN-1
Cut-off frequency	1.0 GHz
Insertion loss	0.6 dB
VSWR	1.5:1 (maximum)
Impedance	50 Ω
Rejection	50 dB (0–0.8 GHz)
Power	25 W
Connectors	SMA

The cut-off frequency of the high-pass filter is the frequency at which the response is 3 dB below the insertion loss of the filter. That is, it indicates the point at which the filter is considered to begin rejecting the unwanted signals. Above this frequency the filter is said to be in its passband and exhibits the low insertion loss shown on the data sheet.

Other terms to take note of, other than the important ones such as insertion loss and rejection, are the power handling capability of the device and the type of connectors with which it is supplied. You should also note the available impedance since different figures cause drastic changes in the filter characteristics.

The high-pass filter does not find as many applications as do the more popular bandpass and low-pass filters. One particular instance does stand out, however, as being very useful when you are concerned with low frequency signals disrupting the operation of your system. The high-pass filter removes any low frequency signals present at the input of a device or section; thereby preventing spurious responses or a microwave transistor breaking into low frequency oscillations. This model also holds true for test setups; the filter can be placed at the output of a generator to ensure a clean output signal which is free of low frequency problems.

To eliminate the low frequency portion of a signal, high-pass filters are the appropriate devices.

4.6.4 Tunable Filter

A *tunable filter* is a very narrow bandpass filter whose center frequency, and associated bandpass, can be varied over a specified range. This tuning process may be done manually, digitally, or with voltage. The tunable filter incorporates many features into a single component that would not be possible in a conventional bandpass filter. One such feature is the narrow bandwidth that is obtainable; it is possible to work over an entire octave with a bandwidth of 1.5 to 2 percent around any center frequency you choose. Thus, you have great selectivity and can reduce unwanted signals to an even larger degree.

A bandpass filter that works over an octave must consist of many sections in order to achieve the desired broad band performance. The tunable filter works over the same band but is concerned with only a small portion at a time. Therefore, tuned resonator sections are used, resulting in a lower insertion loss than with a conventional bandpass filter. This characteristic is very helpful at a receiver front end, where the total loss of the system is of great importance.

Figure 4.28 is a picture of a tuned filter. A typical data sheet of a manually tuned filter is shown below. The manually tuned variety is chosen since it is the one most commonly found in test setups. The voltage and digitally tuned filters are used more in specific applications.

Model number	TF-1
Frequency	1–2 GHz
Impedance	50 Ω
VSWR	1.5:1 (maximum)
Number of sections	5
Insertion loss	0.6 dB
Shape factor	2.2/1
Dial accuracy	±1%
Power	50 W
Connectors	SMA

Fig. 4.28 Tunable filter.

Many of the terms have been explained and, so, are not repeated. However, there are two specifications which need clarification—*dial accuracy* and *shape factor*.

Dial accuracy—As you can see in Figure 4.28, on the front of the tunable filter there is a knob with a dial (on which are scribed the filter's operating frequencies) attached. The operating range has been calibrated very accurately by the manufacturer. However, due to mechanical variations within the filter and internal gears between the filter and the dial, specific frequencies cannot always be reset exactly. Therefore, dial accuracy specifies how close the dial can be set and reset.

Shape factor—The shape factor, very simply, is the ratio of one bandwidth to another, usually the 3 dB and 30 dB. (These values are assumed unless otherwise specified.) In the data sheet above the shape factor is called out as 2.2/1—meaning that the 30 dB bandwidth is 2.2 times greater than the 3 dB bandwidth. Alternately, a 3 dB bandwidth of 50 MHz would yield a 110 MHz 30 dB bandwidth. Another popular shape factor is 3.5/1; it is usually associated with a three-section tunable filter.

Tunable filters, whether manually, digitally, or voltage tuned, find wide applications as preselectors at the front end of many receiving systems. They select a frequency, pass it with a minimum loss, and reject all frequencies above and below the desired one; their characteristics make them ideal for such an application.

They are also well suited for laboratory work where precision measurements are needed, and where noise and harmonics must be kept at a minimum. Their narrow bandwidth and steep skirts ensure that harmonics and spurious (or noise) components are rejected sufficiently.

Thus, the tunable filter's main recommendation is its ability to provide a narrow bandwidth over a wide range of center frequencies.

4.6.5 YIG Filter

One great disadvantage of tunable filters is that they vary greatly in size as their range of operation is changed. The tunable filter used as an illustration occupies a volume of 67.9 cubic inches; one designed to operate at lower frequencies could take up to 200 cubic inches, and some have been known to fill from 500 to 1,000 cubic inches of space. Many times these dimensions cannot be tolerated.

What is needed is a type of filter which exhibits a narrow bandwidth, is tunable over an octave or more, has a reasonable insertion loss, and occupies a small volume regardless of the frequency of operation. A unique variation of a tunable filter fills this order and is much more linear in operation, as well—the *YIG filter*.

The YIG filter (yttrium iron garnet) makes use of an RF interaction between YIG spheres in the filter and their coupling structures. This interplay causes a resonance which, in turn, produces a center frequency dependent only on the magnitude of the dc field and entirely independent of physical dimensions. Since the frequency is determined by the current in the coil of the electromagnet (the magnitude of the dc field), the drivers used to tune the YIG filter are of great importance to both the frequency and linearity of the device. The YIG driver is essentially a voltage-to-current converter; that is, it transforms an input control voltage to an output current in order to vary the field strength, and thus, the frequency of the filter.

Frequency is a linear function of the coil current; in the following data sheets, the deviation from linear (accuracy) is less than ±0.1 percent. The average driver for a YIG filter has an output current, as a function of driver voltage, which is linear within less than 0.01 percent. As a result, the frequency of the filter is a linear function of the drive control voltage to an accuracy of approximately 0.1 percent. This linearity is outstanding for such a compact device and which is controlled by such a small dc voltage.

To aid in understanding the YIG filter, we list a data sheet for a typical one and its driver. Many of the terms on these data sheets need explanation; when most of them are understood, you should have a clearer picture of the YIG filter and its function.

	Filter Data
Model number	YF-1
Frequency range	1–2 GHz
Bandwidth (3 dB)	20 MHz (minimum)
Insertion loss	3.0 dB
Off-resonance isolation	40 dB
Off-resonance spurious	25 dB
Passband ripple	1.0 dB (maximum)
Passband spurious	2.0 dB (maximum)
VSWR	2.0:1 (maximum)

Limit level	+10 dBm
Selectivity	12 dB/octave
Sensitivity	17 MHz/mA
Coil resistance	10 Ω
Coil inductance	110 mH
Time constant	1.0 ms
Deviation from linear	±2.0 MHz
Hysteresis	4.0 MHz
Connectors	SMA
	Driver Data
Operating band	1–2 GHz
Input power	+12 V at 135 mA
	−12 V at 20 mA
Control input	0 to +10 V dc
Sweep time	9 ms (minimum)
Sweep flyback time	5 ms (maximum)
Sweep linearity	±100 kHz

To aid in our discussion of the individual terms, we refer to Figure 4.29. All of the terms on each of the data sheets are listed; if they have been explained sufficiently earlier in this section, that fact is pointed out.

YIG Filter

Frequency Range—This term indicates the total span of frequencies over which the filter can be tuned. Each frequency in this range is the center of the passband of each individual response as the filter is tuned.

Bandwidth (3 dB)—The bandwidth of the response where the loss is 3 dB below that of the filter insertion; it is indicated by point one in Figure 4.29(A).

Insertion Loss—The insertion loss (point two in Figure 4.29(A)) is the internal transmission loss of the filter; it is the minimum loss point on the filter response.

Off-Resonance Isolation—This phrase indicates the amount of rejection (as defined in previous sections) of the filter. Off-response isolation is measured on the skirts of the response, as shown at point three in Figure 4.29(A).

Off-Resonance Spurious—Shown at point four in Figure 4.29(A) is the off-resonance spurious—the attenuation of any spurious signals outside the tuned response.

Passband Ripple—The peak-to-peak variations in the passband response of the filter are the passband ripple, shown at point six in Figure 4.29(A).

Passband Spurious—The amount of additional insertion loss in the passband caused by spurious modes set up in the filter is called the passband spurious and is indicated in Figure 4.29(A) at point five.

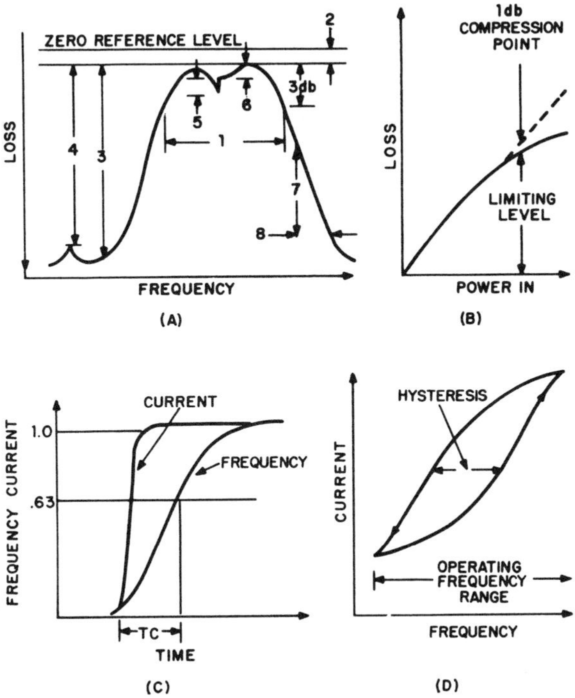

Fig. 4.29 YIG filter characteristics.

VSWR—VSWR is a measure of the match of the filter within the desired passband. Usually, the maximum value across the band is determined and is given in the data sheet.

Limit Level—The input level to the filter which causes the input/output curve to be compressed 1 dB from linear, is the limit level as illustrated in Figure 4.29(B).

Selectivity—Selectivity is actually a measure of the steepnesss of the skirts of the filter response; it is the amount of attenuation from the 3 dB bandwidth for a given increment of frequency. It is a dB/octave figure, shown as item seven in Figure 4.29(A) divided by item eight.

Sensitivity—The number of MHz which can be tuned for each milliamp of control current is the sensitivity. As the term implies, it indicates how sensitive the filter is to control current.

Coil Resistance and Inductance—These values are the inductance and dc resistance of the coil (when de-energized) at room temperature.

Time Constant—Figure 4.29(C) shows what this term indicates—the time it takes for the frequency response to reach 63 percent of its final value when the current step shown is applied. The time constant sometimes is referred to as the switching speed of the filter.

Deviation from Linear—This specification is the amount that the frequency response varies from a theoretical straight line (coil current *versus* frequency) over the operating range of the filter.

Hysteresis—Figure 4.29(D) is an illustration of a hysteresis curve, which is generated by increasing the control current—starting at the low frequency end of the filter—and plotting the frequency as its function. When the upper limit of the filter is reached, the control current is reduced to zero and the frequency is plotted once again. Because a magnetic circuit is involved, the two curves are not identical; there is a frequency difference as shown, called the hysteresis of the filter.

Before discussing the driver data terms, the importance of the last specification on the filter sheet—the connectors—must be emphasized. It is crucial to have the proper connectors on your filter; the insertion loss is already 3 dB and any needed adapters would increase this figure.

YIG Driver

The operating band is the same as that described among the filter specifications. Care should be taken to match the operating ranges of the filter and its driver.

Input Power—This figure is the dc power needed for the proper operation of the driver; usually, both positive and negative voltages are necessary.

Control Input—Control input indicates the range of control voltages which create the current changes out of the driver needed to tune the YIG filter over its assigned frequency range.

Sweep Time and Flyback Time—These two terms need to be considered when the filter is to be swept automatically, as in a swept receiver. The sweep time is the shortest period of time it takes to get from the low end of the frequency band to the high end; flyback time is how long it takes to get back.

Sweep Linearity—Once again, this specification is how the overall response of the driver varies from a theoretical straight line; it also is a plot of current *versus* frequency.

The YIG driver is sometimes mounted with the filter; sometimes, it is a separate component. In the latter case, you should be sure that compatible connections are present on the filter and driver.

The primary application for the YIG filter is in a receiver front end where a range of frequencies must be covered but the selectivity for each received frequency must be high.

From our previous discussions it is clear how it would be possible to cover a complete octave of frequencies but have a 3 dB bandwidth of only 20 MHz for each tuned center frequency. You can also see, by studying the filter's rejection response, how close-in signals to your desired frequency would be rejected.

Before leaving the subject of YIG filters we should make note of the fact that we have made a pretty strong case in their favor. Everything said is true. But it should be remembered that, though the YIG has the advantages we have noted, it also has its share of disadvantages which sometimes make it necessary to use conventional, tuned filters. For example, when a 3 dB insertion loss cannot be tolerated, where dc power for the YIG driver is not available, or with an RF power greater than +10 dBm, the YIG filter cannot be used. As in everything else, the YIG filter has some fantastic advantages but there are cases when its drawbacks eliminate its use.

We have covered five basic types of filters, each of which is designed to eliminate some undesirable portion of a frequency spectrum. We have discussed what specifications are shown on a data sheet and the meaning of those terms. For each filter type we discussed a few applications and, in some cases, advantages and disadvantages. All in all, we have taken a fairly comprehensive look at filters in use in microwaves. Occasionally, there are additional important terms on a data sheet, such as number of sections, peak power, shock, vibration, and temperature. The first is for general information and could be used if you wanted to check various available charts to see how your response will look; the second is of importance if you are working in a pulse system where a large peak power is available; and the final three are environmental specifications which you would check against your own system requirements.

In summary, the microwave filter is, simply, the component designed to eliminate unwanted elements in order to keep a system working.

4.7 CIRCULATORS AND ISOLATORS

In the beginning of this chapter we discussed the directional coupler. We stressed its outstanding quality of being able to pass microwave energy in a forward direction while providing isolation in the reverse; the component was designated as a *directional* device.

Such a quality also is exhibited by the circulator. However, a circulator (or isolator) is a *nonreciprocal* device; that is, it provides a one-way street for the microwave energy.

Figure 4.30 shows the nonreciprocal characteristic of the circulator. You may have seen something similar to this illustration before using water pipes and valves.

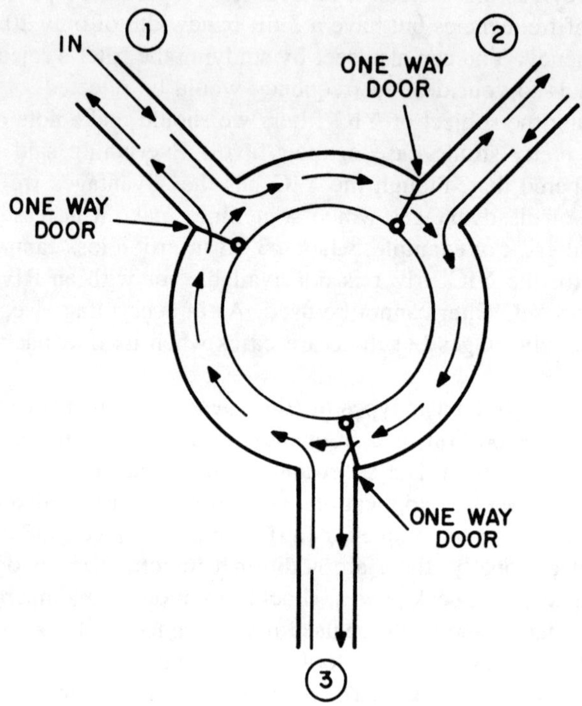

Fig. 4.30 Circulator illustration.

If you were a plumber or plumber's apprentice, such a drawing would be of great interest to you. However, the figure is more understandably related to an every day situation. At one time or another, you may have gone to a football or baseball game at a stadium and experienced the crowds and their random movements—it would be so much easier if everyone were going in only one direction. In Figure 4.30, consider port one to be the main entrance or exit for the stadium, with ports two and three as aisles to individual sections. Each port provides movement in two directions (the main flow and any small amount referred to as "leakage") each of which is controlled. We want to fill section two before section three; when two is full, the overflow goes to section three. From Figure 4.30 you can see that it is impossible to go directly from the main entrance to section three because of the one-way doors at each section. With microwave energy, the action is the same. With the energy input at port one, the easiest path is out of port two. Any reflections (overflow of the crowd) at port two are sent back and end up at port three—they cannot go directly to port one because of the basic, isolation-providing construction of the device. The same is true of any mismatches at port three—they are not felt at port two but appear at port one.

In our illustration we obtain a nonreciprocal device by means of one-way doors or gates. In an actual circulator, a nonreciprocality is produced by a gyro-magnetic field. To understand this term, you must first have an idea of how a circulator (or isolator) is constructed.

Figure 4.31 is a structure diagram of a typical three-port circulator; its basic construction is in the form of a sandwich. The circuit is at the center of the device with pieces of ferrite material on either side. These pieces are sometimes referred to as "pucks," because of their physical resemblance to hockey pucks. Moving outward, a ground plane is placed next to satisfy the requirements for stripline construction. The successive items (pole pieces and magnets) are what produce the gyro-magnetic effect referred to above, and thus aid in the nonreciprocal action of the circulator. The magnetic field of the assembly moves in either a clockwise or counterclockwise circular direction. The interaction of the rotating (gyro-magnetic) field and the ferrite material produces low insertion loss at specified frequencies in one direction, and isolation between ports in the reverse. A shield is placed around the entire assembly to protect it from outside effects and to keep any external circuitry from being affected by the internal magnetic field.

The circuit in Figure 4.31 is basically three arms joined at the center; if you dc check the circulator, all three ports have dc continuity. The nonreciprocal action takes place because of RF, rather than dc, isolation between ports. Interestingly, the directional coupler has both dc and RF isolation between its ports.

The isolator is simply a circulator with its third port internally terminated with a 50 ohm load—a very important factor when working with high power circuits. A manufacturer's building in a high power termination increases the time and cost greatly. A more practical alternative is to buy a three-port circulator and to employ a separate termination for it—a far more economical move when higher powers (25 W and up) are being used.

A data sheet for a typical circulator is given below. Let us examine the data sheet and its terms.

	Circulator	**Isolator**
Model Number	CIRC–1	ISOL–1
Frequency	1–2 GHz	1–2 GHz
Isolation	18 dB	20 dB
Insertion loss	0.5 dB	0.6 dB
VSWR	1.25:1	1.25:1
Power	50 W (CW)	2 W (CW)
	1 kW (peak)	100 W (peak)*
Connectors	SMA	SMA

*higher powers available upon request.

Two terms are worth examining further—*isolation* and *power*.

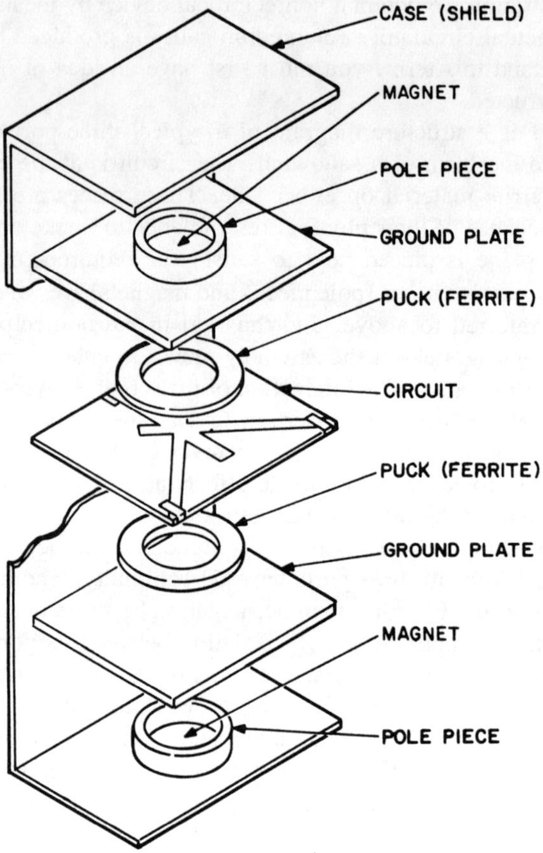

Fig. 4.31 Circulator construction.

Isolation—Isolation, when applied to a circulator or isolator, is not a set figure; for the former, it depends on the quality of a load attached to the third port of the circulator; in the case of the isolator, how good the input and output match is.

A circulator does not provide isolation until it is terminated. Figure 4.32(A) is a plot of expected isolation *versus* third port VSWR; obviously, the better the load VSWR, the better the isolation.

Figure 4.32(B) illustrates why it is necessary to have good input and output VSWRs when an isolator is used. The curve assumes perfect internal termination of the third port. The values printed on the curve (1.2:1 match, and so forth) are the various isolator VSWRs, as shown on the sample data sheet. A good internal isolator match is clearly of great importance, since it allows the input and output to vary to a greater extent, and thereby permits greater flexibility. The better the isolator match, the higher the value for overall isolation.

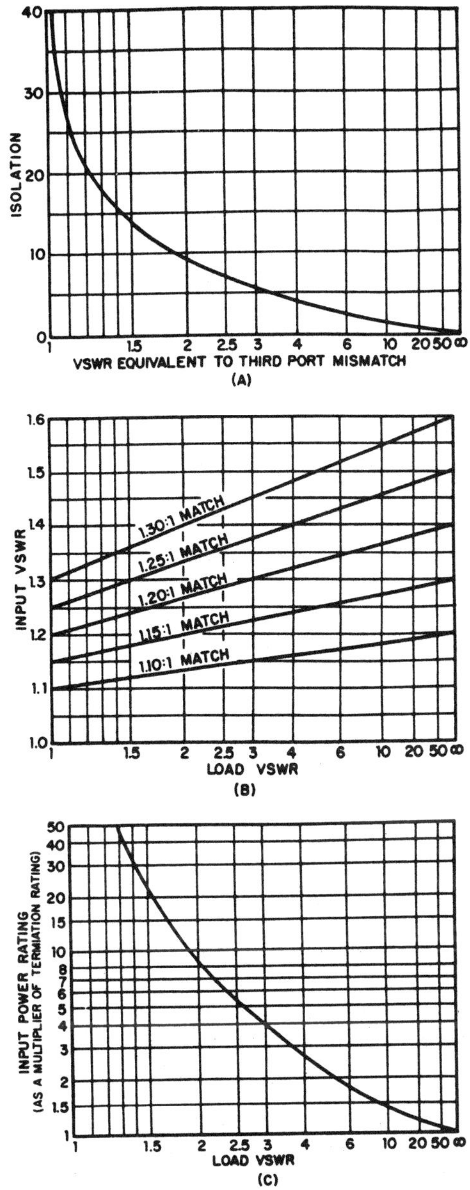

Fig. 4.32 Circulator and isolator isolation curves.

Power—Note from our data sheet the wide variation in power between the circulator and the isolator, resulting from the latter's internal termination which is rated much lower than the overall device. Figure 4.32(C) is a curve of power rating *versus* load VSWR; if your load has a low VSWR, you can extend the range of the isolator to a very high degree. As an example, suppose that you have a need for an isolator that can handle 25 W of power. If you are certain that your load VSWR is never worse than 1.5:1, just about any conventional isolator available (provided the device itself can handle that much power) is usable. If the ratio deteriorates past 1.5:1, your power handling capability drops rapidly. Figure 4.33a shows the schematic symbols for circulators and isolators; Figure 4.33b illustrates a variety of actual devices.

Circulators and isolators find many applications in microwaves today. The primary use of circulators is in what is called a duplexer, a device used in transceivers. Port one is the transmitter output, two is for the antenna, and three goes to the input of the receiver. Clearly, the advantages of low insertion loss and high isolation (with good load match) are put to maximum use. To extend this idea, two circulators can be put in series to form a four-port circulator with increased isolation; this process can be extended to five-, six-, and even seven-port circulators.

In our previous discussion on the construction of the circulator we noted that it contains permanent magnets. However, the magnets used could be electro-magnetic, controlled by an external current. Circulators of this second variety make it possible to create a microwave switch, with the switching accomplished by reversing the electromagnet's coil current. Thus, you change the previously illustrated "right-handed" circulator to being "left-handed"; that is, you have changed the gyro-magnetic direction from clockwise to counter-clockwise. The advantages of this type of device over mechanical or diode switches should be readily apparent. They are smaller, lighter, and free from relay chatter; they require less dc power and have a lower insertion loss; and they are more reliable, and able to handle more RF power, than some other types of switches.

The isolator's most prominent use is for protection from large mismatches in a load. Many times they are used at the output of a power amplifier to protect a power transistor from large antenna variations; any mismatches at the load are dumped into the built-in termination and adsorbed before they can return to the power amplifier. The power amplifier sees an essentially constant load no matter what the load is doing in actuality.

An important advancement in the field of ferrites is the availability of plug-in circulators and isolators, which do exactly what their names imply—plug into a circuit. They have no connectors; instead they have tabs at all ports which can be soldered to a stripline or microstrip circuit. Ground connections are made usually by means of fingers on the device and mounting screws. The plug-in concept has increased the use of circulators and isolators and takes them into the more compact world of stripline microwaves.

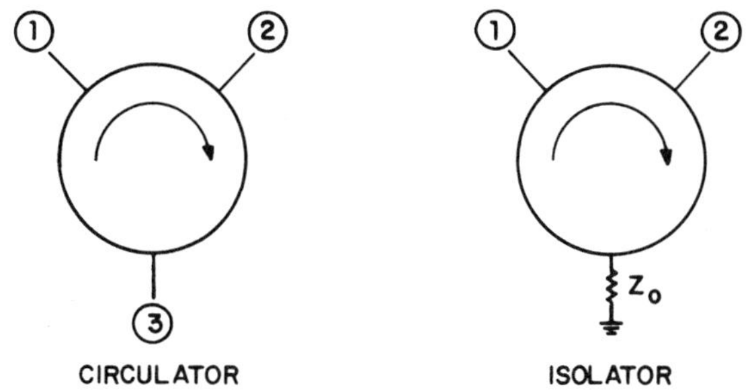

Fig. 4.33(a) Circulator and isolator schematic symbols.

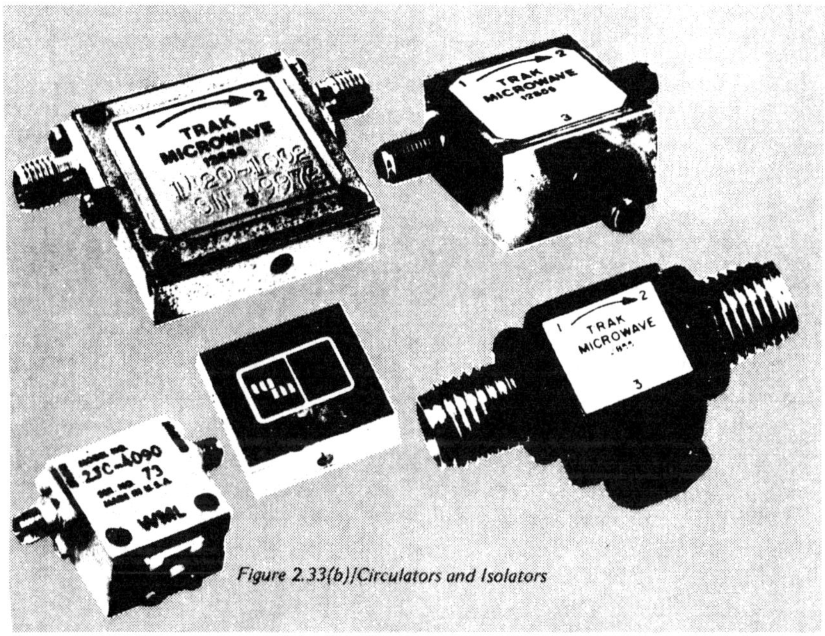

Fig. 4.33(b) Circulators and isolators.

And so the circulator and isolator join the group of microwave components available for system use. In summation, a circulator is a three (or more) port ferrite device which exhibits the characteristic of being non-reciprocal. It provides a low RF insertion loss, a high isolation between ports, and (contrary to the directional coupler) has dc continuity at all ports.

The isolator is, very simply, a circulator with its third port internally terminated.

4.8 ATTENUATORS

To attenuate is to lessen the value of something; in microwaves, that something is the RF level applied to a circuit. There are many ways to accomplish this end—continuously, in various steps, or by a permanently set constant value of attenuation. Irrespective of method, the end result must be to lessen the value of the RF energy without disturbing any other portion of the system—the attenuator must have a good input and output match as it operates.

Our study of attenuators involves three basic types—the fixed, variable (both step and continuous), and *pin* (which makes use of the variable resistance characteristic of the *pin* diode) attenuators. For each of these types we provide a description, explain a typical data sheet, and list applications.

4.8.1. Fixed Attenuators

Probably the most familiar type of attenuator is the fixed coaxial attenuator shown in Figure 4.34; most likely, you have used it many times in your test setups and have come to take it for granted. This attenuator usually is constructed in either of the two ways shown in Figure 4.35; it may be a T section or a π section. You can easily see why there are severe restrictions on the power input to this type of attenuator—each case utilizes a series film resistor whose power handling capability is limited to approximately two W. Provisions are made in some models to handle higher powers, through use of finned heat sinks to dissipate the excess heat produced by the greater amounts of power. Attenuators (or pads, as they are called sometimes), are available that are capable of handling up to 50 W (CW) and 2 kW peak power, but they grow very rapidly in size as their power ratings increase. An attenuator that can handle 50 W would be more than 7 inches long and nearly $1^3/_4$ inches in diameter—a far cry from the miniature, 2-W attenuator which is around $1^1/_4$ inches long and less than $^1/_2$ inches in diameter. Once again, there is a trade-off. If you need high power, you must have the space to put the component. True miniaturization is more easily accomplished at lower power levels.

A typical data sheet for a coaxial attenuator follows. Most of the terms are self-explanatory. However, we examine the most important more closely.

Model number	ATTEN–20
Frequency	dc–12.4 GHz
Attenuation	20 dB
Accuracy	±0.5 dB
VSWR	1.2:1 (dc–8 GHz)
	1.3:1 (8–12.4 GHz)
Input power	2 W (CW)
Connectors	Type N

Terms to be considered are *accuracy, attenuation, input power,* and *VSWR*.

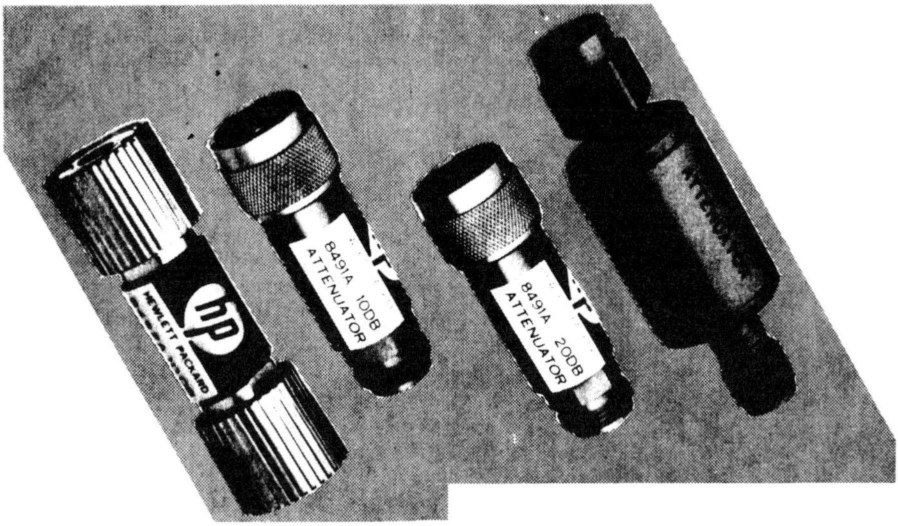

Fig. 4.34 Fixed attenuators.

Accuracy—Accuracy measures how closely the attenuator stays to the attenuation figure (20 dB in the above example) over the specified frequency range. Most coaxial attenuators have calibration marks imprinted on the body of the component giving the attenuation as a function of the frequency, which is an actual reading of the accuracy of the component.

Attenuation—Attenuation is the figure, given in dB, which indicates how much RF power is lost between the input of the attenuator and its output. In other words, it is how much the input power is "lessened" by the time it gets to the output. Expressed mathematically, it is

$$\text{attenuation (dB)} = 10 \log P_{in}/P_{out}$$

Input Power—It should be evident from our previous discussion why this term is of great importance. Many a \$100 pad has been thrown away exhibiting a very sharp, burnt odor because someone took it for granted that the pad would handle the power of their system of test setup. The importance of matching your particular application with specifications of each individual type of attenuator cannot be stressed too strongly.

VSWR—This term is mentioned only to clarify when the VSWR holds true. It is the standing wave ratio (or input match) of the attenuator when the output port is terminated in the characteristic impedance, Z_0 (usually 50 Ω).

Although our discussion of fixed attenuators was of coaxial types, a waveguide variety is also available. Its basic construction is that of a resistance card in the guide

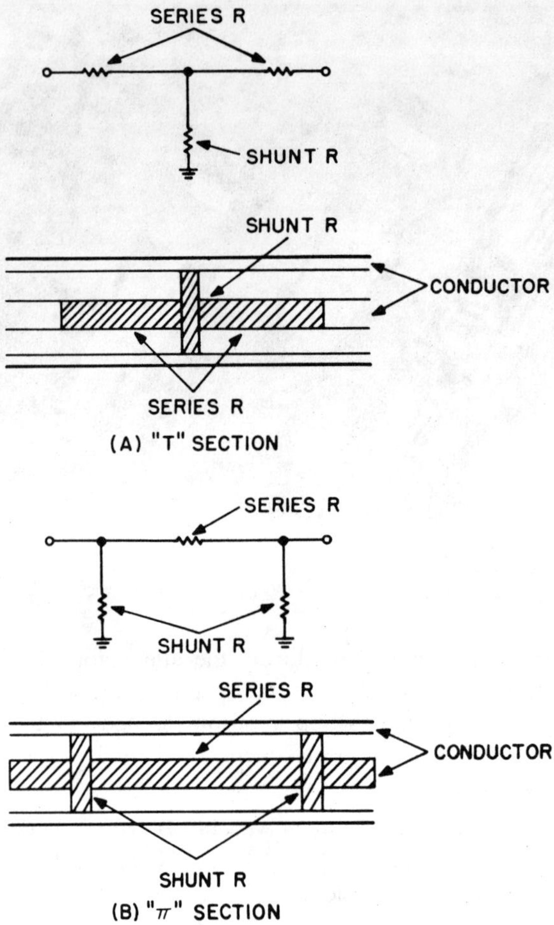

Fig. 4.35 Construction of fixed attenuators.

to absorb a predetermined amount of energy and thus lessen the amount of energy applied to an RF circuit.

Applications of the fixed attenuator are covered at the conclusion of this section.

4.8.2 Variable Attenuators

Many times, you either do not need a fixed value of attenuation or are not sure what value you do need. At this point, the variable (either continuously so, or in discrete steps) attenuator can be used very effectively.

Continuously Variable

There are many different types and configurations of continuously variable attenuators, such as piston, variable card, T, variable coupler, low-loss cut-off, and lossy wall. We look briefly at each of these types and learn how they attenuate continuously over a specified range. One point which is evident with some types is that the attenuation range does not start at zero. The insertion loss for these types of attenuators may be anywhere from 4 to 6 dB. This fact does not, however, cut down on the effectiveness of their operation.

The six types of continuously variable attenuators are shown in Figure 4.36. Each has a unique way of lessening the RF energy at its output while maintaining a good VSWR for the driving circuit.

The initial type is the piston (or waveguide below cut-off) type shown in Figure 4.36(A); the diagram makes it clear how both names apply. First of all, it operates just like a piston in a car, moving in and out of a cylinder; the only difference is that in a car the piston is solid, in the attenuator, it is hollow (because it is made of cylindrical waveguide). The waveguide is of a size so as to be operating at a frequency lower than that for which it was designed (below cut-off); the term "waveguide below cut-off" is justified by this fact. The use of cylindrical waveguide has advantages and disadvantages. A much higher power handling capability favors it over a coaxial configuration. However, since it is operating outside of its intended band of frequencies the initial region is very non-linear and lossy; minimum insertion loss may go as high as 15 to 20 dB, but once this section is passed, the piston attenuator is one of the most accurate available—0.001 dB/10 dB attenuation over a 60 dB range is a plausible figure. The input and output connectors are attached to inductive loops that pick up energy in the cylindrical waveguide; this method of loop coupling produces a good input/output match.

A second type of attenuator is shown in Figure 4.36(B)—the variable card. This device is composed of a resistance card attached to a substrate. The wiper arm of a movable contact varies the resistance between output and input ports, and thus changes the amount of attenuation. You could relate this type of attenuator to a setup in which a variable resistor (potentiometer) is placed in series with a generator and the circuit. As you increase the potentiometer resistance you increase the attenuation; the same is true of the variable card attenuator. Due to variations in resistance, the input and output match is not as good for this type of attenuator as for the piston type. However, the mismatch is not enough to cause either input or output problems. Minimum insertion loss on the variable card attenuator is on the order of 4 dB. Figure 4.36(C) shows the T type of variable attenuator. Recognizable from our discussion of fixed attenuators, the configuration is the same. However, the series and shunt resistors are varied simultaneously to preserve the input and output match while increasing or decreasing the overall attenuation. The three previous types of attenuators do their job by dissipating power within the unit. The variable coupler attenuator

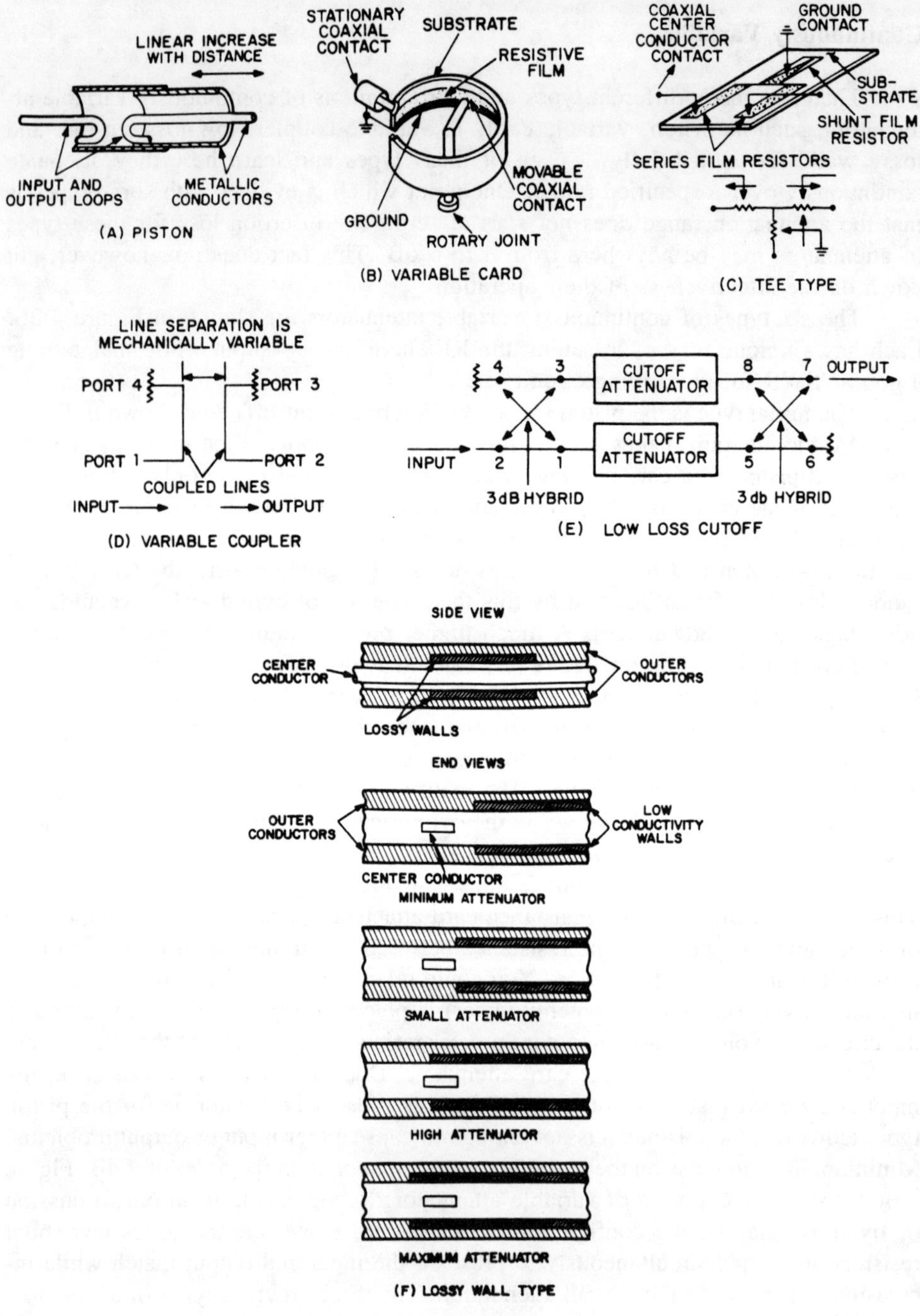

Fig. 4.36 Types of continuously variable attenuators.

relies on differences in the amount of coupling from one line to another. It is composed of the same basic directional coupler discussed earlier in this chapter; the major difference is that the value of coupling is mechanically varied. A very high accuracy attenuator is produced with an excellent input/output match. With an input at port one there is a certain amount of energy coupled to port two, an amount which depends on the spacing between the two lines. The minimum value of attenuation is limited to about 5 dB, due to restrictions on how close the two lines can be. The power handling capability basically is determined by that of the external load at port four. This attenuator type has the additional feature of exhibiting all the properties of a directional coupler while being a good variable attenuator.

The low-loss cut-off (Figure 4.36(E)) is another type of attenuator that does not rely on dissipation within the unit itself. Instead, the device uses two quadrature hybrids (to divide and re-sum the RF energy), between which are sections of mechanically coupled variable cutoff waveguide. By mechanically varying the cut-off point of the waveguide, the attenuation at the desired frequencies can be varied accordingly. The insertion loss of such a device is very low since both legs of the system are balanced; with a zero dB setting, the insertion loss should be 1 dB or less. Due to the use of quadrature hybrids, the input and output match is excellent in this type of attenuator. The configuration shown in Figure 4.36(E) is sometimes referred to as a "constant impedance" attenuator.

The last type of continuously variable attenuators is shown in Figure 4.36(F). Perhaps this variety is the one most easily understood. The main idea of an attenuator is to introduce a loss into the system—what better way to accomplish this function than to have a lossy wall attenuator? This attenuator is a stripline construction with a section of its outer conductor replaced by a lossy material, the position of which is varied to cause the attenuation. As illustrated in Figure 4.36(F), with zero attenuation there is simply a stripline conductor with a low loss dielectric above and below. As the lossy material is moved over the conductor, attenuation increases; maximum attenuation occurs when the conductor is completely covered. At that point, it is surrounded by a lossy material rather than its original low-loss dielectric.

To further understand this attenuator, consider the following analogy. Replace the center conductor by a fluorescent light tube. The low-loss dielectric can be replaced by transparent panels; the lossy material, by dark pieces of cardboard. Figure 4.37 shows this transformed device. With the light on and the cardboard panels completely back, all of the illumination is visible due to the transparent panels (zero attenuation). As the dark panels are moved so as to replace more and more of the transparent, a decreasing amount of light is visible until the panels completely obscure it (maximum attenuation). The dark panels (lossy material) have replaced the transparent panels (low-loss dielectric) and all light is blocked (maximum attenuation of the RF signal).

An attenuator of this sort exhibits many interesting features. First, because of the availability of lossy material at that frequency, it is generally used above 2 GHz.

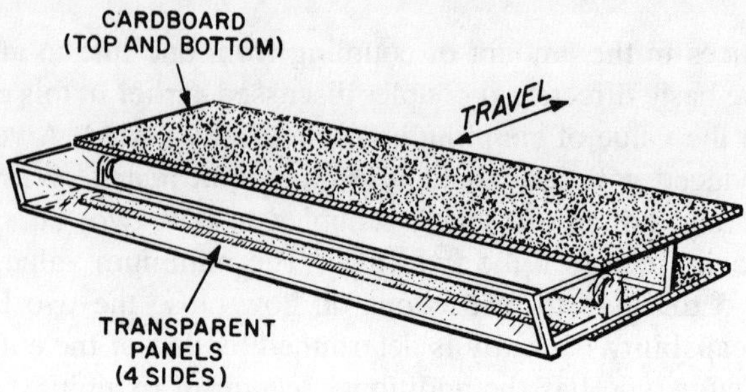

Fig. 4.37 Lossy wall attenuator analogy.

The attenuator once again operates by means of dissipation within the unit, limiting the power handling capability to around 10 W. The insertion loss is generally below 1 dB since, with zero attenuation, the attenuator is composed of nothing more than a single piece of transmission line. The VSWR is also very low because of the construction and the dissipation properties of the transmission lines involved.

A typical data sheet for a general type of continuously variable attenuator is shown below. We examine each of the terms individually after a discussion of the entire sheet.

Model number	VAR–1
Frequency	1–2 GHz
Attenuation range	6–120 dB
Accuracy	±1.5 dB
Power	100 W (CW)
VSWR	2.0:1 (6–15 dB)
	1.5:1 (15–30 dB)
	1.2:1 (30–120 dB)
Connectors	Type N

Accuracy—Accuracy could be referred to as setability or resetability; that is, how precisely can you set the attenuator. For example, if you need 10 dB of attenuation and set the attenuator to 10, how close is the actual result to 10 dB.

Attenuation Range—Very simply put, this range runs from the minimum to the maximum amount of attenuation possible with the individual attenuator. A safe assumption for the attenuator shown would be that it is a variable coupler, based on the insertion loss (6 dB) and total attenuation possible.

Connector—This term is listed to emphasize the need for the proper connectors, which would eliminate both any additional loss due to adapters, and the problem of connector types that do not handle any amount of power. Care must be taken to have the right connectors on your unit. (Types and applications of connectors were covered in the Chapter three).

Frequency—The frequencies indicated by this specification are those over which the attenuator characteristics hold true. Above or below this range, the figures change. Some attenuators, such as variable couplers, have a limited frequency range (usually one octave). Others, like the variable card or T type, are less sensitive to frequency, and thus have a much wider range of operation.

Power—Power is a very important term. There is wide variation in the handling capabilities of the different types of attenuators—some are limited by external terminations while others are restricted by their internal construction. Great care should be taken when choosing an attenuator to make sure that you have not exceeded any of the power limitations.

VSWR—The true reading of the input match of the attenuator over its range of operation is the VSWR. The data sheet above lists more than one figure for VSWR; these different values are at different attenuation levels. As attenuation increases, more power is absorbed by the attenuator and, therefore, is not reflected back to the input. As a result, the VSWR is lowered.

Step Attenuator

When you think of a step you think of a distinct increment by which you move or progress. This idea also applies to step attenuators (see Figure 4.38). This device is a set of fixed attenuators arranged in a turret to obtain discrete steps of attenuation. Attenuators are available in 0.1, 1.0, and 10 dB steps. The 0.1 dB step attenuator usually has a total range of 10 dB, the 1.0 dB step usually has 10 to 12 dB of range, and the 10 dB step can have as high as 120 dB of total range.

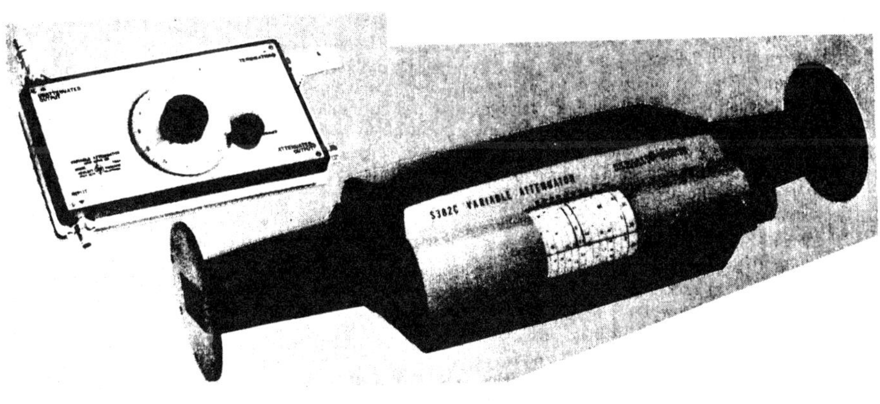

Fig. 4.38 Step attenuators.

The step attenuator has many outstanding features, among which are flatness (over a frequency range), repeatability (0.03 dB is possible), wide range, long life, and either manual or programmable capability. These characteristics are readily apparent from the data sheet shown below.

Model number	STP–1
Frequency	dc–12.4 GHz
Attenuation range	0–60 dB (in 10 dB steps)
Accuracy	±2 dB
Insertion loss	1.5 dB
VSWR	1.5:1
Power	2 W (CW) 100 W (peak)
Connectors	Type N

To help understand the step attenuator more fully, let us examine each of these terms.

Frequency—This term indicates the range of operation over which the attenuator specifications apply. An outstanding feature of the device is its very wide range (dc–12.4 GHz); it is possible because the construction of each fixed attenuator is not frequency dependent. The attenuator is only specified to 12.4 GHz because the connectors begin to limit performance above that point.

Attenuation Range—This specification is the total attenuation (at the maximum and minimum settings) and the increments you can take in between (0.1 dB, 1.0 dB, or 10 dB steps).

Accuracy—Accuracy is how close each step actually is to its specified value. For example, 10 dB steps in the attenuators above could be anywhere from 8 to 12 dB. This information is useful when making precise measurements.

Insertion Loss—This loss is that through the attenuator when its front dial is set at zero dB.

VSWR—VSWR is a measure of the match of the attenuator when it is stepped through its range and terminated in its characteristic impedance (Z_0), usually 50 Ω.

Power—Power is the maximum amount that the attenuator can handle without damage. Notice that we are back to a 2 W CW power capability, as we were when we discussed the fixed attenuator. This similarity is understandable because the step attenuator is actually a series of fixed attenuators.

One feature not yet discussed is repeatability. If you think of switching between each component in a series of fixed, and accurate, attenuators in a row, the concept of repeatability can be understood. An attenuator remains at exactly the same value no matter how long you are switched to another value attenuator; when you return to a specific device, you can be sure that it has exactly the same value as before.

Another attribute is long life, though it has one string attached—its longevity is dependent upon observation of its power handling specifications. If those capabilities are ignored, you end up with a $300 paperweight for your bench. If you observe the specifications religiously, you have one of the most useful test components available.

4.8.3 *pin* Attenuator

The *pin* attenuator is a variable attenuator (see Figure 4.39) that provides either step or continuous attenuation without the use of pistons, card resistors, couplers, waveguide, or lossy walls. Instead, it uses a *pin* diode, which is essentially a highly efficient variable microwave resistor. The *pin* diode is distinguished from a normal *pn* junction diode by an area (called by intrinsic region) sandwiched between the p^+ and n^+ doped silicon layers. (Doping as you may remember, is the term used to describe the addition of impurities to a semiconductor crystal to increase conductivity.) The intrinsic layer has almost no doping and, consequently, a very high resistance. When forward bias is applied across the diode, the conductivity of the *l* layer is increased and you have, at microwave frequencies, an electrically variable resistor. As the bias current is increased, the resistance of the diode decreases. Therefore, if this diode, or a combination thereof, were put in series or in shunt on a line, you would have a very efficient and relatively simple variable attenuator.

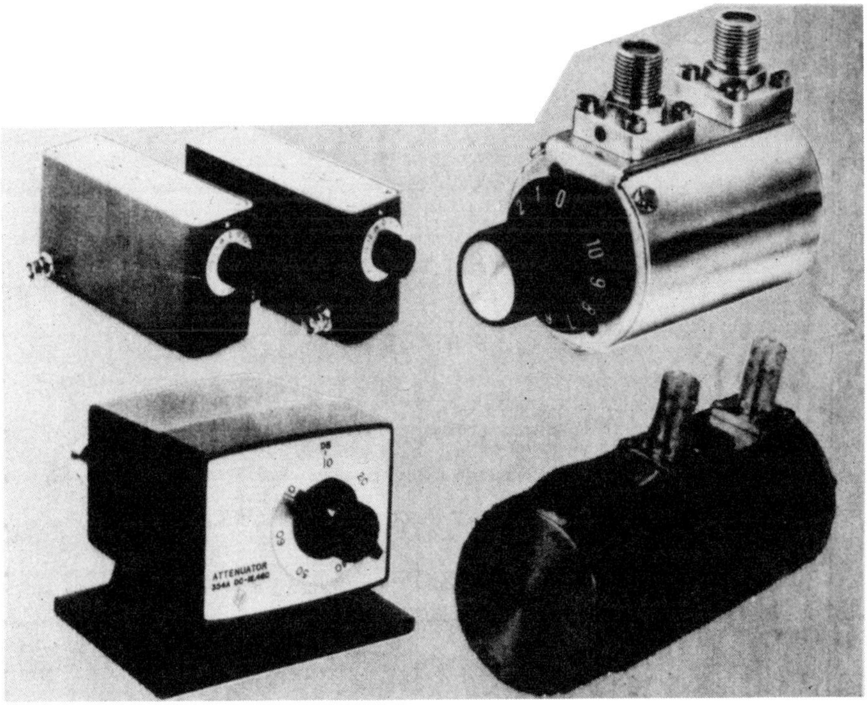

Fig. 4.39 Continuously variable attenuators.

Below about 10 MHz the *pin* diode can be expected to exhibit some rectifying behavior, similar to an ordinary *pn* junction diode. At frequencies much above this point, the *pin* diode is better represented as a variable resistor. The *pin* diode is truly a microwave device as its variable resistance properties are present only at microwave frequencies.

Figure 4.40 shows six common configurations for *pin* attenuators, ranging from a simple series diode to a double hybrid arrangement.

In the series attenuator (A), attenuation decreases as the bias current through the diode increases because the RF resistance is lessened. In the shunt attenuator (B), the opposite occurs; attenuation increases with increased bias current because most of the RF energy is absorbed in the diode.

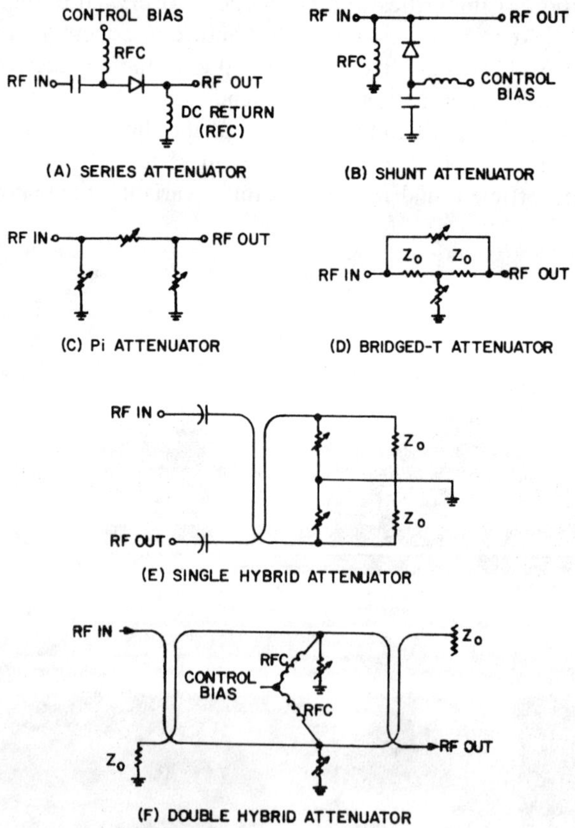

Fig. 4.40 *pin* attenuator configurations.

The π or T attenuator (C and D), operate on the same principle as their fixed attenuator counterparts. By biasing the *pin* diodes properly, a constant impedance is maintained while providing a means of increasing or decreasing the attenuation.

The quadrature hybrid adds a new dimension to the variable attenuator by providing outstanding impedance matching and directivity, no matter if the attenuator is reflective (E) or absorptive (F). The terms *reflective* and *absorptive* are derived from the following: in the first case, attenuation is achieved by the diode causing the RF energy to be reflected; in the second, the diode absorbs the energy as a function of bias current.

To further aid in our understanding of the *pin* attenuator, let us examine a typical data sheet for a commonly available component. This particular attenuator is an absorptive type very similar to that in Figure 4.40(F).

Model number	*PIN*–1
Frequency	1–2 GHz
Attenuation	50 dB
Bias current	100 mA (maximum)
Flatness	±2 dB
Insertion loss	1.5 dB
VSWR	1.5:1 input/output
Power	1 W (CW), 50 W (peak)
Connectors	SMA

Let us examine the terms of the data sheet; they are arranged in alphabetical order for easy reference.

Attenuation—This amount is that by which the input signal is decreased (or lessened) when the *pin* diodes are fully biased (in other words, when maximum current is drawn). This figure is significant in that it indicates that quadrature hybrids are being used. The high value of attenuation achieved is typical when more than one set of diodes is used in each arm connecting the quadrature hybrids.

Bias Current—This amount of current is what is required to obtain the value of attenuation shown above. This figure varies depending on the type and number of diodes used in the attenuator.

Connector—The connectors, of course, are the RF connectors used to connect the attenuator to other circuitry in the system. Great care should be taken to be sure that the type of connector on the attenuator matches the others in the system.

Flatness—This term indicates the variation in attenuation allowed over its frequency range. When accurate measurements are to be made, the attenuator should be calibrated precisely to compensate for these variations.

Frequency—This specification notes both the range of frequencies over which the entire set of specifications apply, and if a frequency dependent device (such as a quadrature hybrid) is used in the attenuator.

Insertion Loss—This loss figure is the minimum amount of attenuation when the bias current is zero and the *pin* diodes are *off* (in the case of shunt diodes used in absorptive attenuators). The loss is made up of any internal losses and the individual insertion losses of the quadrature hybrids.

Power—This term, again, is very significant. The specification for power handling, in this case, is only one watt; very seldom does the level rise above this value with conventional *pin* diodes. Some *pins* can handle 5 to 10 W of power; most, though, are limited to much lower figures. Be very careful when using an attenuator of this type. Check the power rating *before* putting it into your setup.

VSWR—VSWR is a reading of the input and output match of the device. In the beginning of this section, we said that an attenuator must do its job of attenuating without disturbing any connecting circuits. Or, simply put, it must have a good input and output match (or VSWR).

It should be pointed out that *pin* attenuators can serve a double function—consider how the attenuator in the above data sheet would work. With RF energy at the input and no bias applied, there would be a 1.5 dB loss through the device. As the bias is increased gradually, the attenuation increases until we have 50 dB of attenuation (or isolation) between the input and the output of the component at the maximum bias point. If we were to apply the maximum bias value of the attenuator instantaneously, the device would go to its maximum attenuation (or isolation) value in some finite amount of time, depending upon the characteristics of the diode. If we remove the bias instantaneously, the insertion loss figure would be present once again. It should be quite clear by now that what we also have in our *pin* attenuator is a *pin* diode switch. As described above, it would be a simple on-off switch. By re-arranging ports and terminations on different models, it is possible to make a switch which operates between two or more circuits.

The *pin* attenuator proves to be a very versatile component for microwave work. It can be operated remotely by means of a dc line running to bias the diode, or it may be controlled at the component itself. Either way, it is a very efficient attenuator (or switch) when important specifications, such as power handling capability, are observed.

4.8.4 Attenuator Applications

Without a list of applications for each device, the information we have gathered thus far about microwave components would be utterly useless. The attenuator is no exception to the rule; the variety of its uses are what make it such a valuable component.

The primary application for the attenuator is in the control or measurement of RF energy.

Acting as a standard by which to measure loss or gain, in what is called "the RF substitution method," is one area where the attenuator is employed widely. With this method, an initial calibration run is made after removing the particular device being tested from the circuit. It is then replaced, and attenuation is added or removed until the level is back to that of the original measurement. The change in attenuation determines either the loss or gain of the component, whichever the case.

Fixed attenuators are used often to extend the dynamic range of power meters, spectrum analyzers, and other equipment. For example, if you had a power meter with a maximum reading capability of 10 mW, but need a meter to read one watt, the insertion of a 20 dB attenuator (or pad) would allow you to make your reading while staying well within the input power restrictions. This type of padding also can be used at the input of a spectrum analyzer to extend its range while protecting the input circuitry.

Another important use of attenuators is to provide a different sort of protection. Many receivers and amplifiers are very sensitive to high input power and go into "compression" easily; that is, they can be overdriven so that their output flattens out rapidly. To prevent such an occurrence, the input is "padded" with attenuators so that the amplifier cannot be overdriven. An additional benefit of this padding is the reduction of the effects of a variable or a mismatched impedence on the amplifier input. Any RF energy being reflected toward the input is attenuated; thus there appears to be a low return loss, and a good VSWR.

Wherever a reduction in power is needed, the attenuator fulfills that requirement. At the input of an amplifier, a power meter, or a spectrum analyzer, or at the output of a device, the attenuator is a component that can equal or better the overall match of a system.

One application reserved for the *pin* attenuator is in an automatic level control (ALC) circuit; it is usually placed at one of the lower power stages. The system output power is sampled by means of a directional coupler; a dc voltage proportional to the output level is produced and fed to the bias terminal of the *pin* attenuator. If the output power goes up, the dc voltage increases and bias is applied to the *pin* diodes. Attenuation increases and cuts down the drive level. As a result, the output level is brought down. The reverse applies as the power level drops. Thus, the output power level is held virtually constant. Although the *pin* attenuator is used in ALC circuitry, it is not the only type that works—many programmable variable attenuators do the job just as well as the *pin* and in a similar operational manner.

The applications listed above are only general descriptions of what attenuators can do; you can find new and different uses for them everyday. Their versatility and adaptability, whether fixed, variable, or *pin,* allow your imagination to run wild and create the most individually useful systems and test setups.

4.9 ANTENNAS

Usually when people think of microwave antennas, the first thing that comes to mind is the type used by the telephone company in microwave relay systems—an impressive, huge tower with a series of dish antennas on it. However, the parabolic reflector is not the only type of antenna used in microwave systems. There are a great many varieties; each should be evaluated as to its appropriateness to your specific application. Before such an investigation however, it would make sense to have an understanding of just what an antenna is.

In any system that transmits or receives electromagnetic energy into or from a surrounding medium, an antenna is perhaps the single most important component. Its function is to couple energy from a transmission line into its environment (air, for instance), and *vice versa*. Basically, an antenna is a component which transfers energy either from transmitter output circuitry to a medium, or from the medium to a receiver's circuitry. In a broad sense, an antenna is an impedance matching component—effectively, it matches the impedance of the transmitter or receiver circuitry to that of the medium used so as to maximize a transfer of power.

All antennas are described by various, general parameters of their radiation patterns and impedance; familiarity with the terms associated with these parameters greatly aids obtaining an understanding of the function of an antenna. We look into some of these specifications and see how they and others apply to microwave antennas. Those we discuss are *3 dB Beamwidth, polarization, directivity, bandwidth, impedance,* and *VSWR*.

3 dB Beamwidth—This specification is very similar to the 3 dB bandwidth used in describing the operation of filters. It is a number, expressed in degrees, which indicates where the strength of the radiation pattern of an antenna decreases to half power (or 3 dB), as shown in Figure 4.41. With filters, this term measured the number of MHz between two half-power points; referring to antennas, it is the number of degrees between the points.

Polarization—An electromagnetic wave consists of just what the name implies—two components, one electric (E) and one magnetic (M). Polarization refers to the orientation of the E component, or E-vector. As an aid to understanding, consider polarization to be the property of the antenna that allows only waves of a certain specified orientation to be transmitted or received, much the same way as a filter passes only certain frequencies. Three types of polarization are considered commonly—linear, circular, and elliptical. The easiest of the trio to understand is *linear* polarization. The E-vector is associated with only a single plane, (either vertical or horizontal); that is, the electric field (E) is in a plane oriented vertically or horizontally.

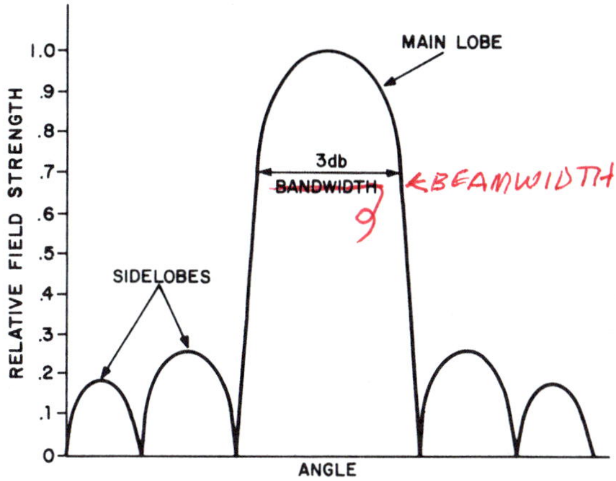

Fig. 4.41 Two-dimensional antenna radiation pattern.

Circular polarization involves more than one plane and is characterized as being involved with the tri-plane atmosphere shown in Figure 4.42. The E-vector rotates in either a clockwise or counter-clockwise direction—if it moves clockwise it is said to be right-hand circular; counter-clockwise, left-hand circular. For circular polarization, the E-vector must remain at a constant amplitude. If it changes magnitude as it rotates, its tip traces an ellipse and the wave is said to be *elliptically* polarized.

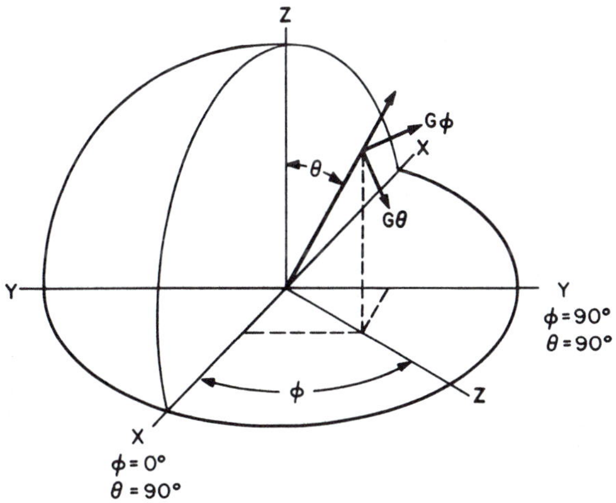

Fig. 4.42 Triplane circular polarized system.

Directivity—This specification is a measure of the ability of an antenna to concentrate energy in a preferred direction. Antennas with narrow beamwidths have a greater directivity than do those characterized as being omnidirectional. The gain of an antenna is closely linked with its directivity—if the antenna has no heat losses and is perfectly matched to the source or load, then the gain is equal to the directivity. However, the gain decreases proportionally to the losses. For example, if the antenna is 50 percent efficient then the gain is 50 percent (or 3 dB) less than the directivity.

Bandwidth—When applied to other microwave components, this term was called frequency; it is the range over which the listed specifications apply. Outside this region, the antenna does not behave as predicted.

Impedance and VSWR—These two terms almost always appear together. Impedance is a prime determining factor in the amount of power transferred from the component to its associated circuitry and *vice versa*. However, when referring to antennas we must know how well energy is transferred between free space and a transmission line.

A more convenient term is VSWR. A high VSWR indicates a large measure of reflected power and a poor transfer of power; a low VSWR characterizes a good power transfer. One consideration, though, when specifying VSWR, is your application. The VSWR of an antenna used in a receiving step can be set as high as 3:1. But use of a 3:1 VSWR in transmitting would cause losses that would put a great burden on the final power stages. Thus, a suitable VSWR for one application may be intolerable in another.

These are the five important terms used to describe antennas. With them in mind we now investigate five common types of microwave antennas: *dipole, helical, log-periodic, horn,* and *parabolic reflector*.

4.9.1 The Dipole Antenna

Very strictly speaking, a *dipole* is defined as a pair of oscillating or stationary charges, or magnetic poles with opposite polarity, which are separated by a distance that is very small in comparison to a wavelength. That definition is discouraging one; let us narrow it down to fit the microwave antenna application and make it more understandable. In our case, the dipole (or half-wave dipole) is a center-fed half-wave antenna connected to balanced resonant lines. More simply, it is a conductor which is nominally a half-wave long at the frequency of use. Figure 4.43 shows the basic half-wave dipole; part (A) is a schematic diagram that illustrates the voltage and current relationship. Figure 4.43(B) is a radiation pattern of a dipole; it shows the very familiar "doughnut" pattern. From the diagrams in Figure 4.43 you have probably recognized the type of antenna as one you have seen before, either by itself, or associated with some type of reflector to improve its directivity.

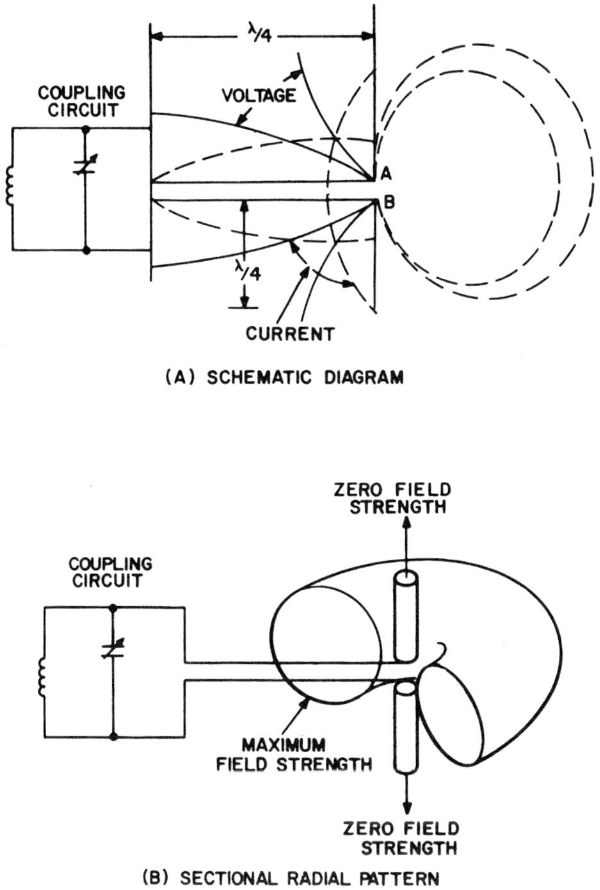

Fig. 4.43 Half-wave dipole.

The voltage and current relationships referred to above are of great importance to understanding the dipole antenna and its operation. From Figure 4.43(A) you can see that the voltage on the antenna is maximum at the ends of the dipole and at the coupling circuitry (the input to the antenna); minimum, at the center of the two elements. It also can be seen that the current is minimum at these points. At the ends of the elements there is no longer a conductor in which the charges can flow; thus, the current goes to zero. At the coupling circuit, minimum current is also logical since it is one half wavelength from the end of the antenna—conditions repeat every half wavelength, giving an estimate of the antenna radiation pattern. Actually, the radiation pattern is a surface symmetrical in all directions around the elements and not simply the curve shown by the dashed lines in Figure 4.43(A). If the lines in part A are imagined to rotate out of the paper so that they revolve around a point midway between A and B, they generate the "doughnut" pattern shown in part B—the true antenna radiation pattern of the dipole. It shows very readily the locations of maximum field strength and a zero area.

The dipole has been described as being nominally one half wavelength long. But under normal circumstances, the correction factor used to develop a working antenna results in a length that is usually 10 to 20 percent shorter. The larger the diameter of the stock used to build the dipole, the shorter the elements must be; such shortening eliminates any inductive reactance present in the antenna impedance. This fact, however, does not keep the antenna from being referred to as a half-wave dipole.

Thus far, we have confined our discussion to a single dipole arrangement. Such a setup has both advantages and disadvantages. Many of the drawbacks can be overcome by using more than one dipole; these multiple arrays usually are termed bent or deformed dipoles, because they are bent back toward their feedpoint to improve their performance. Figure 4.44 shows a multiple deformed dipole, an antenna which (unlike that in Figure 4.43(B)) does not have pattern nulls at the ends of the elements. Since a single straight dipole does have a maximum only in a direction perpendicular to the length of the conductor, the addition of more conductors (bent so that perpendicular lines drawn to them have no unique direction) creates a maximum in nearly every direction. These antennas find wide use on aircraft where it is important that the radiation pattern extend in all horizontal directions.

Another variation of the dipole principle is the slot antenna. If a narrow slot-like opening is cut in a large flat sheet of metal and properly connected to a source of RF power, the resulting radiation bears a certain resemblance to that of a dipole of the same dimensions as the slot. If the conducting sheet is a plane of infinite extent, the radiation pattern is exactly the same shape as that of the corresponding dipole except that the electric and magnetic components are interchanged. Thus, a linear polarization opposite that of the dipole results. Of course, there is no such thing as an infinite plane conducting sheet in the real world. But if the sheet is very large compared to the slot, behavior very closely approximating that of an idealized infinite surface is realized. In fact, slots cut into sheets of even moderate size radiate effectively, but their exact behavior cannot be predicted readily.

The slot described radiates on both sides of the sheet. If radiation is desired on one side only, the "back" region may be enclosed in a box or cavity and the fields along the slot are determined by the dimensions of such a covering. An example of such an arrangement is the cavity-backed, T-fed slot antenna shown in Figure 4.45. It is an efficient, linearly polarized antenna that covers an octave band. The cavity itself is essentially a short section of open-end waveguide with a coaxial input, thereby permitting the device to be a compact element in an array. It can be

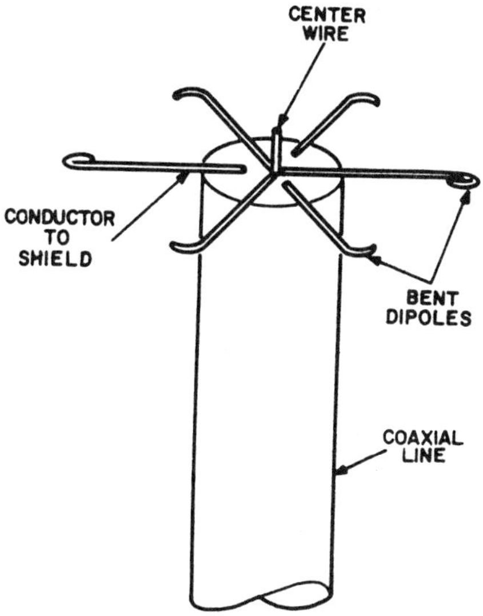

Fig. 4.44 Multiple deformed dipole.

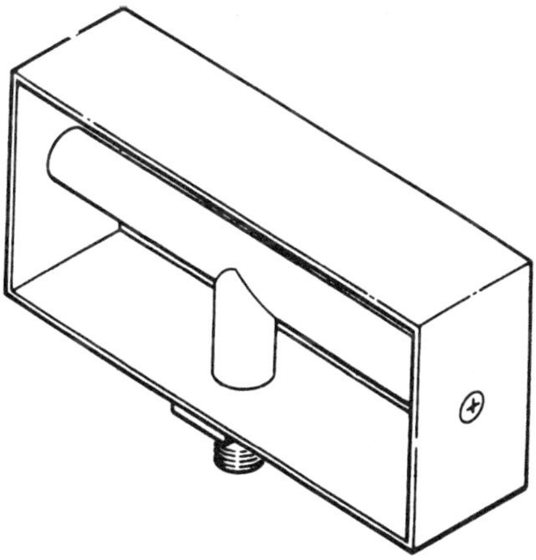

Fig. 4.45 Cavity-backed T-Fed slot antenna.

completely flush-mounted in a ground plane, even in the skin of an aircraft or missile; its power handling capability is determined by the coaxial input. The beamwidths (in both the E- and H-planes) are basically those of an openended waveguide of the same dimensions—around 140° for the E-plane and 35° for the H-plane. The gain of this type of antenna is approximately 3 to 5 dB, but higher figures can be attained by adding an appropriate horn to the slot.

Practically, slot radiators are restricted to the higher frequencies by the requirement of a conducting sheet considerably larger than the slot. They are very useful when a radiator must be devised that does not project from a surface, as when used in an aircraft.

The slot antenna and the dipole, having similar characteristics, find applications in the microwave field where small, efficient radiators are needed.

4.9.2 The Helical Antenna

By definition, a helix is a spiral; a *helical* antenna features a wire wound in the shape of a screwthread (or spiral) and is used in conjunction with a ground plane. Most often, they are used at relatively high frequencies, in order for their dimensions to be large compared to a wavelength.

Figure 4.46 shows a helical antenna. The helix, or spiral, is fed at one end and is, usually connected to the center of a coaxial transmission line whose outer conductor is attached to the ground plane. The basic geometry of the helix is described in terms of its diameter, D, and its turn spacing, S. For an N turn helix, the total length of the antenna is equal to NS, and its circumference is, of course, πD. The length of the wire per turn is $L = \sqrt{S^2 + \pi D^2}$. The pitch angle, α, (an important parameter) is that angle a line tangent to the helix makes with the plane perpendicular to the helical axis. It is equal to $\tan^{-1} S/\pi D$.

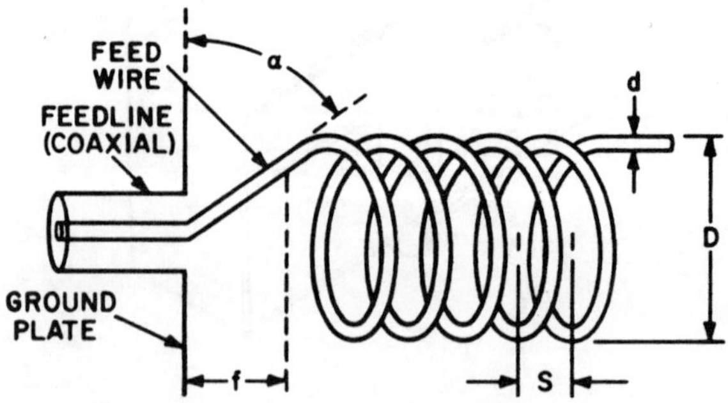

Fig. 4.46 Helical antenna.

The feed wire which connects the coaxial center conductor to the beginning of the helix, lies in a plane through the axis and is inclined, with respect to the ground plane, by approximately the pitch angle. Variations of the geometry effect the input impedance of the antenna. When α is very small, a loop type of radiation is present; when α becomes large, the helix is essentially a short dipole. In these two limiting cases the radiation is linear: in the first case, it has loop polarization and in the second, dipole polarization. For values in between, the polarization is elliptical; at particular values of α, it is circular. The latter case occurs when the pitch angle's tangent is $\pi D/2\lambda$.

When the diameter and spacing (D and S) are appreciable fractions of a wavelength, an entirely different radiation pattern is obtained. In such a case, the maximum intensity is radiated in the direction of the helix axis, in the form of a directional beam with minor lobes at oblique angles. The radiation in the main lobe is circularly polarized. The features of the helix in this mode of radiation (called the *axial* mode) are what account for most of the practical applications of this antenna.

A typical helical antenna operating in the axial mode has a circumference (C) of approximately one wavelength (λ), and a spacing (S) of approximately one-quarter wavelength ($\lambda/4$). The antenna operates quite well over a range of frequencies; the term "approximately" is appropriate in that the dimensions are not very critical. The ground plane (either a solid sheet, a wire grid, or mesh) should be at least one-half wavelength ($\lambda/2$) in diameter. The pitch angle ($^{\alpha}$) may range from 12° to 18°; 14° is considered optimum. The gain and beamwidth depend on the number of turns and the spacing between them, that is, on the total length of the helix. The feedpoint impedance is resistive and on the order of 100 Ω at the frequency where the circumference (C) equals one wavelength. Above and below this point, the resistive component varies and a reactive element appears.

In some applications it is necessary to note if the helix is wound with a right-hand or a left-hand pitch; the orientation, of course, determines whether the wave is right- or left-hand circularly polarized.

Helical antennas have found considerable use in space telemetry at the ground end of the link with ballistic missiles, satellites, and space probes. The circular polarization in the helix is useful in space transmission since the ionosphere causes the waves to be rotationally polarized.

4.9.3 The Log-Periodic Antenna

Log-periodic antennas are some of the most versatile of frequency-independent antennas; they provide a very wide range of patterns, impedances, and polarization characteristics. They achieve bandwidths of 10:1 with ease, and even larger values are possible with careful design. The physical appearance of log-periodic antennas can vary widely, with some being quite bizarre in comparison to conventional devices. But despite their assuming different shapes, they all operate on basically the same principle. Figure 4.47 is a drawing of a log-periodic dipole array which describes the basic operation of the antenna. It is fed at the apex of its structure by a two-wire line. Energy propagates down a transmission line (which generally also supports the radiating elements) until it reaches a region where the elements are nearly resonant; at this point, the energy is extracted from the line and radiated back toward the apex. This "backfire" characteristic is due to the phase relationship among adjacent radiating elements.

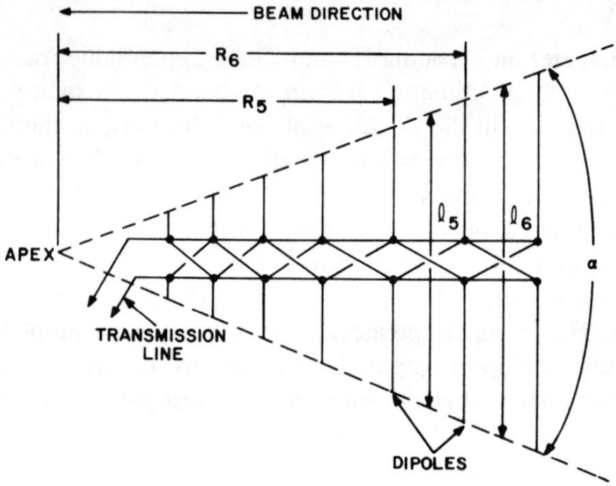

Fig. 4.47 Log-periodic dipole array.

The basic principle of log-periodic antennas can be understood more clearly by referring once again to Figure 4.47. All frequency-independent antennas have a special kind of repetitiveness to their physical structure, thereby a repetitive behavior of electrical characteristics. That is, the standard design involves a geometric pattern that is repeated while changing in size by a constant scale factor. Thus, the shape remains the same, but the structure "grows" (or "shrinks," depending on the direction considered).

Referring back specifically to the log-periodic dipole array, we can define it completely by three basic parameters—$^{\alpha}$, τ, and σ. The angle $^{\alpha}$ is formed by lines coming from the apex, or small end, of the antenna, and travelling along the tips of the elements. The smaller the angle, the higher the gain (within limits)—for a given region of radiation, the elements are more nearly of the same length. The ratio τ (design ratio) relates the lengths of adjacent elements. It can be expressed as follows:

$$\frac{R_2}{R_1} = \frac{R_3}{R_2} = \frac{R_{n+1}}{R_n} = \frac{\ell}{\tau} = \frac{\ell_2}{\ell_1} = \frac{\ell_3}{\ell_2} = \frac{\ell_{n+1}}{\ell_n}$$

(where n = the number of elements in the antenna).

The σ ratio is the relative spacing (in wavelengths) between adjacent elements. It is expressed as

$$\sigma = \frac{R_{n+1} - \mathrm{R}_n}{2\ell_n}$$

If a plot were made of the input impedances of the antenna described above as a function of frequency, the variations would be found to be repetitive. If the plot were against the logarithm of the frequency (rather than the frequency itself) its variations would be *periodic*—the impedance would go through cycles of variations. This behavior is what gives rise to the name "log-periodic," the impedance is a logarithmically periodic function of the frequency. (It is *not* implied, however, that the variations are necessarily sinusoidal—only periodic.) Moreover, all of the electrical properties of the antenna undergo a similarly periodic variation. In particular, the radiation pattern varies in this way, along with parameters such as the directive gain, beamwidth, side-lobe level, and beam direction.

If the structure shown in Figure 4.47 were extended both all the way to the apex (the beginning both all the way to the apex (the beginning of the angle $^{\alpha}$) and indefinitely in the other direction, the periodic action described above would continue to the extremes of zero and infinite frequencies. This situation, of course, is not physically possible—extension to zero frequency (infinite wavelength) would require an infinitely long structure; the extension to infinite frequency (zero wavelength) would necessitate a device of microscopic proportions at the apex. Also, the transmission line feeding the antenna would have to be of infinitesimal conductor size and spacing. None of these conditions is even possible—the structure must be terminated at some point in each direction. These end points determine the high and low cut-off frequencies of the antenna.

When a log-periodic antenna is fed at a particular frequency, the radiation occurs from only a certain portion of the structure. This "active" region is where one or more of the dipoles is nearly one-half wavelength long; its boundary frequencies are those at which either the shortest or longest dipole is approximately one-half wavelength. Thus, the active region is near the apex for the highest radiated frequencies; near the large end for the lowest; and at intermediate positions for those frequencies in between. This behavior also may be expressed by saying that the *phase center* of the antenna progresses from its large end to its apex as the frequency goes from minimum to maximum.

For the antenna shown in Figure 4.47, typical values of the design ratio and apex angle are, respectively, $\tau = 0.7$ and $^{\alpha} = 30°$. For any given value of α, there is a minimum permissible value of τ (which is always between 0 and 1). Larger values of $^{\alpha}$ and smaller values of τ go together, resulting in a more compact design for a given bandwidth. On the other hand, smaller values of $^{\alpha}$ and larger values of τ improve performance—smaller variations in impedance and patterns, and higher gain—at the cost of a larger structure. The reason for this effect may be understood by realizing that for small α and large τ at any given frequency there are more dipoles of nearly a half-wavelength. Thus, the active region of the antenna contains more dipoles and, as is always true, the gain increases with the number of radiating elements and the size of the radiating region. Also, the impedance and pattern variation are smaller because the transitions in the radiating region are smoother when it contains several dipole elements.

The log-periodic antenna can take many different forms, two of which are shown in Figure 4.48.

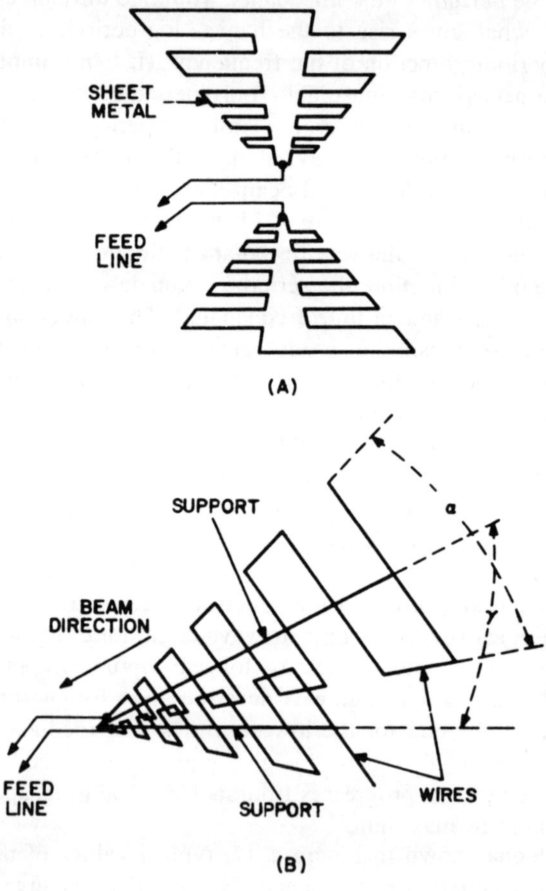

Fig. 4.48 Log-periodic antennas.

The wire-trapezoidal-tooth antenna in part (B) has two arms or branches originating at the apex of the antenna, each of which has a characteristic angle, $^{\alpha}$, that plays the same role in the antenna design as does the angle $^{\alpha}$ in the log-periodic dipole. The other angle, γ, effects the directional characteristics. When $\gamma = 180°$ (as it is in the trapezoidal-tooth structure of Figure 2.48A), the radiation is bidirectional and perpendicular to the plane of the antenna. When γ is 90° or less (acute), the radiation is unidirectional and toward the apex of the structure. At first glance, the dipole array previously discussed does not seem to conform to these observations. However, it can be shown that it is actually just a specialization of the trapezoidal-tooth structure—the "teeth" have been reduced to practically zero width and the angle γ has been reduced from 180° to 0°. That is, the two branches have been folded over until they are in the same plane but not in electrical contact; another name for such an arrangement is a co-planar log-periodic antenna.

The log-periodic is a very useful and versatile antenna. It may be used by itself, in conjunction with a reflector, or accompanied by a cavity in much the same manner as the slot antenna previously discussed. The great versatility of the log-periodic antenna tends to substantiate the title given it by many persons in microwaves—"the workhorse of the broadband antenna family."

4.9.4 The Horn Antenna

Horn antennas usually are waveguide fed, making them ideal for high power transmitting applications. Figure 4.49 shows several types of horn antenna.

The simplest variety of horn radiator is a piece of waveguide "sawed off" in a plane perpendicular to the axis of the guide. The open end radiates energy in a pattern that can be calculated from the known configuration of the field for the particular waveguide mode used. It is not surprising that radiation occurs in this manner—the fields inside a waveguide are propagating in essentially the same way they would in free space. The only real difference is that the walls of the guide constrain them for spreading. When the field reaches the guide mouth, it continues to propagate in the same direction; without the wall constraints, it also can spread laterally. There is a changeover region between the guide and free space where phase velocity and impedance characteristics are altered. The values go from those of the guide to free space figures of $c = 3 \times 10^8\ m/s$ and $Z_c = 377\ \Omega$.

Because the waveguide impedance is ordinarily different from the free space value, the radiating end does not usually present a matched impedance load to the guide. To provide a good match, some sort of transformation is needed. The best method is to flare the walls of the guide; if done properly, this action not only results in a matched impedance, but in a more concentrated radiation pattern (that is, a narrower beamwidth and higher directivity), as well. The flaring of the guide must

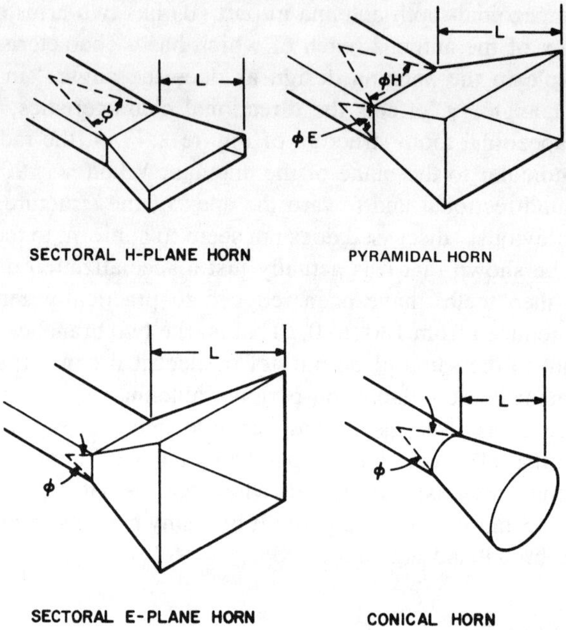

Fig. 4.49 Horn antennas.

be accomplished, as in any other matching network, in a very gradual way. Any abrupt changes in the end of the guide would cause reflections and a larger standing-wave ratio. Referring once again to Figure 4.49, there are two important areas to consider in obtaining a gradual flare—the flare angle, ϕ, and length, L.

If the flare angle is too great, the wavefront at the mouth of the horn becomes curved rather than planar, and there is a nonuniform phase distribution at the aperture. The results are decreased directivity and increased bandwidth. If the flare angle is too small, the aperture area becomes too small for a given length L. The directivity is proportional to the aperture size for a given distribution; therefore, an optimum flare angle is given by the approximation

$$\cos\frac{\phi}{2} = \frac{L/\lambda}{\delta + L/\lambda}$$

where δ is the permissible aperture phase angle variation expressed as a fraction of 360° (for a phase variation of 36°, $\delta = 0.1$). Typical values of δ for different horns are 0.25 (E-plane sectoral horn), 0.4 (H-plane sectoral), and 0.32 (conical horns). For a pyramidal horn the separate flares are calculated using different values in the E- and H-planes.

There are many special types of horn antenna; one that finds wide use is the biconical horn shown in Figure 4.50. It consists of two conical metal surfaces with their axes in line and their vertexes opposed. In other words, it is two cones with their points facing one another and almost touching. As with sectoral horns, there is an optimum flare angle that depends on the mode used. With the coaxial feed shown in the figure, the polarization is vertical and the optimum flare criterion is

$$h = \sqrt{2s\lambda}$$

Horizontal polarization may be excited in this setup by means of a small loop antenna between the vertexes of the cones. The optimum flare angle in this case is

$$\sin\frac{\phi}{2} = \sqrt{3\lambda/4s}$$

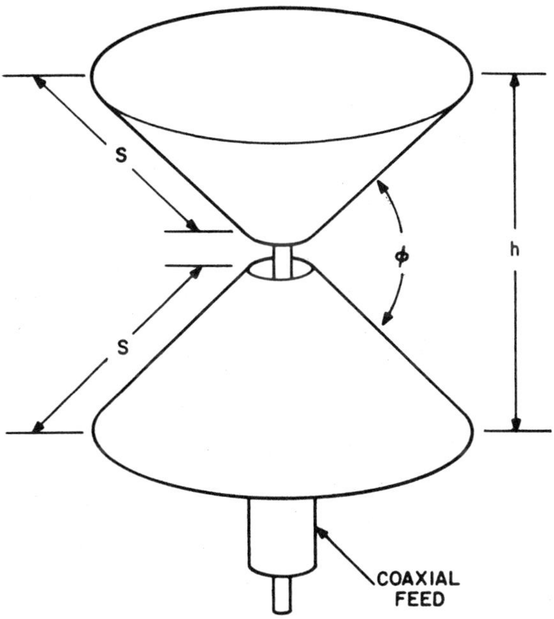

Fig. 4.50 Biconical horn.

A multiple diople feed can also be installed at the apexes of the cones. Such an arrangement produces an aperture which is the same in any azimuth direction, yet it prevents the vertical propagation of any appreciable amount of energy. The horns do not act like horns in the normal sense, inasmuch as the energy usually is not contained inside the horn. Rather, the combined outside surfaces of the two cones act like reflectors. It is even possible to shape the lengthwise dimensions of the cones so that they are parabolic and thus obtain certain advantages. End-to-end biconical horns are also feasible, so as to provide even greater horizontal gain than is obtained easily from a single unit. Stacks of as many as seven or eight such units are not impractical to build and use in microwave ground installations.

To complete our discussion, we show a typical data sheet for a horn antenna. All of the terms in the data sheet were defined at the beginning of the antenna section, but a brief description is given to refresh your memory.

Model number	HRN-1 (pyramidal)
Frequency	1–2 GHz
VSWR	1.4:1
Gain	15 dB
3 dB Beamwidth	35° (E- and H-planes)
Polarization	linear
Connector	Type N

The *frequency* term is related to the bandwidth term expressed previously—it indicates the range of frequencies over which the antenna specifications apply. *VSWR,* in this case, actually is telling you that the flare angle is close to its optimum value and the impedance match is good. The *gain* term measures both the efficiency and *directivity* of the antenna. The *3 dB bandwidth* applies to both the E- and H-planes in this particular case; there are antennas where the term has different values for the discrete planes. Its *polarization* is described as linear; it could be horizontal or vertical, whichever is required for a particular application. The last term, *connector,* is one that has been emphasized throughout this chapter. Once again, careful note should be taken so as to do away with adapters that would detract from the antenna's overall performance.

Horn antennas are highly predictable and can be designed easily to give a choice of radiation patterns over octave bands of frequencies. This high degree of predictability is the major reason that they are found in so many microwave applications.

4.9.5 The Parabolic Reflector

This type of reflector is what we referred to in the beginning of this section as the dish, which is what most people picture when you mention microwave antennas. There is a very good reason why this reflector comes so readily to mind—since it can be used with just about any type of antenna as a feed, it achieves a wide utilization.

The problem faced when constructing a reflective system is how to get the larger amount of transmitted energy to travel forward in a narrow band. The difficulty is similar to that encountered when building a searchlight reflector or a reflecting telescope. Whenever light or radio waves of small length (microwaves) strike a reflecting surface, they are reflected so that an old relationship learned in high school physics holds true—"The angle of incidence equals the angle of reflection." In other words, if a wave hits a curved surface at an angle made with an imaginary line perpendicular to that surface (a normal line), the wave reflects in the opposite direction at an angle exactly equal to that with which it hit. There is, however, a difference between spherical and parabolic surfaces, as can be seen in Figure 4.51.

In each of these cases we assume a point source in a transmitting-antenna reflector or a point receiver for a receiving-antenna reflector—a very small beamwidth that is virtually a single point. In Figure 4.51(A) it is shown that, if a spherical surface is used to intercept microwave energy, the energy reaching various parts of the surface is not reflected to a single point where an antenna could be positioned to collect and carry it to the receiver. The figure is, of course, an extreme case; in practice it might be possible to use a shallower spherical reflector, with the rays achieving an approximate focus. In actual antennas the feed is always of some finite size; thus, an exact focus is not imperative and some deviation from an accurate paraboloid is possible.

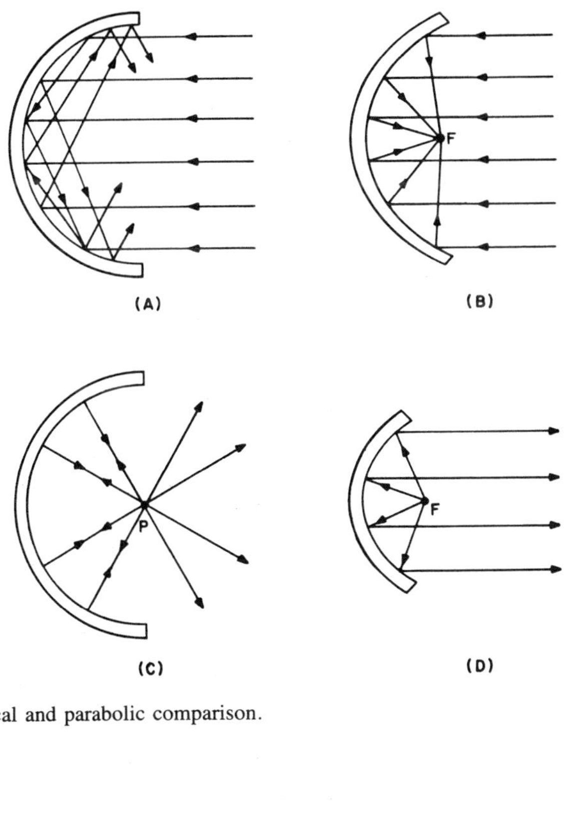

Fig. 4.51 Spherical and parabolic comparison.

The lack of an exact focus always is found with any reflector shape except paraboloid. Therefore, it is common practice to use dishes of the latter sort in forming sharp and symmetrical microwave beams. Although a manufactured parabolic reflector is usually quite accurate, such very high precision is not really necessary. If a reflector is of a form which is approximately paraboloid, there is little enough variation of the focus point so that the feeds can be arranged effectively. Figure 4.51(B) shows how all energy incident to the parabolic surface is reflected back through a common focal point (F). A transmission line may be set at F to collect the incident energy and transport it to the receiver.

A spherical reflector is shown again in Figure 4.51(C), but is acting as a transmitter surface from a point source feed (P), located at the center of the semicircle from which the reflecting surface is generated. In this case, not only is a widely divergent beam obtained, but a reinforcement/cancellation effect is also present. Radiation from the point source to the reflecting surface must, after reflection, travel back through the source again. If the radius is $\lambda/4$, the reaction on the source is to produce resonance; if $\lambda/2$, the result is an unloading. This type of difficulty is overcome partially when a parabolic reflector (D) is used. If the source is exactly at the focus, the exact reverse of the receiving case (B) is obtained and there is no convergence or divergence of the radio beam. Either convergence or divergence, which could occur if the feedpoint were located inside or outside of the focal point, causes a lower gain in the antenna.

Figure 4.52 shows a typical feed arrangement for a parabolic reflector. An open-ended waveguide enters the paraboloid from the back through an appropriate hole cut at the center; a dipole is mounted at the open end and is energized by coupling to the electric and magnetic fields moving down the guide. The dipole then radiates energy which travels to the main parabolic reflecting surface through a thin dielectric support. A small reflecting plate mounted nearby forces all the energy to be reflected from the parabolic surface before becoming part of the beam. This action avoids any difficulty which might otherwise arise—there could be a problem because the beam energy coming directly from the feed has a considerable path difference compared to that travelling from the dipole to the main reflecting system before moving on out into space. The small reflecting plate shades the dipole so that all energy from it must ultimately strike the parabolic surface. In actual practice, it is usually desirable to locate the dipole somewhat outside the focal point, typically at a distance 10 to 20 percent greater than the focal length.

The parabolic reflector can also have other antenna types besides the dipole at its feed point. Depending on which characteristics are most important for a specific application, it may use a helical, log-periodic, or conical antenna. Horn antennas, or variations of them, also are used extensively with parabolic reflectors.

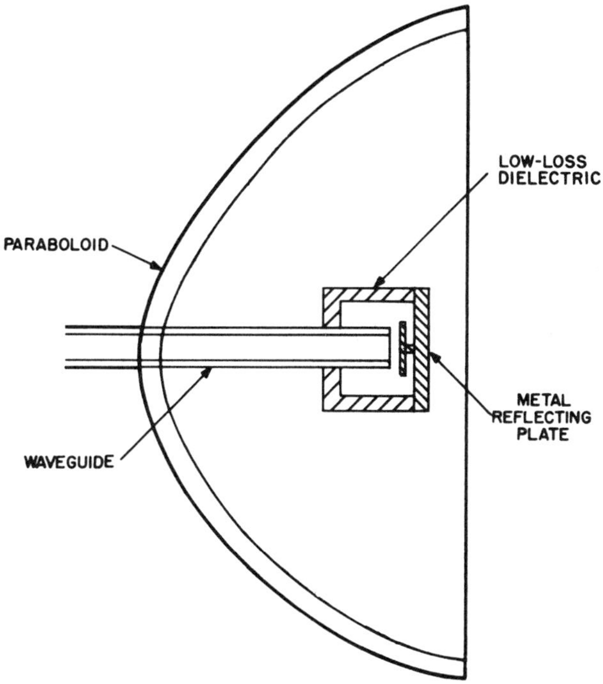

Fig. 4.52 Waveguide feed on a parabolic reflector.

The data sheet shown below is for a parabolic reflector with a log-periodic feed. The terms on this sheet are discussed after the sheet is investigated.

Model No.	PR-1
Frequency	1–11 GHz
Diameter	3 ft
3 dB beamwidth	23° at 1.0 GHz
	2.5° at 11.0 GHz
VSWR	2.5:1
Sidelobe level	15 dB
Gain	14 dB at 1.0 GHz
	33 dB at 11.0 GHz
Connector	Type N

The *frequency* term serves to substantiate that a log-periodic antenna is used with the parabolic reflector. The log-periodic is frequency-independent; the wide frequency range attests to its presence in the system.

The *diameter* is given in most data sheets to take into account any mechanical restrictions that the antenna may produce. The *VSWR,* as in other systems, indicates how good a match is at the antenna port. The *3 dB beamwidth,* described in detail earlier, has two values since the beam gets narrower at the top end of the frequency range.

To understand the *sidelobe* term you must refer back to Figure 4.41. An antenna has characteristics which produce a main lobe—a blast of power—in the desired direction. On either side of this main area smaller patterns—bursts of radiation called sidelobes—are generated. If these secondary levels are appreciable, they interfere with the intended operation of the antenna. They can inhibit the main lobe's ability to effectively increase the beamwidth and thus cut down the gain and directivity of the antenna. Once again, the *gain* is given at two frequencies; the wide variation is understandable since the beamwidth decreases so radically and produces much higher gain. *Connectors* are stressed once again, to eliminate any need for adapters and undesired losses.

The antennas covered in this section include those constructed of tubular sections, one built like a corkscrew, another flared from waveguide, and a reflector utilizing the unique properties of a parabola. Each has specific qualifications which make it useful in the microwave field; each has done its job there for many years and promises to continue to do so.

Chapter 5
Microwave Semiconductors

5.1 INTRODUCTION

The microwave semiconductor has progressed the fastest of any microwave component. This progress, however, has been a very well controlled phenomena. It has been monitored by the industry so that advances and changes are not made just for the sake of advances and changes. Every new device—from higher frequency bipolar transistors, to GaAs FET's, to IMPATTS and TRAPATTS—has been monitored for its most productive usage for the industry. When more advanced models are needed, they are developed. The semiconductor device fits the needs of the industry. It is not developed and then forced to fit an application.

We will cover three major areas of semiconductors in this chapter. First, the very important junction will be presented, represented by the *pn* and the Schottky junctions. Secondly, the microwave diode will be studied. In this area, the *Schottky, pin, Varactor, Gunn, Tunnel, IMPATT,* and *TRAPATT* will be covered. We will show how each fits into the ever-growing and expanding field of microwaves. The third, the microwave transistor, will also be presented. The evolution of the transistor from bipolar to GaAs FET, to HEMT (High Electron Mobility Transistor) will be presented, with applications.

These three areas will give you, the reader, a good background in microwave semiconductors and will result in a more effective use of them.

5.2 SOLID-STATE JUNCTIONS

The heart of any solid-state device, whether it is a microwave device or a simple dc switching transistor, is the junction (or combination of junctions) within that device. It makes no difference if we are dealing with a two-element (diode) or three-element (transistor) device, they all consist of junctions of material of one form or another. It is the characteristics of these junctions that make each device unique from all others.

If you look up the word *junction* in the dictionary you will find that it is: *The line or point at which two bodies are joined*. This definition very accurately defines a solid-state junction, since we are concerned with what takes place at the point where the two semiconductor materials come together, not necessarily what the individual materials are doing by themselves. It is, in actuality, the operation of the two materials *after* they are brought in contact with one another that results in the desired operations. That is, the characteristic of the junction given the overall operation of the device. Nowhere is this junction operation more important than in solid-state microwave devices. Time and distances are of the utmost importance when dealing with devices which will have the capability of operating from 1 GHz upwards into the 100 GHz ranges. Thus, understanding of how a semiconductor junction operates in a quiescent state and when forward and reverse biased is of paramount importance in understanding how the various diode and transistor configurations perform and how they can do the best job for your application.

We will cover two types of solid-state junctions in this chapter. They are the *pn* junction and the Schottky junction. Each of these junctions will be presented; theory of operation will be outlined in detail; and their operation under both forward and reverse biased conditions will be shown. Understanding of these junctions will aid greatly in being knowledgeable of microwave solid-state devices.

5.2.1 *pn* Junctions

When most people think of a junction in a semiconductor device, the *pn* junction is the one most commonly mentioned. This is the type of junction that was used by Dr. Karl Lark-Horovitz at Purdue in 1942 when he did extensive work with germanium for use in rectifiers; it is also the type of junction used by Brattain and Bardeen at Bell Telephone Laboratories in 1948 when they demonstrated the first transistor, a point contact device; and it is still the junction used for many microwave diodes and bipolar transistors available today.

A *pn* junction is formed when a *p*-type material and an *n*-type material are placed in physical contact with one another. When an atom from the third column of the periodic table (boron, for example) is added to a germanium or silicon atom (Column IV), there will be too few electrons available to make a covalent bond. This causes a void, or *hole,* which constitutes a condition termed an *acceptor* atom. In this condition the element formed will accept an electron if one is available. This is now *P*-type material with the *p* representing a positive charge—the charge of the vacated hole. Similarly, an *n*-type material is formed when an atom from the fifth column of the table (arsenic, for example) is added to a germanium or silicon atom. Covalent bonding does occur, but there would also be an excess electron as a result of this combination. This is because arsenic has five electrons in its outermost orbit as opposed to only four for either germanium or silicon. This extra electron is termed a *free electron* and results in a *donor* material which is classified as *n*-type. The *n* in this term is a designation for *negative,* the charge of the electron, which is the primary current carrier in this case.

Thus, the *p*-type material (acceptor, positive charge, holes being the primary current carriers), and the *n*-type material (donor, negative charge, electrons being the primary current carriers), form a junction when they are physically placed together. This allows an operator to conduct a variety of control functions depending on a specific application. The operation of this *pn* combination is what will be investigated.

As stated previously, a *pn* junction is formed when a *p*-type material and an *n*-type material are joined (bonded) together. Immediately after formation, carrier diffusion results in some electrons from the *n* region crossing the boundary, while some holes from the *p* region migrate to the other side. After crossing the junction, an electron which originated in the *n* region is in a high hole concentration environment. In this situation recombination is highly probable, and the electron as a carrier is thus annihilated. Similarly, when a hole crosses the junction from the *p* region it encounters a high electron concentration, recombines, and is annihilated as a carrier. Since each region was originally electrically neutral, electron-hole recombinations on both sides and in the vicinity of the junction result in layers of ionized acceptors in the *p*-region, and donors in the *n*-region. It can be said that the *p*-region has experienced a net accumulation of negative charge buildup, continuous until an equilibrium condition occurs and further carrier diffusion across the junction is negated by the repelling force between the carrier and charge concentration across the boundary. The *pn* junction in an equilibrium state is shown in Figure 5.1. The "built-in" electric field that is produced at the junction in an equilibrium condition is of such a direction and magnitude that, as previously stated, it opposes the diffusion of any mobile carriers across the junction. Thus, with no bias applied to the formed crystal, there is no net flow current. A dynamic equilibrium therefore exists within crystal in which a small current of minority carriers diffuses to the junction and is drawn across by the built-in field referred to earlier. Similarly, an equal current of majority carriers diffuses across the junction in the opposite direction. The potential difference within the crystal resulting from the built-in field is called the *barrier potential*. In most *pn* junctions this is in the order of 0.6 V.

An important aspect of the *pn* junction is the energy levels involved as the two materials are joined to form the junction. Figure 5.2 shows the energy level diagram for a *pn* junction. Notice that the *p* region is at a higher energy level than the *n* region. This is because the *p*-type material has lost some of high energy holes and gained some high energy electrons, as described earlier. This section is no longer neutral at this point, and thus will be displaced above the energy level of the *n*-type material by an amount equal to qVj. In this quantity, q is the charge of the electron and Vj is the electrostatic potential difference across the junction. In order for an electron to move from the *n*-region to the *p*-region, it needs to have sufficient energy to climb the resulting *potential hill* (the electron would have to invade the *p*-region which has more negative charge than the *n*-region). Also, the holes in the *p*-region

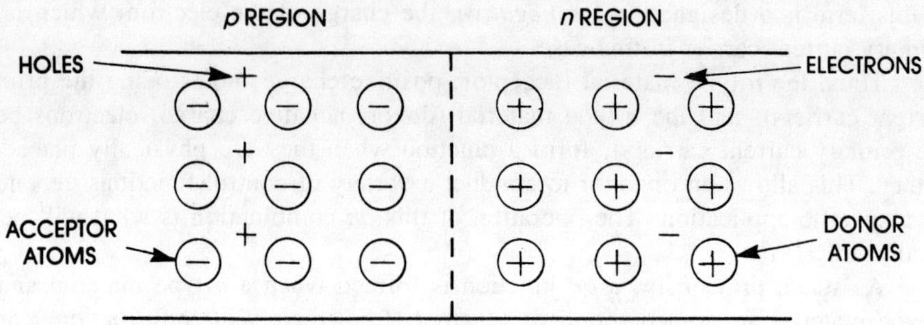

Fig. 5.1 *pn* junction at equilibrium.

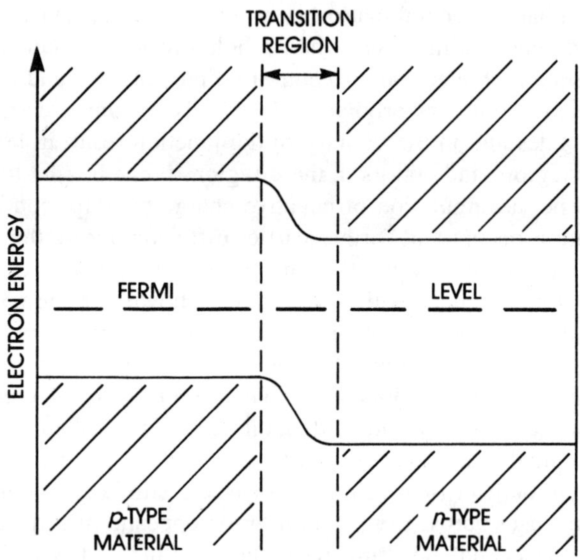

Fig. 5.2 Energy level diagram.

would need to be supplied with external energy in order to climb the potential hill in the other direction. Thermal agitation accounts for a small amount of carrier flow across the junction, but this is minimal. In general, all of the movement of majority carriers is completely balanced by minority carriers.

One term that is shown on the energy diagram in Figure 5.2 is the *fermi level,* which simplified, is a reference energy level at the *pn* junction. There is, however, much more to defining a Fermi level than saying it is a reference level. The filling of energy bands by electrons is accomplished by a simple rule: states of lowest energy are filled first, the next lowest level, and so on. Finally, all of the electrons have been accommodated. The energy of the highest filled state is the *fermi level* or *fermi energy* (W_f). The magnitude of W_f depends on the number of electrons per unit volume in the material.

Discussions thus far have centered around either the *p*-region or the *n*-region and how they are combined at a junction. There is, however, an area on both sides of this junction which must also be defined. This area was termed the *transition region* in Figure 5.2. It is also referred to as a *depletion region* or a *space-charge region*, although the term depletion region is used when the *pn* junction is in a reverse bias condition rather than a no bias condition, as we are now considering. (Forward and reverse conditions will be discussed later in this chapter.)

If we look up the term *space-charge region*, we will find a definition as follows: The region around a *pn* junction in which holes and electron recombine leaving no mobile charge carriers and a net density difference of zero. What this is saying is that as the two materials come together to form the *pn* junction, they are taking on their own individual charge and characteristics. During this "equilibrium" process the carriers in the neighborhood of the junction are swept away from the junction in their appropriate direction. This causes the junction area, and a minute distance on either side of it, to become an area of neutral charge. This can be seen in Figure 5.3. The top portion of the figure shows the *pn* junction with its appropriate acceptors, donors, holes, and electrons. The junction follows down through to parts (b) and (c) of the figure, and the distance from the junction on the right and left are distances through the semiconductor away from the junction. Figure 5.3(b) shows the charge density at and around the junction. It can be seen, as mentioned previously, that the net charge in the space-charge region is zero. Also shown is a *potential hill* in Figure 5.3(c). This is an electrostatic potential which is present as you progress from the junctions into the semiconductor material. The one shown in solid is a formidable obstacle for holes from the *p*-region to overcome. Similarly, the dashed line also shows that there is a hill of equal magnitude. This presents a large obstacle for the electrons from the *n*-region. These potential hills further substantiate the neutrality of the space-charge region when there is no bias applied to a *pn* junction.

As a result of the diffusion process within the region of the *pn* junction, a potential gradient is established across the space-charge region. This gradient is represented by an imaginary battery, as shown in Figure 5.4. The battery is termed "imaginary" because it is shown only to represent the internal effects of the junction and the space-charge region. The actual potential is not a measurable quantity. In the absence of any external voltages (no bias applied), the potential gradient discourages further diffusion across the *pn* junction because electrons from the *n*-type material are repelled by the very slight negative charge induced in the *p*-type material. Thus, the gradient prevents total interaction between the two materials (*n*-type and *p*-type) and consequently preserves the differences in their individual characteristics.

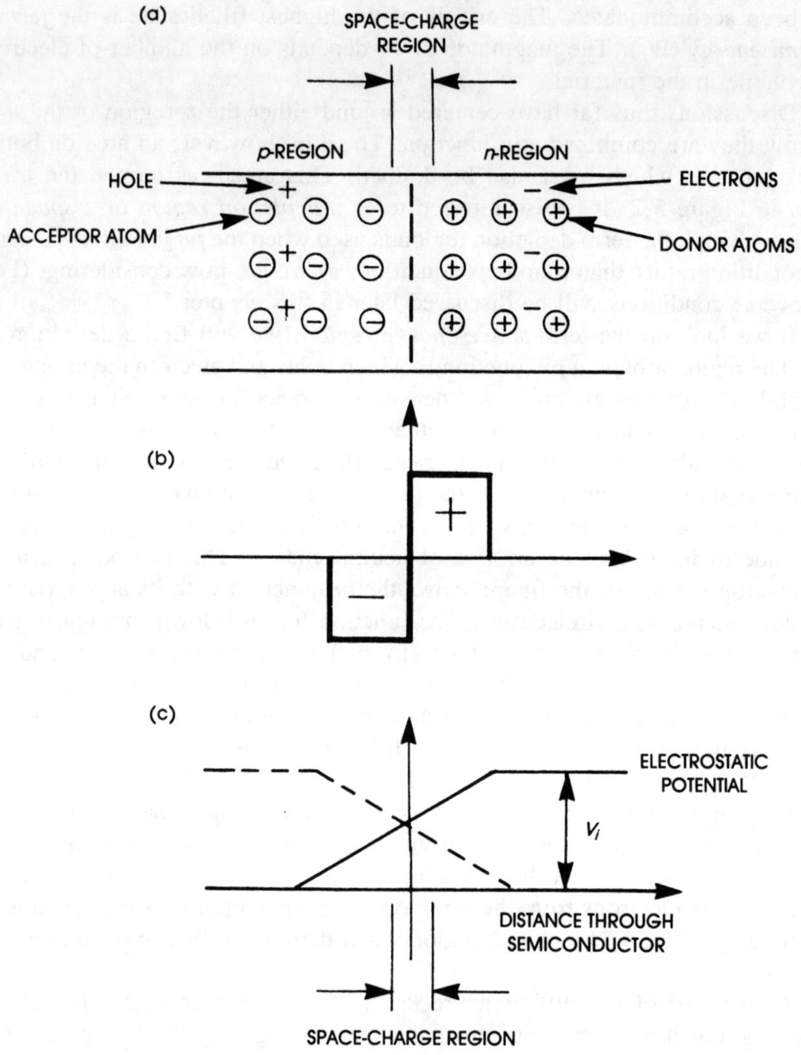

Fig. 5.3 Space-charge characteristics.

Two junctions which can be analyzed as discussed above are the *abrupt* junction and the *graded* junction. The abrupt junction is one in which the impurity concentration (assuming that the inpurities are fully ionized) is constant on both sides of the junction. Figure 5.5 shows the charge density for this junction.

The *linearly graded junction*, on the other hand, is one in which the impurity concentration is a linear function of the distance across the junction. This is shown in Figure 5.6.

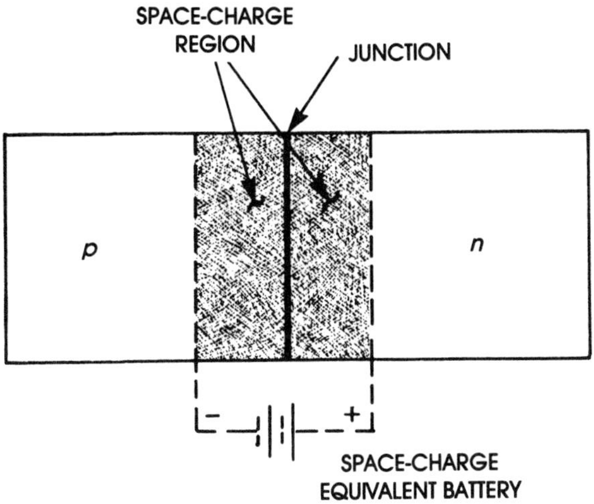

Fig. 5.4 Imaginary space-charge battery.

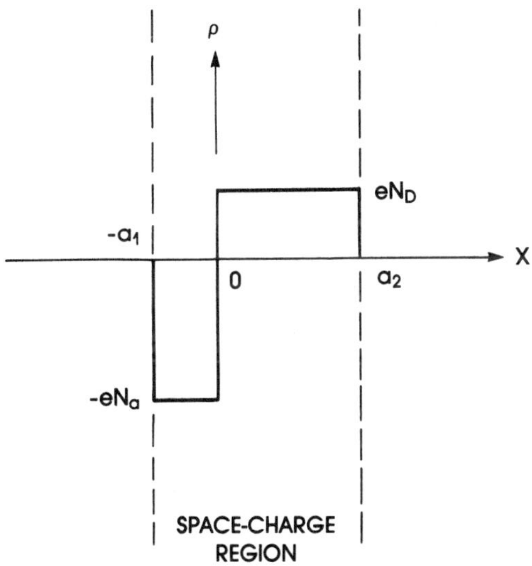

Fig. 5.5 Abrupt junction.

The space-charge approximations described in the text above allow you to calculate the width, w, of the space-charge region. Knowing this width makes it possible to determine what the junction capacitance of the *pn* junction is. This is of prime importance when dealing with higher frequency devices such as those used for microwave applications. Junction capacitance will be covered in detail following

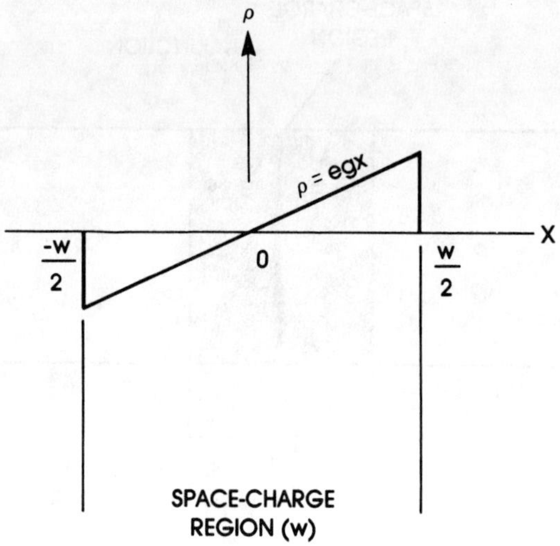

Fig. 5.6 Graded junction.

the discussions on biased *pn* junctions. This is because the capacitance has much more meaning when the junction is either forward or reverse biased.

All of our discussions on the *pn* junction have thus far had everything in an equilibrium state. That is, one side has holes and acceptor atoms (*p*-region) while the other has an equal number of electron and donor atoms (*n*-region). There were no external voltages applied to disturb this state. To understand the operation of the junction, however, we must apply voltage since, in actuality, all microwave solid-state devices are active devices (they require voltage for operation). It is now time to apply a *bias* to the *pn* junction we have worked so hard to create.

There are two conditions we can achieve by biasing a *pn* junction. One condition is *forward* biased and the other is *reverse* biased. Both conditions must be understood if an individual is to comprehend the capabilities of the *pn* junction devices available for circuit applications. For you to gain the proper perspective and grasp the operation of a biased *pn* junction, we will compare it to the unbiased *pn* junction, which we discussed above.

You will recall from previous discussions that the unbiased *pn* junction energy diagram, as shown in Figure 5.2, indicated *p*-type material energy levels slightly higher than that of the *n*-type material. This is shown again in Figure 5.7(a) and is due to the *p*-type material losing some of the high energy holes while gaining high energy electrons. The amount of the difference, as previously stated, is qVj with q being the charge of the electron, and Vj the electrostatic potential difference across the junction. This results in a potential hill which must be overcome when conduction is to be accomplished. In the unbiased *pn* junction case, however, conduction is not what the designer is looking for primarily, although when the electrons are raised

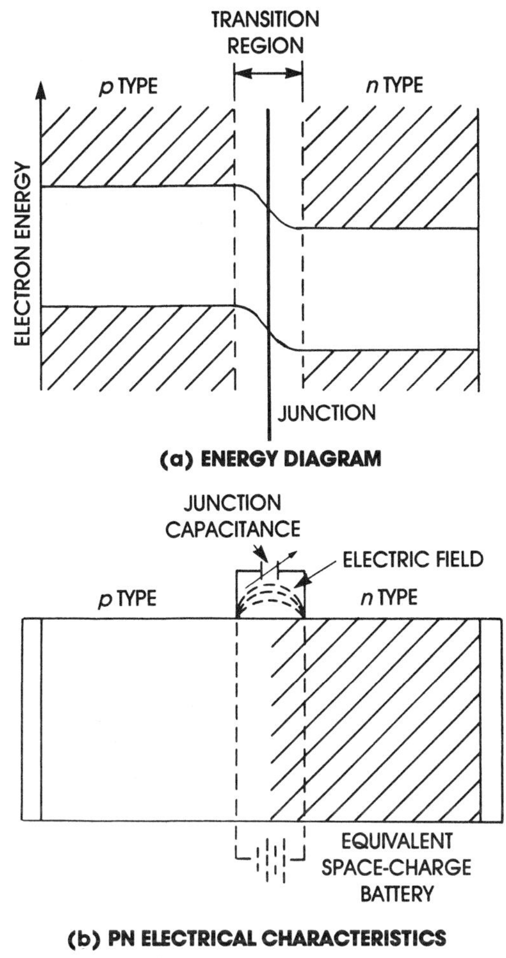

Fig. 5.7 Unbiased *pn* junction.

to a sufficient level conduction will occur. This is used when unbiased diodes are used for detection or mixing purposes. In these cases the microwave energy is responsible for allowing the electrons to achieve sufficient energy to cross the junction.

Figure 5.7(b) shows the unbiased *pn* junction's electrical characteristics. There is a junction capacitance, dependent on the width of the transition region; an electric field which is generated because of the difference in the materials which have been joined; and then there is the equivalent (imaginary) space-charge battery which was described earlier. The importance of these characteristics lies *not* in any of their specific values. Rather, the importance is in the relative change in each of these characteristics as the junction is biased in a forward or uneven direction.

An important condition to have in a *pn* junction device is to have it forward biased. In this condition current can flow across the junction very freely. A look at Figure 5.8 will show why this current flows so freely. Figure 5.8(a) shows that when the forward bias is applied, the potential hill is reduced dramatically so that carriers can cross the junction with great ease. Further reduction of the hill by increasing the bias will result in an even greater number of carriers crossing the junction. There is, however, a maximum amount of voltage that can be applied. Exceeding this value will destroy the junction and its original characteristics.

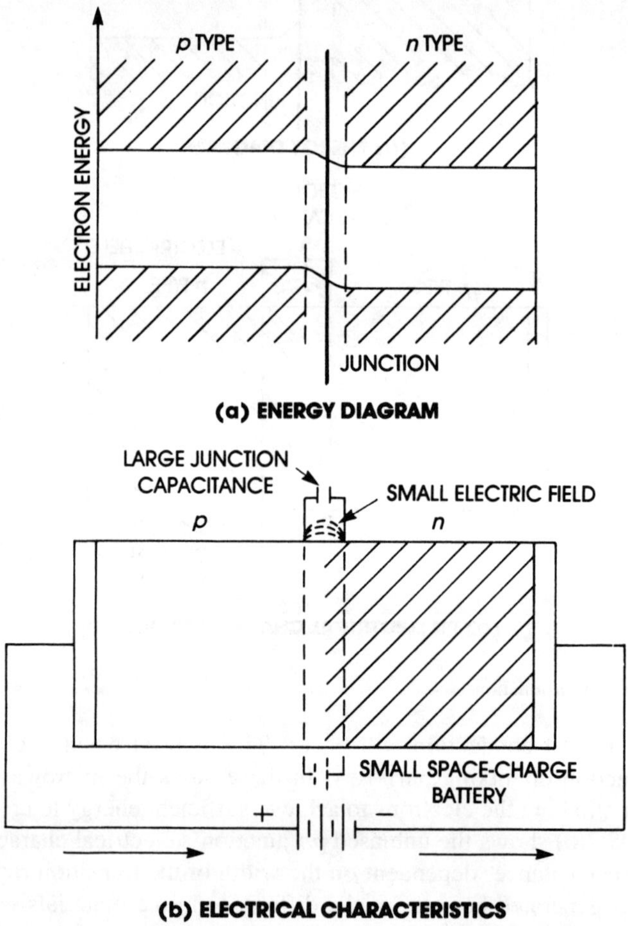

Fig. 5.8 Forward biased *pn* junction.

Figure 5.8(b) shows how the junction reacts electrically with a forward bias applied. Notice how the transition, or space-charge region is closer together than that shown with no bias applied to a *pn* junction. This is because, with a forward bias applied, the area around the junction has many more carriers in it and is, thus, less of a "dead" or neutral zone. There obviously is more conduction in this region than with no bias, so naturally the region will be smaller. You will also notice that the junction capacitance is shown to be large as compared to an unbiased junction; the electric field is small; and the space-charge battery is small. These are all functions of the space-charge region.

The junction capacitance is larger because the *effective* plates of a capacitor (the edges of the space-charge region) are closer together. Refer back to basic ac theory and you will recall that this does increase the capacitance. Similarly, the electric field across the region is smaller since there is less region across which to develop a field. As a consequence, the space-charge equivalent battery will be smaller. Since the electric field is across less of a junction (its *width* is decreased) and its intensity, or amplitude is less (*height* is decreased), it follows naturally that the equivalent (imaginary) battery will be smaller.

A curve used to characterize a biased junction is the *I-V* curve (current-voltage curve). This is shown in Figure 5.9(a) for a forward-biased *pn* junction. Curves are shown for both germanium (Ge) and silicon (Si). It can be seen that when voltage is applied in a forward direction, there is a point when very little forward current (I_F) is being drawn. For germanium, there is very little current present until approximately 0.3 V is applied and 0.6 V for silicon. This is because the space-charge equivalent battery has an approximate potential of 0.3 V for germanium and 0.6 V for silicon. This battery voltage must be overcome before any current can be drawn. A negative voltage applied to the junction will cause an entirely different reaction. This will be covered in the discussions on reverse bias *pn* junction. (I_R is reverse current in the junction.)

Figure 5.9(b) is an *I-V* curve for a *pn* junction with no bias. Notice how it differs drastically from that of Figure 5.9(a), a forward biased junction. Since no bias is applied, there is no positive or negative voltage on the device. Thus, the forward current, I_F, equals the reverse current, I_R, and the net result is no current flowing through the device. This verifies what was discussed in earlier sections of this text on unbiased *pn* junctions.

There are times when zero bias or a forward-biased condition is not what a designer desires. One such time is when a *pn* junction is being used as a switching diode. To ensure that the diode does not conduct current, a *reverse* bias is applied to the junction. With this bias applied to the junction, Figure 5.10, the potential hill (previously discussed) which the electrons must climb in order to produce current flow, becomes very high and difficult to climb. This is shown in Figure 5.10(a).

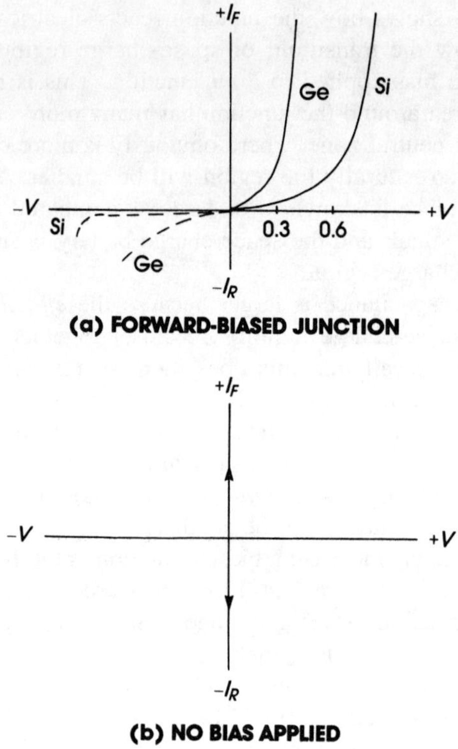

Fig. 5.9 *I-V* curves.

The only current possible under these conditions results from a few minority carriers on each side of the junction (electron and hole leakage current). In Figure 5.10(b) you can see how the width of the space-charge region has increased dramatically. This increased width results in a small junction capacity, since the plates of the capacitor have effectively been moved further apart; there is a high value of electric field over this wider area; and the space-charge battery is a larger value, since the electric field intensity is high. This, obviously, has created a much different condition than when the junction was forward biased.

As discussed above, the only current possible under these reverse biasing conditions is the result of a few minority carriers on each side of the junction. This leakage current is almost completely independent of the magnitude of the reverse-bias voltage until the *avalanche*, or *zener*, voltage is exceeded. This is shown in Figure 5.11. At this point, reverse current (I_R) increases very rapidly. The avalanche breakdown, as it is called, occurs when the electric field across the junction produces ionization as resulting high energy carriers collide with valence electrons. The breakdown appears to be a "field-emission" phenomenon with the strong field in the junction region pulling carriers from their atoms. It should be clear that the junction

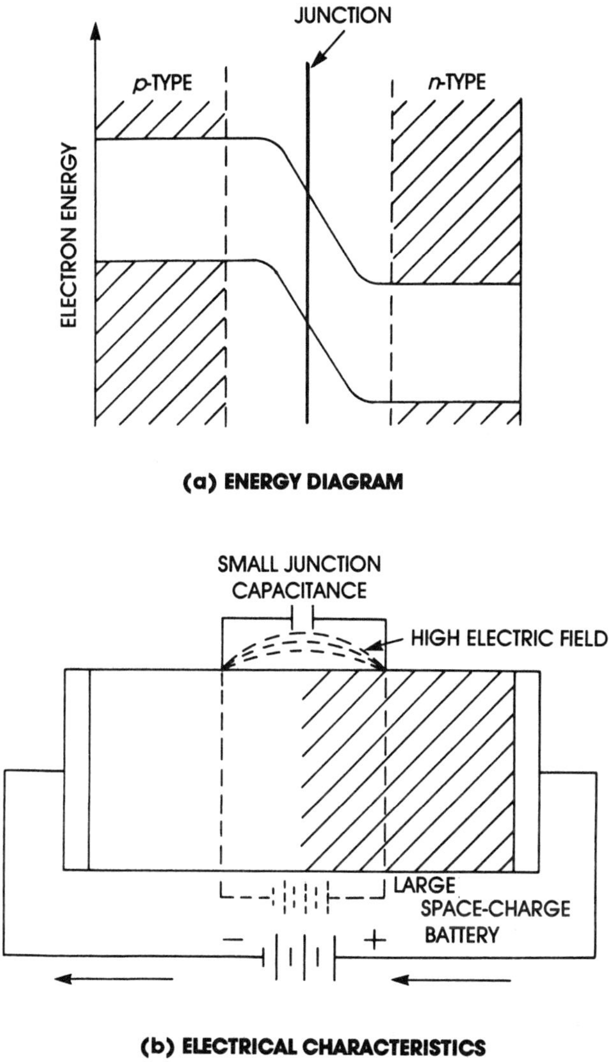

Fig. 5.10 Reverse biased *pn* junction.

should not be left in this state for any period of time or severe damage or destruction of the junction can occur.

With all of the biasing conditions covered for the *pn* junction, it is now time to name the junction we have been discussing. This two-element device is actually a *diode*. That is, it will conduct current in one direction and not conduct it in the other. Extensive coverage of diodes will be presented later in this chapter. For now,

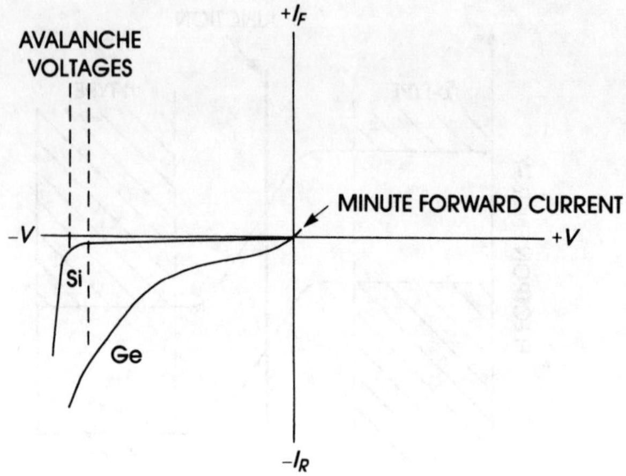

Fig. 5.11 *I-V* curve for reverse bias.

we will be concerned with the diode, not as a microwave device, but as a two element *pn* junction.

Operation of the *pn* junction diode can best be summarized as:

- Joining *p*-type material to *n*-type material establishes a contact-potential hill which limits random movement of majority carriers across the junction;
- Minority carriers, usually generated by thermal energy, can easily cross a reverse-biased junction. Their flow is essentially independent of the magnitude of the bias, unless breakdown occurs; and
- With forward biasing, the height of the potential hill is reduced and majority carriers cross the junction in great numbers.

Before we leave our discussion on the *pn* junction, two terms need to be presented and defined. Both terms are critical for microwave operation of solid-state devices. The terms are *junction capacitance* and *transit time*.

The first term, *junction capacitance*, has been mentioned frequently throughout our discussion of the *pn* junction. You will recall that we stated that junction capacitance has more meaning when the *pn* junction is either forward or reverse biased. It should now be clear what was meant by that statement, following the discussion presented on unbiased-, forward-, and reverse-biased junctions.

Junction capacitance is, basically, what the name implies. It is the capacitance set up by placing a *p*-type material and an *n*-type material in contact with one another and then either applying a bias voltage to them, or letting the junction remain in an unbiased state. You will recall from basic ac theory that a capacitor was produced when two conductive plates were separated by a dielectric. The spacing of the plates (thickness of the dielectric) determined the amount of capacitance. A relationship for this definition is

$$C = 0.0885\ \varepsilon_r \frac{(N - 1)A}{t}$$

where

C = the capacitance in farads (F);
ε_r = the dielectric constant;
N = number of plates;
A = area of one side of one plate in cm;
t = dielectric thickness (plate spacing) in cm.

You will recall from our discussions of unbiased-, forward-, and reverse-biased *pn* junctions that the junction capacity varied exactly as predicted by the relationship alone. That is, an unbiased junction had a value of capacitance determined by the material joined and the resulting space-charge region; the forward-biased junction had a much smaller space-charge region (t decreased) and also a resulting higher junction capacitance; and the reverse-biased junction had a very wide space-charge region (t increased) and a corresponding lower value of junction capacitance. Thus, the previous theory has been substantiated. There are two statements that can be said of junction capacitance:

1. It is present across *every pn* junction.
2. It is caused by the uncovered charge layers that come about as a result of the ionization of impurities.

For microwave applications the junction capacitance must be very low. Large values of junction capacitance will limit the high frequency operations of devices or may cause positive feedback within a device that will result in oscillations. When choosing devices for microwave operation, therefore, be sure that the junction capacitance of the device is minimal or very small.

The final term to be investigated is the one which made solid-state devices much more desirable than vacuum tubes—*transit time*. In a semiconductor this is the time it takes for carriers to cross the depletion or space-charge region. It is a *delay* (τ) which is encountered as a carrier travels from one material to another. The effect of transit time is identical to that in vacuum tubes, although its operation is different, since one has a vacuum medium and the other has a dielectric medium. In traveling across a solid-state junction, the holes or electrons drift across with velocities determined by the resulting ion mobility, bias voltages, and junction construction. When more than one junction is concerned, as in a transistor, there are a combination of conditions which add up to form the total transit time. This total is

$$\tau_{ec} = \tau_e + \tau_b + \tau_{s1} + \tau_c$$

where

τ_{ec} = emitter-to collector delay;
τ_e = emitter barrier changing time (delay);
τ_b = base transit time;
τ_{s1} = collector transit time;
τ_c = collector charging time.

- The emitter barrier charging time is an *RC* product of r_eC_e, where r_e is the emitter series resistance (determined by temperature, charge of electrons, and emitter current) and C_e is the emitter barrier capacitance.
- The base transit time is an expression determined by W_b (base layer thickness), *D* (minority carrier diffusion coefficient), and *n* (a factor related to the drift field in the base region associated with the amount of impurities).
- Collector transit time is determined by the time of carriers crossing the depletion layer, the drift velocity of the carriers, and the total width of the depletion layer in the collector region.
- The collector charging time is once again an *RC* product. This time it consists of r_c, the collector resistance and C_i which is the collector depletion-layer capacitance

All of these parameters combine to produce an overall transit time (delay) through the device. Further discussions of two-junction, three-element (base-emitter-collector) devices will be presented later in this chapter on microwave transistors.

Thus, the *pn* junction has been formed and characterized. It can be seen that when two types of materials (*p*-type and *n*-type) are joined, a junction is formed which exhibits electrical characteristics which can be modified by application of bias voltages. Such parameters as junction capacitance and transit time are parameters which are critical when operating solid-state devices (with *pn* junctions) at microwave frequencies. Thus, when choosing the device for your particular application, be sure to consider these and other parameters which will aid the *pn* junction in doing the best job for your microwave application.

5.2.2 Schottky Barrier Junction

If you were to look up the definition of a Schottky barrier in an electronics dictionary, you would find words which say:

> "A simple metal-to-semiconductor interface that exhibits a nonlinear impedance."

A large portion of that definition is absolutely true. One part is questionable. That part is the word *simple*. The Schottky barrier junction is not simple in terms of operation, it is only simple in the fact that a metal and a semiconductor are joined together and produce one of the most useful junctions ever devised for microwave applications.

As stated above, the Schottky barrier junction is set up by contact between a metal and a semiconductor. It gets its name from early analysis in an area termed *majority-carrier rectification* by W. Schottky in 1938. The semiconductor used is generally *n*-type silicon or *n*-type GaAs. This metal-semiconductor junction has many advantages over a *pn* junction when used at microwave frequencies. When the junction is forward biased, current flows because of majority carriers moving from the semiconductor to the metal. Minority carrier effects are virtually eliminated with this type of junction. This means there is no reverse recovery time in a Schottky junction (since only majority carriers are used) and, then, there is no charge-storage capacitance. Low capacitance, as we have discussed previously, is an excellent property for microwave devices, since it allows them to operate at higher frequencies. The reliance on only majority carriers to perform the necessary operations within the junction classifies the Schottky, and any devices using it, as *unipolar*. This is as opposed to the *bipolar* effects of the *pn* junction.

Let us now take a look at the energy levels of a Schottky barrier junction and find out more about its operation. Figure 5.12 shows energy diagrams for metal (a) and a semiconductor with no surface charge (b). In the case of the metal, ψ_m is the work function of the metal. This is considered a unit of negative charge for this discussion. It denotes the energy required to remove an electron from the Fermi level (W_f) of the metal and place it at rest in free space (vaçuum). The factor e is used to convert the charge ψ_m to energy units. This work function is a constant value for a given atomically clean metal, but varies as surface contamination increases. This is the only factor which governs its value. This is as opposed to the work functions of a semiconductor, ψ_s, which not only varies with surface contamination but also depends on the positions of the Fermi level and upon carrier density. These factors can be seen in Figure 5.12(b). (W_c is the wave function in the conductance band and W_v is the work function in the valence band.) One parameter which is not affected by semiconductor doping (or contamination) is termed X, the electron affinity. This is the amount of energy released or absorbed when an electron is added to a neutral atom. Its unit of measure is electronvolts and will increase with smaller (decreasing) atoms. This indicates that the electron affinity is greater for higher group numbers (Groups V, VI, VII).

The semiconductor depicted in Figure 5.12(b) is one without a surface charge, that is, a neutral state of the semiconductor exists. This, however, is a condition which is *not* a common occurrence, and, when it does occur, does not last for any period of time. What is a common occurrence in a semiconductor is shown in Figure 5.13. This is a semiconductor with a net negative charge. In this state, positively ionized donors are formed just below the surface and distort the band edges as shown. Since the material is still at equilibrium, there is no net current and the Fermi level is still constant throughout. The distance of the Fermi level from the conduction load is designated as the amount of bending of the band edges, and is shown as ψ'_{ns}. These are shown in Figure 5.13.

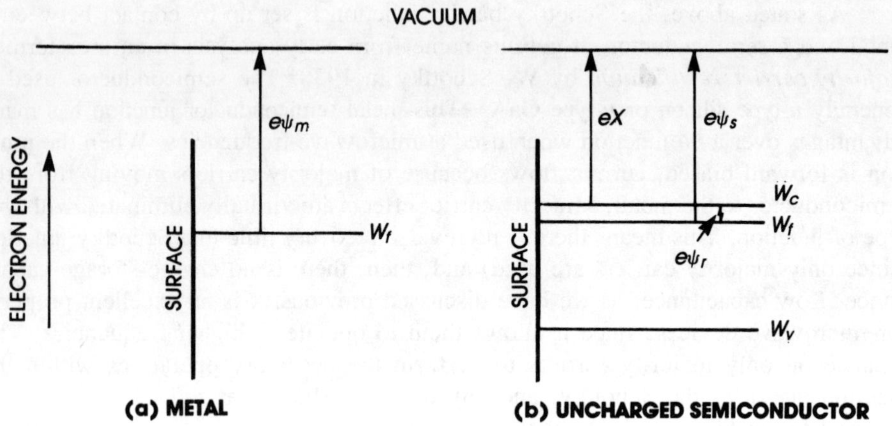

Fig. 5.12 Energy diagram for metal and uncharged semiconductor.

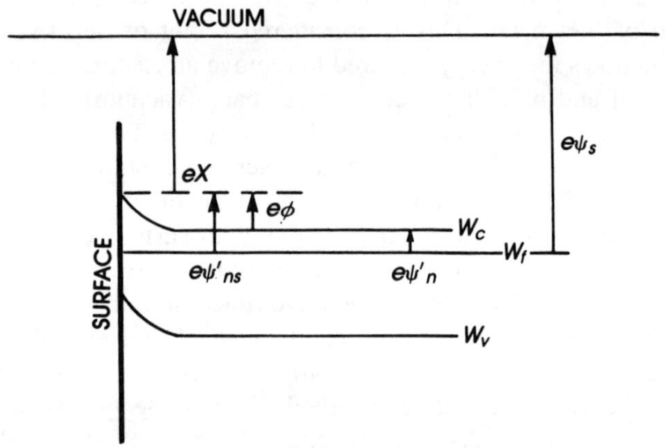

Fig. 5.13 Semiconductor with net negative charge.

With the basic concepts of both the metal and the semiconductor presented, it is now time to place the two in contact with one another and form the junction. Figure 5.14 shows the two individual materials (metal and semiconductor) prior to their attachment to one another. It is important to notice the individual energy levels and work function placements before contact. Notice that the Fermi levels are not coincident in both materials. This is because, at this point, the two materials are not in equilibrium with one another. The figure also shows that there is some net surface charge present in the semiconductor, since there is band bending present.

Figure 5.15 shows the energy levels of the Schottky barrier as the metal and semiconductor are brought into contact with one another. Equilibrium has not been

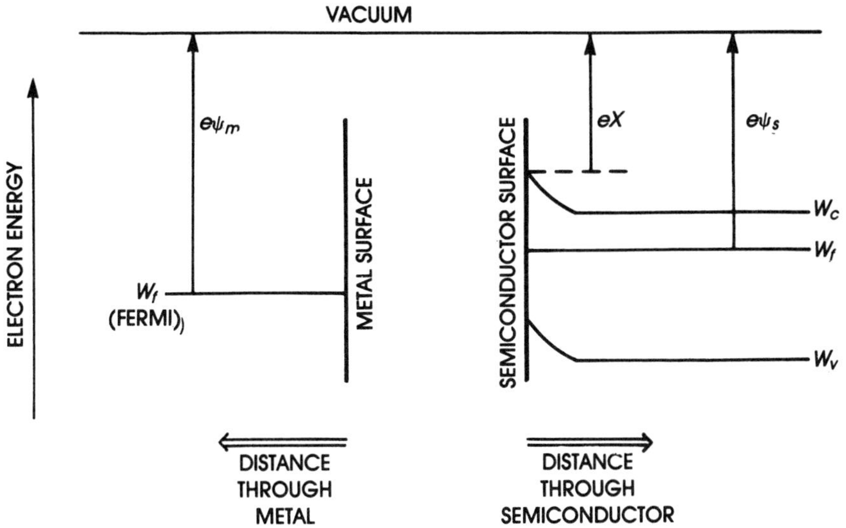

Fig. 5.14 Metal / semiconductor before contact.

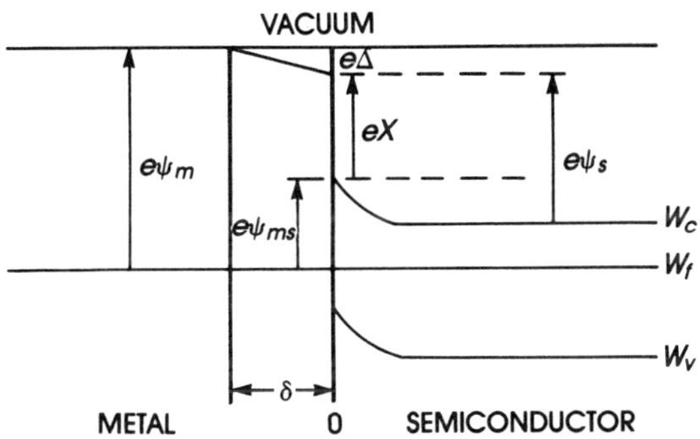

Fig. 5.15 Metal and semiconductor in contact.

established, and the Fermi levels are not in line. A change in potential designated $\psi_m - \psi_s$ occurs across the gap. This is termed Δ in the figure. This charge is called the *contact potential*. As the surfaces are brought closer together, the field between the two surfaces, designated by ψ_m—ψ_s/δ will increase. The gap distance δ is in the order of 0.5 to 5Å, depending on surface preparation and method of contact. This is a distance of 0.5×10^{-8} cm (1.96×10^{-9} in) to 5×10^{-8} cm (19.68×10^{-9} in). This thin layer is virtually transparent to carriers coming between the metal and the semiconductor. At its maximum height, the barrier which is seen by the carrier is known as the *metal-semiconductor barrier height* and is designated as

$$\psi_{ms} = \psi_m - X - \Delta$$

where

ψ_{ms} = work function of the metal-semiconductor
ψ_m = work function of the metal
X = electron affinity
Δ = change in potential across the gap

Table 5.1 shows values of ψ_{ms} for various metals when contacted with Si (silicon) and GaAs (gallium arsenide), both *n*-type semiconductors. (Values are in volts.)

Notice the heights of the barriers using silicon as opposed to those in gallium arsenide. All values are lower for silicon which is why gallium arsenide is used for transistor application when the Schottky junction is advantageous for high frequency operation. Generally, diodes will use silicon for the Schottky junctions, since barrier heights need not be as high for diode operation. (The maximum width of the barrier is that of the depletion layer. This is usually on the order of thousands of angstroms.)

Table 5.1

Metal	*Si*	*GaAs*
Au (Gold)	0.81	0.90
Ag (Silver)	0.69	0.88
Al (Aluminum)	0.67	0.80
Cu (Copper)	0.71	0.82
Pt (Platinum)	0.85	0.86
Ni (Nickel)	0.66	—
W (Tungsten)	0.69	0.80

With the metal-semiconductor Schottky barrier junction analyzed and formed, it now remains to place bias voltages on the junction to produce the desired operation. Figure 5.16 shows the three conditions possible for a Schottky junction; no bias, forward bias, and reverse bias. Figure 5.16(a) is the familiar energy diagram we have been discussing throughout this session. It is the zero-bias condition and is reproduced in this figure to be used as a reference. It can be seen from the bias conditions presented that the bias voltage causes changes in the energy band bending and, more importantly, a change in the Fermi level (W_f) within the junction. In Figure 5.16(b), the forward bias condition, the bending is not as severe as the zero bias condition and the Fermi level is raised slightly above that of the metal. This allows the junction to conduct majority carriers more easily. (V_A is positive for an applied forward bias—the positive voltage being applied to the metal.)

In Figure 5.16(c) the reverse-bias condition is shown. In this case the energy band bending is extensive and the Fermi level of the semiconductor is much lower

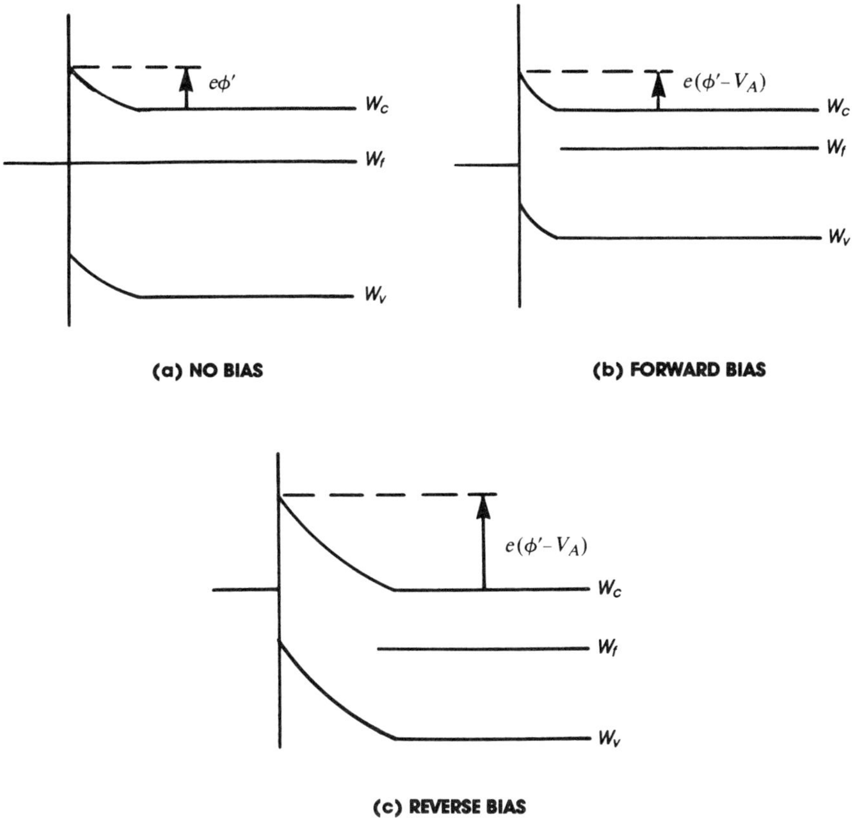

Fig. 5.16 Schottky barrier bias condition.

than that of the metal level. This basically eliminates any conduction across the junction. (V_A is negative for this condition.)

Thus, the Schottky junction has been shown for applications in the microwave area. Its operation as a *unipolar* device which relies only on majority carriers for conduction allows it to be used at much higher frequencies than its *pn* junction counterpart (a *bipolar* device).

The junction is used for the diodes used in detection and mixers and has become the heart of GaAs field effect transistors and the gate junction. For this Schottky gate, certain metals such as aluminum, tungsten, platinum, and molybdenum will form the desired diode suitable for a gate junction when evaporated on GaAs. (All of these metals were described in Table 5.1 when we were discussing the work function, ψ_{ms}, of the metal-semiconductor combination.) When using any of the last three metals (molybdenum, platinum, tungsten) a layer of titanium is first put down to act as an adhesive, since they will not stick to GaAs on their own. A layer of gold gives the gate area the needed conductivity. Thus, gates are much simpler to fabricate with aluminum than with other metals, even though other materials, such as titanium-tungsten-gold (TiW-Au) gates have advantages which are:

1. Not subject to *purple plague* (a brittle alloy which is formed when an aluminum-gold (Al-Au) bond is exposed to extended heat—as in the gate formation process).
2. Less subject to failure from electric discharge.
3. Can be plated an arbitrary thickness to reduce resistance (and reduce the device noise figure).

Thus, it can be seen that, as in all devices, trade-offs are necessary when utilizing the Schottky barrier junction.

5.3 MICROWAVE DIODES

When most people think of a diode, they picture a small glass or plastic-packaged device with two leads on it, used for rectifying ac in a power supply or clamping a pulse to a specific dc level. This representation, of course, is very accurate for many of the low-frequency, dc, and logic applications encountered throughout electronics. When we speak of *microwave diodes,* however, we are speaking of a two-element device which will do much more than rectify or clamp a dc voltage.

To understand the complexity and versatility of microwave diodes, consider what function they can perform. This two element-device is able to switch, attenuate, mix frequencies, detect, amplify, oscillate, and modulate signals in both the microwave and millimeter bands of frequencies. This small device which is defined by the dictionary of electronics as a *two terminal electronic device that will conduct electricity much more easily in one direction than in the other,* can perform functions which were unheard of with three-terminal, or any other, devices as little as thirty years ago. Obviously, we are talking about devices which have characteristics that involve much more than "conducting electricity much more easily in one direction than in another." We are speaking of devices which use silicon or gallium arsenide as a semiconductor; which may use a *pn* junction or Schottky junction for operation; may have an intrinsic layer added during construction to control the resistances of the device; will exhibit negative characteristics; or may not even rely on semiconductor junction properties for their operation. These are the characteristics of a variety of microwave (and millimeter) diodes. Not, as you can see, the typical characteristics used in the basic definition of a diode. This is because a microwave diode is much more than just a two-element device which has limited capabilities. It is a complex device which is an integral part of many sophisticated microwave circuits and systems and, as such, deserves close inspection and study to determine how these tiny marvels can be used to their greatest potential.

We will investigate seven types of microwave diodes in this chapter. They are: *Schottky, pin, Varactor, Tunnel, Gunn, IMPATT,* and *TRAPATT* diodes. Each of these devices will be presented with an analysis of the device, theory of operation, construction, and applications. Where applicable, a typical data sheet will also be presented to illustrate common parameters, how to interpret them, and where (and how) they can be useful for your applications.

5.3.1 Schottky Diodes

Like the conventional point contact diode, the Schottky-barrier (or hot carrier) type is a majority carrier—its conduction is by means of a predominant carrier (in this case, electrons, although *p*-type silicon has holes for majority carriers). The energy, or Schottky barrier, exists at the metal to semiconductor interface because of the difference in the work functions of the two materials. The barrier is unaffected by a reverse-bias, but is decreased by a forward-bias—the familiar characteristic of a rectifying diode. Therefore, when the device is forward biased, the majority carriers (electrons) can be easily injected from the semiconductor material into the metal, where the energy level is now much higher. Once in the metal, the electrons give up this excess energy in a very short time, on the order of 10^{-15} seconds (or one *Fento-second,* which is one thousandth of a pico-second), after which they become a part of the free electrons of the metal. Of course, when reverse biased, the energy level of the barrier is too high for the electrons to overcome and the device does not conduct. These characteristics indicate the Schottky-barrier's potential for being very efficient high frequency mixers and detectors.

Although there is little difference in the theory of operation of both point contact and Schottky-barrier diodes, they differ in the manner by which the rectifying junction is formed. The point contact junction is realized by the pressure contact of a metal whisker against the semiconductor material, establishing a metal-semiconductor rectifying barrier. Instead of the pressure contact, the Schottky-barrier diode has its rectifying barrier formed by a metal contact deposited on the silicon epitaxial layer as shown in Figure 5.17. This technique achieves a more nearly ideal metal to semiconductor interface. The use of a conductive pad above the junction allows the whisker wire to be thermal-compression bonded, thus eliminating the need for a pressure contact while increasing the mechanical reliability of the device.

To obtain a diode which approaches the ideal junction conversion loss, it is necessary to minimize the loss factors R_s, C_j, and L_s, shown in Figure 5.18. The metallic junction (R_j) is shunted by the barrier capacitances (C_j). The series resistance (R_s) is a result of the bulk resistivity of the heavily doped silicon substrate. L_s is the inductance of the gold whisker wire which is bonded to the metallized area over the junction and C_c is the case capacitance.

Two parameters limit the flow of energy to the barrier. Power can be dissipated in R_s and shorted across the barrier by C_j; however, this dispersion can be reduced by minimizing both parameters. The RF mismatch caused by L_s and C_c can be eliminated by external matching circuitry. However, it is important to minimize L_s both

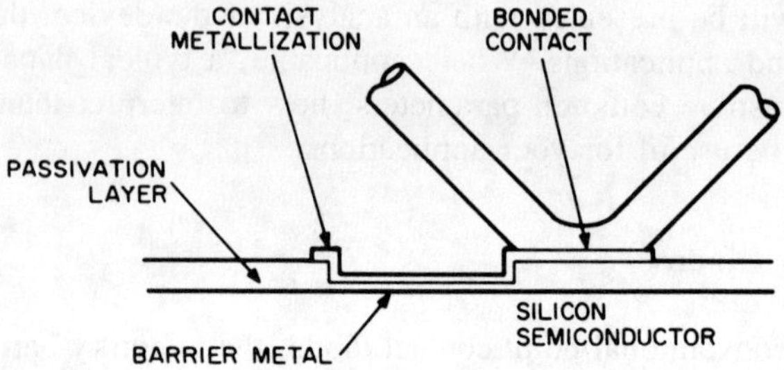

Fig. 5.17 Schottky-barrier diode.

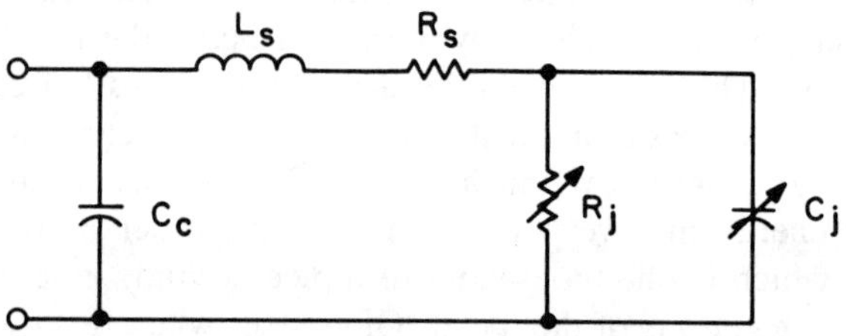

Fig. 5.18 Schottky-barrier diode electrical representation.

for its effect on operating bandwidth and on conversion loss. So when looking for a Schottky diode, the important things to remember are that R_s, C_j, and L_s should always be as low as possible. Typical values of diodes available today are R_s = 4–6Ω, C_j = 0.3–0.5 pF, and L_s = 0.4–0.9 nH.

Two types of packages for Schottky's, along with their internal construction, are shown in Figure 5.19.

Two common but very important applications for the Schottky diode are as mixers and detectors. To understand what part the diode plays in each, the following investigation is offered.

In general, the term mixer applies to a circuit that is used for converting an incoming high frequency signal to one having a lower frequency. This conversion is accomplished by means of a local oscillator (LO) signal and a non-linear resistive element. The LO signal and the incoming RF signal are coupled into the non-linear element, which becomes a time varying resistance under the influence of the LO. Under these conditions, the Schottky diode, insofar as it functions in the linear region, generates a difference frequency, called an intermediate frequency (IF), the amplitude of which is proportional to the RF signal amplitude and is independent of the amplitude of the LO signal. The non-linear element, referred to above as the most important portion of the mixer, is of course the Schottky-barrier diode. Different types of mixers are single-diode, balanced and double balanced, and image-rejection.

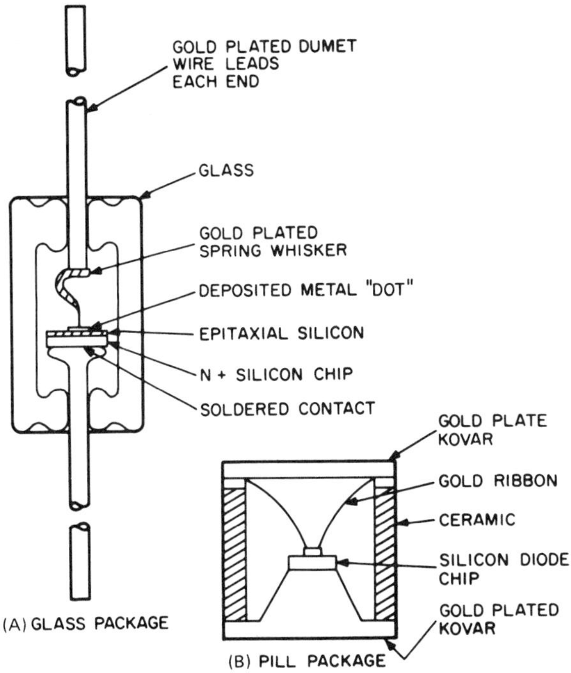

Fig. 5.19 Schottky diode package.

Detectors are essentially low-sensitivity receivers which function on the basis of direct rectification of the RF signal through the use of a non-linear resistive element—the Schottky diode. Generally, detectors can be classified into two distinct types: the small-signal type, also known as square-law detectors; and the large-signal type, or linear or peak detectors.

The former operation is dependent on the slope and curvature of the *I-V* characteristic (current-voltage) of the diode in the neighborhood of the bias point. The output of the detector is proportional to the power input to the diode, that is, the output voltage (or current) is proportional to the square of the input voltage (or current), hence the term *square law*.

The large-signal detector operation is dependent on the slope of the *I-V* characteristic in the linear portion, consequently the diode functions essentially as a switch. In large-signal detection, the diode conducts over a portion of the input cycle, and the output current of the diode follows the peaks of the input signal waveform with a linear relationship between the output current and the input voltage.

So one can see how the characteristics of the Schottky-barrier diode can be used to provide low conversion loss mixers and very efficient detectors. They are far superior to the old point contact diode and can be used in these applications through X-band.

5.3.2 *Pin* Diodes

The *pin* diode is distinguished from the normal *pn* junction type by an area called an intrinsic region, sandwiched between the p^+ doped and the n^+ doped silicon layers. (Doping is the term used to describe the addition of impurities to a semiconductor material to increase conduction). The intrinsic layer has almost no doping and is consequently a very high resistance. When forward bias is applied across the diode, the conductivity of the *i* layer is increased, creating an electrically variable resistor at microwave frequencies. As the bias current is increased the resistance of the diode decreases, making the *pin* diode ideal for use as a variable attenuator.

Figure 5.20 is the conventional *planar* construction used widely throughout the industry. Figure 5.21 is the *Plesa*™ construction used by Varian Associates.

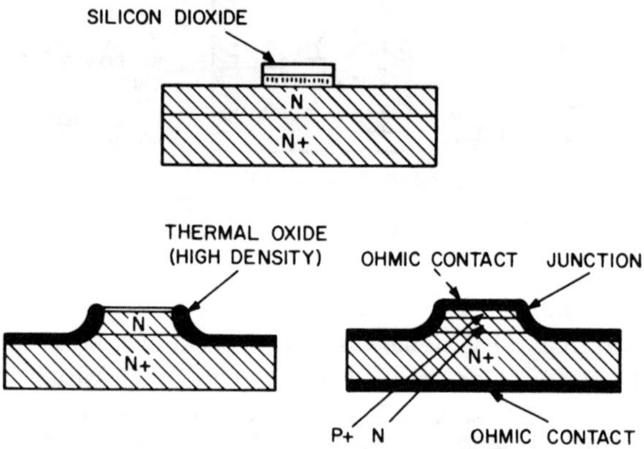

Fig. 5.20 Planar *pin* diode construction.

Planar processing is a well-established, high-yield method for producing reliable devices—especially for lower frequencies. Application of the process to an epitaxial *pn*-junction diode is shown in Figure 5.20. An *n* on the n^+ epitaxial silicon slice identical to that of the mesa process is selected, on the surface of which a very dense thermal oxide is grown. Circular holes, which serve as diffusion windows, are opened in the SiO_2 layer. The slice is then exposed to a boron dopant at high temperatures and diffusion occurs through the window openings, with the SiO_2 layer acting as a mask. The *pn*-junctions are formed in the bulk silicon under the dense layer of SiO_2.

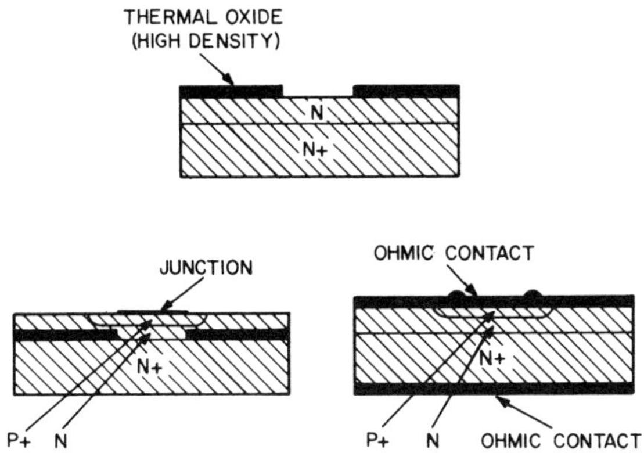

Fig. 5.21 *pin* diode Plesa™ construction.

The absence of exposure of the *pn*- junction to external contaminants is the major technical advantage of this process; thus, a highly uniform device with excellent leakage properties and inherently high reliability results. The dominant limitation occurs because the junction formed with the planar process is not a simple parallel plate junction capacitance, as in mesa construction. The diffusion through windows into bulk silicon forms a more complex near plane junction in the center with significant edge effects. Breakdown voltage that can be obtained in the planar structure is less than with the mesa equivalent. Abruptness of the diode, as described by the capacitance-voltage relationship, is also lower. Generally, planar microwave diodes are less suitable for high performance than mesa diodes, if breakdown voltage or tuning ratio are the primary criteria for selection. When used on *pin* diodes, the planar construction produces properties that are unique from mesa construction and, for certain applications, are very desirable.

Before going any further, a few details should be offered about the mesa method of fabrication, the type of *pn* junction and fabrication usually found in basic diode construction.

Normally, to make a microwave *pn*-junction diode, a slice of n on n^+ epitaxial silicon of the proper doping density and thickness is selected. Then boron is diffused to produce a *p*-type layer on the front surface. (The slice at this point is illustrated in Figure 5.22(A).) The next step is to metallize the front and back of the slice to produce low-resistance ohmic contacts to the silicon, as in Figure 5.22(B). Circular areas of the proper diameter are marked on the p^+ side, using photolithography, and mesas are etched on the slice, as seen in Figure 5.22(C). After mesa etching, one must scribe and break apart the individual devices, then mount them in packages. Now the constructions of the mesa and planar types of fabrication can be compared.

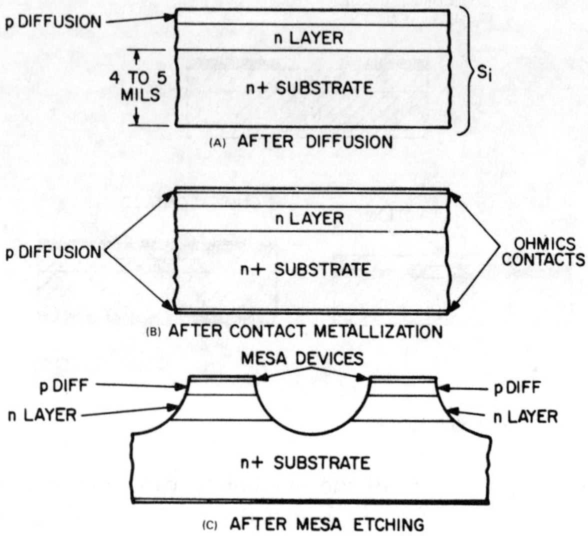

Fig. 5.22 Mesa fabrication.

The *Plesa*™ fabrication process is a hybrid of the classical mesa and planar processes, extracting the advantages from each. To produce a *pn*-junction epitaxial diode, an n on n^+ epitaxial silicon slice is selected as with mesa or planar technologies. Prior to any diffusions, silicon mesas are formed. Masks are formed for etching by first pyrolytically depositing a silicon nitride (Si_3N_4) layer and an SiO_2 layer on the n layer surface. Suitable etches result in an array of masking dots on the epitaxial silicon as shown in Figure 5.21. The mesas are then etched in the silicon slice and a very dense thermal oxide is grown on the sides. An important consideration is that the *pn* junctions have not yet been formed, allowing high temperature passivation without unwanted contaminants. At this point in the *Plesa*™ processing, the die shown in Figure 5.21(B) is passivated with a window prepared for the diffusion. This die is the same as the planar form illustrated in Figure 5.20(A), except for the geometry of the epitaxial silicon.

Conventional diffusions through the windows produce junctions under the dense passivation, as in the planar method. Because of the unique processing, the junction is in a mesa geometry, resulting in a near parallel plate capacitance and excellent electrical characteristics. Final metallization for ohmic contacts completes the devices.

Plesa™ devices have an integrity of passivation not achieved in any other type of mesa diode. The low-reverse-bias leakage currents, and exceptional leakage current stability under time-temperature stress, are evidence of a new level of reliability achieved with this process.

The parameters that characterize a *pin* diode on a data sheet are V_B, C_j (−50V), R_s (100 mA), τ_L, and θ_j. The voltage breakdown (V_B) is controlled by the width of the *i* layer and directly limits the RF voltage swing that may be applied to the device. C_j (−50V) is the junction capacitance in the fully depleted state. For most *pin* diodes, 50 V is enough to fully deplete the *i* layer. This capacitance is chosen to be a value compatible with the circuit in which the device is to be used. C_j (−50V) is controlled by the device area for a fixed voltage breakdown. R_s (100 mA) is the total RF resistance of the device (R_f) when 100 mA of forward current is flowing through it. R_s (100 mA) is measured at a frequency of 1.0 GHz, unless otherwise specified. At 100 mA of forward current, the resistance of the i layer (R_i) is usually very small. Therefore, R_s (100 mA) is a good indication of R_s, the minimum series resistance of the diode. Carrier lifetime (τ_L) influences the resistances of the intrinsic layer and the switching properties of the diode. Thermal resistance (θ_j) indicates the cw power-handling capability of a *pin* diode.

Because of the unique properties of the *pin* diode, it may be used in the following applications:

1. Voltage or current-controlled attenuators (discussed in Chapter Four).
2. Modulators.
3. Low and high-power switches.
4. Phase shifters.
5. Limiters (passive and active).

The basic small signal equivalent circuit of the *pin* diode as used in circuit analysis is shown in Figure 5.23(B). This circuit differs from that in (A) in that there is an inductance (L_s) in series with R_s and C_j, due to the ribbon or wire used to connect the *pin* chip (Figure 5.23(A)) to the package or circuit (stripline or microstrip). A parasitic capacitance (C_p) caused by the package in which the diode is mounted, shunts both the elements of the chip and the inductance. (This parasitic capacitance is not present in unpackaged chip diodes). Another parasitic capacitance, (C_f), found between the contact ribbon or wire and the chip mounting surface that shunts the active part of the *pin* chips, has been omitted from Figure 5.23(B) since its value is ≈0.02 pf. With the equivalent circuit in mind, the applications listed above for *pin* diodes are described.

The parameters (all listed above) of importance for determining RF performance of *pin* attenuators or switches are series resistance (R_S), junction capacitance (C_j), and carrier lifetime (τ_L).

Series resistance is measured utilizing a transmission loss technique at a frequency of 1 GHz with a junction capacitance at 1 MHz. The resistance of a *pin* diode is varied by applying a dc forward current, allowing RF energy passing through the device to be controlled. The capability of high levels of charge storage permits the application of high power without serious harmonic problems, as the device cannot follow the high frequency RF in equilibrium. As long as charge is stored in the

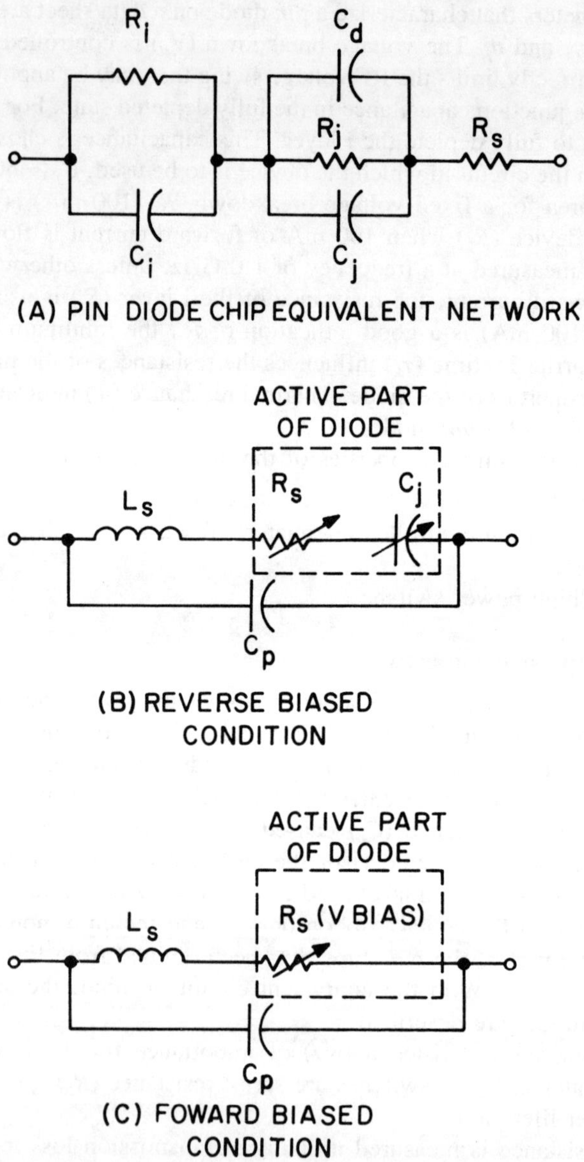

Fig. 5.23 *pin* diode equivalent circuits.

pin diode, an RF signal removes only a little of the charge, rather than change the conduction state of the diode. Values of series resistance can vary from 100Ω to 1Ω as forward current is increased from zero to approximately 100 mA. Upon examination of a plot comparing a *planar* and the *Varian Plesa*™ device, note that for the former device the series resistance changes almost linearly on a log-log plot as the forward current is varied. Since a *pin* diode biased in the reverse direction is a high impedance and, in the forward direction, is a low impedance, it is ideal for RF switching. The devices may be used in either series or shunt-type circuits, depending on the switching properties desired.

For a single diode, series mounting provides broadband performance because resistance in series with the transmission line changes at different bias levels. The upper limit of operation is governed by parasitic reactances of the chip mount. The insertion loss and isolation are determined by the lower and upper values of resistance at the switch drive levels. At high frequencies, the insertion loss remains constant, but the isolation decreases approximately 6 dB per octave. This decrease in isolation with increase in frequency occurs because the capacitive reactance decreases with increasing frequency.

For a single stage, shunt mounting provides more narrow band operation, since the parasitic elements are tuned out. Insertion loss is reduced and, since the device can be directly heat sinked, higher power operation is possible. To increase the bandwidth of operation for shunt-mounted devices, many properly spaced chips may be cascaded.

The junction capacitance is measured at 1 MHz and at a reverse voltage beyond the punchthrough voltage. The RF capacitance is independent of bias voltage if the frequency of operation is higher than the dielectric relaxation frequency (which is inversely proportional to intrinsic layer resistivity and is plotted in Figure 5.24).

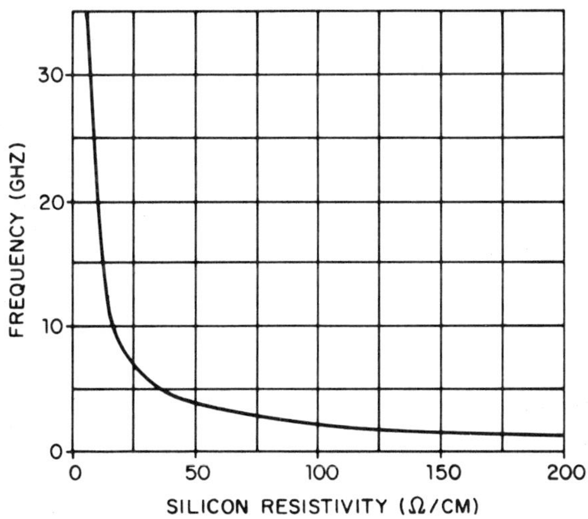

Fig. 5.24 Dielectric relaxation frequency *versus* resistivity.

Another important fact to consider is the difference in minority carrier lifetime between *planar* and *Plesa*™ *pin* devices. It has been found that *planar* devices have lifetimes that are 25 percent faster than *Plesa*™ devices with identical series resistance and junction capacitance. The differences are important in how they influence switching.

(A more detailed description of *pin* attenuators and switches is in Chapter Four under *Attenuators*.)

If a dc current is applied to the attenuator mentioned above with a 1 kHz signal superimposed on the current, the attenuation varies from some minimum to a maximum value. If the RF signal is detected, a waveform similar to the low-frequency (1 kHz) modulation signal can be observed. Any discrepancy in these waveforms is a measure of distortion, caused by two major factors. First, if the modulation frequency is too high, the diode cannot follow the signal because of stored charges. Secondly, the attenuator is not linear over all bias ranges; hence a range must be chosen where linearity is present. This information, of course, leads to the use of the *pin* diode as a *modulator*.

Pin diodes are frequently used as high-power digital phase shifters in phased-array radars, of which three major types are employed: *switched-line, periodically-loaded-line,* and *hybrid-coupled-bit*.

The switched-line phase shifter (Figure 5.25(A)) consists of two single-pole double-throw switches, and two line lengths per bit. RF energy applied to the first SPDT switch can be diverted either through the reference arm or through the first phase-shift bit. A minimum of four diodes per bit, regardless of bit size, must be used; both the number and size of the bits may vary according to the application. By adding more bits, the only parameter change that occurs, other than the increase in phase shift, is the increase in insertion loss.

The periodically-loaded-line phase shifter uses switched loading susceptances spaced approximately $\lambda/4$ apart along a transmission line (see Figure 5.25(B)). Adjacent $\lambda/4$-spaced loading susceptances are equal and take on either of two impedance states. The number of pairs in a given state determines the transmission phase shift. If the normalized loading susceptances are small, then matched transmission results for either loading state. The largest reasonable phase shift per diode pair for this type of phase shifter is 22.5°; hence, 32 diodes are required to shift the phase 360° in 22.5° steps.

The hybrid-coupled phase shifter is shown in Figure 5.25(C); it includes a 3 dB hybrid junction with balanced phase bits connected to coupled arms. For each bit at least two diodes are required which, as in other phase shifters, are used as switches.

To compare the three phase shifters mentioned above, let us assume the construction of a 360° phase shifter. The minimum number of diodes needed in the switched-line phase shifter is 16, the periodically-loaded-line type requires 32 diodes, and the hybrid-coupled-bit requires eight. Regarding peak-power-handling capability

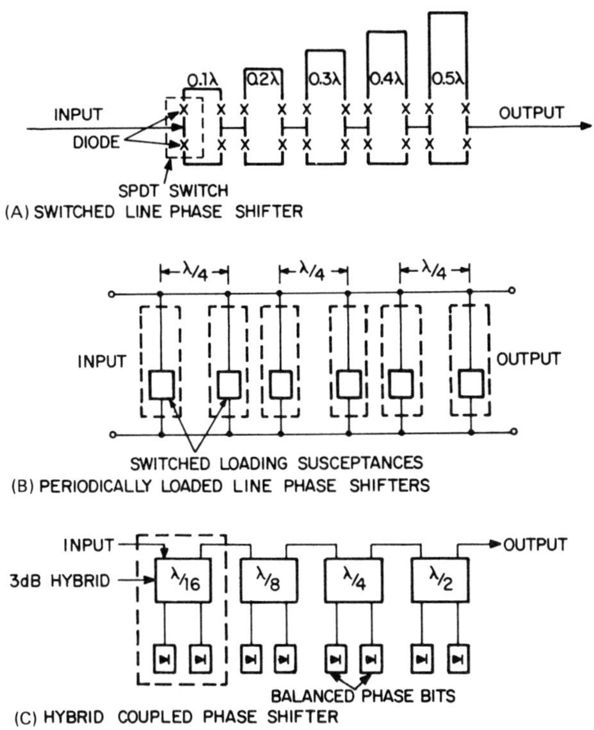

Fig. 5.25 *pin* diode phase shifters.

when equipped with diodes having equal breakdown voltages, the hybrid-coupled unit has twice the capability of the loaded-line version, while the switched-line model has twice that of the hybrid-coupled. The hybrid-coupled phase shifter has the least loss; the switched-line type has the greatest. In the switched-line unit, the loss is constant; with the other two types, it varies with the amount of phase shift required.

Pin diodes can also be used as active or passive limiters. The active limiter requires a bias generated from an external source, such as the rectified current generated from a point-contact detector diode used to sample the RF power in the system. This current is sufficient to drive the device into forward conduction and, in the shunt case, to provide a low impedance to the incident power. A passive limiter operates at zero volts and is activated only by incident RF power.

The varactor-diode limiter is essentially a power-sensitive attenuator. It uses the nonlinear capacitance properties of a varactor or *pin* (thick and thin base) diode to detune a filter-type structure when large amounts of RF power are applied to the device. It also reduces the incident power to levels that do not paralyze the receiver nor burn out the mixer. Even with varying input power, the output power remains constant over a wide range. The limiter must dispose of excess power by either reflecting it, diverting it to another load, or absorbing it. However, diodes are not designed to absorb high power.

The *pin* diode is a very versatile component that has many uses, as shown above, and can be summarized in one statement—it is a voltage (or current) controlled variable microwave resistor that can be either varied gradually or switched rapidly.

5.3.3 Varactor Diodes

Considerable controversy exists as to why a varactor is capable of frequency multiplication. One hears such terms as step recovery, abrupt-junction, punch-through and the like. The difference in these terms is simply the behavior of the junction capacitance (C_j) as a function of bias voltage.

Early multiplier theory assumed that the diode was never driven toward forward conduction; that is, the multiplication process was due entirely to the non-linear reactance of the diode in the reverse-bias region. Circuit design theories advocated by several of the early workers in the field proved to be quite accurate as long as the diode was not overdriven.

The perpetual quest for higher power and higher efficiency led some designers into breaking the rules with considerable success. By intentionally overdriving a diode, much higher powers could be obtained along with increased efficiency. Consequently, design rules fell by the wayside.

The theorist came to the rescue with the stored-charge concept and, almost immediately, two schools of thought developed. The step-recovery theory states that as a diode is driven toward forward conduction by the first half-cycle of the input signal, it stores a very large charge and appears as a very low impedance. As the input signal reverses polarity, this charge is extracted and the diode remains in the low impedance state until all of the charge is removed. At this time, the diode instantaneously snaps to a very high impedance state, an action which generates a spike of power thought to be the source of harmonic energy. Little, if any acknowledgement was given to the nonlinear reactance theory, based on reverse-bias junction capacity alone. Diodes tailored to enhance this characteristic are called step-recovery diodes (SRD) or snap varactors. The available circuit design information based upon the step-recovery theory is quite complete and comprehensive. Multipliers designed and built to this set of rules perform very well and seem to confirm the theory.

The punch-through theory of charge-storage multipliers, however, does not accept the fact that the charge-recovery step is the fundamental mechanism by which harmonics are generated. It is, in fact, argued that should such a fast step exist in any tuned multiplier circuit, the very presence of any inductance would drive the diode into avalanche breakdown. The multiplication process must, therefore, be due to some other phenomenon. We now return to a modified form of the nonlinear reactance theory of old. The type of varactor now considered exhibits an abrupt change in its capacity *versus* bias characteristics. As a direct result of this relationship, the diode's stored-charge capability *versus* drive level also exhibits an abruptness. The bias voltage at which this transition in charge-storage capability takes place is termed the punch-through voltage, thus a diode which displays this characteristic is called a *punch-through varactor* (*PTV*). The above is essentially the same phenomenon which supposedly makes the SRD multiplier work. *Bimode*® and *Super Bimode*® diodes also fall into this general classification of charge-storage diodes. The major differences between any of these various diode types lie in the method by which the manufacturer chooses to measure and specify his device. How the circuit designer applies the device to his frequency multiplication problem is governed almost entirely by which theory provides the best understanding or greatest success.

Although all these devices seem to be generally the same, they are usually specified differently. The SRD, or snap varactor, is specified in the time domain in terms of its pulse-shaping and switching characteristics. The PTV is specified in the frequency domain in terms of capacity *versus* bias (and therefore, reactance *versus* frequency) and diode Q, or cut-off frequency. Bimode diodes, leaving nothing to chance, specify all of these parameters.

The active element of a varactor diode consists of a semiconductor wafer containing a junction of well-defined geometry, usually formed by diffusion. Figure 5.26 shows a cutaway view of a typical varactor wafer and the equivalent circuit; the latter includes the following:

C_j—Junction capacitance, which is a function of the applied voltage.

R_j—Junction resistance, which is in shunt with C_j and is also a function of bias.

R_s—Series resistance, which may be a function of bias, including the resistance of the semiconductor on either side of the junction through which the conduction current passes and the resistance of the ohmic electrical contacts to the wafer.

Varactor diodes are normally operated under reverse bias, where the junction resistance, which is ordinarily 10 MΩ or more, is negligible in comparison with the microwave capacitive reactance of the junction. Therefore, the equivalent circuit of the reverse-biased varactor wafer at microwave frequencies is simply a capacitance and resistance in series. The equivalent circuit of a forward-biased varactor at microwave frequencies is generally more complicated, since it must include the diffusion of the injected carriers, as well as the effect of these carriers on the conductance of the semiconductor material.

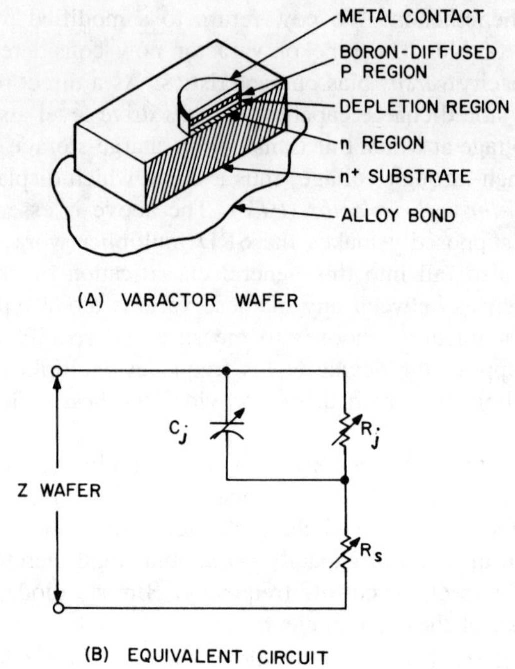

Fig. 5.26 Varactor diode.

At low frequencies, the varactor exhibits the conventional forward and reverse characteristics of an ordinary diode. For forward bias, the diode current increases exponentially with the applied voltage, and for reverse bias, a small saturation current (I_s) flows. When the reverse bias is increased to the avalanche breakdown voltage (V_B), the diode reverse current increases very rapidly, since it is limited only by the small diode resistance and any external resistance that is present in the circuit. Figure 5.27 shows typical dc *I-V* characteristics, microwave series resistance, and 1 MHz *C-V* characteristics for the harmonic generator diode described in Figure 5.26.

Although the variation in junction capacitance is the most important characteristic of the varactor diode, there are parasitic resistances, capacitances, and inductances associated with any practical encapsulated varactor. Part of the parasitic resistance is an unavoidable result of the variation in the junction capacitance, as shown later. All the other parasitics are consequences of the method of contacting and encapsulating the varactor diode. However, all varactor encapsulations have very similar microwave equivalent circuits.

The possible applications for varactors include pulse shaping and broadband spectrum generation (both for which the charge-storage type is best suited), in addition to their better known uses as frequency multipliers, parametric amplifiers, and

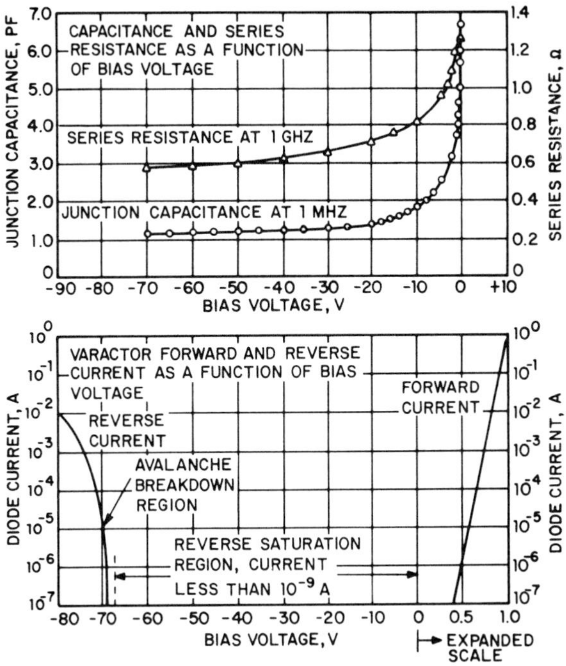

Fig. 5.27 Varactor characteristics.

tuning elements. Figure 5.28 shows some basic pulse shaping circuits which employ the diodes as charge-controlled switches. The fixed forward bias inserts a charge into the diode, making it appear as a low impedance; it remains at that level until all of the charge has been removed by the input signal, at which time it snaps to a high-impedance state. The diode stays at this value until the input signal falls back to zero, at which time the bias re-establishes the charge on the diode for the next cycle. The trailing-edge-sharpener circuit functions in exactly the same manner. These two basic circuits may then be combined into a pulse-squaring circuit as shown, where the last circuit is a cascaded version of the first. This type of multiple-stage operation is sometimes required when a very slow rise-time signal is to be converted to a very fast rise-time (or fall-time) pulse. These basic circuits are only a few of the many possible ways in which charge-storage diodes may be used in waveform shaping.

The pulse-shaping circuits of Figure 5.28 indicate a resistive generator and load. A modification to this circuit is shown in Figure 5.29. The effect of the reactive components on diode operation may be seen in the time domain waveform. The diode now acts as a fast voltage spike generator instead of a fast voltage step generator. This spike gives rise to the spectrum shown in the frequency domain plot. The interval between the teeth of the comb is equal to the frequency of the drive signal. The relative flatness of the spectrum is a function of the pulse width. This

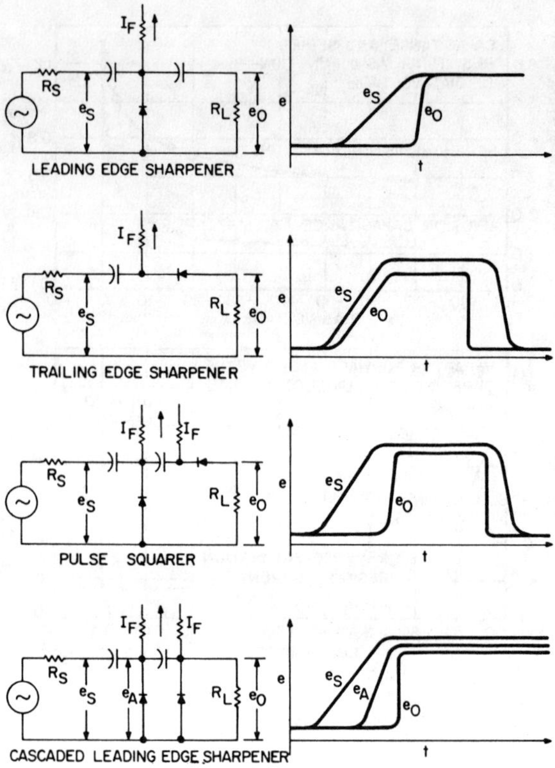

Fig. 5.28 Pulse shaping circuits.

class of frequency multiplier finds application anywhere that "a family of frequencies" is required—in test equipment calibration, multiband transmitters or receivers, phase-lock systems, or birdy markers in swept-frequency test setups.

We now enter the muddy region of multiplier diode applications; that of high-efficiency, single-output frequency multipliers. If one rigidly adheres to the step-recovery theory of multipliers, then all that is necessary to convert a comb generator into a single-frequency multiplier is to add the output network and filter as shown in Figure 5.30. The output matching network serves two functions—a transformer which matches the diode impulse generator to the output filter, and a low-Q resonator which converts the impulse into a damped *sinusoid*. The output filter requirements for this type of multiplier are no different than for any other. The Q and skirt-selectivity requirements are determined from the allowable insertion loss and required adjacent sideband rejection, and may be realized in any of the conventional forms—coaxial, lumped element, stripline, or waveguide.

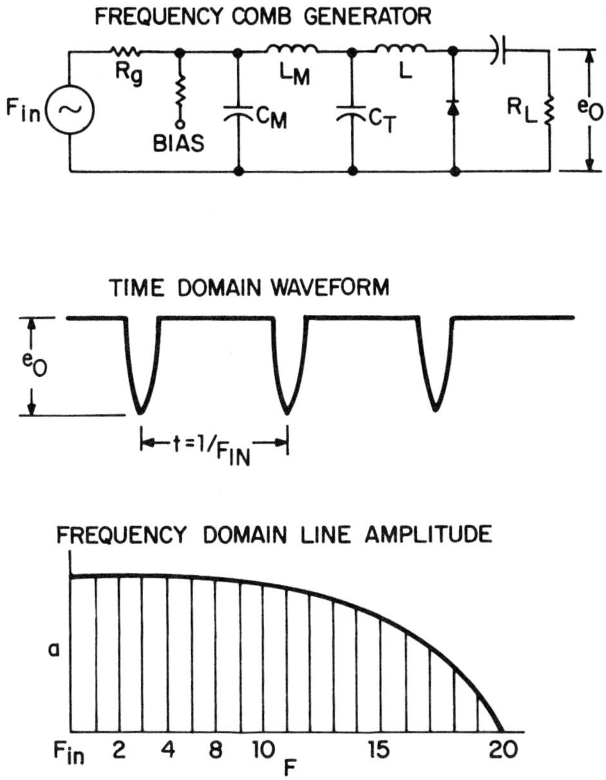

Fig. 5.29 Comb generator.

The original versions of SRD theory stated that idlers were not needed for high efficiency; in fact, idlers were thought to be detrimental to the multiplier's performance. These beliefs were in direct opposition to the idler concepts employed by designers building multipliers based on the nonlinear reactance theory.

In selecting the proper device for a multiplier circuit, one must, from RF characteristics (input and output frequency, input and output power), determine the device characteristics—breakdown voltage (V_b), minimum junction capacitance (C_j), lifetime (τ_L), transition time (τ_T), and thermal resistance (θ_{Th}).

Typical values for these specifications for a $\times 5$ multiple with an output of 6 GHz are:

$V_b = 50$ V
$C_j = 2$ pF
$\tau_L = 50\text{–}150$ ns

$\tau_T = <200$ ps
$\theta_{Th} = 30°$ C/W

Table 5.2 is a comparison of diodes listing their applications, advantages, and limitations. It can be a very useful table when building a multiplier or evaluating one that someone else has built. With this information one can check to see if a manufacturer has used the proper type of diode for the job to get the maximum output at the maximum efficiency.

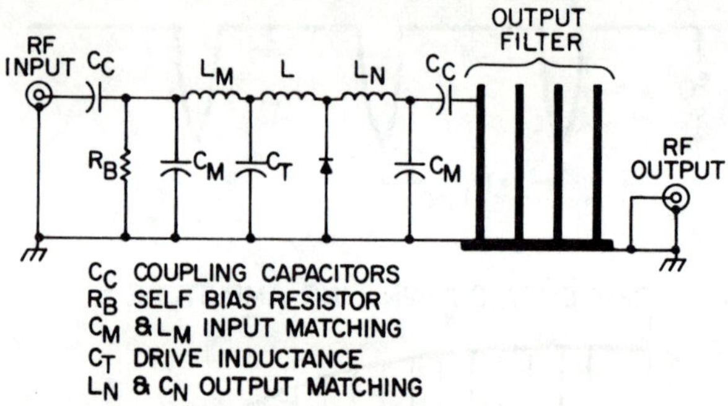

Fig. 5.30 SRD frequency multiplier.

5.3.4 Tunnel Diodes

The tunnel diode is a *pn*-junction device with a very high doping on both sides of the junction. As defined earlier, doping is the term used to describe the addition of impurities to a semiconductor material to increase conduction. The tunnel diode differs from the *pn*-junction diode primarily because of its high doping level. The impurity level in microwave *pn*-junction diodes is in the range of 10^{16} to 10^{18} atoms/cubic centimeters, while values of 10^{19} atoms/cubic centimeters are quite common for the tunnel diode. In electrical specifications, the tunnel diode is highly conductive near zero bias, whereas the *pn*-junction signal diode does not start conducting until it is forward biased to 0.6 V.

Figure 5.31 shows the two curves which characterize tunnel diodes. On top is the *I-V* curve (current *versus* voltage); on the bottom is the *R-V* (resistance *versus* voltage). Two important points concerning these curves should be defined. The first is R_M (minimum negative resistance), which is the terminal minimum negative resistance that occurs at the inflection point on the *I-V* curve, as can be seen in Figure 5.31. Measurement of this value is made at a low frequency so the reactive components present can be neglected. The second point is R_N (negative resistance at a minimum *K*). The value *K* (noise constant) is a term dependent on temperature, diode

Table 5.2 A Comparison of Diodes

Diode Classification	*Applications*	*Advantages*	*Limitations*
Bimodes and Super-Bimodes	Low-order multipliers Fast pulse-shaping Comb generators	High efficiency at any power level with very high power handling capability	Usually requires tuned idlers for maximum efficiency and power handling; useful only thru X-band
Step Recovery Diodes SRD	High-order multipliers Fast pulse-shaping Comb generators	Efficient multiplication at low and medium power levels. Does not require tuned idlers	Not suited for very high power operation. Useful thru Ku band
Gallium Arsenide Varactors Multiplier Types	Low-order multipliers to millimeter frequencies	Efficient multipliers at low- and medium-power levels at X-band and above	Requires tuned idlers for maximum efficiency and power handling
Gallium Arsenide Varactors Parametric Amplitude Types	Low-order, low-power multipliers for X-band and higher	Efficient low-power multipliers for millimeter frequencies	Same as above

Definitions—Low order is ×4 or less
High order is greater than ×4
Low power is less than 200 mW input
Medium power is between 200 mW and 2 W
High power is greater than 2 W

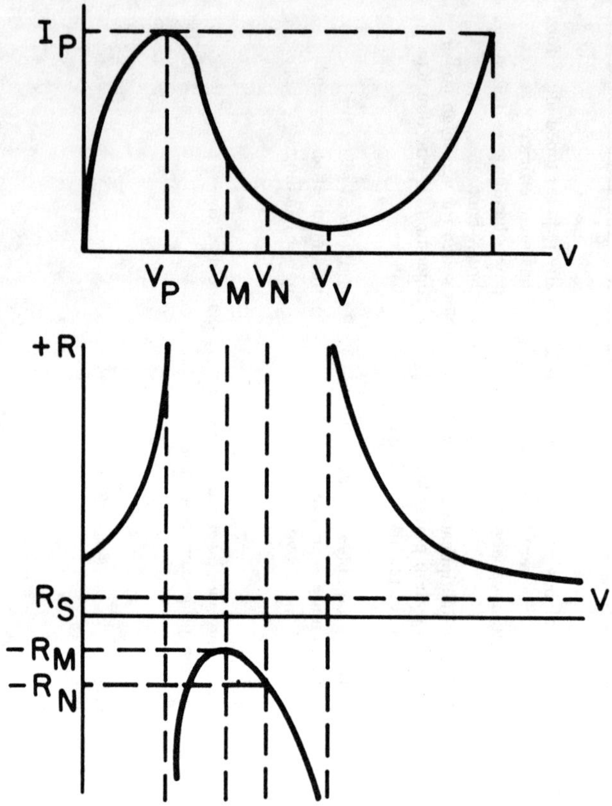

Fig. 5.31 Tunnel diode curves.

current, junction resistance and diode voltage at which the measurement is made. The normal operating point for tunnel diode amplifiers is in the region of R_N and is defined as the value of negative resistance at minimum K. The noise constant does not vary rapidly with junction resistance in this region, so the exact operating point is usually determined by linearity and gain requirements. Values of R_M vary from 35 to 70Ω depending on the value of I_p used. Values of R_N can range from 40Ω to the neighborhood of 120Ω, once again, depending on I_p.

Construction of the tunnel diode, a hermetically-sealed pill type package, is shown in Figure 5.32. It offers high heat transfer efficiency (so the diode can handle higher power), ruggedness, low capacitance, and extremely low inductance. Its rugged construction reduces the possibility of failures due to shock, vibration, and temperature cycling. Figure 5.33(A) is an equivalent circuit of a tunnel diode in this

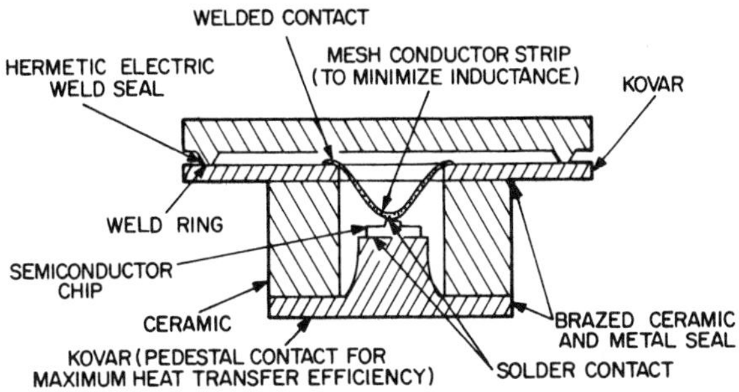

Fig. 5.32 Tunnel diode construction.

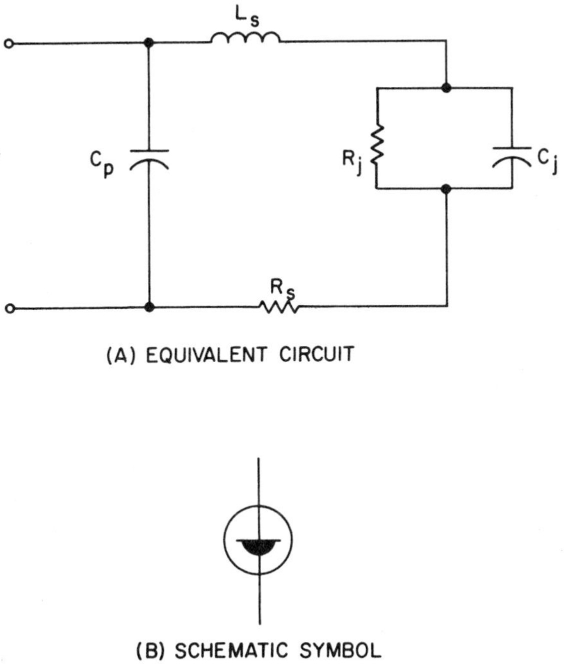

Fig. 5.33 Tunnel diode equivalent circuit and schematic symbol.

package, with Figure 5.33(B) being a schematic symbol. The parameters shown and listed below are important to the diode operation and are dependent on many factors.

Package Capacitance (C_p)—The package capacity is a function of the package type. Specified values are average values for the specific type of package. In high-frequency diodes, the junction capacity is often much less than the package capacity.

Series Inductance (L_s)—The effect of series inductance is somewhat dependent upon the diode mounting in the microwave circuit. The values specified are average values for diodes mounted in series with center conductor of a coaxial transmission line.

Junction resistance (R_j)—The junction resistance is the dynamic (ac) resistance of the diode junction. It varies from almost zero where the diode is heavily back-biased, to infinity at the peak current point, then to a minimum negative value and back to infinity at the valley current point. It is also approximately zero when the diode is biased heavily in the forward direction.

Series resistance (R_s)—The series resistance is a combination of contact resistance and spreading resistance, determined by pulsing the diode far into the forward region (where the junction resistance is very low), then measuring the incremental resistance with a small sampling pulse on top of the biasing pulse.

Junction capacity (C_j)—The junction capacity is determined by measuring the total capacity with the diode biased at the valley voltage, then subtracting the package capacity. The operating point junction capacity is defined to be 75 percent of the valley point junction capacity. (The valley point is designated as V_r in Figure 5.31.)

There is one term that is dependent on all of the above parameters except C_p and L_s—the *Resistive cut-off frequency* (f_{ro}). This frequency is where the resistive part of the parallel combination of C_i and $-R_j$ (Figure 5.33) equals R_s; the diode no longer displays a negative resistance; f_{ro} is specified at R_M (Figure 5.31) where $|R_j| = |R_M| + |R_s|$.

$$f_{ro} = \frac{1}{2\pi |R_j| C_j} \sqrt{\frac{R_i}{R_s} - 1}$$

Microwave tunnel diode amplifiers are light-weight, inexpensive, ultra-low noise components that are useful in the front end of receivers to improve the sensitivity of communications systems, satellites, radar, and countermeasures systems. The retrofit of tunnel diode amplifiers into existing receivers provides performance improvements which might otherwise be obtained only through expensive and complex modifications to the transmitter. Because of the wide bandwidth obtainable, tunnel diode amplifiers provide the low signal distortion required in radar astronomy and communications systems.

A design guide for adequate amplifier gain and low noise is

$$f_{ro} \approx 4 f_o$$

where:

f_{ro} = resistive cut-off frequency of the tunnel diode;
f_o = amplifier operating frequency

Since tunnel diodes in tunnel diode amplifiers are operated in the negative resistance region, adequate consideration must be given to the prevention of instabilities. Switching may occur unless the biasing circuit is well regulated and is of low impedance. Likewise, oscillations may occur if off-band frequencies are not terminated properly, or if the self-resonant frequency is less than the diode resistive cutoff frequency. It is also important to properly isolate the incident from the amplified wave, an act which may be accomplished with the use of ferrite circulators or hybrid combiners, in configurations such as those shown in Figures 5.34 (A) and (B).

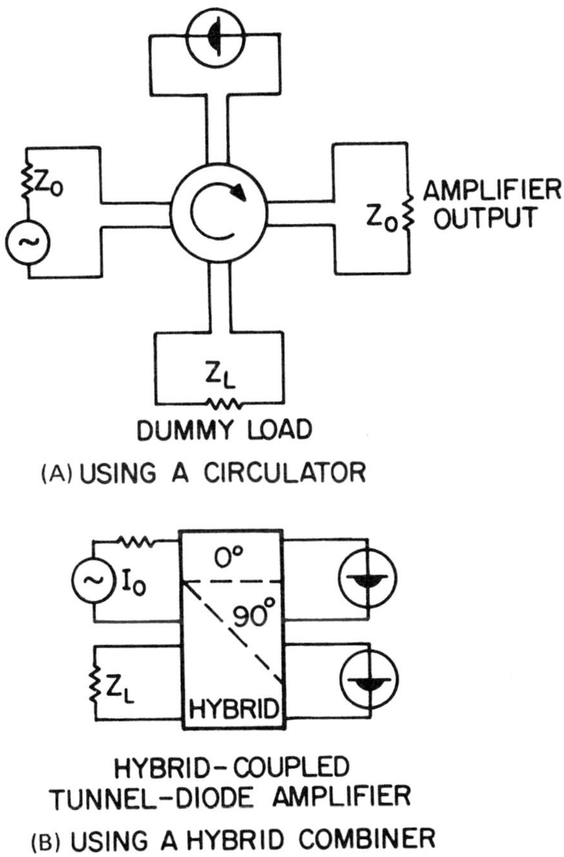

Fig. 5.34 Tunnel diode amplifier.

The materials used in the tunnel diodes of amplifiers are of great importance. The substances can be gallium antimonide, germanium, or gallium arsenide; they have much to do with the amplifier parameters, such as the dynamic range covered below.

The dynamic range of an amplifier is the input signal variation over which the amplifier functions effectively. The lower limit is determined by noise figure; the upper limit by saturation. To compare the effect of semiconductor materials on dynamic range, consider gallium antimonide, germanium, and gallium arsenide diodes with the same values of R_M. The peak current values for equal R_M are in the respective ratios of 1 to 1.5 to 2.5.

The level at which saturation occurs is roughly proportional to 10 log I_p^2. Thus, for a 2.5 to 1 variation of I_p, the saturation level changes by about 8 dB. Fortunately, this variation is considerably larger than are the noise figure variations for each material; a net result of about 5 dB in dynamic range is effected by using a gallium arsenide diode rather than one of gallium antimonide.

It is possible to have increased dynamic range and a low noise figure by constructing dual-stage amplifiers, using a gallium antimonide unit in the first stage, and a gallium arsenide device in the second.

Since microwave tunnel diodes are free from transit time effects, they are a logical choice for use as high-frequency microwave oscillators. Such devices tend to be simple, compact, rugged, and reliable. Gallium arsenide offers the best choice for oscillator applications because of a wide dynamic range in the *I-V* characteristic. As evidenced in Figure 5.35, GaAs has about nine percent greater current swing, and twice the voltage range obtainable with germanium diodes. The power output from a tunnel diode microwave oscillator may be calculated from

$$P_{out} \approx 3/16\ \Delta I\ \Delta V\left(1 - \frac{f_o^2}{f_{ro}^2}\right)$$

where:

$\Delta I = I_p - I_v$;
$\Delta V = V_v - V_p$;
f_o = operating frequency;
f_{ro} = resistive cut-off frequency.

Below self-resonance, the diode reactance is capacitive; therefore, in the lower frequency range, the load at the desired frequency must be inductive.

Conversely, above self-resonance, the diode reactance is inductive; therefore, the load at the desired frequency must be capacitive. Thus, the tunnel diode can be considered as the device which changes all of the preconceptions about diodes. Those who believe the sole purpose of a diode to be rectification of RF energy are very surprised upon discovering the extended uses of the *tunnel diode*.

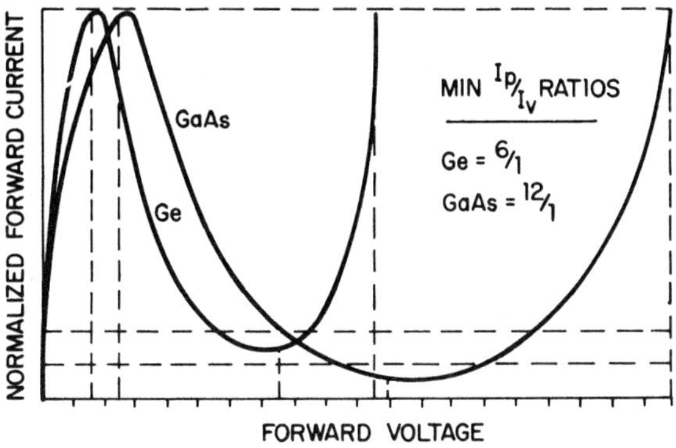

Fig. 5.35 Tunnel diode forward characteristics.

5.3.5 Gunn Diodes

In the early 1980s it was said that as long as the properties of semiconductors depended on junctions, and that those junctions must be made thinner as the frequencies increased, high powers using semiconductor devices at microwave frequencies would not be possible.

Devices have been developed however, which do not depend on a junction for operation. One such class of devices exhibits microwave power properties that depend on the behavior of *bulk* semiconductors, rather than junctions. The *gunn effect* is the main representative of this group, and the *gunn diode* the main device. The term *diode* may be a confusing term here, since there is no junction, and no rectification is involved. The device is called a diode only because it has two terminals.

In 1963, J.B. Gunn discovered the *transferred electron effect,* now known as the Gunn effect. The effect was found to be exhibited by gallium arsenide (GaAs) and indium phosphide (InP). Subsequently, cadmium telluride (CdTe) and indium arsenide (InAs) were also found to possess this property. If a relatively small dc voltage is placed across a thin slice of GaAs, a negative resistance will be present under certain conditions. The one primary condition is that the voltage gradient across the GaAs slice exceed 3300V/cm. A GaAs Gunn slice is shown in Figure 5.36.

As previously stated, the Gunn effect is a *bulk* property of semiconductors and does not depend on either junction or contact properties for its operation. The effect is independent of total voltage or current and is not affected by magnetic fields. Furthermore, it occurs in *n*-type material only, so that it must be associated with electrons rather than holes. The frequency of oscillation produced corresponds closely to the time electrons take to cross such a slice of *n*-type material (such as in Figure 5.36) as a result of the voltage applied. This suggests that a bunch of electrons,

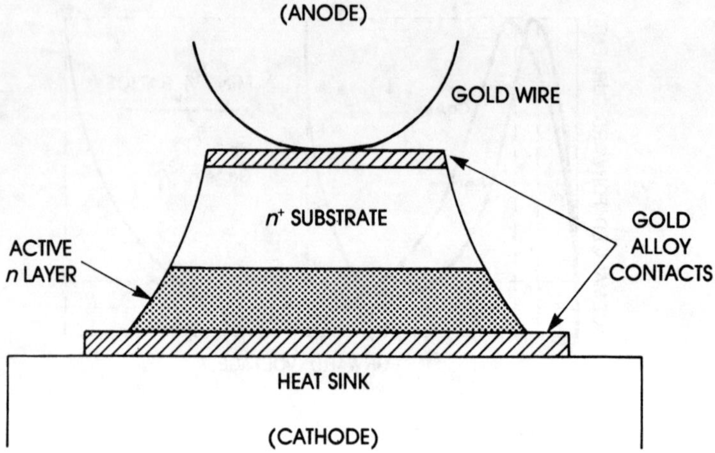

Fig. 5.36 GaAs Gunn slice.

called a *domain,* is formed somehow, once per cycle, and arrives at the positive end of the slice to excite oscillations in the associated tuned circuit.

The idea of how this device obtains its negative-resistance characteristics requires further investigation. Gallium arsenide is one of a fairly small number of semiconductor materials which, in an n-doped sample, has an empty energy band higher in energy than the highest filled band. Also, the size of the forbidden energy gap between these two is relatively small. When a voltage is applied across a GaAs slice which is n-type, these electrons flow as current toward the positive end of the slice. The greater the potential across the slice, the higher the velocity with which electrons move toward the positive end and, thus, the greater the current. Thus far, this appears much the same as any positive resistance device. In this case however, there is so much energy imparted to the electrons by the extremely high voltage gradient (positive resistance devices will never have a gradient anywhere near this high), that instead of traveling faster and increasing the current, they actually slow down. This is because the electrons have acquired enough energy to be transferred to the higher band, which is normally empty. This process produces an effect called the *transferred electron effect*. Electrons have thus been transferred from the conduction band to a higher energy band in which they are much less mobile. Thus, the current has been reduced as a result of a voltage rise. This phenomena, a classic representation of negative resistance, can be seen in the *I-V* curve for a typical Gunn diode in Figure 5.37. It can be seen from the figure that as the voltage rises past what is termed the *threshold negative resistance value,* the current also rises to this point, and then falls. If the bias voltage was to be increased further this phenomenon would cease, because eventually the voltage would be sufficient to remove the electron from the higher-energy low-mobility band, and the current would once again

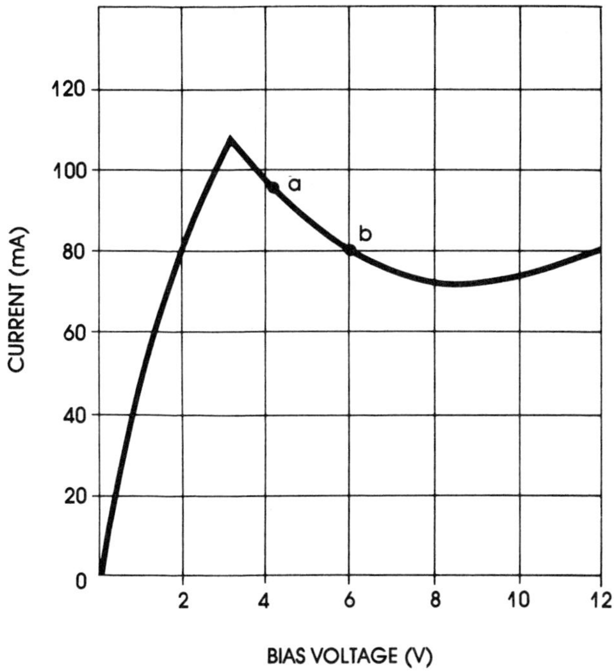

Fig. 5.37 *I-V* curve for a Gunn diode.

increase. For this reason, the ideal area to operate in is the linear region just past the threshold value (the area a to b in Figure 5.37). When the Gunn diode is biased in this negative-resistance region, the field distribution within the sample rearranges itself to form a thin, high-field domain, with the rest of the sample experiencing a sub threshold field. The domain drifts through the sample at the saturated drift velocity and collapses when it reaches the anode. The bias current then momentarily increases until a new domain (or bunch of current carriers) forms at the cathode. This periodic current increase results in high frequency power being made available to the external circuit. The output frequency, thus, is determined by transit time effect (the time it takes for one bunch of current carriers to get from the cathode to the anode), but, in practice, to vary the frequency over about one octave is possible by tuning the external resonant circuit (because of modulation of the bias voltage by the RF voltage developed in the resonant circuit, which can cause the instantaneous diode voltage to fall below the diode threshold). Formation of subsequent domains is then delayed, giving rise to the name *delayed domain* for the mode of operation.

Limited space-charge accumulation (LSA) operation in Gunn diodes is obtained by tuning the diode in a high-impedance circuit at a frequency such that a full domain cannot form during the period of an oscillation. The RF swing necessary is such that the pulsed operation is normally used. The principal advantage of this mode is that, since the material is displaying bulk-negative resistance, the dimensions of the diode are no longer limited by transit-time considerations, meaning that millimeter-wave frequencies can be achieved without the use of impossibly small active regions. A disadvantage of LSA operation is that, if domains do start to form, the RF voltages involved are so high that the domain fields may cause an avalanche breakdown, leading to the destruction of the device. Since domains can be created by even the slightest disruption in the consistency of the material, the requirements of the material for LSA operations are extremely strict. Also, it is possible to use Gunn diodes in other modes (quenched domain, multiple domain, and hybrid, for example), but these suffer from some, or all, of the disadvantages of being little understood, being difficult to control, or requiring diodes of special design.

Certain precautions should be observed when operating Gunn diodes. They should be operated from a constant-voltage power supply. The required operating voltage range is determined by the threshold field, power dissipation, and breakdown considerations. The data sheets should be very carefully checked for maximum voltage ratings. Although in some circumstances it is possible to obtain a greater output by operating at higher voltages (pulsed conditions), no guarantee of reliability is given. The quoted ratings are maximum values, and the power supply should be free of any overvoltage transients (which are most likely to occur during switch on). Bias polarity must, of course, also be heeded very carefully. Since the frequency-pushing characteristics of these devices can be as high as 30 MHz/V, noise and ripple on the power supply can produce a significant amount of FM noise on the oscillator output. A well stabilized power supply is, therefore, necessary to reduce frequency-noise modulation.

Bias-circuit oscillations may occur in certain circumstances due to a slightly negative resistance in the diode dc characteristics just above threshold, and they are found generally in the range of one to 100 MHz. Although they are not damaging to the diode, they can result in a poor output spectrum. These oscillations can be suppressed by connecting a 0.01 μF to a 1.0 μF capacitor across the oscillator bias terminals.

The InP (indium phosphide) Gunn device is developed specifically for low-noise amplifiers from 26 GHz to 100 GHz, and for medium-power low-noise oscillators from 40 GHz to 140 GHz. Typical oscillator powers so far obtained are shown here.

Indium phosphide (InP) has advantages over gallium arsenide (GaAs) for high-frequency millimeter-wave oscillator applications. Two of these advantages are

- Higher peak-to-valley ratio,
- Reduced scattering time due to a high-threshold field.

Peak-to-valley ratio is illustrated in Figure 5.38. Figure 5.38 is a plot of particle velocity *versus* electric field for *n*-type GaAs and InP. InP has a peak (maximum)-

Table 5.3 Oscillator Powers

Frequency (GHz)	*Power Output (mW)*
56	200
90	100
95	68
100	44
55 (pulsed 10% duty)	860

to-valley (minimum) ratio of approximately three, and the GaAs shows less than 2.4. This ratio must be high because the basic efficiency of a transferred-electron oscillator (Gunn device) is strongly influenced by this ratio of the material used. This InP ratio has been shown to be higher than GaAs; but, even better is the fact that when the materials are heated, the ratio in GaAs is significantly reduced, and the InP reduces both peak-and-valley velocities nearly the same amounts, preserving the high resultant ratio.

The scattering process defines the movement of charges (or particles) from one point to another. Each of the scattering processes within the Gunn device takes a certain amount of time to move between these points. (The points referred to are termed the *central valley,* which is the main valley of the curves, and the *satellite valley,* which is a secondary valley.) Times considered are those required to move from the central valley to the satellite valley, and the reverse time. These times have been shown to be a factor of two smaller in InP than in GaAs. Thus, if we have a GaAs Gunn device operating at 90 GHz, we can expect similar performance of InP at 180 GHz, which is a sizeable improvement in overall performance.

A third characteristic that makes InP more desirable at millimeter frequencies should also be mentioned. That characteristic is the higher impedance on InP. The design of millimeter-wave oscillators is significantly simplified when a large-device negative resistance and small reactance is present, which is the case for InP Gunn devices. The obvious higher-peak velocity of InP (shown in Figure 5.38) permits an increased active layer length, maintaining a transit-time relationship with a low bias-to-threshold ratio.

It should also be pointed out that the wideband negative resistance characteristics of an InP Gunn device, coupled with its low-noise properties, make the device an excellent choice as a reflection amplifier. The applications are suitable for millimeter-wave narrow and broadband circuits. To sum up the Gunn device applications, we can say:

- Microwave application—GaAs,
- Millimeter-wave application—InP.

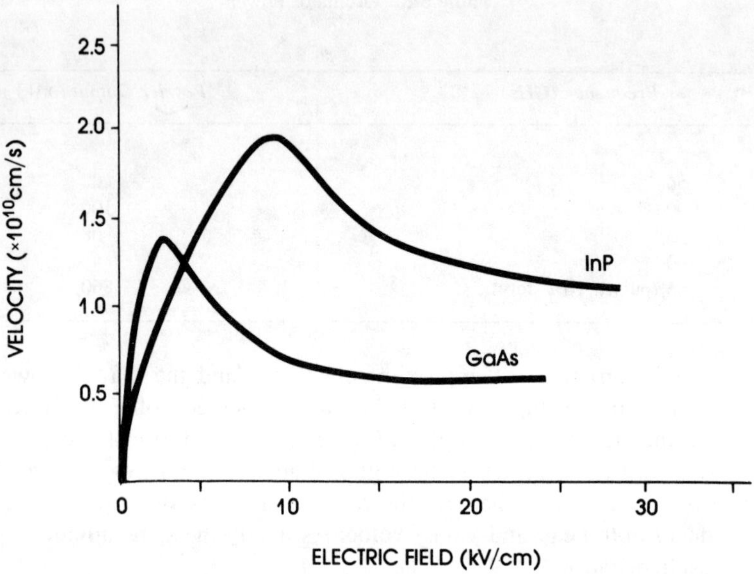

Fig. 5.38 Peak-to-valley comparison.

5.3.6 IMPATT Diodes

The name IMPATT stands for *IMPact Avalanche and Transit Time* diode. Its name and what it stands for will become more apparent as our discussions are presented. In our previous discussions we have said that the Gunn diode was a negative resistance device. It was said to have a dynamic dc-negative resistance. This meant that, over a certain range, current decreased with an increase in voltage, and *vice versa*. This particular point was pursued no further, it being taken for granted that any device which exhibits a dynamic negative resistance for direct current will also exhibit it for alternating current. This is to say, if an alternating voltage is applied, current will rise when voltage falls, at an ac rate. We may thus now redefine negative resistance as that property of a device which causes the current through it to be 180° out of phase with the voltage across it. The point is important here, because this is the only kind of negative resistance exhibited by the IMPATT diode.

A combination of delay involved in generating avalanche current multiplication, together with delay due to transit time through a drift space, provides the necessary 180° phase difference between applied voltage and the resulting current in an IMPATT diode. The cross section of the active region of this device is shown in Figure 5.39. Note that it is a diode, the junction being between the p and the n layers.

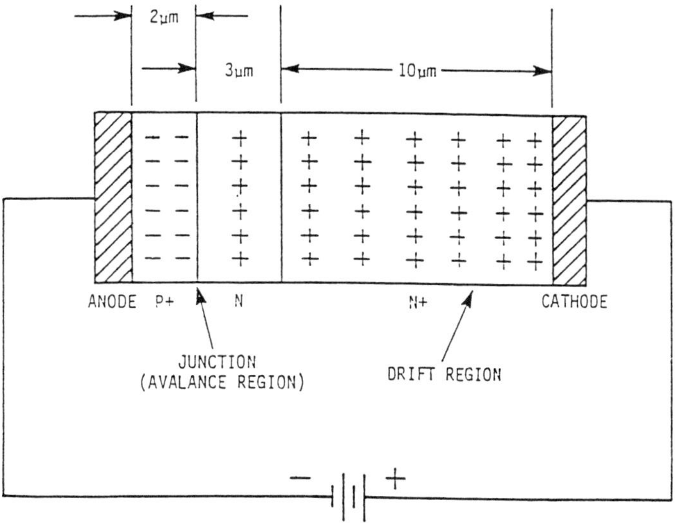

Fig. 5.39 IMPATT schematic.

An extremely high voltage gradient is applied to the IMPATT diode, of the order of 400 kV/cm, eventually resulting in a very high current. A normal diode would very quickly break down under these conditions, but the IMPATT diode is constructed so as to be able to withstand these conditions repeatedly. Such a high potential gradient, back-biasing the diode, causes a flow of minority carriers across the junction. If it is now assumed that oscillations exist, we may consider the effect of a positive swing of the RF voltage superimposed on top of the high dc voltage. Electron and hole velocity has now become so high that these carriers form additional holes and electrons by knocking them out of the crystal structure, by so called impact ionization. These additional carriers continue the process at the junction, and it now snowballs into an avalanche. If the original dc field was just at the threshold of allowing this situation to develop, this voltage will be exceeded during the entire RF positive cycle, and avalanche current multiplication will be taking place during this entire time. However, since it is a multiplication process, avalanche is not instantaneous. Indeed, as shown in Figure 5.40, the process takes a time such that the current pulse maximum, at the junction, occurs at the instant when the RF voltage across the diode is zero and going negative. A 90° phase difference between voltage and current has thus been obtained.

As so far described, the current pulse in the IMPATT diode is situated at the junction. However, it does not stay there. Because of the reverse bias, the current pulse flows to the cathode, at a drift velocity dependent on the presence of the high-dc field. The time taken by the pulse to reach the cathode depends on this velocity, and of course on the thickness of the highly doped (n) layer. Also, the thickness of

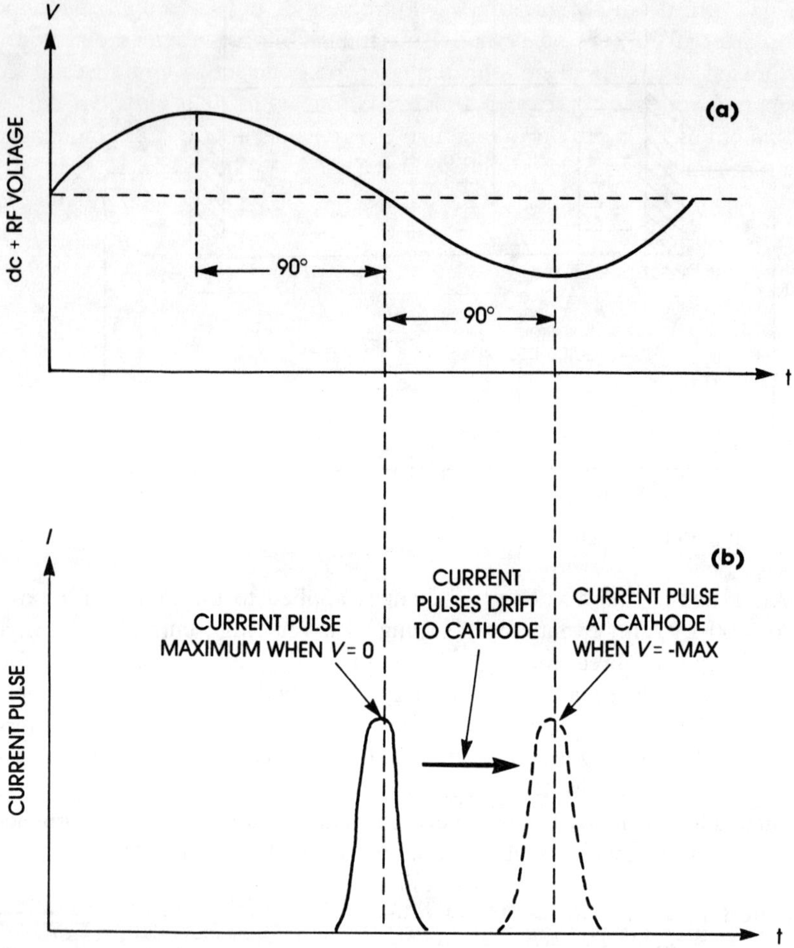

Fig. 5.40 IMPATT behavior.

the drift region is selected so that the time taken for the current pulse to arrive at the cathode corresponds to a further 90° phase difference. Thus, as shown in Figure 5.40 when the current pulse actually arrives at the cathode terminal, the RF voltage there is at its negative peak. Accordingly, voltage and current in the IMPATT diode are 180° out of phase, and a dynamic RF negative resistance has been proved to exist. As with other devices studied in this chapter, such a negative resistance lends itself to use in oscillators and amplifiers. Because of the short times involved, these can be microwave.

Commercial IMPATT diodes have been available for quite some time. They are made of either silicon or gallium arsenide, invariably epitaxial and mostly mesa;

some have Schottky-barrier junctions. Gallium arsenide is technically preferable. It gives lower noise, higher efficiencies and higher maximum operating frequencies. It is also more difficult to work, however, and more expensive. Silicon diodes were developed first, and even now give higher output powers than GaAs for commercial diodes. It is expected that silicon diodes will continue to be used for less exacting applications.

The IMPATT diode shown in Figure 5.41 is typical of what is manufactured, and could house a chip made of either GaAs or silicon. The construction is deceptively simple. A lot of thought and development have gone into its manufacture, however, particularly the contacts, which must have extremely low ohmic and thermal resistance. In a practical circuit, the IMPATT diode is generally embedded in the wall of a cavity, which then acts as an external heat sink.

It should be noted that, until recently, practical IMPATT diodes were unlike Read's original proposal which called for a double drift region. Whereas Figures 5.39 and 5.40 show diodes with single (n) drift regions. The reason for the departure from what was theoretically a high-efficiency structure was difficulty in fabrication. This has been solved, and GaAs RIMPATT (*read*-IMPATT) devices are beginning to show efficiencies approaching 30 percent in the laboratory, more than three times as high as could be expected from a commercial IMPATT diode.

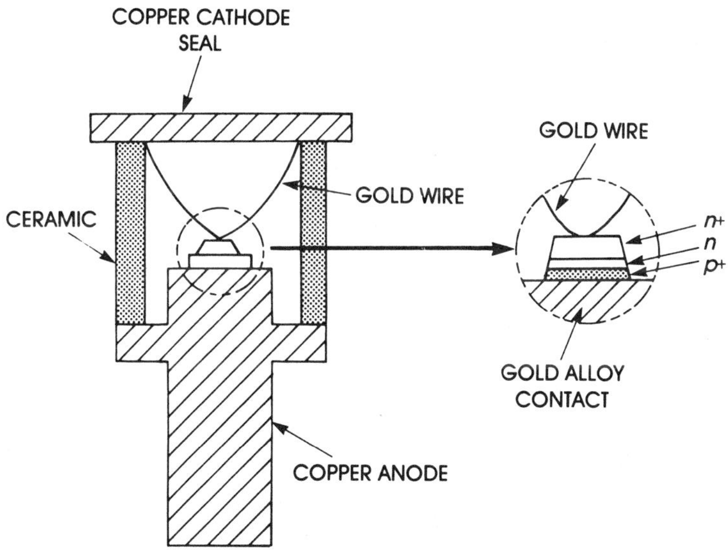

Fig. 5.41 IMPATT diode.

5.3.7 TRAPATT Diodes

The term TRAPATT stands for *Trapped Plasma Avalanche Triggered Transit* diode. The TRAPATT diode is derived from the IMPATT diode, and is closely related to it. Indeed, at first it was merely a different, "anomalous," method of operating the IMPATT diode. A greatly simplified operation will now be described.

Consider an IMPATT diode mounted in a coaxial cavity, so arranged that there is a short circuit a half-wavelength away from the diode at the IMPATT operating frequency. When oscillations begin, most of the power will be reflected across the diode, and thus the RF field across it will be many times the normal value for IMPATT operation. This will rapidly cause the total voltage across the diode to rise well above the breakdown threshold value used in IMPATT operation. As avalanche now takes place, a plasma of generated electrons and holes is generated, placing a large potential across the junction, which opposes the applied dc voltage. The total voltage is thereby reduced, and the current pulse is, as it were, trapped behind it. When this pulse travels across the *n*-drift region of the semiconductor chip, the voltage across it is thus much lower than in IMPATT operation. This has two effects. The first is a much slower drift velocity, and, consequently, longer transit time, so that for a given thickness the operating frequency is several times lower than for corresponding IMPATT operation. The second point of great interest is, that when the current pulse does arrive at the cathode, the diode voltage is much lower than in an IMPATT diode. Thus dissipation is also much lower, and efficiency much higher. The operation is similar to class C, and indeed the TRAPATT diode lends itself to pulsed instead of CW operation. It must be reiterated that the foregoing is a much simplified description of the operating mechanism. It is, however, adequate to emphasize the similarity to the IMPATT as well as the individual features and applications (mainly pulse operation) which are unique to the TRAPATT.

Commercial TRAPATT diodes are constructed of silicon and GaAs with structures corresponding to those of IMPATT diodes, but with gradual, rather than abrupt, changes in doping level between the junction and the anode. Furthermore, they use complementary n^+-p-p^+ structures as shown in Figure 5.42 instead of the p^+-n-n^+ IMPATT chip of Figure 5.39 for reasons of better dissipation. The two figures should be examined in conjunction with each other.

Because the drift velocity in a TRAPATT diode is much less than in an IMPATT diode, either operating frequencies must be lower, or the active regions must be made thinner. In fact, both these considerations are borne out by results obtained thus far. On the one hand, most good TRAPATT results have been for frequencies under 8 GHz. On the other hand, it has been found that by the time 5 GHz is reached, the width of the depletion layer is only 2 μm, which is just about the lower practical limit. Since the TRAPATT pulse is rich in harmonics, however, amplifiers or oscillators can be designed to tune to these harmonics, and operation above *X* band in this manner is possible.

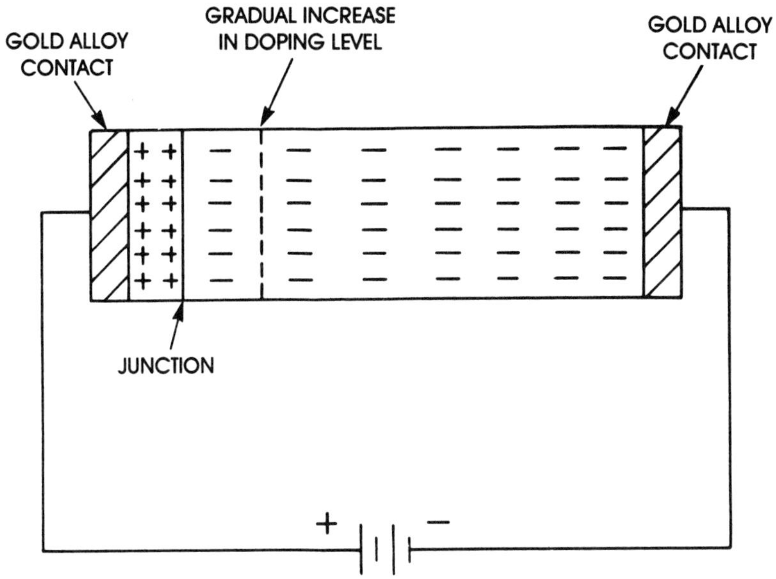

Fig. 5.42 TRAPATT diode.

5.4 MICROWAVE TRANSISTORS

The microwave transistor is the device which has probably advanced and improved the most of any microwave device over the years. This statement may be challenged after reading the previous sections on diodes, but, even though many new diodes have been introduced over a period of time, the transistor is the device which has advanced and improved on its basic designs and structure while introducing basically only two new types.

To aid in understanding just how far the microwave transistor has advanced we will give a brief history of the transistor (*transfer resistor*). The first transistors were fabricated by Drs. W.H. Brattain and J. Bardeen in 1948 while they were studying the properties of germanium semiconductor rectifiers at Bell Laboratories. During these studies they observed that the flow of current through the rectifier could be controlled if a third electrode was added to the device. This electrode would serve much the same purpose that the control grid was then doing in triode vacuum tubes. What resulted from this observation was the *point-contact transistor*.

To fabricate such a device, the collector and emitter wires are pressed into an *n*-type germanium material similar to the method used in early crystal sets which used "cat's whiskers." This arrangement is the basic germanium rectifier that Brattain and Bardeen were investigating when they discovered the transistor characteristics. With this basic rectifier configuration established, the *p*-regions are introduced into the device by a process called *forming*. This process involves the passage of a large current through the wire and into the germanium for a short period of time. The *p*-type germanium forms almost immediately at the contact point as a result of the diffusion of some acceptor impurities from the wire into the germanium. Thus, the name *point-contact transistor* is derived and the resulting device is a *pnp* transistor.

This device, of course, was a major breakthrough for the electronics industry, because it promised to replace many of the bulky, hot, and high power consuming vacuum tubes of the day. The point contact transistor, however, had drawbacks of its own. This revolutionary device generated much more noise internally than the vacuum tube it was designed to replace; since it was not hermetically sealed, it could not tolerate temperature or humidity; and it was a very fragile device to handle. These problems were rather serious if one was to use the transistor for any useful and demanding applications other than sitting on a bench and controlling the flow of current. In 1949, these problems were addressed when W. Schockley published a paper describing the possibilities of producing a *junction transistor*. Today this concept is one that is widely known and used. These devices, as proposed, were much more rugged and had much improved noise performance.

The first of the junction transistors were produced using a process called a *grown-junction technique*. The term, *to grow,* is one which is very common in semiconductor discussions and, as we can see, it is one which goes back to the earliest days of transistors. The *growing* of junctions is accomplished by starting what is termed a *seed* of material (this is a small quantity of material used as a foundation for the finished product). The seed is touched to the surface of a bath or *melt* of molten semiconductor, then is slowly pulled away. (The semiconductor is germanium, silicon, *et cetera*.) The size of the seed is increased, or grows, because the melt in the area of the seed adheres to the seed and effectively "freezes" when withdrawn from the melt solution. Characteristics of the solution can be changed simply by adding more impurities to the melt, or by changing the impurities which are added. When the crystals are completely "grown," they are cut to the appropriate size, then leads are attached to the proper areas. These devices are also called *double-doped transistors*.

Alloy or *fused-junction* devices were the next to appear on the transistor scene when they were developed in 1951. These turned out to be primarily *pnp*-type transistors, whereas the grown junction was primarily *npn*. A wafer of base material (usually *n*-type) germanium is held in a jig between two pellets of impurity (possibly indium). The structure is held until the impurity melts (155° C for indium), and the molten impurity penetrates the base material. During cooling, crystal growth results in a completed three-layer structure. This device is characterized by a very high junction capacitance due to the fusing process during construction. Thus, this type of device is not used for high frequency applications.

The disadvantage, and thus the limitation, of the alloy-junction transistor was overcome in 1953 by the Philco Corporation when the *surface-barrier transistor* was developed. This type of transistor still finds many applications in today's electronic circuits. Modifications have been made to the original 1953 version, but the basic device, and its principles, are still what are needed and used. Construction of this device was achieved by a manufacturing process that subjected a small piece of germanium to two jets of electrolyte, (such as a solution of indium chloride), which strikes the opposite faces of the germanium-base slab. Etching of the semiconductor was achieved by passing a current through the electrolyte stream and the germanium. When the desired base thickness was achieved, current polarities were reversed, and the jets plated the metal emitter and collector contacts on either side of the base. Improvements were made to this procedure by the addition of impurity diffusion techniques. In this process, impurities of one conductivity type are made to diffuse at high temperatures into the surface of a bar of the opposite type. This results in the formation of a junction we discussed in detail in section 5.2.1, the *pn* junction.

The next generation of transistor was the *mesa* or *diffused-base transistor*. In this device, phosphorous pentoxide (P_2O_5) may be diffused into an *n*-type silicon collector region to form the base of the device. Acid etching of the proper areas results in a plateau, or mesa.

The final two types of transistors are the *planar* and *epitaxial*. These are the transistors used in microwaves today. The planar construction was shown earlier in Chapter Five when we discussed *pin* diodes and, as you will recall, uses diffusion and surface passivation that protects the surfaces and junction edges from contamination. Leakage occurs in this type of construction because silicon dioxide (SiO_2), which is an insulator, is formed on the exposed surfaces. You will recognize the planar construction once again as it is shown for a transistor in Figure 5.43.

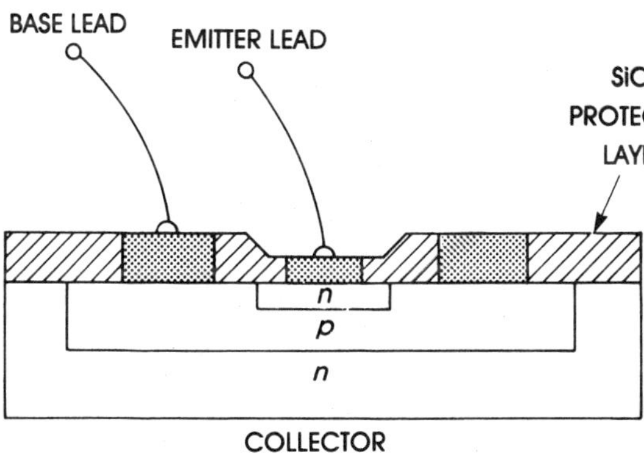

Fig. 5.43 Planar construction.

The *epitaxial* transistor is similar to the planar construction, except that it has a thin layer of low-conductivity material for its collector, with the remainder of the collector region of high conductivity. It is for this reason that an epitaxial device is sometimes referred to as an epitaxial-collector type of construction. This type of construction, as we have previously stated, is the main type used for microwave bipolar transistors today. Its construction is shown in Figure 5.44.

The next phase of our history lesson involves the *field-effect transistor* (FET). Although it was known in theory as far back as 1926, and analyzed completely by Schockley in 1952, the FET was not successfully fabricated for general usage until 1961. Beyond this, it was not used for microwave applications until the mid 1970s when gallium arsenide pushed the frequency operating area for microwave solid-state devices well up into the high microwave frequency spectrum. The FET has certain advantages over a junction device. These will be investigated as we discuss both the bipolar and FET in detail in the following sections.

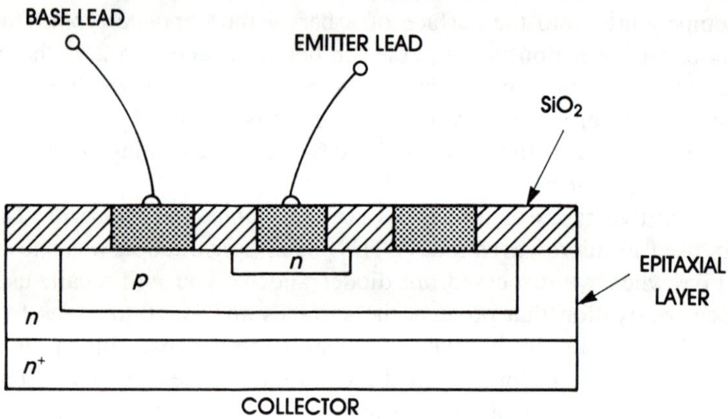

Fig. 5.44 Epitaxial construction.

The final chapter to this point in the transistor story is the *high electron mobility transistor* (*HEMT*). This device, which found its way to the microwave industry around 1984, extends the frequency range even higher due to its main property—high electron mobility. This device will be discussed in detail later in the following sections.

With the preceding history lesson and the discussion presented in this chapter on junction theory, you are now ready to investigate the three types of transistors presented: *bipolar, FET,* and *HEMT*.

5.4.1 Bipolar Transistors

If you were to look up the term *bipolar* in the dictionary, you would find it would say it was "having two poles, or extremities." The dictionary of electronics defines it as:

Having to do with a device in which both majority and minority carriers are present.

This is the one we will be concerned with, because it involves the two types of carriers, electrons, and holes. With most devices we are concerned only with one energy carrier; those are electrons. The bipolar, however, which consists of two *pn*-junctions, operates because of two types of carriers. Both electron and hole movement are of vital importance in the operation of the bipolar transistor.

The bipolar transistor finds many applications in microwaves, usually at 4 GHz and below. They are used for low-noise devices, low power linear amplifiers, and higher power class-C power amplifiers. Each of these classifications: *low-noise, linear power,* and *high power,* will be discussed in this chapter.

5.4.1.1 Low-Noise Transistors

The low-noise bipolar transistor finds its most prominent application as an amplifier in the front end of a microwave receiver. In this application, a device that will amplify the low-level input signal and introduce a minimum amount of noise in the process is needed. To obtain this low-noise characteristic requires special construction techniques. As we describe this construction, two terms will be coming up, *planar* and *epitaxial*. It would be beneficial for us to get a detailed explanation of them at this point.

Recall, from our previous discussions, that planar construction uses diffusion and surface passivation that protects surfaces and junction edges from contamination. This construction makes the protected areas less prone to surface problems that sometimes occur when structures such as the *mesa* construction are used. The term, *planar,* is one that denotes that both emitter-base and base-collector junctions of the transistor intersect the device surface in a common plane. Perhaps a better term for the structure is *co-planar*. The real significance of the *planar* structure, however, is that the fabrication technique of diffusing dopants (or impurities) through an oxide mask results in the junction being formed beneath a protective oxide layer. Thus, the previously mentioned isolation from surface problems.

The term *epitaxial* is actually a shortening of the term epitaxial collector. This term means that the collector region of the transistor is formed by the epitaxial technique, rather than by the diffusion method as mentioned earlier for the base and emitter. The epitaxial layer is formed by condensing a single-crystal film of semiconductor material on a wafer of substance, usually the same material. Thus, an epitaxial device is one in which the collector region is formed on a low-resistivity silicon substrate. Note in Figure 5.44, which is the epitaxial construction of a transistor, where the epitaxial layer is located. Location is the important factor, not its material. It is actually the same material as the collector.

The base and emitter regions on the epitaxial device are diffused into the EPI layer, as it is termed. This diffusion also can be noticed in Figure 5.44, as well as the fact that the epitaxial structure can be thought of as the planar construction with an extra layer that separates the collector from the base and emitter structures. The epitaxial technique lends itself to precise tailoring of the collector-region thickness and resistivity that improve device performance and uniformity.

The internal "geometry" of most low-noise transistors is an interdigital type of construction. This interdigital construction is illustrated in Figure 5.45. This geometry achieves the lowest base resistance without sacrificing gain. This low-base resistance high-gain combination results in a low-noise figure. To understand why this comes about, consider what the noise figure of a transistor is.

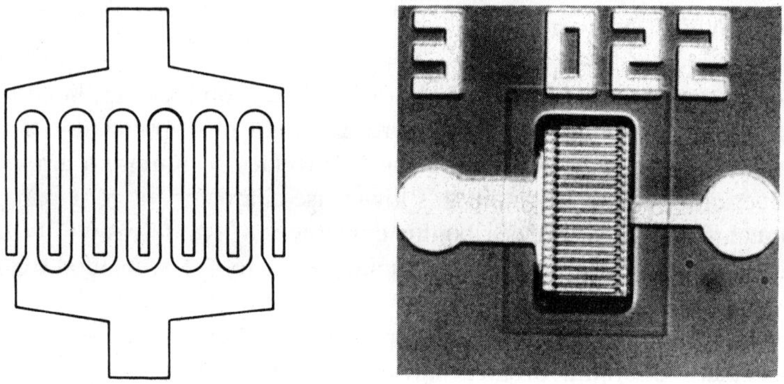

Fig. 5.45 Interdigital construction.

The noise figure of a transistor is a measure of how much the signal-to-noise ratio (comparison of signal level to noise level) is degraded as the signal passes through the transistor. The degradation that does occur is due to the presence of three noise sources within the chip itself. These are:

- Shot noise in the emitter-base circuit,
- Shot noise in the collector circuit,
- Thermal noise generated by the base resistance.

Shot noise is the noise generated at a semiconductor junction due to the current that is passing through it. It is directly proportional to the square root of the current applied to the junction. It is, therefore, logical to keep the current through the device as low as possible. Notice that when a low-noise transistor is presented later, the lower noise figures are achieved when the collector current (I_c) is low.

Thermal noise is the noise generated by the random thermal motion of charged particles (or electrons within the base region). Remember that we referred to the fact that the interdigital structure achieved the lowest base resistance without sacrificing gain. If we have a low resistance in the base region, we will have a small voltage dropped across that region, and thus lower power dissipated. This lower power will mean very slight thermal activity in the base region and, therefore, low thermal noise.

With the information above, we can say that the best low-noise transistor should have a construction that offers an "efficient" junction structure and a low base resistance. By efficient junction, we mean one that will provide the *pn* junction action with a minimum amount of current or loss across that junction.

With the low-noise transistor defined, we will now look at a data sheet which would be associated with such a device. This data sheet is for a general low-noise microwave bipolar transistor. The primary parameters to be concerned with are *frequency, noise figure,* and *gain*. Each of these will be discussed with a variety of gains covered.

The first parameter, *frequency,* should be just that—the first parameter at which you look. This is because you may find a transistor with an excellent noise figure but it only operates to 2 GHz. If your application is from 3.0 to 3.5 GHz, the excellent noise figure will do you no good at all. So, the first thing to consider when selecting a transistor is its frequency range. Frequency range is usually one of the first items on the data sheet, and it will appear in an opening paragraph termed "Description" or "Description and Applications." It also may appear in a summary of specifications termed "Features." Wherever it appears, be sure to examine this first. A paragraph entitled *Description* would read something like this:

> The ABC-442 is a silicon bipolar transistor designed for use in low-noise, small-signal amplifiers up to 4 GHz. This device features excellent gain characteristics while maintaining its specified low-noise figure over a broad range of frequencies. It is available in chip form as well as in our 70-mil and 100-mil packages.

Check this paragraph on the transistor data sheet first. It will save time and aggravation later on.

The next logical parameter to look at on the data sheet is, of course, the *noise figure*. This figure is sometimes called *spot noise figure* or *minimum noise figure*. When reading this number, be sure to read the conditions that go along with it. Simply saying that a transistor has a noise figure of 2.8 dB means little or nothing. If, however, the manufacturer says the device has a noise figure of 2.8 dB at 4 GHz with V_{CE} = 10 V and I_C = 5 mA, the device is characterized so that we can determine whether we can use it in a particular application. (V_{CE} is the collector-to-emitter voltage, and I_C is the collector current.) This type of detailed characterization is absolutely necessary, since the design resulting from the use of a particular device depends on the *s*-parameter data that are taken at specific conditions (V_{CE}, I_C). (*s*-parameters are the parameters which define the transistor in reference to a 50Ω system.) To have the proper matching circuits and ensure that the transistor will operate with the expected noise characteristics, these conditions must be duplicated. Figure

5.46 shows the two conditions governing the noise figure of a device. Figure 5.46(A) is noise figure *versus* frequency and Figure 5.46(B) is noise figure *versus* collector current at a fixed frequency (3.5 GHz in this case). If you compare I_C = 5 mA in Figure 5.46(B) to 3.5 GHz in Figure 5.46(A) you will see that the same noise figure results.

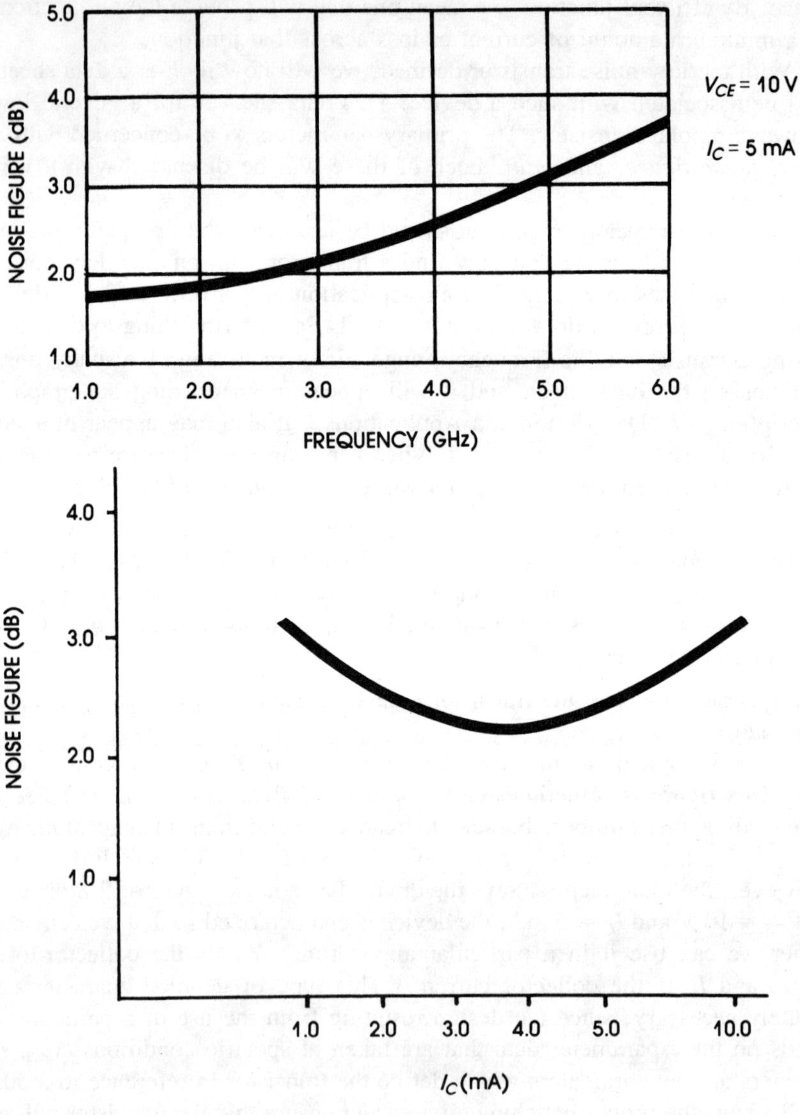

Fig. 5.46 Noise figure *versus* frequency and collector current.

With the noise characterized, the next parameter to be investigated is the *transistor gain*. With this parameter we are looking at two values: G_{MAX} and G_{NF}, G_{MAX} is the gain from conjugately matching the device for a specific gain. The noise figure obtained with this condition depends on the characteristics of the particular device used. A conjugately matched transistor will produce the maximum gain available.

G_{NF} is the gain achieved when the transistor is matched specifically for the optimum noise figure at that particular frequency. This gain is significantly lower than the gain achieved at G_{MAX}. To illustrate, consider the following comparison of G_{MAX}, G_{NF}, and noise figure for three devices in Table 5.4.

Table 5.4 A Comparison of G_{max} and G_{NF}

Device	*Frequency*	*NF*	G_{max}	G_{NF}
A	4 GHz	2.8 dB	11.0 dB	7.0 dB
B	4 GHz	2.7 dB	11.0 dB	8.0 dB
C	4 GHz	3.0 dB	13.0 dB	10.0 dB

This data is from three commercially available devices designed to provide basically the same performance. When designing for gain with these devices, considerably more gain is received than when designing for optimum noise figure. In most cases there is a difference of 3 to 4 dB in gain. This difference, of course, is a logical result, since the design is around a conjugate match for power gain, and around noise admittance parameters for a noise match. This latter type of matching does not guarantee a maximum transfer of power and thus will not guarantee a maximum amount of gain through the device.

One additional gain figure is sometimes given on transistor data sheets. This figure is the *insertion power gain,* designated as $|S_{21}|^2$. This gain is obtained by dropping the device into a circuit with a 50-Ω line at the input, and a 50-Ω line at the output. This gain is useful for wideband operations where a wide range of matching is not feasible using a conjugate match. The value of $|S_{21}|^2$ for the three previous devices is 5.5 dB, 6 dB, and 7 dB, respectively. About the same gain is received as with optimum noise figure with $|S_{21}|^2$. However, the good noise figure will not be received as when performing a specific noise match.

5.4.1.2 Linear Bipolar Transistors

When transistors are biased for a particular application they are usually biased either class A, B, or C. Class A is biased well above cutoff, so that the transistor is conducting over the entire 360° of the input cycle. Class B is biased at cut-off and the device conducts for only one-half (180°) of the input cycle. Finally, class C is biased well below cut-off (usually 2 1/2 time cut-off) and only conducts for 120° to 150° of the input cycle. Generally, class A operation is reserved for low power levels where noise is the primary concern; class B is used for push-pull operation of power stages; and class C is almost always used when high power is needed.

It is possible, however, to get a higher power output by biasing some transistors class A. These are the microwave linear transistors. Devices are available that produce 7 W at 2 GHz, or as much as 2 W at 3 GHz. This capability may be very impressive, or it may mean nothing without the understanding of what a linear power transistor does.

To aid in understanding the linear device, consider a low level transistor stage. It may be operating in the microwave range or in the audio range. The device is built with its associated components, proper bias is applied, and a signal is placed at the input. With these conditions satisfied, a certain output will be obtained. If the input level is increased, the output level will follow according to the characteristics of the transistor and its circuitry. This increased output will increase until a point called the *compression point* is reached. This is where the output does not increase at the same rate as it previously did. A further increase in the input level will not result in a corresponding increase at the output. If a 10 dB increase is applied to the input and there is only a 9 dB increase at the output, the 1 dB *compression point* has been reached. This situation is one which all class A-biased circuits encounter. The compression point is what limits the amount of power available in most class A circuits.

In class C operation the device is biased well below cutoff (usually two and a half times), as we previously stated. With this type of biasing, low-level inputs will have no effect on the output, since the input level is not large enough to overcome the cut-off voltage applied for class C. As the input power level is increased more and more, it will approach the cut-off point of the transistor. Once the bias has been overcome, the input level will be sufficient to turn the device on. The transistor will then snap on and display a constant power output which will increase very little, if any, as the input power is increased further. If you have an input power which is required to vary over some range and needs the power output to vary, neither the class A device (which is limited in power level) nor the class C diode (which does not vary the output power) will do the job. A different device must then be used to do this task—the linear power transistor.

The linear power transistor is the bridge needed between the high power class C output stage and the low-level class A input stage. It enables a designer to use class A operation (linear operation) right up to the final power stages. It should not be assumed that the linear power transistor competes with the class C power transistor. On the contrary, the linear device complements the power stage by providing very good drivers for the high power stage. The linear-power stages are unique in that they are able to operate class A, but still are built rugged enough to survive the high collector currents needed for power amplification. Methods used to accomplish this are, of course, proprietary and thus cannot be discussed at this point. Suffice it to say that the transistors are a unique blend of low-level operational characteristics combined with the stout construction needed for higher power.

Items from a typical linear power transistor data sheet are shown below:

Frequency	1–2 GHz
Power output	1.6 W at 2 GHz
VSWR mismatch	∞
Linearity	−0.2 dB, + 1.0 dB

All of the parameters listed above have specific conditions associated with them, just as the low-noise transistor did. An example of these conditions may be V_{CE} = 20 V, I_C = 200 mA, and frequency = 2 GHz. These conditions indicate how the data you are viewing was taken and what it takes to duplicate it. A brief explanation of each of these parameters is shown below.

The first parameter, *frequency,* gives the range over which the transistor will operate. Frequency does not indicate, however, that the device will cover this entire band in one shot and have a response that will be flat within 0.5 dB. It only says that the frequency range given is the one the transistor was designed to handle. Some devices will be characterized as operating from 1 to 2 GHz, for example, but will not have a constant output power over this band. They will only have the required output over a narrow band—1.0 to 1.15 Hz, 1.6 to 1.7 GHz, and so forth. The power output *versus* frequency curve in Figure 5.47 is a typical example. It shows that approximately 1.9 W is available from 1 to 2 GHz. In reality you will not be able to match the device over this broad a band. You will have to settle for something less so check the data sheets carefully to see just what the frequency is.

Power output is the next important parameter to be checked when choosing a linear transistor. As we have said previously, certain conditions are spelled out when specifying parameters. A device cannot be said to deliver 1.7 W of power at 2 GHz without defining the conditions under which the 1.7 W was obtained. What is needed is such data as V_{CE}, I_C, and power input. You will recall that these parameters were shown for FIgure 5.47 where the power output *versus* frequency curve was shown. The power on that graph was only valid for those conditions.

Another curve which illustrates how important these parameters are is shown in Figure 5.48. If we take the case of V_{CE} set at 15 V, we can get anywhere from 0.5 W to approximately 1.6 W by varying the emitter current (I_E) from 100 to 300 mA. If we increase V_{CE} to 20 V, the same range of current yields from 0.6 W to about 2.25 W. So you can see how important it is to specify the operating voltage and current to make the parameter repeatable. Therefore, the power output on the data sheet should always include test conditions, including frequency.

VSWR *mismatch*: One of the most important properties of a device handling power is its ability to perform under mismatch conditions. If, for example, a device is producing 2 W and has a large mismatch for 2 W, the power (or a large majority of it), will be reflected back to the device. If the device is not equipped to handle

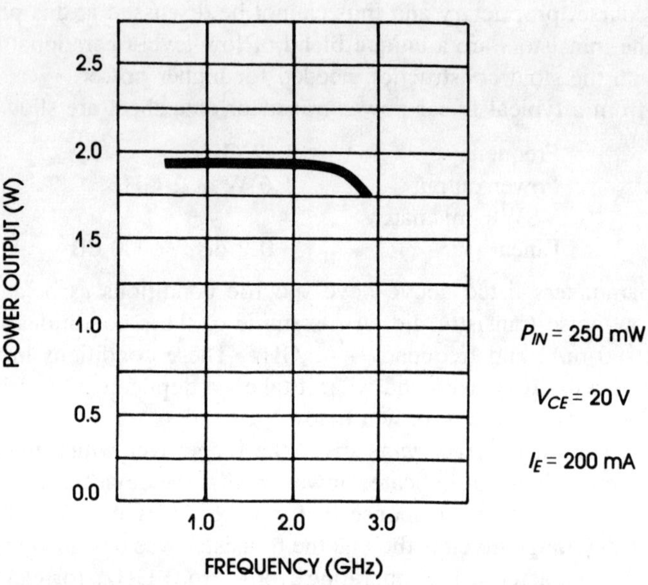

Fig. 5.47 Power out *versus* frequency.

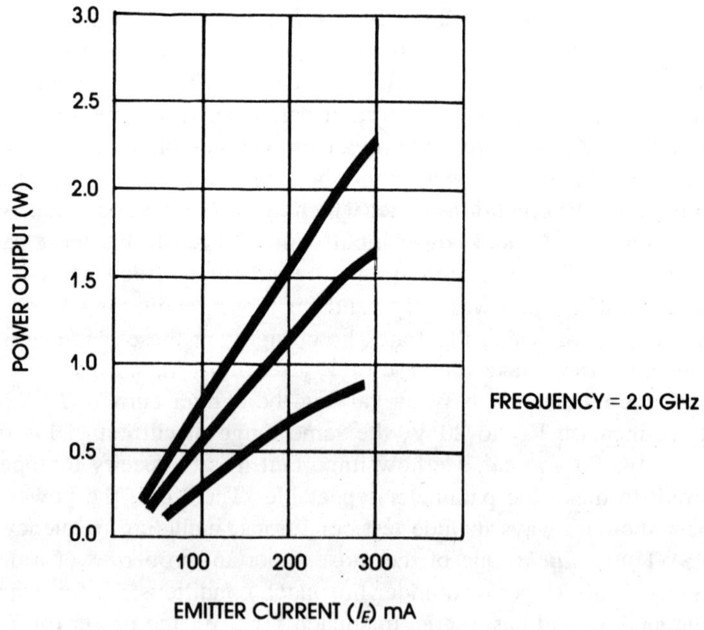

Fig. 5.48 Power out *versus* current.

this reflected power, it will be destroyed. This will result in lost time waiting for a replacement device plus the cost of a new device, neither of which are conductive to schedule and budget constraints. So always check to see what sort of a mismatch tolerance the device exhibits. The example we have presented shows infinity as the specification for mismatch, meaning that a VSWR of infinity (or an open circuit) will not affect or damage the transistor. If there is any chance of a large mismatch present at the output of a power stage, specify a transistor which will take an infinite mismatch.

The final parameter on our typical data sheet is *linearity*. Linearity is a natural parameter to investigate since we are talking about linear devices. Linearity of a device is how close its output response tracks a straight line drawn between a low- and high-power and of its operating range. The data sheet device showed −0.2 to +1.0 dB of variation over this range. The conditions given for this parameter are usually frequency, power number 1 (low end) and power number 2 (high end). Typical numbers might be 2 GHz, P_1 = 0.15 W, P_2 = 1.5 W. If the response of P_{IN} *versus* P_{OUT} is plotted, and a straight line is drawn between P_1 and P_2, the variation from this line will be the linearity. The ideal figure, of course, would be ±0 dB.

The microwave linear power transistor, then, is the device that produces an intermediate level of power amplification while operating in a linear, or class A mode. Whenever a range of linear-output power levels is needed, this transistor is the type to use. After investigating the parameters above for your particular application, choose the linear-power device which will perform up to these parameters.

5.4.1.3 Power Transistors

It is not possible to talk about microwave power transistors today without encountering arguments concerning gold and alumimum metallization and their respective advantages and disadvantages. Some people say that gold gives better performance and reliability; others make the same claim for aluminum. Each has very sound reasons to recommend its choice and the particular application influences the decision. For this reason, both aluminum and gold are covered here.

As a means of understanding microwave power transistors, let us construct one. To start, the wafer is made of silicon (Si) which, in a transistor, does the same job as a vacuum in a vacuum tube—it provides a medium with certain properties which allow current to flow. The wafers used in the manufacture of RF power transistors are epitaxial. Recall from the section on low-noise transistors that epitaxial is defined as the growing of a substance on a substrate of the same substance; that is, the technique for growing a layer of silicon on a silicon substrate. Figure 5.49 shows this process. The silicon itself is grown in ingots 1 to 3 inches in diameter and from 6 inches to several feet in length. A single crystal "seed," one fourth inch square by several inches long, is used to start the growth. This seed is rotated and slowly withdrawn from a crucible of molten silicon, which is doped *n*-type or *p*-type by the

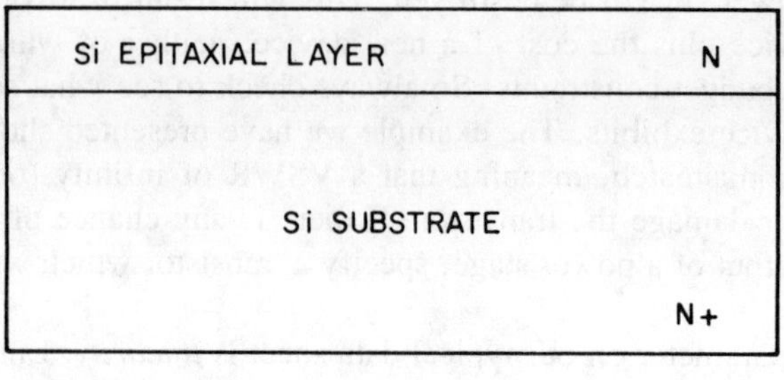

Fig. 5.49 Epitaxial wafer.

addition of basic elements. Ingots of *n*-type doping, for example, use antimony or arsenic; they are grown at temperatures of approximately 1,500° C. All of the dopant used must be distributed uniformly. The ingots are sliced with a diamond saw into wafers to a thickness of approximately 15 mils, then are mechanically and chemically polished to a mirror finish (reducing the wafer to approximately 12 mils).

The polished wafer substrates, discussed above, are next placed in an epitaxial reactor at about 1,200° C and a doped film of silicon is grown by chemical vapor deposition on top of them. The doped film can be as thin as one tenth of a mil (0.1) or as thick as two mils, and can be either *n*-type or *p*-type in nature. This extensive process results in a *silicon epitaxial wafer* which is used as the beginning of microwave power transistor. This section is a very important part of the transistor because the device is constructed totally in the silicon epitaxial layer (Figure 5.49). For all intents and purposes, the substrate is only a carrier for the epitaxial silicon, and has little or no effect on the electrical parameters of the device.

With the epitaxial wafer in hand, let us proceed with the building of our power device. The next step in the process is that of oxidation and masking.

High frequency power transistors require a large number of narrow emitter images in order to obtain a high emitter perimeter to base area ratio. A minimum practical emitter width is one micron. The base, emitter, and contact patterns are defined in an oxidated silicon wafer by a photoresist and etch operation. This process, of course, must be capable of obtaining and maintaining one micron geometries. Photoresists (polymers sensitive to exposure to ultraviolet light) are used to define the necessary patterns; they can be either positive or negative. With the positive type, the exposed area is developed and the resulting image is the same as that which was exposed. Thus, to etch a hole in the oxide to permit diffusion of the impurities, the mask has clear areas with the required pattern, but its major portion is a dark field. In the past, such a dark mask was difficult to align because the operator could not see through the dark emulsion. However, see-through masks, which are opaque to ultraviolet light, are now available and have made the use of positive resists practical. Negative resists react in the opposite manner, and the ultraviolet light polymerizes them so that, in this case, the unexposed area is developed. Organic solvents are used to develop these resists, causing them to swell; thus the openings shrink and edge definition is lost. Various rinses are employed to reduce this swelling, but as the resist shrinks, its adhesion to the substrate is degraded. Therefore, it is desirable to use positive resists for small geometries. Their developer does not cause swelling, so a much thicker resist can be used for a given image. Ease of processing is another advantage of positive resists, because they do not have to be post baked at high temperatures to ensure adhesion; they are stripped easily in many different organic solvents.

The oxidation process is performed to protect the surface of the device. All diffusion takes place through "windows" cut in the oxide to allow the *n* dopants to diffuse into the silicon. As described above, photo-sensitive material is applied to the top of the wafer. The photoresist is exposed when a light source is shown on the wafer, in a manner similar to the development of a photograph. The mask shields the window from exposure; when developed, the unexposed window is removed since light is required to harden the photoresist.

Exposing the silicon for dopants requires an etchant that attacks only the SiO_2 and not the silicon or hardened photoresist—hydrofluoric acid (HF). Any photoresist that remains is removed with a solution of chronic acid.

Figure 5.50 shows how the complete process works for multiple oxidations. This figure also serves to illustrate the next process—base and emitter diffusion.

Base diffusion precedes emitter diffusion. The base of an *npn* device is *boron* while its emitter uses *phosphorus*. Diffusion temperatures are on the order of 1,000° C, causing the dopant (boron or phosphorus) to diffuse from areas of high concentration to those of low. Either higher temperature or higher concentrations accelerate the diffusion. The farther the dopant travels into the transistor, the lower the concentration. Whenever a high dopant concentration region exists, a low resistance (resistivity ρ) also exists, resulting in a high concentration of carriers (electrons or holes).

The action taking place during diffusion determines the quality of the transistor. Once the base is diffused, the emitter/base junction must overtake the collector/base junction, in order to control base thickness. Transistor parameters are greatly affected by the location of the base region. A thin base is the desired goal. Control of this region is best over the emitter and base depth when doping densities are large (low ρ). A rule-of-thumb emitter depth should be one half to three quarters that of the base.

Next comes the process where the gold and aluminum proponents no longer agree—transistor metallization. Aluminum metallization is covered first, simply because of alphabetic order; an effort is made to totally avoid partiality.

The aluminum metallization technique is illustrated in Figure 5.51. Aluminum is evaporated on the wafer to a target value of $<\pm 2$ percent, using an electron beam

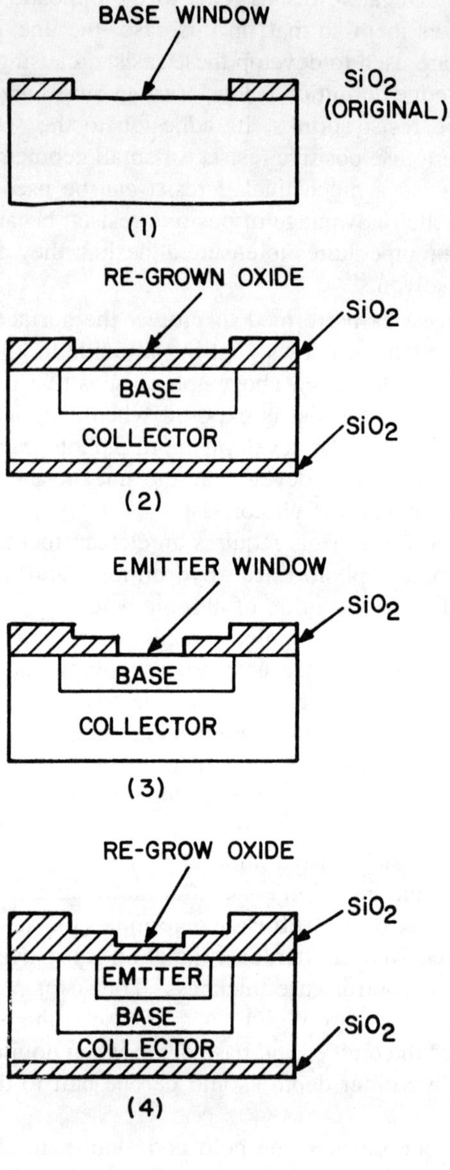

Fig. 5.50 Multiple oxidations.

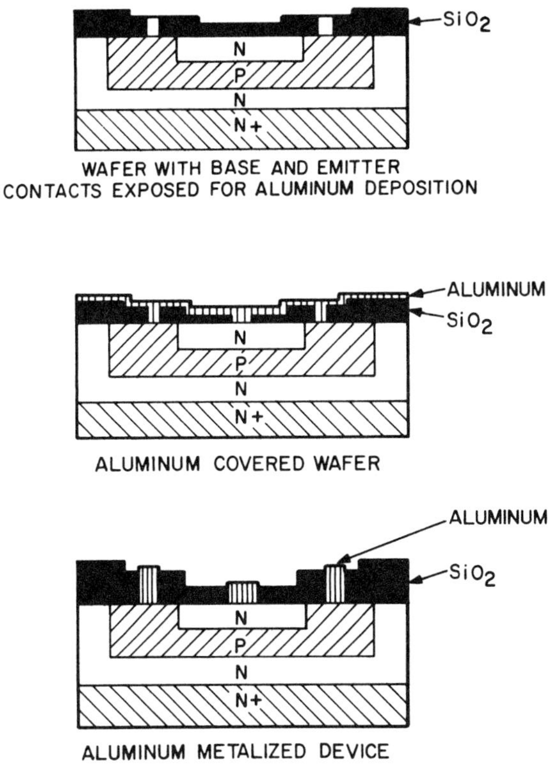

Fig. 5.51 Aluminum metallization technique.

technique to achieve a uniform and pure aluminum layer. A photolithography process, described earlier, is used to define the final metal pattern needed. A patented etch process is used to etch away the unwanted aluminum. This technique is very powerful in that it allows extremely fine aluminum lines of less than 0.1 mil to be cut. In addition the process is inherently very high yield and defect free. Its major advantages are finer line geometries and very high yield on large devices (permitting very high-power, high frequency units).

Aluminum silicon metallization is another process employed, in which there is an addition of 1 to 4 percent silicon to the aluminum. The advantages are the same as with the aluminum system, but there is the additional benefit of vastly improved yield and reliability on very high frequency devices. Usually when a transistor is said to have aluminum metallization, it in fact has aluminum silicon metallization.

To those who prefer gold, the above discussion of improving aluminum would seem to be more "patching the wound" than "curing" the aluminum problem. Gold, it is said, offers distinct advantages over any aluminum system. It is less prone to electromigration (15 to 20 times better); it eliminates the corrosion problems encountered with aluminum; it can be etched easily (it does not undercut); and it virtually eliminates the microcracking associated with relatively brittle aluminum.

All of the advantages mentioned above are applicable to pure gold, but pure gold cannot be deposited on bare silicon. Silicon and gold form a eutectic around 400°C, a temperature not compatible with transistor processing and assembly. Thus a problem arises because the gold migrates into the silicon and no clear boundary is formed between the two. A high temperature metal therefore must be used with gold to act as a barrier separating the gold from the silicon. This barrrier metal should be a good barrier for gold (gold should not diffuse through it); it should make good, uniform electrical contact to the silicon; and it should adhere to the oxide of the device to allow for subsequent assembly.

In actual practice there is no single element that satisfies all of the above conditions, so a multi-metal system must be used. Some of the systems most commonly used as multi-metal barriers are:

Platinum-Silicide/Titanium/Platinum/Gold—Platinum-silicide is used to provide good electrical contact, titanium is used for adherance to the oxide, platinum is used to prove a barrier to gold diffusion and gold to carry high current densities. This system is primarily used in *beam lead* applications. It is very complex and impractical for very small geometries because of the etching difficulties of both titanium and platinum. This system, if it does work, is very reliable, if care is taken to passivate the titanium which is worse than aluminum in terms of corrosion.

Platinum-Silicide/Molybdenum/Gold—This metallization system has been used for the last several years in small signal/high frequency transistors. The main problem in this system is the high electrical contact resistance. In high-power transistors, low input resistance is one of the essential features for achieving good performance. Platinum-silicide is used to minimize this problem. However, getting consistent low contact resistance still remains a problem to be solved.

Platinum-Silicide/Tungsten/Gold—This system has been introduced recently for use with high power/high frequency transistors. Tungsten does not make good electrical contact to silicon, therefore, platinum-silicide has to be used to lower the contact resistance. Tungsten also does not adhere well to oxide. To overcome this problem a very high sputtering energy, which tends to degrade the E-B junctions and thereby results in very poor yields, is required. To solve this problem a fourth metal is required to "glue" the tungsten to the oxide. Tungsten, while better than titanium, is not an acceptable barrier to gold, as accelerated tests at 600° C have proven. Moreover, tungsten is very hard to etch, as it undercuts the gold during etching; but it is impossible to quality inspect the seriousness of the damage. How far it has been undercut is clear only when the gold falls off the wafer. Obviously, this type of metallic barrier is not used very often.

Auromet®—This gold metallization system has been introduced recently and answers most of the problems encountered in gold metallization systems. It has excellent adhesion to oxide; it is a very good barrier to gold, as accelerated tests at

700° C for 24 hours have shown no significant inter-diffusions, and it has excellent step coverage with no microcracking.

Despite many claims from several manufacturers of the superiority of one system over another, unfortunately there has not been enough time to pile up long hours of operating life tests to prove the absolute value of long term reliability of any one of these systems. The only meaningful data is that of comparison of different systems to each other. Where some of the gold systems have shown to be at least an order of magnitude better than aluminum in electromigration induced effects, other gold systems have shown worse MTBF, because of the gold diffusion through the barrier metal.

The package is perhaps as important as the chip itself in achieving device performance. Moreover, the technology involved in designing the one is as complex as that of designing the other. The concept behind a good package is outlined as follows.

1. Reduce the lead inductance to a minimum. Priorities are the common lead first, the input lead second, and the output lead (or least significant lead inductance) third.
2. Increase the thermal dissipation to a maximum. *Get the heat out!*
3. Make it a reliable, easy-to-use package by considering such important factors as package type, shape, size, material, and environmental resistance.

As the frequency of operation increases, the package becomes more and more important. This concept should be obvious since the inductance factor, for example, becomes much more critical in a transistor that has to operate properly at 2 or 3 GHz. Also, the materials must be selected with greater care. Figure 5.52 shows the internal construction of three microwave transistors and shows how some of the factors listed above are incorporated. Figure 5.52 (a) is a low power linear device (1 W), Figure 5.52 (b) is a 10 W device, and Figure 5.52 (c) is a 40 W transistor. Notice how the density of the device increases with power. The number of cells increases to handle the current that the increased power produces. Notice also how the bonding wires are all of equal size and shape to ensure that the lead inductance is kept to a minimum and that its devices operate properly. Figure 5.53 is a picture of six different styles of cases available for microwave transistors. As you can readily see, virtually any style needed for a particular application is available.

One of the most important factors governing the proper operation of power transistors cannot be ignored—the heat removal capability of the chip. The thermal characteristics of the chip are largely dependent on the particular design. From heat transfer theory, a point heat source (the chip) dissipates heat through a solid in the shape of a cone; that is, the heat is concentrated at the origin and fans out as it travels farther from that point. One cell of a transistor could be approximated as a collection of point heat sources next to each other. Therefore, when heat is dissipated, the point heat source closest to the center of the cell is subjected to "heat

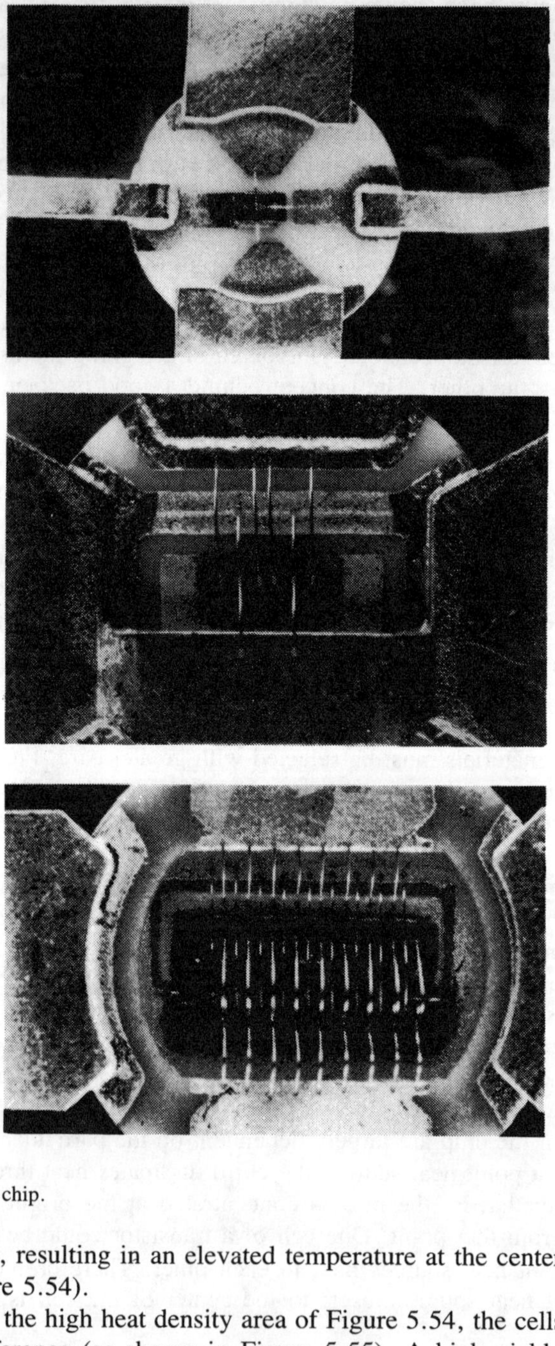

Fig. 5.52 Transistor chip.

cone" interference, resulting in an elevated temperature at the center of the cell (as illustrated in Figure 5.54).

To eliminate the high heat density area of Figure 5.54, the cells can be spaced to minimize interference (as shown in Figure 5.55). A high yield construction is

Fig. 5.53 Transistor package.

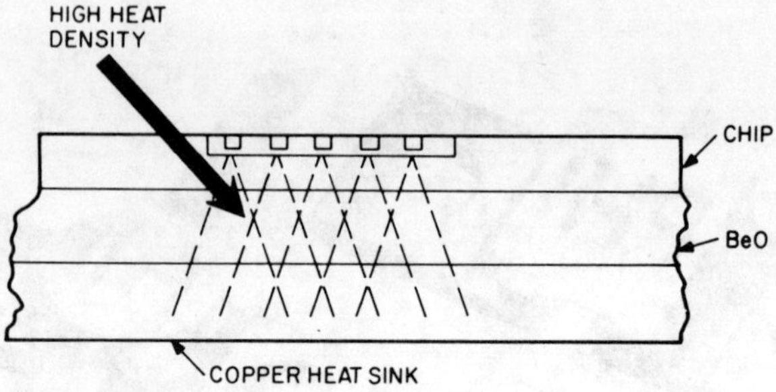

Fig. 5.54 Thermal dissipation in the chip—"heat cone" interference.

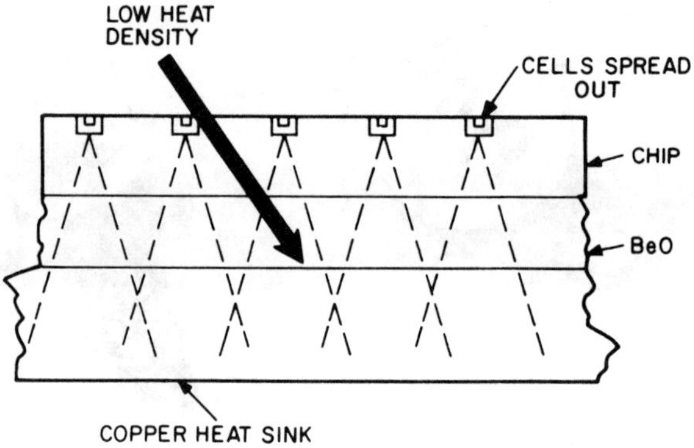

Fig. 5.55 Thermal dissipation with cells spread out.

used in this case and a geometric symmetry arises from the use of cellular and interval matching techniques, virtually assuring thermal balance. As an example, thermal measurements were taken in a 32 cell, 125 × 75 mil chip and slow thermal gradients (temperature differences) of only 3 percent from cell to cell. Single cell transistors often show temperature variations of 50° C from center to edge. Uniform heat distributions allow the junctions to operate at moderate temperature assuring long life reliability of the device.

An RF power transistor is a reliable device capable of operating in excess of 100,000 hours without failure when electrical specifications are observed. Without proper mounting, though, the transistor may be both mechanically and thermally stressed beyond reliable limits.

Of prime importance when using a transistor with a stud is the torque put on that stud. Too much torque can sheer the stud off; too little torque results in inadequate contact for electrical ground and thermal conductivity. A stud torque of 8 ± 1 in/lb should be used when installing a 3/8 inch stud transistor; 5 ± 1 in/lb for a 1/4 inch stud; and 10 ± 1 in/lb for a 1/2 inch stud. A rclease type torque wrench should always be used.

Flange packages must be mounted on a flat (±2 mil) surface if the proper heat transfer is expected. It is most important that the flange not be twisted or bent before or during installation.

Studless packages may be installed using epoxy or solder. When soldered properly, thermal resistances equivalent to the stud package can be expected. Some degradation in thermal resistance is noted when epoxy is used; the extent depends on the conductivity of the epoxy employed.

A high quality silicone grease should always be used on both stud and flange devices; it improves the interface thermal resistance by at least 0.2° C/W and as much as several ° C/W (if the heatsink surface has a poor flatness).

One of the most important aspects of transistor mounting is the relationship of the leads to the circuits. The transistor should fit into the circuit without stressing the leads when torqued to the heatsink. Never solder a transistor into a circuit before torqueing. The two most important points to remember when mounting a transistor are not to stress the leads excessively and keep the leads short.

Figure 5.56 shows three improper transistor mounting techniques; Figure 5.57 shows the proper methods to use.

Thus, the bipolar transistor for microwave applications has been discussed in three distinct and individual areas: *Low-noise, linear power,* and *power*. Each has specific areas where it can be used and has found areas where only microwave tubes were used previously. Truly, the bipolar microwave transistor has a prominent place in microwaves today.

5.4.2 Field Effect Transistors

The overwhelming choice for microwave transistors in the mid-1970s was the gallium arsenide field effect transistor (GaAs FET). You could not pick up a microwave magazine without seeing more and more advances being made in these devices. Their popularity and usage grew and their acceptance increased with good reason. They were, and are, excellent devices for microwave applications.

The field effect transistor differs from the bipolar in more than name. You will recall from our discussion of bipolar transistors that the device operated with two types of carriers, electrons and holes. The field effect transistor, on the other hand, is what is termed a *unipolar* device, that is, it operates with only one type of carrier—electrons. As discussed in Section 5.2.2., this reliance on only one type of carrier allows the device to operate at significantly higher frequencies than their bipolar counterparts. This, of course, makes them an ideal choice for microwave applications.

We have stated that the mid-1970s saw the blossoming of the FET on the microwave horizon. These devices actually have a much longer history than that.

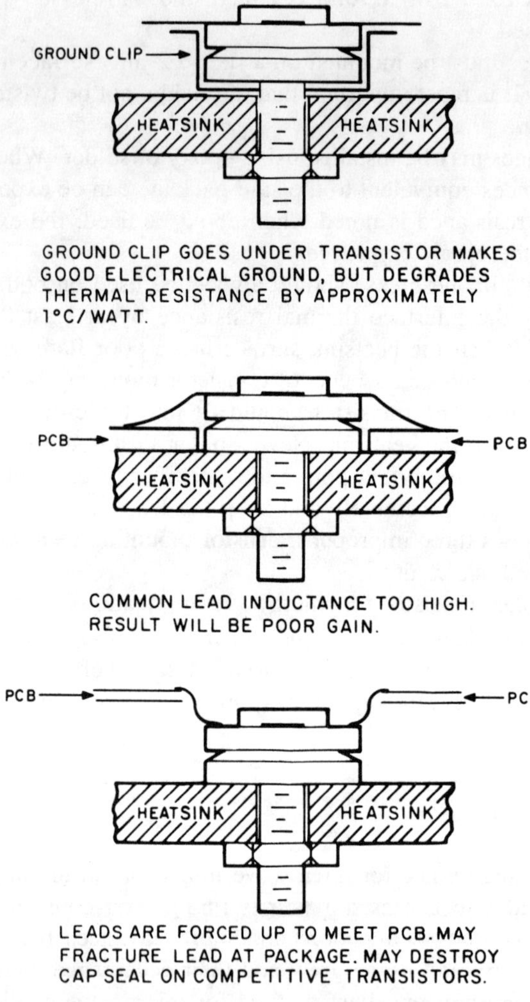

Fig. 5.56 Improper mounting techniques.

The theory for FETs goes back to 1926, and William Shockley first proposed them in 1952. At that time, however, there were many technological and fabrication difficulties that kept the FET from arriving on the scene until the early 1960s. At this time the silicon bipolar device was so well defined that it pushed FET development further in to the 1970s.

The earliest FET was the junction FET (JFET), which became available about the same time as the first microwave bipolar transistors. The JFET construction is shown in Figure 5.58 (a). Advances in techniques and a need for low power resulted

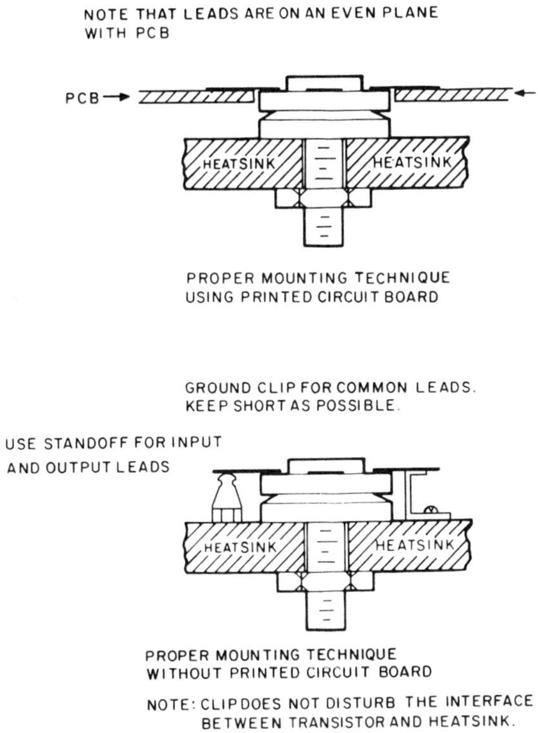

Fig. 5.57 Proper mounting techniques.

in the metal-oxide-semiconductor FET (MOSFET). This device also may be referred to as an insulated-gate FET (IGFET) because of the oxide layer between the gate and the substrate as shown in Figure 5.58 (b).

These FETs offered no competition to the well-established bipolar transistor in either of the forms shown in Figure 5.58. However, when the metal-semiconductor FET (MESFET) came on the scene, the bipolar had competition. This device now introduced the Schottky barrier as the gate of the device and completely departed from the conventional *pn*-junction. Figure 5.59 shows a representation of the HESFET. With the construction came the now familiar gallium arsenide field effect transistor (GaAs FET). This was the device which was the buzz word of the 1970s and on to the 1980s.

Designation of the elements are different from those of a bipolar device. When using a bipolar device you put the 10 V or 12 V on the collector, biased the base, and attached the emitter to ground. In a FET you put the 10 or 12 V on the drain, bias the *gate,* and attach the source to ground. This is illustrated in Figure 5.60.

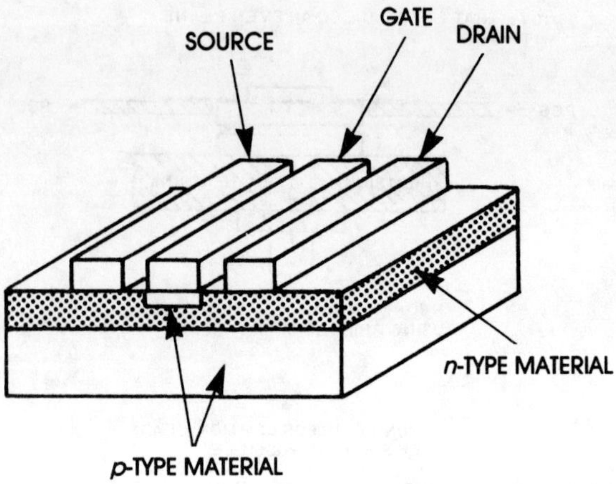

(a) JUNCTION FET (JFET)

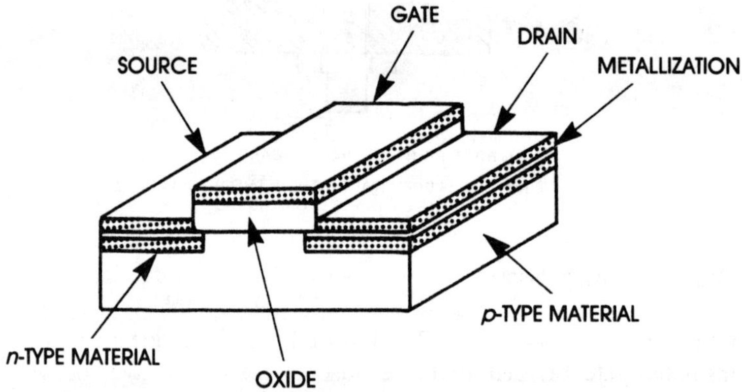

(b) METAL-OXIDE-SEMICONDUCTOR FET (MOSFET)

Fig. 5.58 Field effect transistor (FET).

Presently GaAs FETs are in use in many areas of microwaves. Their range of frequency is upwards to 40 GHz and above in many cases. Noise figures of 3 dB are possible in this frequency range also.

As was discussed in Chapter 2 gallium arsenide is a semiconductor compound (number III-IV) which has many advantages over the conventional material, silicon. Some of these advantages are:

1. The drift mobility is higher and, therefore, unwanted resistances which reduce gain and increase noise are lower.

2. Peak velocity of electrons is higher—thus, the high frequency gain is increased.
3. It can be made in an insulating form which allows the construction of low-capacitance, high-gain devices on it.

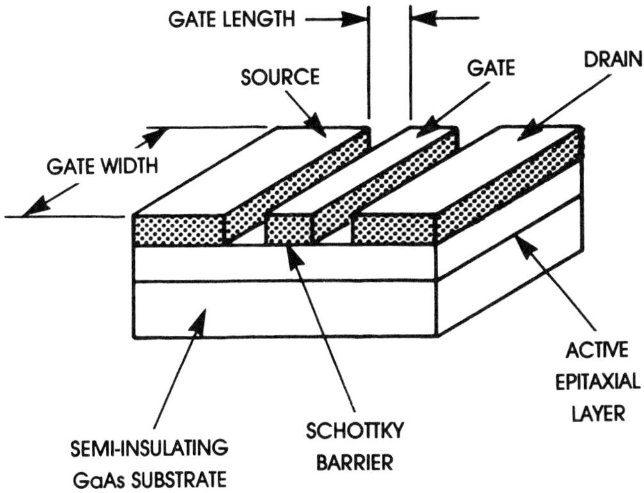

Fig. 5.59 MESFET.

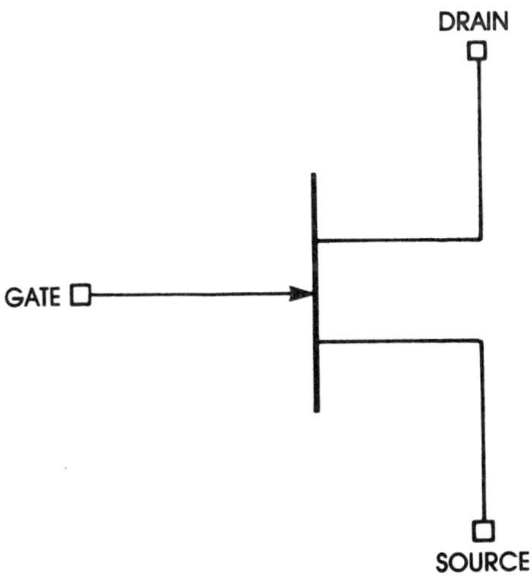

Fig. 5.60 GaAs schematic.

Figure 5.61 is a diagram of GaAs FET geometrics. Figure 5.61 (a) is a low noise device, while Figure 5.61 (b) is a power device. The source, gate, and drain of each device is marked. Notice the difference in appearance of the two geometries shown. First, both are oriented the same. That is, the gate is always between the source and the drain. This allows the gate to perform its modulating function and produce gain through the device. Second, notice the different types of structures for power and low noise. The power device has many areas for the element so that each may share the current and, thus, carry more total current. The low-noise device must be a structure which carries very little current and also has small gate resistance to keep the internal noise level low.

A point should be brought up before proceeding on with further information. That point is the importance of the gate when specifying GaAs FETs. You notice that with different manufactures, GaAs FETs may be called 0.5 μm, or 1 $-\mu$m, or some other value device. This number refers to the gate length and is the most important single FET element that determines high frequency gain. Commercial, high-performance FETs have the 0.5 μm gate lengths and will operate to around 18 GHz. Lower-frequency FETs have 1.0 μm gate lengths and will operate in the range of 8 to 10 GHz. Some FETs have been built that have 0.2 μm lengths and will operate to somewhere around 30 GHz. Gate length, then, is the term which determines the frequency of a GaAs FET. By referring back to Figure 5.59 you can see the location of the gate length of a device.

A characteristic of the gate that comes into play when discussing power FETs is the width. Small-signal FETs may have a gate width of 0.15 mm, while a power device might have a width of 26 mm. This is also shown in Figure 5.59. The differences in width can be distinguished in Figure 5.60 where both the low-noise and power devices are shown.

With the FET explained, we now can look at its operation. Normally, a positive voltage is placed on the drain relative to the source. This is usually between 2 and 10 V, but can be higher for power FETs. If the gate is first connected to the source ($V_{GS} = 0$) and the drain voltage gradually increased, the source-drain current will increase and then remain almost unchanging. Figure 5.62 (a) shows a plot such as would be obtained. If the voltage is increased beyond a certain point (BV_{DS}), a catastrophic failure will occur. This voltage is somewhat higher than the maximum given on the data sheet. The current level at which the current saturates is called I_{DSS}, current-drain-to-source, with gate shorted to the source. In microwave FETs, the current saturates because the electrons reach a velocity which is limited by collisions with atoms in the GaAs crystal lattice. Increasing the voltage further simply causes more violent collisions but no increase in current. This velocity is about 1×10^7 cm/s. In lower frequency FETs current saturation occurs in a different way. The gate is too long to allow building up the necessary field to cause velocity saturation, but the voltage drop under the gate will eventually become as large as the pinchoff voltage and saturation will occur.

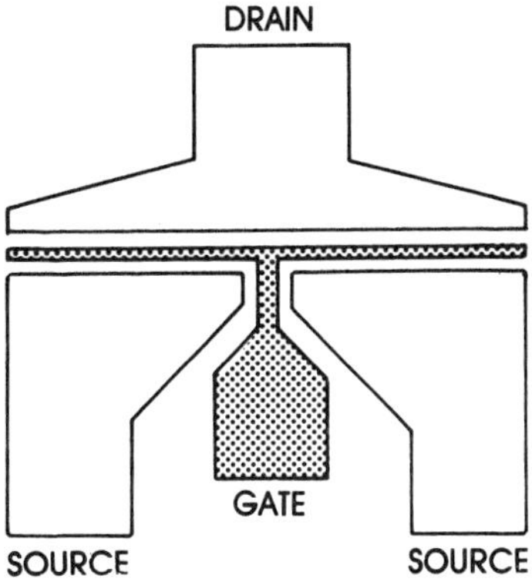

(a) LOW-NOISE GaAs FET GEOMETRY

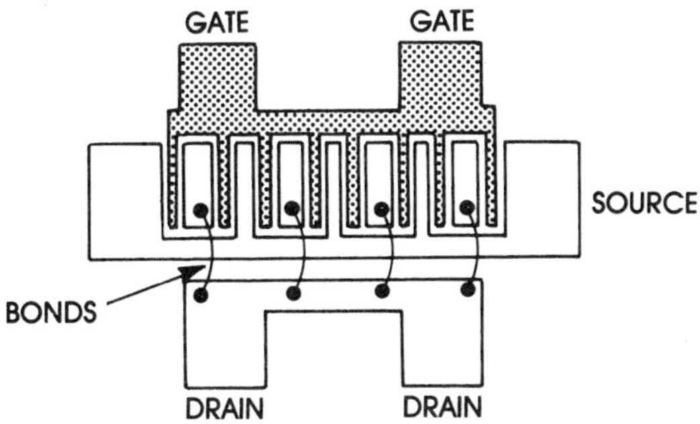

(b) POWER GaAs FET GEOMETRY

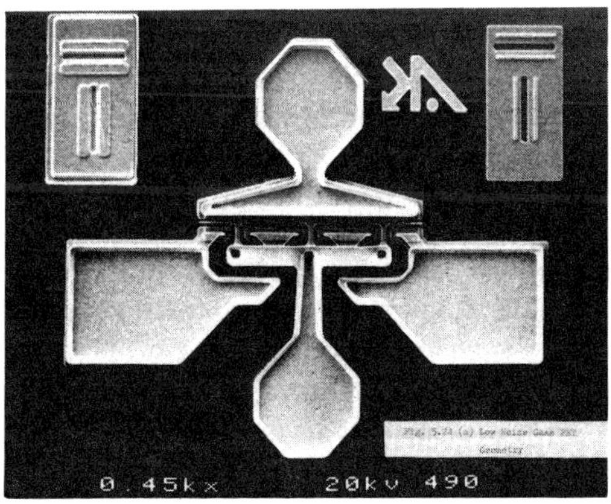

Fig. 5.61 GaAs geometry.

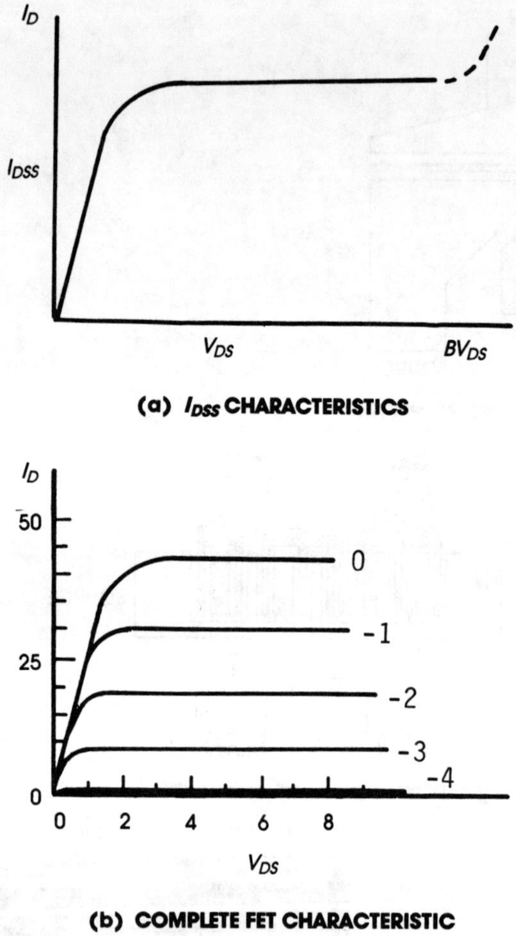

Fig. 5.62 FET characteristics.

If we now refer to Figure 5.63 we will see a configuration which should be familiar. In this situation, electrons travel from the source to drain through the undepleted part of the channel. It should be noted that, even if the external gate-to-source voltage is zero, the gate-channel diodes have a "built-in" voltage equal to about 0.8 V. This means that the channel will always be partially depleted by the built-in voltage, even if no external voltage is applied. Next, the gate is disconnected from the source, and a negative voltage is applied to the gate. Figure 5.62 (b) shows the characteristic of a FET with 0, −1, −2, −3, and −4 V applied to the gate. As the gate is made more negative, there is less of the channel left to carry current, and the saturated current is lower. Also shown in the example is the pinchoff voltage,

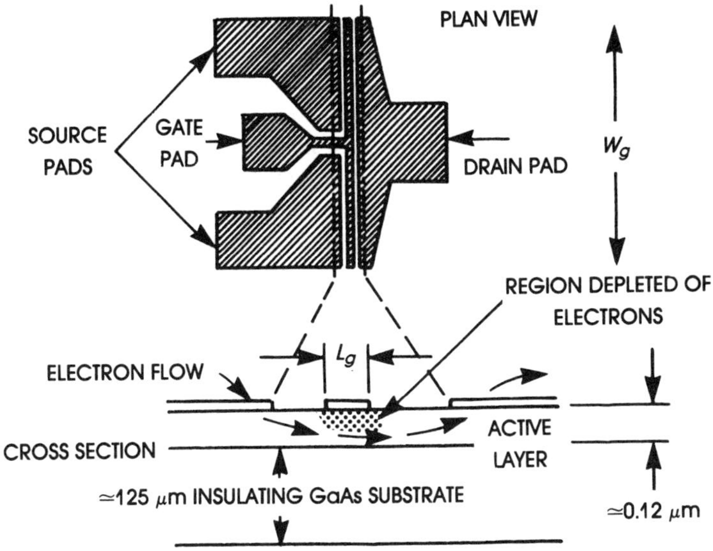

Fig. 5.63 Typical low-noise GaAs FET.

V_p, which is approximately −4 V. Since the FET current never really reaches zero at pinchoff, an arbitrary drain current is selected for defining V_p. For example, define V_p as the gate voltage necessary to reduce the drain current to 1 percent of I_{DSS}. If I_{DSS} = 100 mA, then V_p is the gate voltage to reduce I_d to 1 mA. As a matter of practicality, V_p is defined by a drain current of 0.1 to 1 mA for small FETs, and at a larger current for power FETs.

The preceding description of FET action (*dc*) is also applicable to *ac* or RF conditions. First, an appropriate positive voltage is applied to the drain. Next, a negative voltage is applied to the gate which will give the required drain current. After this, a signal can be applied to the gate. The signal voltage will alternately widen or narrow the depletion layer under the gate, also causing the drain-to-source current to vary. Note that when the gate is positive, the current is increased, and when negative, the drain current decreases. When the drain current is positive and high, the voltage across the load will be maximum and the voltage on the drain minimum. Thus, at low frequencies a common-source FET exhibits a 180° phase shift from gate to drain. At higher frequencies this changes, due to reactive elements and internal phase shift.

There are two additional parameters which are of importance. These are the breakdown voltages, gate-to-source and gate-to-drain (BV_{GSO} BV_{GDO}) with the unused terminal unconnected. When measured, these voltages will be almost the same. In an actual circuit however, it will be noted that the gate-to-drain voltage will be the sum of the gate and drain voltages. The gate breakdown voltages are "soft"; that is, the current change near breakdown is gradual. This is very desirable in some circuits, for it means that if the gate is driven slightly into breakdown by a large signal, no damage occurs. On the other hand, if BV_{DS} is exceeded by using too high a drain voltage, breakdown is instantaneous and catastropic.

GaAs FETs are excellent devices for low noise characteristics. Their frequency range can be from 0.5 GHz up to 30 GHz. Below 0.5 GHz, the $1/f$ noise dominates performance and the noise figure increases.

Since the FET has several noise sources and these are correlated, there is an optimum input tuning for best noise figure which is not the same as that for gain. As a result, when the FET is tuned for best noise figure, some gain will be lost. The gain obtained when optimum noise figure tuning is used is called the "associated gain," G_A. (This and other terms will be covered in the next section when we present a low-noise data sheet.)

In order to utilize GaAs FETs to their full potential and choose the right devices for your particular application, you should be familiar with the terms present in a low-noise device data sheet. We will now present these.

The first thing to look for when considering a low-noise GaAs FET is the frequency of operation. If possible, try to choose a device that operates at a higher frequency than needed, so that the portion of the curve that is increasing in noise figure is avoided. For example, in Figure 5.64, operation at 8 GHz or less would be needed, if the device were used with the characteristics shown. Above the frequency the noise figure increases, and the associated gain decreases, which is not an ideal situation. So be sure your frequency of operation is below the maximum frequency of the device used. One note on Figure 5.64 is that this curve holds for the condition given—V_{DS} = 3 V, and I_{DS} = 10 mA; that is, a drain-to-source voltage of 3 V and a drain current of 10 mA. If your FET is biased differently than these conditions, it will not perform as shown in Figure 5.64.

The noise figure on a data sheet can be termed optimum noise figure (NF_{OPT}), minimum noise figure (NF_{MIN}), or spot noise figure. Any one of these terms may be used, but they all mean that this figure is the lowest noise figure when the conditions listed (V_{DS} and I_{DS}) are used and the corresponding *s*-parameters for these conditions are used.

Gain is a very important parameter for a low-noise circuit. There are actually two gain numbers involved when talking of low-noise GaAs FETs. One is the maximum available gain (G_{MAX} or *MAG*) and the other is the gain at optimum noise figure (G_{NF} or *Ga*).

The maximum available gain figure is not really concerned with low-noise performance. It is concerned with obtaining the most output available from a device. Conditions are once again important when obtaining this gain figure. The conditions for the maximum gain are usually spelled out as V_{DS} = 3 V, I_{DS} = 30 mA. Notice the greater amount of current necessary to obtain the increased gain.

The gain at optimum noise figure numbers are those obtained when the conditions are duplicated for the best noise figure obtained. That is, we have said that

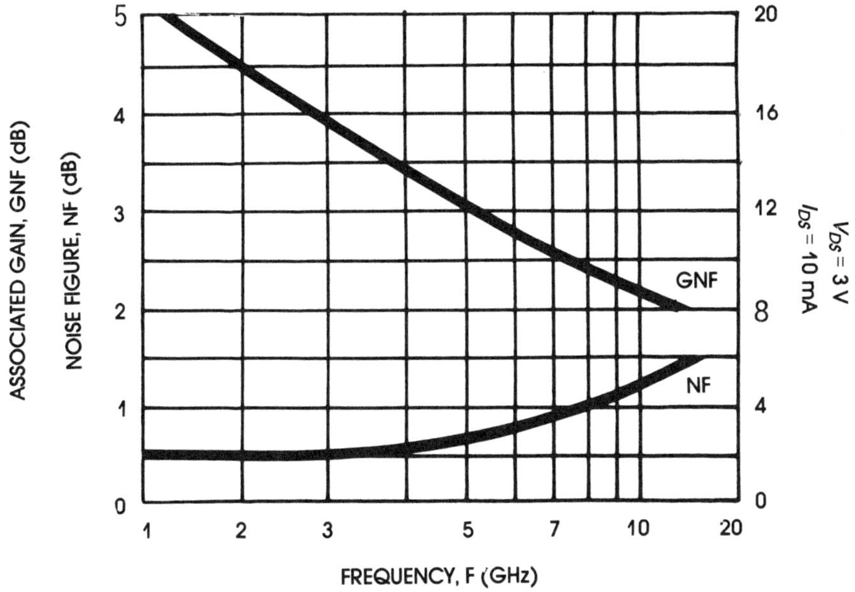

Fig. 5.64 Noise figure and gain *versus* frequency.

the best noise figure was obtained when $V_{DS} = 3$ V, and $I_{DS} = 10$ mA. We, therefore, will duplicate the specified gain when these conditions are met. This gain, as might be imagined, is lower than the maximum available gain, since we are operating with less drain current in order to get a low noise figure. Less drain current will automatically reduce the gain. The gain figure can be on the order of 2 to 3 dB difference from the maximum available gain figure. This difference however, can be made up in following stages, if increased gain is really important. The prime function of the first stage is low noise.

Additional parameters which will appear on low-noise device data sheets are shown below. Most have been defined previously.

I_{DSS} = Saturated drain current (maximum);
I_{GS} = Gate to source leakage current;
I_{GO} = Gate to drain leakage current;
I_{DS} = Drain current;
V_{DS} = Drain to source voltage;
V_{GS} = Gate to source voltage;
gm = Transconductance (ratio of change in drain current divided by the change in gate voltage as previously defined).

When FETs are made very large, a design such as shown in Figure 5.61 (b) becomes impractical because of the extreme width of the gate. To avoid this problem, the gate is divided into segments which are connected in parallel. Figure 5.61 presented a power FET with eight gate segments. When the gate is broken up into segments, a topological problem arises. One of the elements (the drain in this case) becomes enclosed by the other elements and must be brought out by bridging. Metal straps over air (air bridges), or metal straps over silicon dioxide (dielectric bridges), are sometimes used. Figure 5.61 shows wire bridges.

As the signal travels down the gate from the end feed point, the input signal undergoes a loss and phase shift which are worse at higher frequencies. The gate metal resistance, combined with the gate-to-source capacitance, form what is the equivalent of a lossy RC transmission line. As a result, the gates near the feed point and at the far end receive signals which are different in amplitude and phase. To compensate for this, the gate is broken up into narrower and narrower (W_g) segments for higher frequency FETs.

A figure of merit for power FETs is watts per millimeter of gate width. For 4 GHz lab FETs the number is about 1 W/mm; at Ku band this drops to 0.3 to 0.4 W/mm. A conservative operating level for X band is 0.4 to 0.6 W per mm.

As FETs are made increasingly larger to obtain higher power, the matching problem becomes more acute because the input impedance becomes much less than 50Ω. At higher frequencies the problem becomes even more severe, because the impedance level also drops with increasing frequency. Some of the larger packaged power FETs employ matching or partial matching at the input and output, using inductors and capacitors inside the package. This alleviates the problem considerably but limits bandwidth to a specific frequency range.

Thermal resistance is as important in power FETs as it is in power bipolar transistors. While FETs do not "run away" as do bipolars, the thermal resistance of GaAs is higher than that of silicon. The safe operating region for a FET is bounded by BV_{DS} and I_{DS} at low dissipation and about 180° C channel temperature at high current and voltage, as shown in Figure 5.65. The correct operating temperature should be based on the MTTF desired.

As in the case for low-noise devices, items from a data sheet are presented to acquaint the reader with important terms.

The first parameter to consider, of course, is the frequency of operation. If the short paragraph at the top of the sheet does not call out a specific frequency range, it will usually call out a band. If you are not familiar with the frequency bands, look further down the data sheet and notice frequencies for specific parameters and determine the operating range of the device.

The second parameter to be concerned with is the output power, which will be presented in two ways: power output and power output at the 1-dB compression point. The power output figure is the one that tells the linear power output; that is, the power output of the device when operating within a specified input power range. This range does not drive the device into compression. For this power output figure,

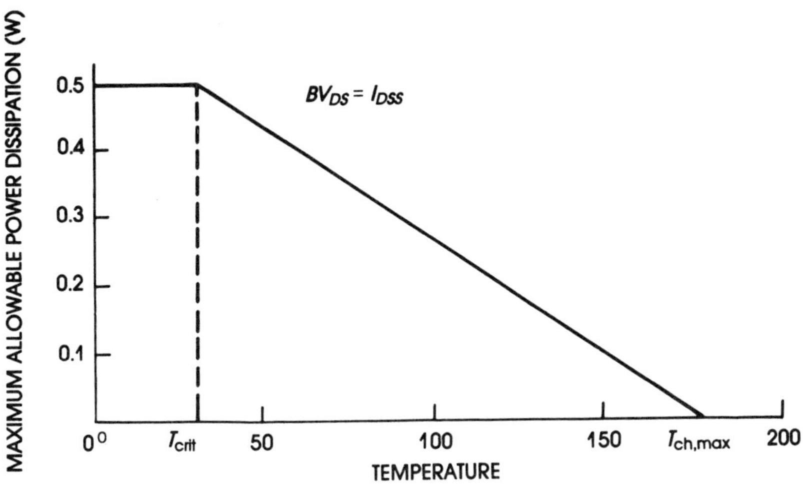

Fig. 5.65 Power dissipation.

there must be conditions set down that will result in this power output. These conditions are power input, V_{DS}, and either I_{DS} or V_{GS} (V_{DS} is drain to source voltage, I_{DS} is drain to source current, and V_{GS} is gate to source voltage).

Power output at the 1 dB compression point is found by increasing the power input in 10 dB steps and noting that the ouput also increases by 10 dB. When a 10 dB input results in a 9 dB increase in output, the 1 dB compression point has been reached. To increase the output further would not make any sense because you will get no more power at the output; the limit has been reached. Many times the 1 dB compression output power is used to keep the output of an amplifier constant over a wide range of environmental conditions, or over a varying range of input powers.

Gain, of course, is also an important parameter to consider in a power device. It would be foolish to supply dc and RF power to a device and only produce 1 or 2 dB of gain. Most power FETs will exhibit from 7 to 10 dB of gain up to around 8 GHz with very little effort. Be sure to check the particular device to be used to be sure you know what gain it should produce.

The use of power devices brings to light a whole new set of parameters. These parameters are thermal. First, check the operating temperature range of the device to be used. Most devices will operate from around −65° C to +175° C. These temperatures are somewhat less than the temperatures of their bipolar counterparts that will operate up to +200° C. When making the transition from bipolar to GaAs FETs, you should be aware of this temperature difference.

The other thermal parameter, thermal resistance, is expressed in degrees Celsius per watt and tells how well the device gets the heat generated within the device out of that same device. This figure should be as small as possible, since as little rise in temperature per watt of power as possible is wanted.

The final parameter to be investigated is the impedance of the device to be used (S_{11} and S_{22}). Figure 5.66 shows a typical plot of S_{11} and S_{22} for a power device that produces a 1-W output at 8 GHz. Notice how much lower the impedance is at the input than for the low-noise device. This lower impedance will make the input matching to a 50-Ω circuit a little more difficult, and may require multisection matching if a broad band of frequencies is to be covered.

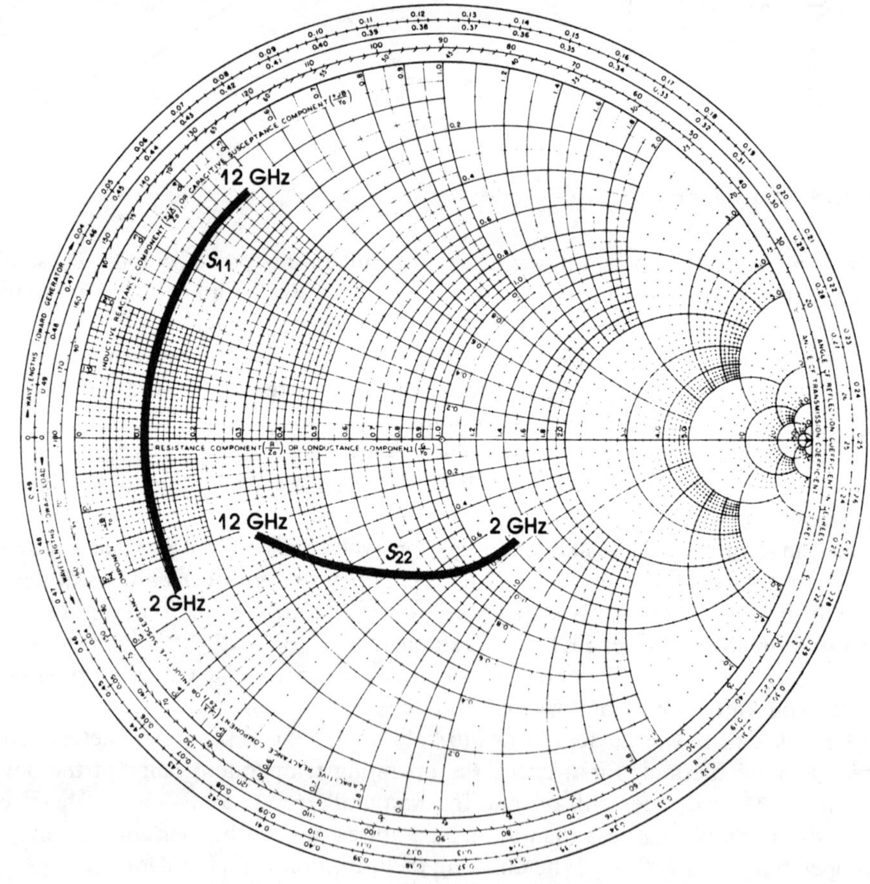

Fig. 5.66 S_{11} and S_{22} for a power device.

One area that is of prime importance, but is not covered on data sheets, is handling precautions for GaAs FETs. These precautions are for both low-noise devices and power FETs.

When handling the packaged unit, never pass the device from one person to the next by the leads, especially on a dry day, since static charges will destroy it. Although the MESFET is not as sensitive as a MOSFET, it is still quite delicate.

When soldering the device into a circuit, the soldering iron must be properly grounded. Soldering times should not exceed 20 s at 260° C. Never insert a FET into a prebiased circuit, particularly one where self-biasing is used; the bypass capacitor will charge through the forward-biased barrier and most likely will destroy the gate.

Once the device is soldered into the circuit, do not just "switch on" the gate and drain bias supplies. Many regulated power supplies feature large in-rush transients due to their stability-limited gain-bandwidth product. Adjust V_{GS} slowly to about -1.0 V, gradually increase V_{DS} from 0 V to the desired bias point and, finally, readjust V_{GS} to give the desired drain current.

Remember, for a FET, $V_{DG} = V_{DS} - V_{GS}$, so place no more than 10 V between the drain and gate. If something is suspect, avoid the use of a digital multimeter to check the resistance because the voltage supply within the meter may destroy the gate.

We must also caution against the use of curve tracers. The high-voltage transformers used in most curve tracers may have high leakage currents that can, if not properly grounded, destroy the FET. If a curve tracer test is required, we suggest the following:

1. Ground the curve tracer to earth.
2. Set the gate-source voltage to zero voltage.
3. Increase the drain-source voltage slowly to the desired value ($\approx$3 V).
4. Adjust the gate-source voltage to the desired value.

When handling GaAs FET chips, the most important precaution to remember is that all operations must be performed in a clean environment, by the fewest number of people possible. If incoming inspection is performed, leave the chips in the carrier and store in a dry environment, preferably in dry nitrogen. When removing the chip from the carrier, use a vacuum probe with a PTFE tip. If tweezers are used, extreme caution is mandatory. Gallium arsenide is more brittle than silicon, and dust may break off and become lodged in the small channel.

The back of the chip is metalized with pure gold and can be die-attached with low-temperature epoxy, AuGe or AuSn 0.5 mil preforms. Whatever material is used, the maximum allowable exposure is 300° C for five minutes. The use of a die-attach machine is not recommended because the gate may be damaged.

If AuSn preforms are used, recommendations are heating the stage to 200° C and the probe to about 250° C. Probe pressure should be less than 22 g. For specific chip handling procedures, consult the individual manufacturer for any peculiarities that may be present for their devices.

In general, three simple precautions will prevent the destruction of devices from transient breakdown:

1. Unless it is to be negatively biased, dc-ground the device gate.
2. Decouple the bias ports near the device with a low-inductance capacitor (1μF tantalum) shunted with a zener diode (5.8 V, 1.3 W). In addition to limiting transients, the zener diode gives protection against over-voltage and reverse biasing.
3. Switch on power supplies and set their outputs to minimum before connecting the device.

Thus, two types of GaAs FETs have been presented. It can be seen how they are matured devices for microwave operations. Their structure and operations make them more and more suitable for microwave and millimeter application.

5.4.3 High Electron Mobility Transistor (HEMT)

The latest addition to the lineup of solid-state devices designed primarily for microwave applications is the *High Electron Mobility Transistor* (HEMT). This relatively new technology is going through the typical growing pains that both bipolar and GaAs devices have undergone. When they were first proposed in 1978, they received a rather indifferent reception. They have, however, recently attained a much higher credibility in the microwave industry by demonstrating their performance with commercially available devices. Tests have shown some areas where HEMTs have outperformed GaAs FETs. It should be pointed out that the high electron mobility transistor is designed to be a low-noise device only. It is not designed to compete with high power GaAs FETs at microwave frequencies.

The HEMT device looks very similar to the GaAs FET. Figure 5.67 shows a typical geometry for a HEMT. Notice its striking similarity to the low-noise GaAs FET discussed in the previous section. Also, the structure is shown to be similar as presented in Figure 5.68. These two figures lead one to believe that nothing is really different about the HEMT, so why does it perform as well as it does? The answer is not in the structure, as we have seen, but rather the semiconductor materials it uses. These devices are based on the modulation-doped GaAs/AlGaAs heterostructure epitaxy in which the motion of the charge carriers is confined to a thin sheet within the GaAs buffer layer. By referring once again to Figure 5.68 you can see the layers being discussed.

In the HEMT, a silicon-doped AlGaAs layer grown on top of an undoped GaAs layer brings about the formation of a two-dimensional electron gas on the GaAs side of the heterostructure (Figure 5.68). The gas forms because of the electrons' greater affinity to the GaAs. The two-dimensional gas layer is about 150 Å thick and forms the carrier channel that links the device source and drain.

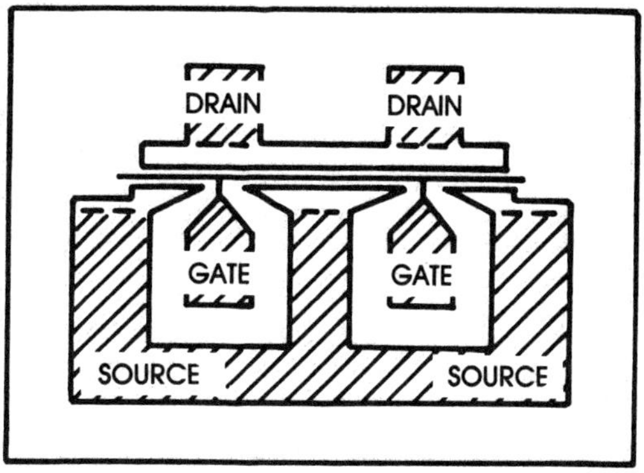

Fig. 5.67 HEMT geometry.

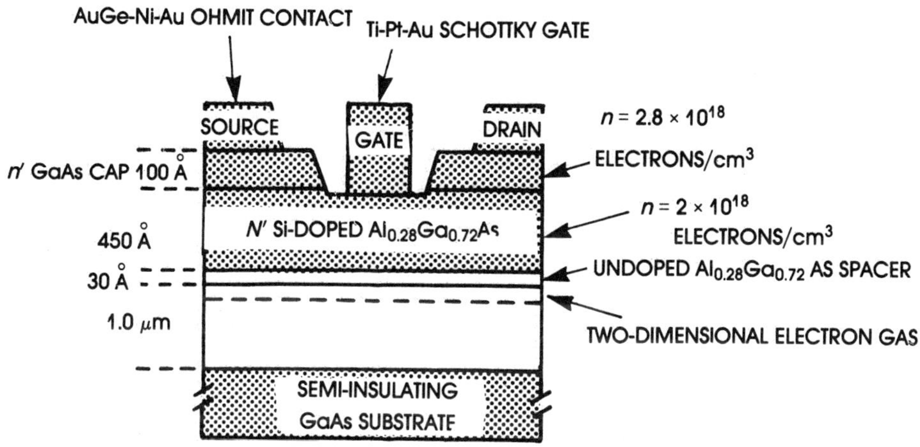

Fig. 5.68 HEMT construction.

When a Schottky barrier gate is placed on top of the AlGaAs layer, a depletion region forms beneath the gate. If the AlGaAs layer is sufficiently thick, the gate and the interface depletion regions will not overlap, and the device will operate as a "normally on" transistor. In this *depletion-mode* device, an application of negative bias to the gate will extend the gate depletion region to the hetero-junction interface, thereby barring electron flow and pinching off the drain-source current.

Etching a recess to seat the gate deeper in the AlGaAs layer creates a device that operates in the *enhancement mode*. Such a device operates as a "normally off" transistor. When a positive voltage greater than the threshold voltage is applied to this type of HEMT, electrons will accumulate at the hetero-junction interface and form the two-dimensional gas, turning the device on. Thus, unlike the MESFET, the HEMT actually functions more like a metal oxide semiconductor (MOSFET), with the Schottky barrier gate controlling the number of electrons in the two-dimensional electron gas by raising and lowering the interface barrier. In a MESFET, bias applied to the gate modulates the channel, and thereby, the number of electrons passing through it. With a HEMT, the channel thickness remains constant; it is the number of carriers that is modulated.

Electrons in a HEMT move in a very thin sheet of material, traveling in a direction that is parallel to the hetero-junction interface, with coordinates x and y (two-dimensional motion). In a MESFET, electrons are allowed three degrees of movement: x, y, and z (three-dimensional motion).

There is no actual proven connection between the degree of electron freedom and a device's noise figure, but certain theories have arisen that the HEMT has a lower noise figure than the MESFET because of the two-dimensional electron freedom as compared to three for the FET. To substantiate these theories, it is true that the average quadratic fluctuation of the electron drift velocity—which is a critical factor in determining noise as a measure of random electron motion—is smaller with two degrees of freedom than with three.

In the HEMT, carrier transport in the two-dimensional electron gas is similar to movement in undoped GaAs. Unlike the highly doped channel of a conventional GaAs MESFET, there is little or no impurity scattering in the undoped GaAs in which the two-dimensional gas resides. As a result, electrons in HEMTs travel at twice the saturated velocity (about 2×10^7 cm/*s*) at room temperature as those in GaAs FETs, with an electron mobility of nearly 8000 cm^2/V-*s* (compared to 4000 cm^2/V-*s* for the GaAs FET channel). At lower temperatures, 77 K for example, the electron velocity increases to about 3×10^7 cm/s while the mobility increases to approximately 80,000 cm^2/V-*s*, depending on the epitaxial layer structure.

We have been making general comparisons of GaAs FETs and HEMTs throughout this section. Now let us show some data. Table 5.5 is an actual test run to compare the two devices side by side.

It can be seen that for these particular devices the HEMT has either matched or surpassed the FET in all catagories. A second set of tests has yielded Figure 5.69. This shows maximum available gain and noise figure for a 0.5 μm gate HEMT and a 0.3 μm gate GaAs FET. It can be seen that the GaAs FET outdoes the HEMT in available gain up to about 14 GHz, but the HEMT is superior in noise figure throughout.

With its inherent electron speed advantages, the HEMT has naturally generated great optimism among digital circuit designers. HEMT ring oscillators have already been fabricated with switching delays of only 12 ps at room temperature.

Only the Josephson junction exhibits lower power dissipation per gate than the HEMT, 4 K (compared to 77 K operation for the HEMT). Research on the application of the HEMT as a digital device continues.

Table 5.5

Parameters and Conditions	*GaAs FET*	*HEMT*
Minimum noise figure (dB) at		
V_{DS} = 3.5 Vdc and I_{DS} = 10 to 15 mA		
at 4 GHz	0.7	0.4
at 8 GHz	1.4	0.8
at 12 GHz	1.9	1.2
at 18 GHz	2.6	1.8
Associated gain (dB) at		
V_{DS} = 3.5 Vdc and I_{DS} = 10 to 15 mA		
at 8 GHz	11.5	12.0
at 12 GHz	9.5	10.5
at 18 GHz	7.5	9.5
Maximum available gain (dB) at		
V_{DS} = 4 Vdc and I_{DS} = 0.5 I_{DSS}		
at 4 GHz	17.0	17.0
at 8 GHz	13.0	15.0
at 12 GHz	11.5	14.0
at 18 GHz	9.5	11.5
Output power at 1 dB compression (dBm) at		
V_{DS} = 4 Vdc and I_{DS} = 0.5 I_{DSS}		
at 12 GHz	13.0	13.0
Saturated drain current (mA) at		
V_{DS} = 3 Vdc and V_{GS} = 0 Vdc	45	36
Transconductance (mmho) at		
V_{DS} = 3 Vdc and V_{GS} = 0.2 Vdc	40	45

Note: All specifications are at +25° C.

The HEMT also brings optimism to analog designers, especially those working at millimeter-wave frequencies. With its low noise and high gain, 0.2 - m HEMTs may one day form the heart of low-noise amplifiers working up to 100 GHz. Such transistors have been fabricated at TRW and are currently undergoing full evaluation. In the meantime, HEMTs with gate lengths ranging from 0.50 to 0.25 μm have completed testing by a number of manufacturers and show good performance up to 45 GHz. (An "average" HEMT, is one having a 0.35-μm gate length. Such a transistor has a nominal gate width of 75 μm and dc transconductance of 240 mS/mm of gate periphery.)

While the HEMT should see the majority of its analog use at millimeter-wave frequencies, the technology has also been explored for lower frequency applications.

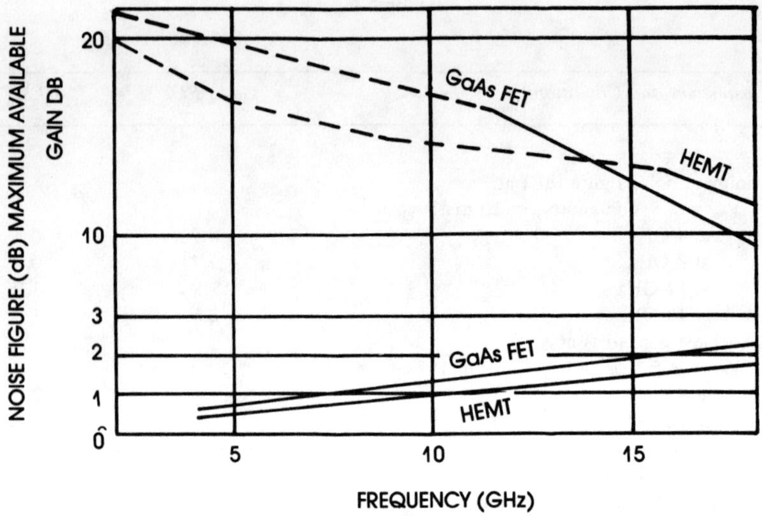

Fig. 5.69 GaAs FET *versus* HEMT.

All of the characteristics that make the HEMT such a viable candidate for millimeter-wave work also make it attractive for low power-consumption, lower frequency designs.

Thus, the HEMT (high electron mobility transistor) joins its predecessors in the microwave field. It has had its triumphs to this point, and it has had its defects. These are the ups and downs that the bipolar and the GaAs FET have been through and, to some degree, are still going through. The HEMT has found many application areas and will continue to find even more.

5.5 SUMMARY

This chapter has presented an area of microwaves which has been an area of mystery for years—that of semiconductors. It has provided an explanation of two types of junctions, *pn* and Schottky. It has presented seven types of diodes useful for microwave applications and has dispelled the idea that a diode is just for rectification in a power supply.

Lastly, a long look was taken at microwave transistors. The bipolar, GaAs FET, and HEMT were covered.

Surely it can be seen that our opening statement of how microwave semiconductors have progressed the fastest of any microwave component is very true.

Chapter 6
The Smith Chart

When Phillip H. Smith passed away in 1987, the microwave industry lost one of its greatest contributors. His *Transmission Line Calculator* developed in 1939, and which later became known as the *Smith chart,* has become the microwave tool of the trade. It is synonymous with virtually every transmission line problem you can imagine.

Since this book is primarily technically oriented, information concerning the theory and derivation of the chart is omitted in favor of more precise concentration on its utilitarian aspects. But first, a preliminary examination of the Smith chart is in order.

The Smith chart shown in Figure 6.1 is called a *normalized chart*. It has a 1.0 at the center. Values on this chart are obtained by dividing the impedance you have by the characteristic impedance of the system (50Ω, for example). If you measured an impedance of 150Ω on a 50Ω transmission line, the normalized impedance would be plotted at point three on the chart. This would be along the center line of the chart, to the right where the number three is located.

If you are working with complex impedances, you have a quantity $R +/- jX$. This shows a real and an imaginary component for each impedance. To normalize this type of impedance you divide each component by the characteristic impedance. That is

$$Z/Zo = R/Zo +/- j\,X/Zo$$

As an example of a normalized impedance, consider an impedance of $60 + j75\Omega$ in a 50Ω system. The normalized impedance would be $1.2 + j1.5\Omega$.

The Smith chart has many unique features which can be used to your advantage. Points along the center line of the chart are all pure resistance points. They have no imaginary (or reactive) component associated with them. Similarly, points along the outer edge of the chart are all pure reactance. They are inductive reactances

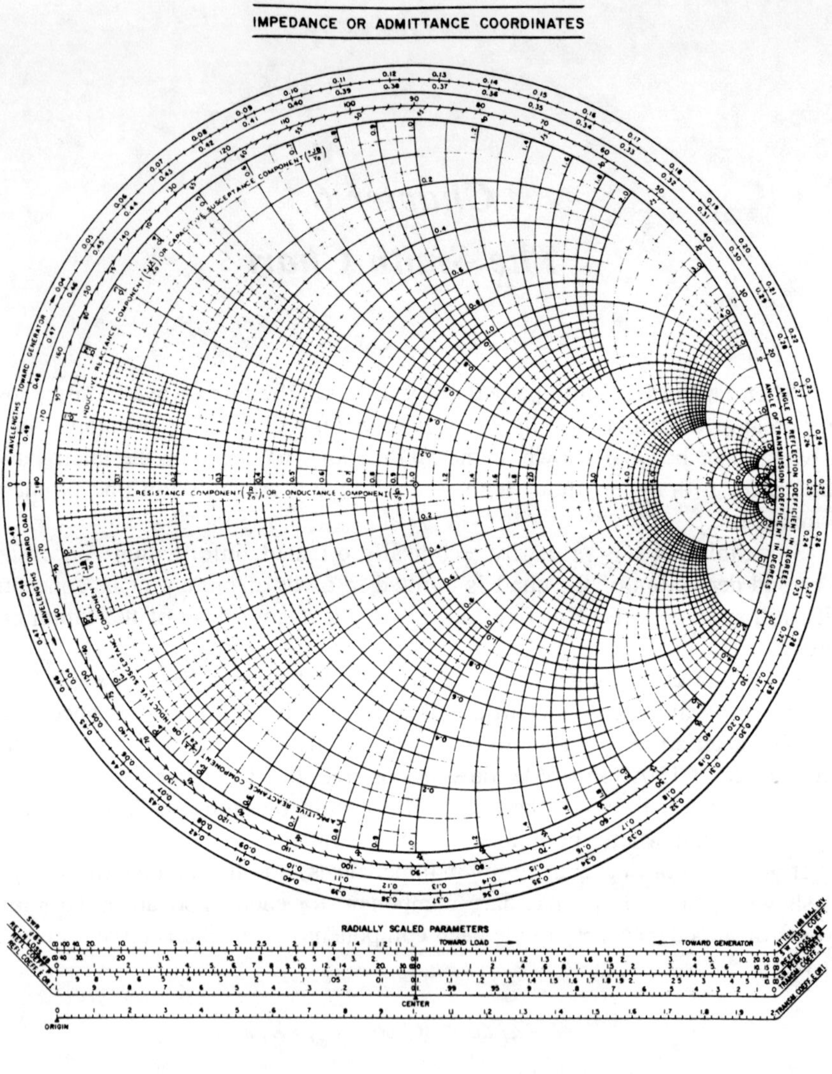

Fig. 6.1 The basic Smith chart.

on the top edge and capacitive reactance on the lower edge. These points have no resistance associated with them.

The two outer "tracks" display *wavelengths toward the generator* in a clockwise direction and *wavelengths toward the load* in a counterclockwise direction. The bottom of the chart displays *radial parameters*. These will be covered in detail later in the chapter.

6.1 IMPEDANCE

Any impedance can be plotted on a Smith chart. To illustrate this, an example will be presented. Suppose we are given an impedance $Z = 5 + j25\Omega$ in a 50Ω system. (the $+j$ indicated a reactance component *above* the 0 reactance line. A $-j$ would be below it) When this impedance is normalized it becomes $5/50 + j25/50$ or $0.1 + j0.5\Omega$. This is plotted at point A in Figure 6.2. It is plotted first by finding the 0.1 point on the horizontal real axis. Then follow the 0.1 circle up ($+j$) until the 0.5 reactance circle is intersected. This is the normalized impedance $0.1 + j0.5$. Draw a line from the center of the chart through this point (A).

The scales on the outside edge of the chart are calibrated both in wavelength and in degrees, and one complete revolution of the chart is equivalent to one-half a wavelength, or 180°. The impedance at any point along the transmission line is that which would be measured if the line were cut at that point and measurements were made looking into the line section connected to the load. The impedance of the generator has no effect on this value. If the transmission line is terminated with the value of its characteristic impedance, then, no matter where the line is cut, the characteristic impedance is measured. However, if this is not the case, and the line is cut at various points, the impedance varies cyclically. If there are no losses on the line, the impedance of the load remains the same at each half-wave point. Thus, as complete revolutions of the Smith chart are made, the beginning point is passed on each turn, representing the cyclical variations of standing waves on a transmission line.

Assume that point A of Figure 6.2 represents a terminating impedance on the transmission line, and that the impedance one-tenth of a wavelength from the load is desired. Using the center of the chart ($1.0 + j0$) as a pivot point, swing an arc equal to the radius from the center to point A toward the generator by an amount equivalent to one-tenth wavelength, as determined by the outer calibration circle. This is from 0.074λ to 0.174λ. At the resulting point B, the normalized impedance is $0.37 + j1.9$. For this specific example, remember that the load remains constant and that conditions along the transmission line must be considered. Therefore, the arc drawn is of constant VSWR, as are all circles drawn from the center of the Smith chart.

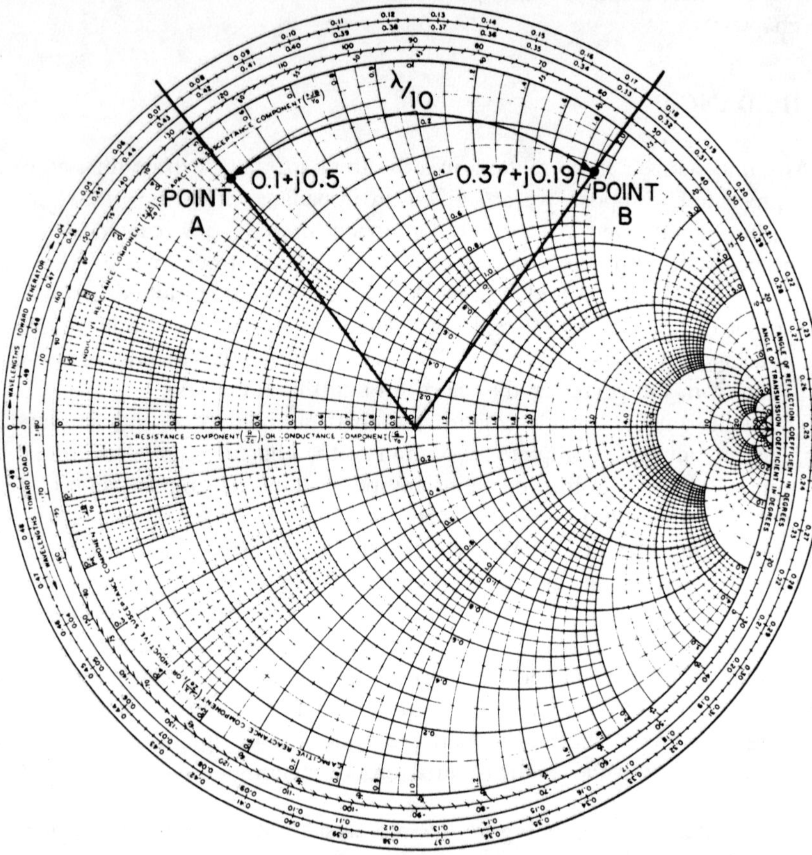

Fig. 6.2 Impedance measurements.

At this point a practical application of the Smith chart is in order. In the problem below, the normalized impedance at a point on a transmission line is $0.1 + j0.5$, and the frequency of the source is 1,000 MHz. What impedance can be measured on the transmission line 3.8 cm toward the source? What is the admittance at this point? What is the VSWR on the line?

Solution—The above is a typical problem that demonstrates the transformer properties of a transmission line. To find the impedance 3.8 cm away, one must know the starting point on the chart, the length of transmission line in wavelengths, and the direction of travel. The frequency of 1,000 MHz corresponds to a free-space wavelength of 30 cm. Thus, the length of the section of line is $3.8/30 = 0.126$ wavelength, traveling toward the generator. A circle is drawn (see Figure 6.3) using as a pivot the $1 + j0$ point at the center of the chart. Using the radius from here to point A, the arc is drawn toward the generator a distance of 0.126 wavelength, at which point (B) a normalized impedance of approximately $0.8 + j3$ is formed. The normalized admittance value, approximately $0.09 - j0.32$, is obtained by moving around the circle exactly one-quarter wavelength from point B, to a point (C) exactly on the opposite side of the chart. The above characteristic of the chart is extremely useful in design work with parallel impedances because the admittance values are far more convenient for parallel circuit considerations. The VSWR may now be found by continuing the arc until it intercepts the 0 reactance line, at which point (D) a constant resistance value (VSWR) of 11.8 is plotted.

The principles of impedance matching are, in general, applicable to both waveguides and transmission lines, although the physical form and the behavior of a given matching structure as a function of frequency may be different.

Suppose a match of a 50Ω transmission line terminated with a load $Z_L = 100 + j50\Omega$ is desired. The normalized load impedance is $2 + j1\Omega$ and is plotted as point A in Figure 6.4. If the reactive component of the load is eliminated by placing a capacitive reactance of $j50\Omega$ in series with the load, only the resistive component remains, as shown by moving along the constant resistance circle to point B. Because there is still a VSWR of 2.0, elimination of the reactive component of the load is obviously not the only requirement for matching the transmission line. In order to achieve this condition, the normalized input impedance must be $1 + j0$; therefore, the reactive component must be eliminated on the $R/Z_o = 1$ circle. A line is drawn from the center of the chart through point A, intersecting the wavelength scale at 0.213λ at D. The $R/Z_o = 1$ circle is located by traveling clockwise around the constant VSWR circle to point C, where the normalized input impedance of the line is $1 - j1$. The distance along the line to this point is $0.338\lambda - 0.213\lambda = 0.125\lambda$, as indicated on the wavelengths toward generator scale. A series reactance of $+j50\Omega$ placed at point C eliminates the capacitive reactance, and the input impedance point

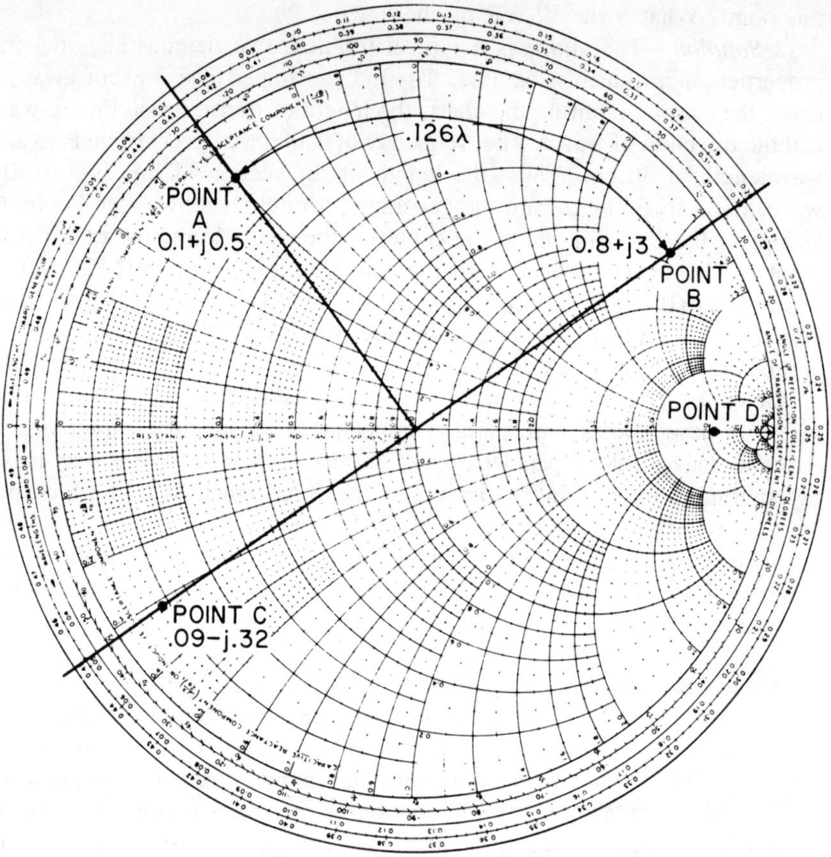

Fig. 6.3 Impedance problem.

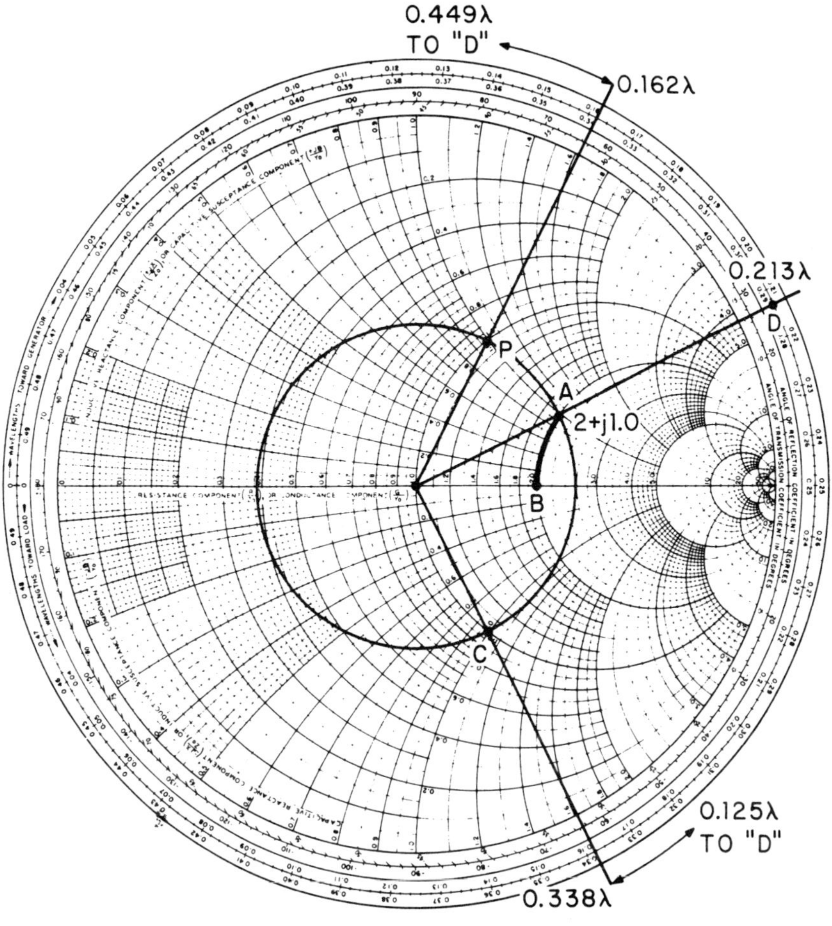

Fig. 6.4 Impedance-matching example.

moves to the center of the chart, where $R/Z_o = 1$ and $X/Z_o = 0$. The transmission line is now matched or, in other words, a VSWR equal in amplitude and opposite in phase has been introduced at point C.

The required value of an input impedance of $1 + j0$ for a match can also be obtained by traveling around the constant VSWR circle to point P. The impedance here is $1 + j1$, and a capacitive reactance must be added in series with the line to cancel the inductive reactance at this point. The distance from the load at A to this new point P is 0.449λ.

6.2 SINGLE STUB-TRANSFORMER

Since admittance values are additive at parallel junctions, transmission line problems are simplified considerably if considered in terms of the input admittance. The *single-stub transformer,* an impedance transforming device often employed with parallel-wire and coaxial lines, consists of a short section of transmission line connected in parallel with the original line and terminated in a short circuit. The unit is shown in Figure 6.5 (a).

The problem of Figure 6.4 is now one of admittance (shunt). In Figure 6.5 (b), the normalized input impedance is converted to normalized input admittance by $\lambda/4$ rotation of the diagram, corresponding to location of the reciprocal of the normalized input impedance at a point diametrically opposite point A. The normalized input admittance is located at point M, and the normalized value of input admittance y_i is read as $0.4 - j0.2$. For a match, $B/Y_o \pm jB/Y_o$ must equal $1 + j0$. The line from the center of the chart through M intersects the wavelengths toward generator scale at 0.4630. The distance measured on the wavelength scale by traveling clockwise from the load admittance point M to point P, where $B/Y_o = 1$, is 0.1990λ. The normalized input admittance at this point is $1 + j1$ (a capacitive susceptance). An inductive susceptance connected across the line at P results in the input admittance point moving to the center of the chart, where $G/Y_o = 1$. The required inductive susceptance is $-j1.0$ and can be accomplished by making the shorted line the proper length. In computing the length, remember that one end is a short circuit—a zero impedance and an infinite admittance ($Y_o = 1/Z_o$). A starting point is thus achieved at the right side of the chart where all values of wavelength are 0.250. The second point is found at $-j1.0$ on the chart (point F = $.375\lambda$), to which a line is drawn from the center of the chart, forming our end point. Thus the shorted line (or stub) is $0.3.75\lambda - 0.250\lambda$ or $0.1.25\lambda$ in length, and the final result is $d_1 = 0.199\lambda$ and $d_2 = 0.1.25\lambda$. This arrangement produces a good 50Ω match.

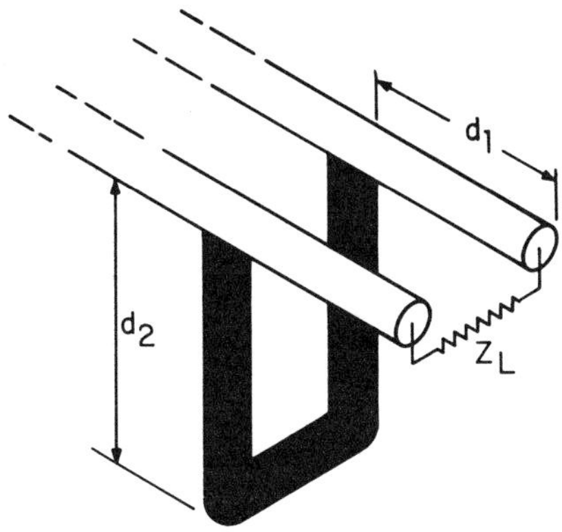

Fig. 6.5(a) Single stub transformer.

6.3 DOUBLE-STUB TUNER

In the previous discussion the matching procedure required a point be located on the line where the reactance or susceptance element is placed either in series or in shunt with the line. As the single-stub method of matching a line is unsuitable for most transmission line structures because of the difficulty of building a single stub adjustable in position, the *double-stub transformer* is used. As shown in Figure 6.6, it consists of two stubs separated by any distance less than one-half wavelength (usually 3/8 wavelength); each stub is an adjustable short circuit. The load admittance to be matched is the impedance as seen at the load side of stub 1, even though the actual load may be any distance from the stub. The separation between the load and tuner should not be great in terms of wavelengths, especially if the source exhibits frequency instability. If the tuner is many wavelengths from the load, a small change in frequency causes large changes in admittance (or impedance) as seen at the tuner—a phenomenon known as the *long line effect*. Even though a given spacing cannot accommodate all impedances, one can be chosen so that the range of impedances is sufficiently wide to allow for any slight variations in load or frequency that may be encountered.

The double-stub tuner system of Figure 6.6 has two adjustable-length shorted stubs with lengths d_l and d_2; and a fixed-length section of line between them which must be other than a half-wavelength long (usually 1/8 or $3/8\lambda$). The lengths d_l and d_2 of the two shorted stubs must be determined to match a given load to the characteristic resistance of the line. This problem is handled most efficiently on a transmission-line chart. On a per-unit admittance form of Smith chart the principle of the double-stub tuner may be summarized as follows:

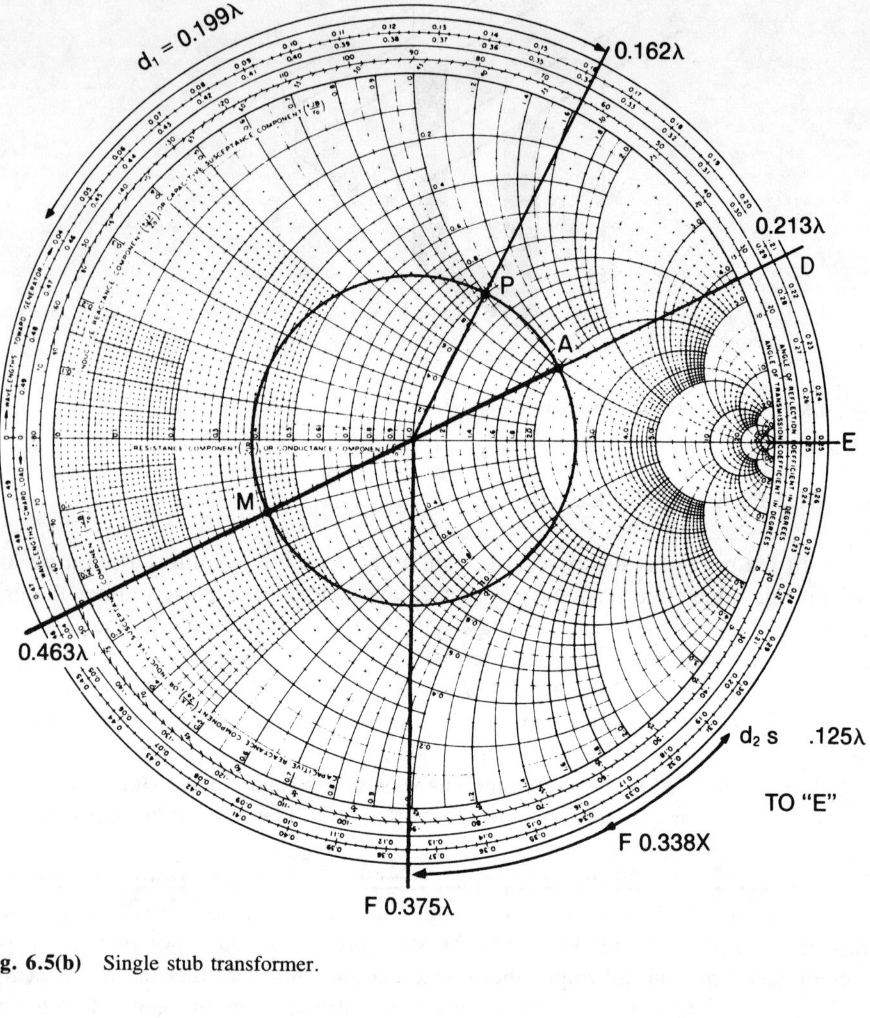

Fig. 6.5(b) Single stub transformer.

The length d_1 must be chosen such that the load and stub in parallel are transformed to the unit conductance circle by transferral down the fixed-length section of line. This choice permits the length d_2 to be chosen such that the pure susceptance which it presents exactly cancels that reflected by the load at its terminals, with the resulting admittance a perfect match of $1 + j0$ per-unit mhos.

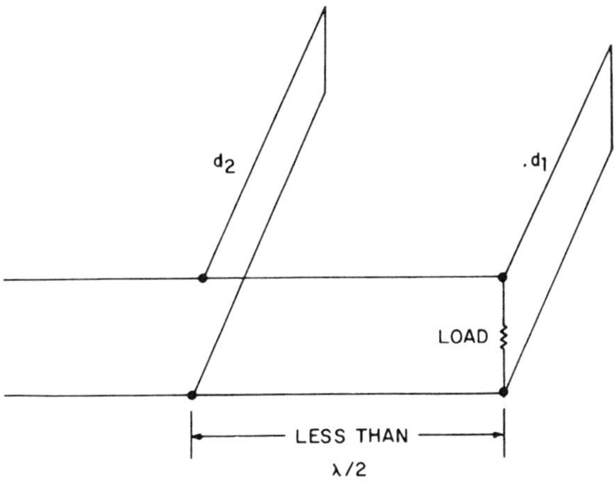

Fig. 6.6 Double stub tuner.

The most difficult step in analyzing the double-stub tuner is choosing the length d_1 such that the load and stub in parallel are transformed to the unit conductance circle by transferral down the fixed-length section of line. Once this step is accomplished, the process continues in the same fashion as that of the single-stub tuner when choosing the length d_2. Although the length d_1 can be chosen by trial and error, such a time-consuming task is not necessary. The secret of success lies in finding the locus of points on the chart such that transferral down the fixed-length line toward the source yields the locus of points represented by the unit conductance circle. The problem is really quite simple because, by going backwards from source to load, the locus is easily seen to be the unit conductance circle rotated toward the load by the length of the fixed section of line. If the fixed length of line is $\lambda/8$ long, the locus appears as shown in Figure 6.7 (a). Transferral of this locus of points toward the source by $\lambda/8$ yields the unit conductance circle, as illustrated in Figure 6.7 (b).

By adding susceptance by the stub at the load so that the combination is on the required locus (shown for the special case of $\lambda/8$ stub spacing in Figure 6.7 (a)), the admittance may then be transferred toward the source by the length of fixed line, with complete assurance that the resulting admittance lies on the unit conductance

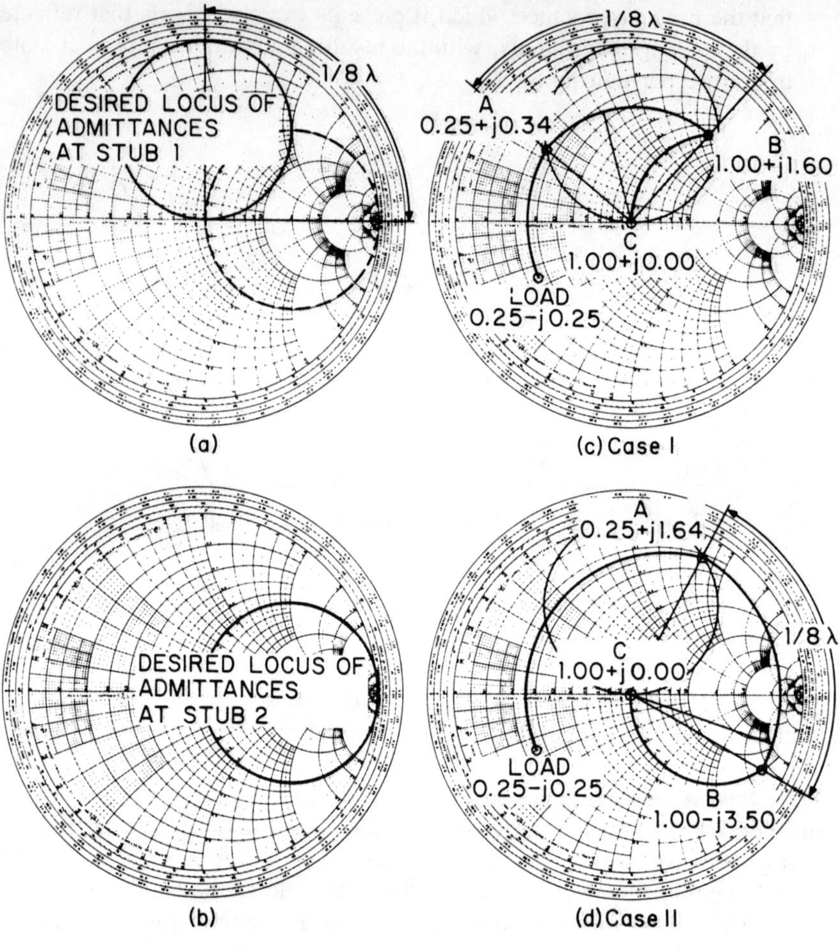

Fig. 6.7 Smith charts of double stub tuner.

circle. The remaining stub thus may be used to cancel the susceptive portion and yield a perfect match.

The following example serves to illustrate how these dimensions are calculated. Suppose the load is $100 + j100\ \Omega$, the lines have a characteristic impedance of 50 Ω, and the spacing between stubs is $\lambda/8$. The per-unit load admittance is easily computed as $0.25 - j0.25$ per-unit mhos. There are two solutions, just as for the single-stub tuner. The steps, illustrated in Figures 6.7 (c) and 6.7 (d), are summarized below:

1. Add $+j0.59$ per-unit mhos ($+j1.89$ in case II) with 0.335λ (0.422 in case II) of line to reach the admittance of $0.25 + j0.34$ per-unit mhos ($0.25 + j1.64$ in case II), which is on locus of 6.7 (a) for $\lambda/8$ stub spacing (the rotated circle).
2. Travel toward the source $\lambda/8$ to reach the admittance of $1.00 + j1.63$ per-unit mhos ($1.00 - j3.50$ in case II).
3. Add $-j1.63$ per-unit mhos ($+j3.50$ in case II) with 0.088λ (0.455 in case II) of line to reach the admittance $1 + j0$ ($1 + j0$ in case II).

A summary of the final results is listed below.

	Case I	*Case II*
Load	$100 + j100\Omega$	$100 + j100\Omega$
d_1	0.335λ	0.422λ
d_2	0.088λ	0.455λ
Spacing	0.125λ	0.125λ

By referring to Figure 6.7 (a), notice that load admittances inside the $g = 2$ circle never intersect the desired locus of admittances at stub 1 for $\lambda/8$ stub spacing. Consequently these loads may not be matched with such a double-stub tuner. In a similar fashion, it is possible to find the loads which are or are not matched for other spacings between the stubs. A summary of these results is presented in Figure 6.8, the shaded areas representing the loads that cannot be matched. It is interesting to note that, except for the special case where the distance between stubs is precisely a half-wavelength, the widest possible ranges of loads can be matched for any distance between stubs approaching a half-wavelength. The smallest range of load can be matched for a stub spacing of a quarter-wavelength; however, the adjustments become exceedingly critical as the distance between stubs approaches $\lambda/2$. Thus, a choice of spacing such as $\lambda/8$ or $3\lambda/8$ represents a reasonable compromise between matching range and critical adjustment.

If the load happens to fall on a portion of the chart where matching is not directly possible, then it is necessary to add a fixed length of line between the load and first stub in order to adjust the load to a value that can be matched. The analysis then proceeds in a conventional manner.

An important laboratory consideration when analyzing the double-stub tuner (unless the advantage of not requiring an adjustable length line is to be lost) is that

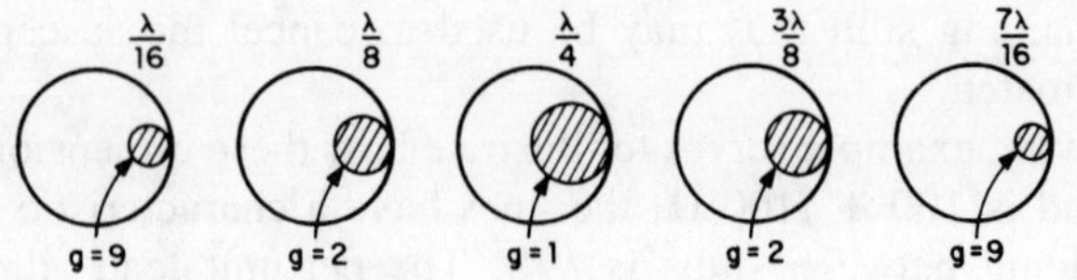

Fig. 6.8 Spacing between stubs.

the distance between stubs is not precise but rather, with fixed sections of line, an odd value like 0.147λ. This fact once again proves that, although theoretical considerations are important, things work out a little differently in the real world.

6.4 RADIAL SCALED PARAMETERS

Figure 6.9 shows the radially scaled parameters found at the bottom of the Smith chart. Parameters on this scale are *reflection coefficient, return loss, transmission loss, standing-wave ratio in dB,* and *mismatch (or reflected) loss*. Let us see how each of these can be read from the radial scale and put to use.

6.4.1 Reflection Coefficient

The *reflection coefficient* is read on the right hand side of the radial scale in Figure 6.9. There are two reflection coefficient scales, labeled voltage and power. The former is above the line, a plot of E_r/E_i, and it varies from zero at the center of the chart (Z_o point) to one at the outside (zero resistance). The power reflection coefficient scale, designated ρ^2, is located beneath the same line. Notice that each value on this scale is precisely the square of the value of ρ noted above the line.

The angle of the reflection coefficient is zero on the Z_{max} resistance line where the incident and reflected waves are in phase. The angle of Γ from the zero point is linearly related to the distance traveled, and its sign is indicated on the outside rim of the impedance coordinate system. The relationship of various parameters to the reflection coefficient can be illustrated on the chart. Use dividers to measure $\rho = 0.5$ on the radial scale (point A). By comparing the measured distance to the VSWR scale reading on the right-hand axis of the chart, a VSWR of 3.0 is found, (B), in Figure 6.9. The power reflection coefficient ρ^2 (at A) is noted to be 0.25 and, by multiplying this value by 100, the percent of reflected power is obtained. Reflected power itself is 25 percent of incident value. When performing measurements on the radial scales and the chart proper, the center of one corresponds to that of the other, while the end of the radial scales corresponds to the outside rim of the chart (zero resistance). Therefore, dividers can be used to compare measured values either from the center out, or from the outside rim toward the center, as long as the

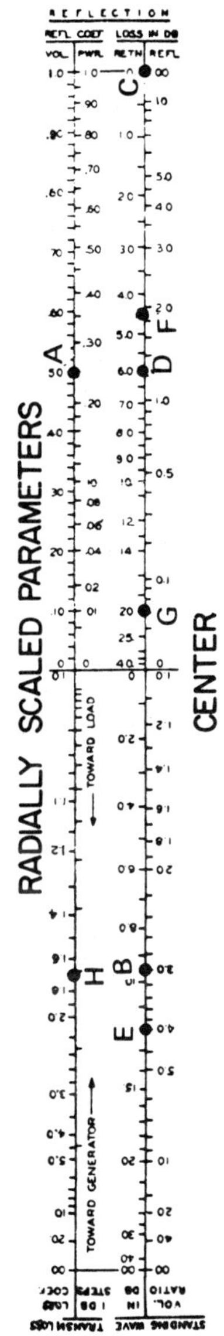

Fig. 6.9 Radial scaled parameters.

chosen reference remains consistent. The relationships of the remaining parameters to reflection coefficient can be obtained by measuring the corresponding distances on the scales of interest.

6.4.2 Return Loss

The *return loss* is also read on the right-hand side of the radial scale in Figure 6.9. As mentioned previously, the term stands for the ratio of the incident power to the reflected power at a point on the transmission line, and is expressed in dB. The return loss scale is headed "loss in dB" with the return loss values indicated above the line. When the reflection coefficient is one, the return loss is zero dB, as indicated at the extreme right side (point C in Figure 6.9). This return loss value indicates that all of the signal incident on the line was returned toward the source. Similarly, as the reflection coefficient approaches zero, the return loss approaches infinity—the more perfect the load, the less reflection from it. (This relationship was stated earlier as a maximum transfer of power when $Z_L = Z_o$).

To illustrate how the return loss principle and the radial parameter can be used, consider the following example:

Assume that the 3 dB attenuator in Figure 6.10 is perfectly matched (input and output VSWR = 1.0). The indicated input power of 100 mW is decreased to 50 mW at the output of the attenuator. This 50 mW is reflected from the short circuit back through the attenuator in the reverse direction. One-half of this reflected power is lost, resulting in a reflected power at the input of 25 mW. By definition, return loss signifies the total round-trip loss of signal, being 6 dB in the present example. Using a set of dividers, measure the distance from zero to the 6 dB line on the return loss scale (point D). Use this setting to mark the distance from the right-hand rim of the impedance coordinate system toward the center of the chart. The VSWR is read as 3.0 at B. This example is of special significance since it shows that the VSWR is decreased when attenuation exists on the line; a high VSWR thus can be decreased by placing attenuation in the line. In many cases, though, this action is not desirable because of the resulting loss of power in the forward direction. Special components of a unidirectional nature are required to provide the necessary attenuation and to dissipate minimum power in the desired direction of power flow. Using the same setting on the dividers, measure the distance from 1.0 on the voltage and power reflection coefficient scales to obtain the voltage reflection coefficient of 0.5 and power reflection coefficient of 0.25. This measurement verifies that the reflected power is 25 percent of the incident power. The voltage reflection coefficient is squared and multiplied by 100 in order to obtain the reflected power in percentage form.

A second example, shown in Figure 6.11, is also of special significance as it explains why an attenuator is placed at a generator output. The problem seeks the value of the attenuator needed to reduce the VSWR from 4.0 to 1.22. Measure the

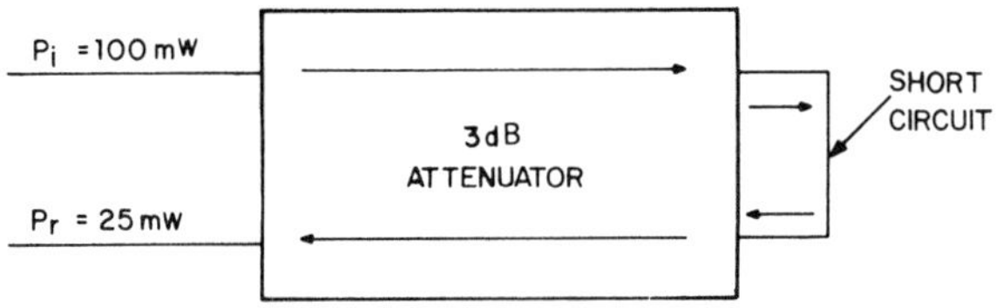

Fig. 6.10 Return loss with a shorted 3 dB attenuator.

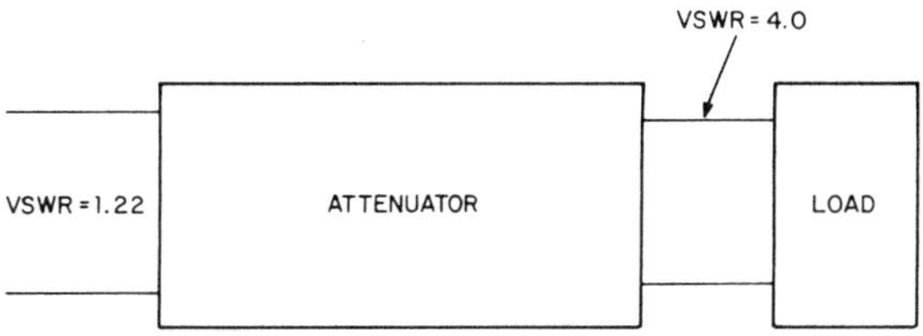

Fig. 6.11 Attenuator used to improve a match.

distance from the left end of the scale in Figure 6.9 (infinity) to the point on the scale that indicates a return loss of 4.4 dB (point F). Measure the scale length for a VSWR of 1.22 and mark off the same distance on the return loss scale. The return loss is found to be 20 dB (point G). Therefore 20 dB − 4.4 dB = 15.6 dB, which is the value of return loss to be added. The attenuator value required to obtain this value of return loss is 7.8 dB, since the round-trip loss must be 15.6 dB. Of course, since 7.8 dB is a theoretical value, no such pads are readily available. In reality a 10 dB pad would be inserted in the line, resulting in a VSWR of 1.13:1. If you recall, we usually insert a pad of at least 6 dB. If the above VSWR of 4.0 is used as a starting point, a return loss of 16.4 dB, or a VSWR of 1.36:1 is achieved—respectable values.

In the above examples the distances are measured from the right end of the radial scales. The measurements can also be made from the center line of the scales, as long as the chosen starting point remains consistent.

6.4.3 Transmission Loss

The transmission loss in 1 dB steps, or the equivalent one-way loss in the signal on the transmission line, is presented on the left hand side of the radial scaled parameters. By measuring the distance to D on the return loss scale, and measuring off the same distance to H on the transmission loss scale, it is noted that the attenuation reading is one half the reading on the return loss scale. (The letters referred to are in Figure 6.9).

6.4.4 Standing-Wave Ration in dB

In many practical applications the standing-wave ratio is measured in decibels, and the corresponding VSWR and other parameters are obtained from this measurement. It is expressed as

$$\text{SWR (dB)} = 20 \log \text{VSWR}$$

A radial scale plot of the standing-wave ratio in decibels is shown in Figure 6.9, and its values (in dB) can be compared to the various parameters by measuring equidistant points on the desired scales as demonstrated in previous examples. For example, the return loss of 6 dB is equivalent to a SWR of 9.5 dB; the return loss of 4.4 dB is equivalent to a SWR of 12 dB; and the return loss of 16.4 dB when using a 10 dB pad is equivalent to a SWR of 2.5 dB.

6.4.5 Mismatch Loss (Reflected Loss)

The mismatch loss is a measure of the loss caused by reflection and is the ratio of incident power to the difference between incident and reflected power, expressed in decibels as follows:

$$\text{Mismatch Loss (dB)} = 10 \log \frac{P_i}{P_i - P_r}$$

$$= 10 \log \frac{E_i^2}{E_i^2 - E_r^2}$$

where:

P_i = Incident Power;
P_r = Reflected Power;
E_i = Incident Voltage;
E_r = Reflected Voltage.

As an example, a VSWR of 3.0 represents a mismatch loss of 1.25 dB. This point is shown, opposite the return loss of 6 dB, at B in Figure 6.9.

6.5 EXPANDED CHARTS

The accuracy that can be obtained from the conventional Smith chart can be improved by simply isolating a portion of the chart and enlarging it. However, the only region that can be expanded exists from the center to the 1.5:1 VSWR circle (actually the outer edge of 1.59:1). Figure 6.12 shows the expanded Smith chart with the outer edge of 1.5:1. Figure 6.13 blows the chart up further by going to an expanded section whose outer edge is a 1.12:1 VSWR.

The expanded charts use the same outer scales as the full size Smith chart; that is,"wavelengths toward generator and load", and angle of reflection coefficient. All of the radially scaled parameters have the same value at the center and increase linearly to the outside edge where they coincide with the chart. In other words, a VSWR of 1.59:1 on the chart is read as a 1.59:1 on the scales. Also a return loss of 13 dB (or 25 dB) corresponds to the outer edge of the chart resulting in a 1.59:1 (or 1.12:1) VSWR.

The value of such charts is obvious when high return loss (low VSWR) devices must be characterized accurately. For example, if 1.25:1 VSWR had to be maintained on a series of components, they would have to be measured with fairly close accuracy. If the standard chart were used, some points would be very difficult to read. To relate this concept to purely physical size, notice that the space between a 1.2 and 1.4 VSWR is 0.25 inches on the full Smith chart; on the expanded 1.59:1 chart, it measures 1.25 inches. By basic mathematics, one can see that there is five times more space in the expanded chart, giving excellent accuracy. It should be obvious at this point to figure how accurately the very low VSWR's (between 1.0:1 and 1.10:1) can be read, by using the chart in Figure 6.13.

6.6 POLAR IMPEDANCE CHARTS

There are other types of charts that relate microwave impedance in polar form. That is, rather than resistive and reactive components $(R + jX)$, the impedance is in the form of a magnitude and phase (A/θ). Figure 6.14 is *a Carter chart* which displays both magnitude and phase angle of normalized impedance or admittance. It makes an excellent overlay for the Smith chart to permit conversion of normalized rectangular components of impedance to equivalent magnitude and phase angle. When rotated 180° from its orientation in Figure 6.14, the Carter chart permits conversion of normalized rectangular components of admittance to equivalent magnitude and phase angle. Thus the Carter chart provides a useful means of converting components of any complex impedance or admittance from the rectangular to the equivalent polar form, and *vice versa*.

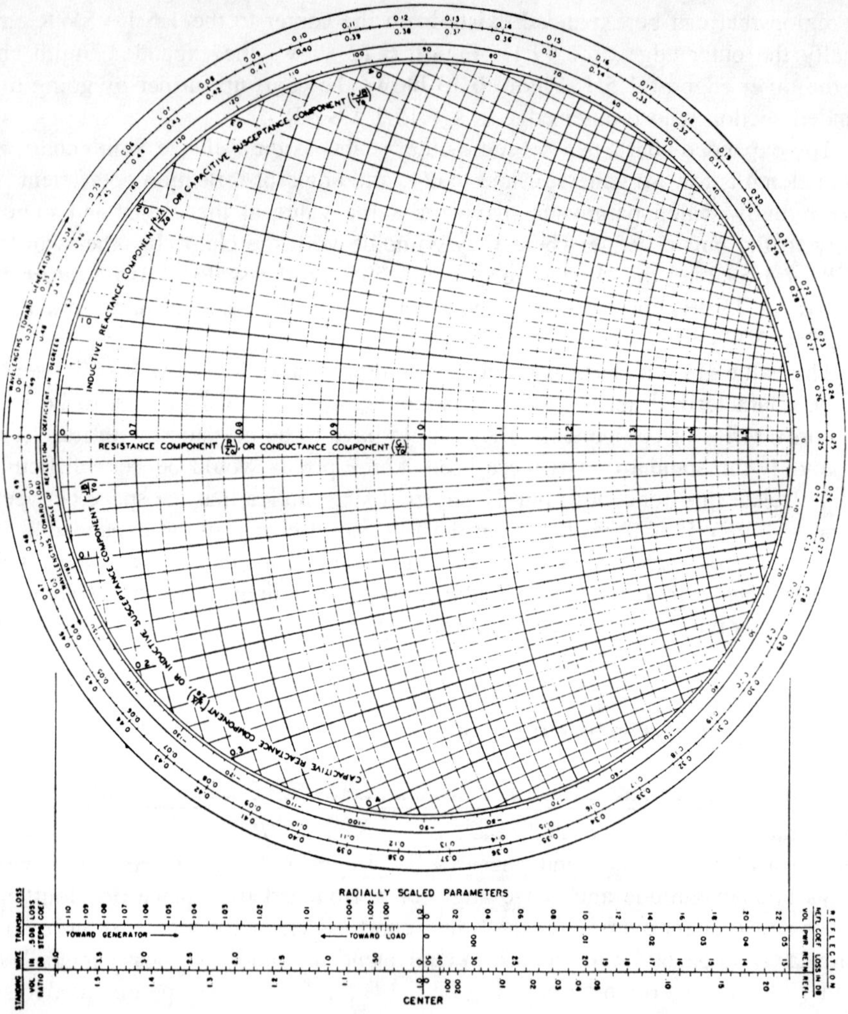

Fig. 6.12 Expanded Smith chart (1.5:1).

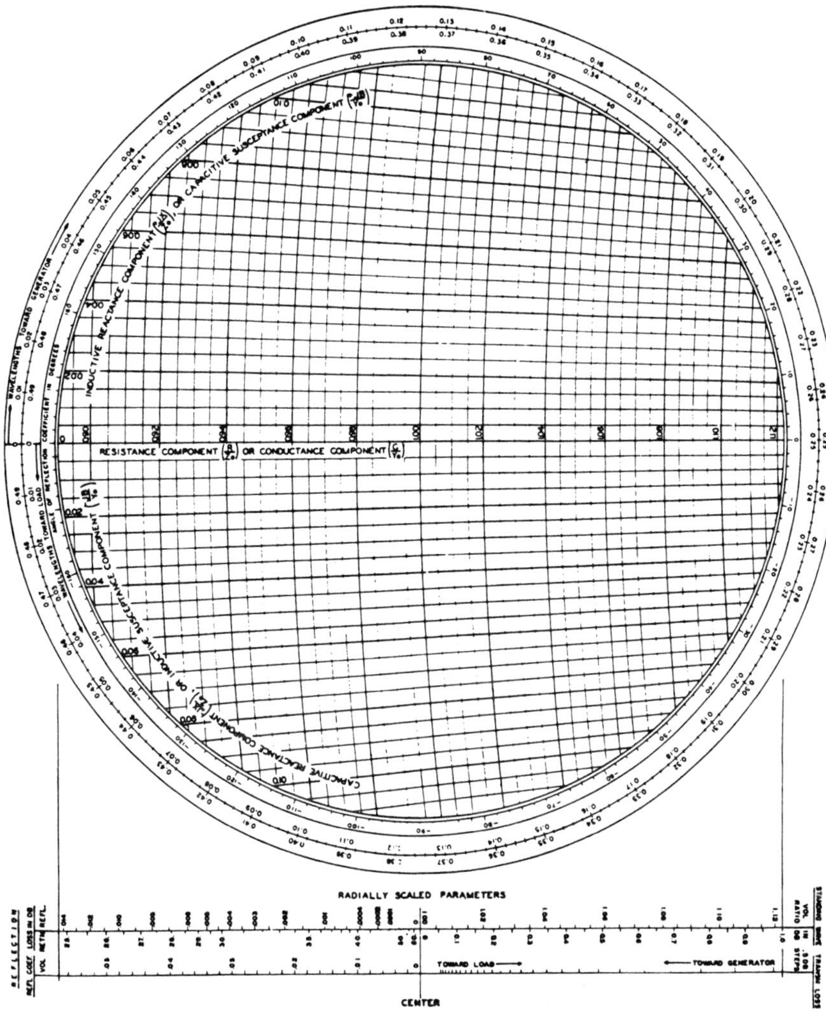

Fig. 6.13 Expanded Smith chart (1.12:1).

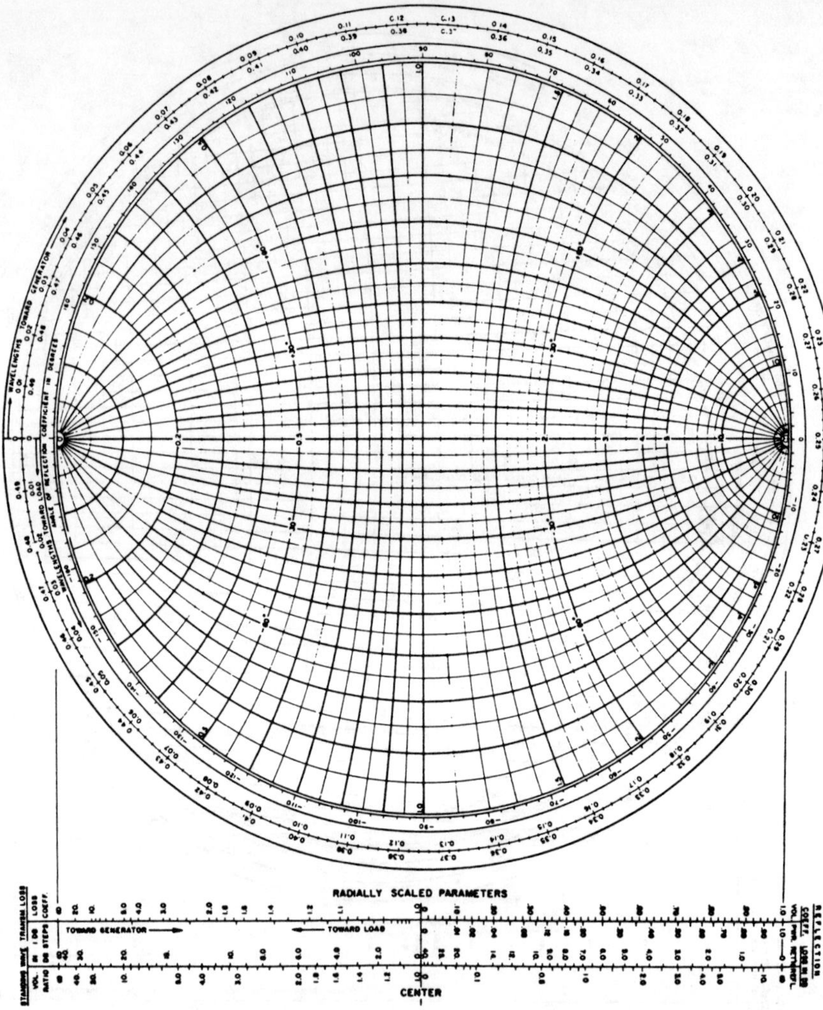

Fig. 6.14 Carter chart for polar coordinates.

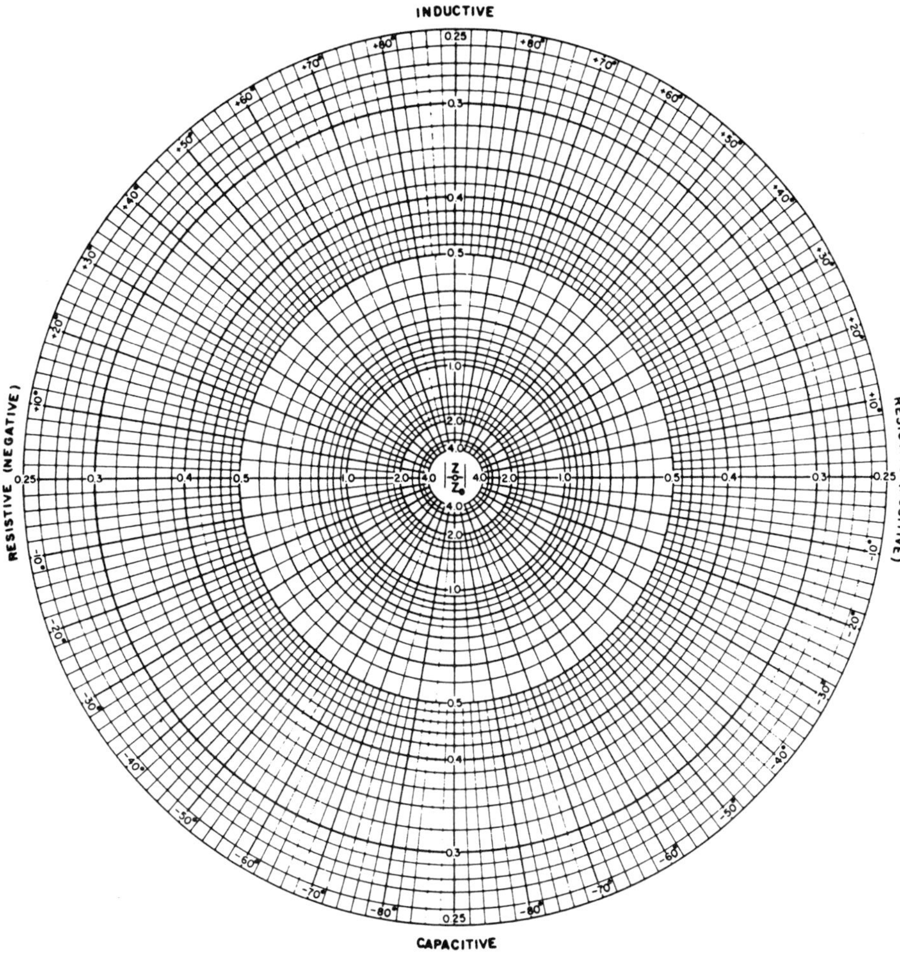

Fig. 6.15 Polar coordinate plot.

Figure 6.15 illustrates another way of showing polar coordinates. Here is a polar coordinate system on which it is possible to plot directly the resultant of two normalized impedance vectors that represent two circuit elements connected in parallel. There are definite quadrants for negative resistance-inductance, negative resistance-capacitance, positive resistance-inductance, and positive resistance-capacitance. Magnitudes are normalized values (Z/Z_o).

6.7 PROBLEMS

The following are examples of problems to be worked on the Smith chart. Answers are included with each.

1. Given that Z (load) = 40 + j42.5 and Z_o = 50 Ω, Find: VSWR, reflection coefficient, angle of reflection coefficient, return loss, mismatch loss, and admittance.

   Solution: (2.5:1, 0.43, + 78, 7.5 dB, 0.9 dB, 0.6 − j0.6)

2. Given Z (load) = 80 + j60 Ω and Z_o = 50Ω, Find: The input impedance (Z_{in}) of a line 4 cm long, the VSWR and output admittance. (Frequency = 2 GHz, dielectric constant = 2.1)

   Solution: (0.46 + j0.49 or 23 + j24.5, 2.75:1, 0.4 − j0.3)

3. Given Z (load) = 60 + j50 Ω, and Z_o = 50 Ω, Find: The impedance matching circuit using a single short circuit stub. Find d_1 and d_2.

   Solution: (d_1 = 0.233 λ, d_2 = 0.129λ)

6.8 SUMMARY

In the beginning of this chapter, the Smith chart was referred to as a very important microwave "tool." Now it is evident that the chart is indeed an *indispensable* tool. We have seen how it facilitates the computation of impedances and impedance-matching networks, and we have used it to justify the use of an attenuator in test setups. *Radially scaled parameters* have been introduced, referring to the nearly invisible numbers and letters at the bottom of the chart. And, finally, we have learned that the tool we need so much has many variations to make our job easier and our results more accurate.

So when "tools of the trade" are discussed over a cup of coffee, the microwave worker has one of his own—the Smith chart.

Chapter 7
Microwave Measurements

The importance of proper microwave measurements can not be overstated. That ideal design that took so much time and effort is completely useless until it is tested and proven to work. It must not only be tested, but it must be tested properly. This involves not only getting the right equipment, but planning ahead as to which parameters you need to test; constructing the setup properly; knowing how to conduct the tests; and, finally, knowing how to interpret the results. A good plan before you begin will greatly enhance the possibility of a successful and fruitful test.

This chapter will first present a series of hints, precautions, and tips to aid in the testing of microwave components or systems. These will help to avoid many testing problems for you and are all based on actual experiences. Secondly, the chapter will present representative test setups for *power*, *transmission*, *impedance*, *frequency*, and *noise measurements*. These are called *representative setups* because you should never attempt to memorize a setup from a book. This is because your requirements may not be the same as those for the setup shown, and thus your results would not be what you need for your particular application. It is much better to learn why each component is used in a setup, and then design the setup which is best for your application. This will save time, money, and effort in the long run.

7.1 PROCEDURES

Everything in life is governed by sets of rules and regulations. There are regulations for traffic and boating, and rules of conduct, etiquette, and order, to name just a few. Basically, each of these groups of "laws" delineates the "right" and "wrong" ways of doing things.

The field of microwaves is no different than any other—it too has its set of rules and regulations to help you find the proper methods by which to produce optimal results at these higher frequencies. The test procedures that follow are intended to prepare you for specific work later on. This area is the preliminary one of microwaves—what you should know before you attempt to make a measurement. The greater your knowledge of standard test procedures, the more meaningful and valuable your tests and their results.

We cover seven general areas, spelling out certain rules and listing helpful hints for each.

1. Microwave component ratings
2. Equipment calibration
3. Low power testing
4. Medium power testing
5. High power testing
6. Noise testing
7. Data sheet information

To clarify topics three, four, and five, we arbitrarily set low power to be everything with an output below 1 mW (0 dBm); medium power, from 1 mW to 10 W (0 dBm to +40 dBm); and high power, above 10 W (from +40 dBm).

7.1.1 Component Ratings

If you were buying a new tire for your car and a friend offered to sell you one of his, the first thing you would want to find out (after the price) is the size—to see if it would fit on your car. The same initial precaution should be taken when obtaining components for a microwave test setup. Be sure that they fit your specific application.

Component ratings form a general classification of individual components and test equipment. These ratings include *maximum input power* or *power handling capability*, *input and/or output impedance*, *frequency ranges*, *polarity*, and *dc voltages*. To have your setup or system perform as expected, you must stay within each of the specifications that applies; in some cases, such control also preserves the component or some part of it. Let us examine each of the above ratings and see how, by observing them, we can operate at peak performance.

Maximum input power can apply to either components or test equipment. The term usually is associated with any type of circuitry in which an active device is used at the input, such as mixers, detectors, amplifiers, and *pin* circuits. Equipment such as spectrum and network analyzers, counters, and power meters have a maximum level printed in red on the front panel to gain your attention. In general, a spectrum analyzer is rated from 1 mw to 2 W maximum input; most counters have a maximum rating around +10 dBm (10 mW); power meters with thermistor mounts can operate to 30 mW while those with power sensors can go as high as 3 W (though most are limited to 100 mW (+20 dBm); and the network analyzer maximum is 50 mW (+17 dBm) to the solid state input circuit of the frequency converter.

Power handling capability applies primarily to components, indicative of how much power it can dissipate. The two devices where specification is a prime consideration are attenuators and isolators. The power handling capacity of a fixed attenuator is usually from 1 to 5 W in continuously variable attenuators, the power rating can go as high as 100 W CW. Step attenuators typically do not have a high power rating because they are made up of individual fixed pads attached to some sort of switching apparatus.

The isolator is, of course, a terminated circulator; this termination is what limits the power handling capability of the device. Some isolators have a maximum rating of 10 W; the majority can handle no more than 5 W. The limits for the power of an isolator termination are size and, consequently, manufacturing price. If techniques were known that could make a small, 100 W termination, the isolator would be in the same power class as its kin, the circulator.

The *input/output impedance* ratings apply to both components and test equipment. For a maximum transfer of power (maximum system efficiency), the impedance of a load should equal that of the driving circuitry. In other words, if you want to test a 50Ω component, your generator should have an *output* impedance of 50Ω. Similarly, if you want to check the response of a 50Ω system, your test equipment should have an *input* impedance of 50Ω.

In the case of microwave devices, it is much easier to interface two or more if they all have the same input and output impedances. If you tried to put a 50Ω coupler with a 75Ω filter, they would not perform as expected because of reflections caused by the mismatch.

The term *frequency range* was used many times when describing the individual microwave components and always was defined as the band of frequencies over which all of the component specifications applied. Again, this meaning is true of any microwave test equipment. For instance, a sweeper or signal generator has a certain specified range of frequencies over which it is designed to operate—it does not work properly above or below this band. Likewise for a spectrum analyzer, only in the range of frequencies over which it was designed to operate does it give you a true picture of your specific signal and its associated harmonics. Outside of this frequency band it gives unreliable readings and even can generate signals of its own.

The rating *polarity* is applicable anywhere a dc bias or dc output is involved. Components such as externally biased mixers, detectors, and *pin* attenuators, and switches, would fall into this category. A YIG filter and its driver also need a dc voltage for proper operation; the polarity of that voltage must be observed. A wrong dc voltage polarity at the inputs of X–Y recorder or plotter would result in completely useless data, as well.

You should become familiar with every component and piece of test equipment you use in a test setup or plan to utilize in a system. Know the maximum ratings (especially power) of each to avoid the waste of time and money caused by burning out a pad, filter, or set of mixer diodes. Neglecting to consult either a data sheet (for a dc voltage value or polarity), catalog (to verify how much power a 20 dB attenuator can handle), or an instruction manual (when your signal generator does not work properly, only to find that its impedance is 600Ω and your system's is 50Ω) can result in just such a loss.

These things may sound quite basic as you read them in a book. But, under the pressures of a contract deadline or the need for a working and demonstrable system or test setup on Tuesday morning (and it is now Monday afternoon), you may have the tendency to ignore precisely the points mentioned above.

Take the extra few minutes to check on component ratings—that action will pay off in terms of a component that lasts many years and provides peak performance.

7.1.2 Equipment Calibration

Very simply, to calibrate your equipment means to set up and adjust it so as to have it conform to standards (such as those for frequency, noise, attenuation, and power). Whichever standard you use, you can be assured that it is highly accurate and supported completely by the reputation of the National Bureau of Standards.

Often, calibration charts are attached to equipment or components; two familiar ones are shown in Figure 7.1. The chart in (A) is for a 20 dB coupler, and is riveted permanently to the device—it indicates the amount of coupling, in dB, as a function of frequency. There are relatively few places over the entire band that are actually at the 20 dB level—some are greater, others are less. This chart is of great value if you need a more accurate reading from your coupler than just the approximate 20 dB figure. As an example, suppose you were using the device to monitor an RF line and needed a reading within 0.2 dB because the following circuit was sensitive to levels. If the operating frequency were 1.5 GHz, the coupling would be actually 19.5 dB instead of 20. All of your readings, in fact, would be 0.5 dB worse than if you assumed a 20 dB coupling. Because of the faulty presumption of a perfect coupler, your circuit really may not be as good as it looks.

A calibration factor curve for a thermistor mount is shown in (B). This mount attaches to a power meter that has a switch which ranges from 88 percent to 100 percent to compensate for errors in the overall power measuring system; thus, the meter is more efficient and accurate. As the frequency of operation increases, more compensation must be added in order to maintain a constant calibration (or, in this case, efficiency).

There are other components that have calibration marks printed on them. For example, some coaxial attenuators have an attenuation figure at dc, 3 GHz, 6 GHz, and 10 GHz, providing exact reading at these frequencies; by interpolating, you can estimate the values in between.

The majority of devices and equipment, however, have only a calibration sticker which indicates the date calibrated, when the process should next be repeated, a calibration stamp, and the signature of the test man. Another type of sticker often found reads as follows:

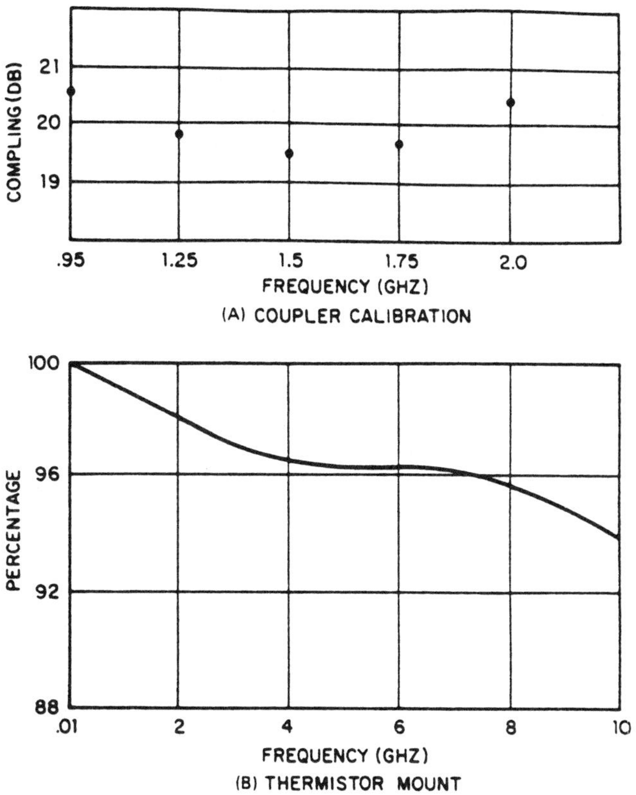

Fig. 7.1 Calibration curves.

Not Calibrated—Do Not Use For Quantitative Data

Equipment or components with the above label cannot be used to provide the absolute data that could be presented to a customer. However, they can be used for relative readings to verify proper operation of a complete system or part of it. The sticker indicates that the equipment is in working order but cannot be traced to a standard.

One other calibration sticker you may have seen says:

No Calibration Necessary

This piece of equipment, usually a coupler, step attenuator, or device of that nature, never needs to be called in for calibration or a calibration check. Data obtained on this sort of component does not vary appreciably in accuracy. In fact, if there was significant variation in performance, little could be done, short of replacing the entire unit, to improve the situation.

It cannot be stressed too strongly, how important it is to have your equipment calibrated. Using uncalibrated equipment is like walking through a maze in the dark—you eventually get somewhere, but you don't know how or where. Take the time needed to have all your equipment checked periodically. It is a great temptation to hang on to a device when its calibration date comes due; yet a couple of hours without it can make all the difference in regard to how meaningful your data will be. To obtain clear, concise, and useful information, *use calibrated equipment.*

7.1.3 Low Power Testing

The low power region has been characterized by output levels below 1 mW (0 dBm). In this area, great care must be taken at the input of any active device not to overdrive it and produce inaccurate data. Figure 7.2 is a basic block diagram of a setup for low power testing; it is not intended to be an actual setup but rather a general outline to help explain the procedures involved. The source could be either a signal generator synthesizer, or a sweeper; its prime requirement must be that it has an output between 1.0 and 10 mW in the proper frequency range, thus ensuring that the source operates satisfactorily at the necessary lower levels.

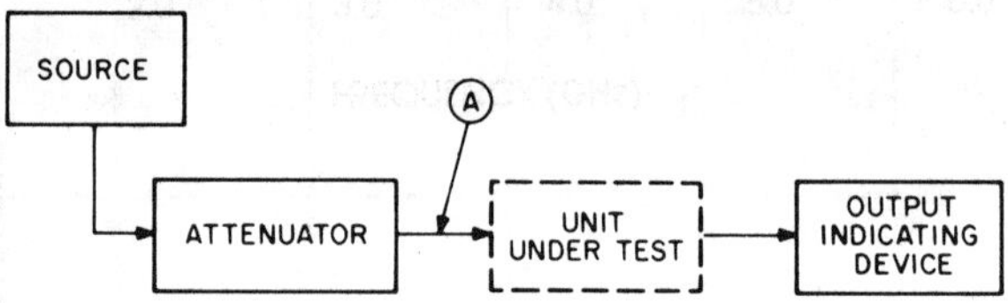

Fig. 7.2 Low power testing block diagram.

The second block in the figure is the one that validates the testing setup. This attenuator, preferably a step or continuously variable type, allows you to set the input level at a safe level so as not to overdrive the unit being tested. The actual input power can be measured at point A.

The output indicating device could be a scope, dc meter, X–Y recorder or power meter, depending on the type of indication desired. There is no need for concern regarding the power rating of any of these devices; most of them can operate a safe distance above what we are classifying as our low power region. In the same vein, there is no need to be concerned with the power handling capability of the input attenuator. Most of these components are rated for at least one watt—in our case, we are operating well below 1.0 mW.

Only one last point need be repeated—*when running low power tests—do not overdrive the unit under test.*

7.1.4 Medium Power Testing

This region of power testing is the most widely used in microwaves—it will encompass over 90 percent of all your checking. Once again, this area is defined from 1.0 mW to 10 W (0 dBm to +40 dBm). Figure 7.3 is a basic block diagram for the medium power region. The source, once again, could be a signal generator or a sweeper, though the latter is not adequate above a power level of around 50 mW (+17 dBm). There are sweepers that can produce 100 mW (+20 dBm), but they are at the lower end of the microwave frequency range and are the clear exceptions to the norm.

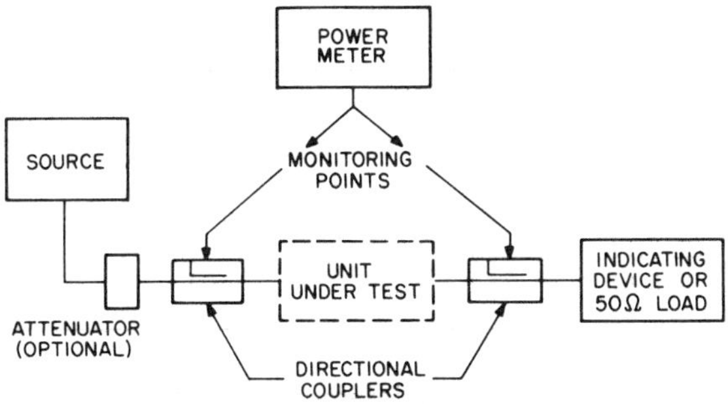

Fig. 7.3 Medium power testing block diagram.

The attenuator in the diagram is listed as "optional" because, at lower levels (1 to 500 mW), the input must be padded to prevent the device from being overdriven. Above the half-watt level the attenuator is removed for two reasons. First of all, many of them (especially step types) have a limited power handling capability. Secondly, at these levels, there is little danger of overdriving a component; usually there is enough tolerance between zero power input and the required drive level to provide adequate control.

The directional couplers shown are used to monitor the input and output of the device; sometimes the input coupler is dual-directional to provide a means of monitoring the reflected power.

The indicating device is usually not at the output for the higher power levels in this range. Rather it is at the coupled port of a directional coupler with a termination, of adequate power capability, at the output.

In this power region, you must begin to be aware of the power handling abilities of every component to be used—all attenuators, whether used at the input, output, or off the coupled port of a coupler, should be double-checked. You must be conscious of the maximum allowable input power to your power meter. If a termination is used at the output, it must have an adequate rating to handle the power dumped into it—were the load underrated, its properties (namely its resistance) would change, power would be reflected back, and inaccurate and useless readings would be given.

The source utilized should also be checked very carefully. The output impedance should correspond to that of your system; a check of harmonic content (by means of a spectrum analyzer) can save valuable time later by determining that a second harmonic is present that would effect (and invalidate) all your readings. The low frequency section of the output should also be investigated—frequencies on the order of 200 to 400 MHz have been known to set up oscillations in microwave transistors and to destroy them completely. When potentially trouble-causing low frequency components are present, a high pass filter can be used to eliminate the problem.

As the power levels get above a few milliwatts, there are more items to take into consideration, and greater care must be taken in choosing both components and test equipment. These levels are where both the power ratings of components and equipment are important, and the VSWR begins to play an ever increasing role in the proper operation of your test—any excessive reflections could invalidate hours of work.

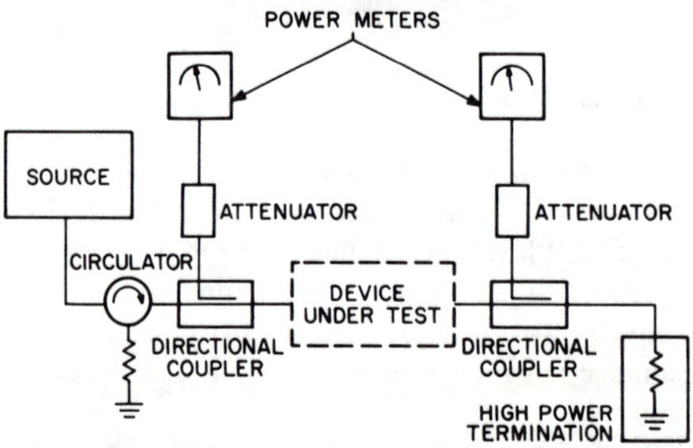

Fig. 7.4 High power block diagram.

7.1.5 High Power Testing

When testing in the range from 10 W and up, all the properties of a component become extremely important. Items such as insertion losses, thermal properties, and connectors join power ratings as critical qualifications.

The one point too often overlooked is the *thermal* properties of the particular component. Very simply put, this specification indicates how efficiently the device dissipates the heat generated by the high power transmitted through the system. When RF power is produced at the levels we are discussing now (+40 dBm and up), a great deal of heat is produced—one fact of life from which we simply cannot escape.

Figure 7.4 is a block diagram for high power testing—there are some obvious differences from Figures 7.2 and 7.3. First of all, a circulator definitely is needed at the output of the source. Any reflections present at these power levels had better be channeled to a load, not back into the generator. Another difference is the addition of attenuators at the coupled parts of the directional couplers. The thermistor mounts of most power meters have a fairly low maximum input level; the combination of coupler and attenuator help to keep below that figure. The third basic point of departure is that the indicating device is not placed in series with the unit under test; rather, it is placed off the coupler, with a termination in series because of both the limited inputs to the microwave indicating devices, and the availability of coaxial terminations rated to 275 W and above (if you are willing to cool them with water).

The parameters to be checked and re-checked in the high power region are, simply, *all of them*—the power rating, coupling values, VSWR, maximum input levels, insertion loss, and every other one pertaining to the individual component. To illustrate just how important one of these parameters could be, let us examine the insertion loss figure. Suppose we had a device producing 100 W of power, a +50 dBm level. If a circulator is placed at the output of such a device to protect it from excessive VSWR, it must have a low insertion loss. With a loss of 0.5 dB the output level would drop to 90 W; if the loss were allowed to go to 1.0 dB, the output level would be down to 80 W. What took so much to generate would be thrown away in a lossy component. This loss figure in some other component or application mught be acceptable, but at high powers, it is disastrous.

In conclusion let us list the rules for high power testing:

1. *Be sure your source is adequate.*
2. *Use a circulator, or high power isolator, to protect the source.*
3. *Check each component's power ratings.*
4. *Be very aware of all parameters, especially VSWR and insertion loss.*
5. *Never hit a device under test with full power—run the input level up gradually.*

One final note of caution is to *never touch high power microwave circuitry.* Whether stripline, microstrip, coax, or waveguide, it can burn. The heat generated by the high power is to be feared, but RF heat created by the combination of high frequencies and high power can burn even more severely. Narrow microstrip lines and center conductors of coaxial connectors are prime areas of RF burn danger. When testing high power microwave devices—*hands off*!

7.1.6 Noise Testing

Every network has a certain amount of internal noise associated with its operation, particularly when an active device is involved. A comparison of this noise level to the level of the desired signal is called the signal-to-noise ratio. There are input and output signal-to-noise ratios for each network or device. The noise figure is a measure of the reduction of a signal-to-noise ratio by a network; it is a term, expressed in dB, which compares the input and output ratios.

The noise figure test is one in which you must pay careful attention to the frequency ranges of the equipment used; Figure 7.5 helps to explain this statement. The diagram is of a typical noise figure setup, with the noise generator taking the place of the signal source in previous illustrations. A noise spectrum either broad or narrow band is produced—the type of noise dictates whether a single noise source for many frequency ranges is sufficient, or if a different one for each region is needed. Thus, the frequency of operation ratings should be checked.

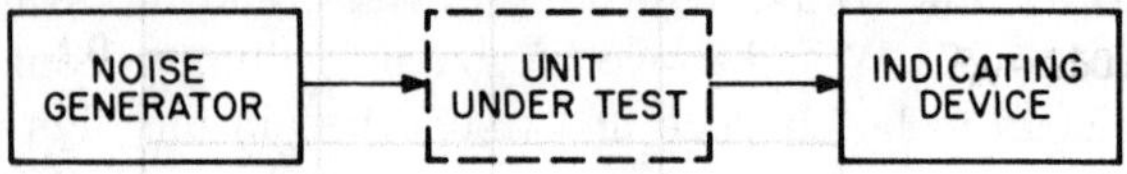

Fig. 7.5 Noise figure block diagram.

The indicating device in the setup is also frequency sensitive. The required input may be at 30 MHz, 60 MHz, 70 MHz, 200 MHz, 1,500 MHz, or some other IF frequency. If your particular system or component does not produce any of these, additional work is necessary. As an example, suppose you need a noise figure reading on a receiver that operates from 8 to 10 GHz. The IF frequency for this receiver is 400 MHz, but your noise figure meter only takes an input of either 30, 60, or 70 MHz. One way to test this receiver is a process called "double conversion," which mixes the IF signal once again to get it to a frequency appropriate for your noise figure meter. In this case, a 460 MHz or 340 MHz signal into a mixer would yield 60 MHz, and the meter would work properly. An alternate method in the example above would be to use a noise figure meter with an IF input that would handle 400 MHz. Instruments are available that tune the IF from 10–1,600 MHz, and thus would require no extra conversion.

It should be remembered that the voltages involved in taking noise measurements are at a very low level; interfering signals from outside sources could be present and would disrupt the readings. Therefore, it is always wise to take your noise measurements in a screen room or shielded area whenever possible. These locations are specifically designed to keep out all outside interference. If you still obtain erratic readings in a screen room, check to see if an unneeded generator might be turned on and radiating.

A noise reading is a very important piece of data; great care should be taken when obtaining it. Checking frequency ratings, noise source impedance, and the input frequency and power level at the noise figure meter (indicating device) enables you to make the most accurate readings possible.

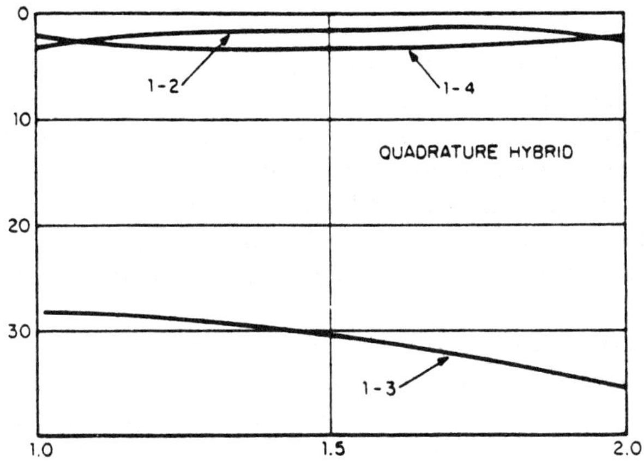

Fig. 7.6 Improper data sheet.

7.1.7 Data Sheet Information

The data sheet from a microwave test can be either a very useful tool to an engineer, or as worthless as a blank piece of paper. Its value depends entirely on what information is included by the tester.

Figure 7.6 shows a data sheet that is practically useless—all it tells you is that someone tested a quadrature hybrid at some time and ran parts 1-2, 1-3, and 1-4 from 1.0 to 2.0 something. Nothing is said as to a model, manufacturer, serial number, or any other details.

Figure 7.7 is the same data sheet, only with some meaning. Again, it is clear that a quadrature hybrid was tested, but there the resemblance between the two sheets ends. Company XYZ made the hybrid, assigned it a model number QH-1, and serial number 401 was tested on 1/9/75 by a person with the initials ABC. The device was run from 1 to 2 GHz with an attenuation range calibrated from 0-40 dB. Port 1-2 is the 0° phase port, 1-4 is the 90° port, and 1-3 is the isolation port. This data sheet is so thorough that only a photograph could possibly increase your knowledge of the hybrid. Clearly, a good data sheet is valuable and can supply great amounts of information. But since an overcrowded data sheet is as useless as one with not enough information, we list the most pertinent information. The items are as follows:

1. *What is it?*
2. *Model number*
3. *Serial number*
4. *What ports were run?*
5. *What test was run?*

6. *Use a diagram where necessary*
7. *Label each axis*
8. *Date the sheet*
9. *Sign the sheet with your full name or initials*

There may be times when more information is needed, and times when less is desired. As a rule, though, if you follow each of the items listed above, you create a very helpful means by which to evaluate a component or a system. Before leaving this section, take another look at Figures 7.6 and 7.7. Do the data sheets you have been turning out look like Figure 7.6? Do you figure that "he'll know what I'm talking about"? If you *have* been producing sheets like that in the first figure, you should study Figure 7.7 very carefully and see its advantages. If you take each set of readings with the idea in mind that no one will even look at it for a month, you will concentrate on putting down all the information so that you do not forget what you have done. It can be embarrassing when an engineer comes to you a couple of weeks after you have made a test with questions about it that you cannot answer. Be thorough; pretend that no one will look at your data for a month.

An area which should be mentioned at this point is a data sheet produced by a plotter. In this age of automatic and computer-controlled systems, it is very easy to let the machine do "its thing." For plotting, it is usually a case where the computer asks you for specific information and it is up to you to give it the proper response. Remember, these systems are only as smart as what you put into them. They will follow your instructions. Be sure these instructions are correct.

Our discussion up to this point could be viewed as pieces of a puzzle—each section is important in itself but is not the complete picture. These sections assemble all the parts to create a whole. By using the microwave components, cables, connectors, adapters, and equipment, and by following our "rules" for microwaves, we can assemble test setups that enable us to check out either an entire system or any individual segment. Five basic microwave measurements can be made:

1. *Power*
2. *Transmission*
3. *Impedance*
4. *Frequency*
5. *Noise*

Each of these is investigated and shown with a basic test setup and any available alternative methods. Included are the theory behind each measurement, the use of every component in the setup, and the procedures to follow for a successful test. When these items are understood for each of the five determinations above, there should be no microwave measurement that you cannot perform. Remember, do *not* memorize these setups—study and learn from them.

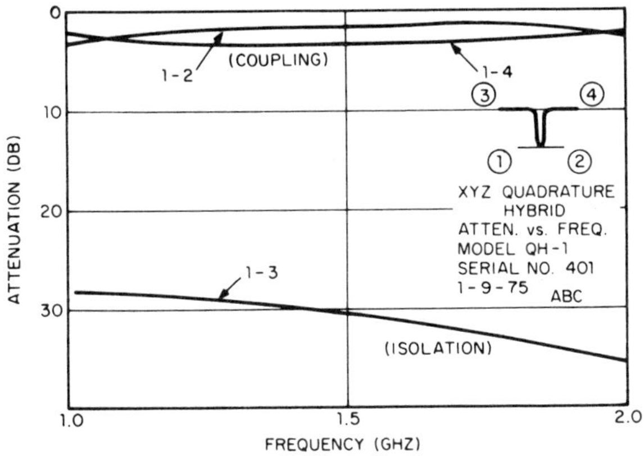

Fig. 7.7 Proper data sheet.

7.2 MICROWAVE POWER MEASUREMENTS

When we talk about power at low frequencies, we refer to voltage, current, and power factor. The measuring instruments used in these areas are designed to gauge these parameters. However, power measurements made at microwave frequencies are by means of thermally sensitive devices; that is, they measure the heat generated by the RF energy present in a particular circuit. Such devices fall into one of three common categories: *thermocouples, bolometers* (*barretters* and *thermistors*), and *calorimeters*. Each has a unique way of measuring microwave power; let us examine them and determine the area of the power spectrum for which each is best suited. Remember our classification of power levels: low (1 mW and below), medium (1 mW to 10 W), and high power (10 W and up).

7.2.1 Thermocouple

The *thermocouple* is formed when two wires of different metals have one of their junctions at a higher temperature than the other. The difference in temperature produces a proportional voltage. The device is used to measure the temperature rise of a load which dissipates microwave power; by appropriate calibration, the temperature change is converted to an indication of power. Thermocouples have excellent sensitivity and are especially useful as power monitors; their primary applications are in low and medium power areas.

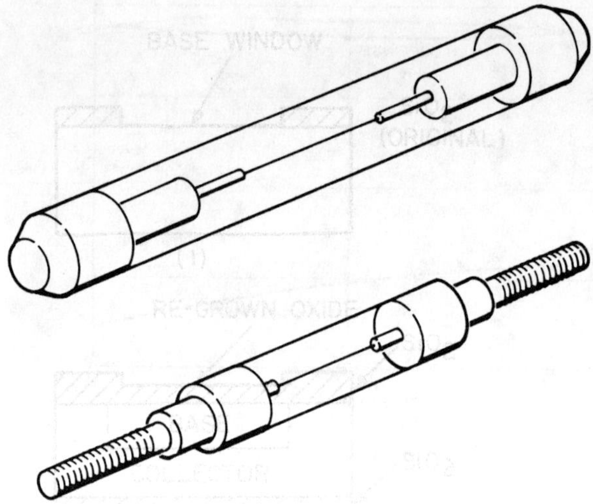

Fig. 7.8 Microwave barretters.

7.2.2 Bolometer

Bolometers are classed as *barretters* or *thermistors*. The most common usage of the term bolometer refers to a barretter; however, we describe both classes to provide a familiarity with all types of microwave power measuring devices.

The *barretter* consists of an appropriately mounted short length of fine wire, usually platinum, with sufficient resistance to enable it to be impedance matched as a termination for a transmission line. It has a positive temperature coefficient; that is, the resistance of the barretter increases as the temperature rises. It is usually biased to an operating resistance of from 50 to 400 ohms; 200Ω is the most common. Figure 7.8 shows two common barretters.

The *thermistor* consists of a tiny bead of semiconducting material which bridges the gap between two fine, closely spaced, parallel supporting wires. This construction is shown in Figure 7.9. Effectively, all the resistance of the thermistor is concentrated in the bead material which has a negative temperature coefficient of resistance (an increase in temperature results in a resistive decrease). They usually are made of complex metallic-oxide components using oxides of manganese, nickel, copper, cobalt, and other metals.

The changes in resistance resulting from the dissipation of microwave power in the bolometer (barretter or thermistor) are commonly measured by using the devices as one arm of a dc or audio frequency Wheatstone bridge circuit.

The barretter and the thermistor are sensitive power detectors, and are capable of measuring as little as a few microwatts of power when used in properly designed bridge circuits. The thermistor is the more flexible device, in that its resistance may be varied over an extremely broad range, depending on the magnitude of bias current. This feature is often a decided advantage in the broadband impedance matching of the detector as it leads to excellent overload and burnout qualities, and its operating resistance is confined to a considerably smaller range. However, the latter can be made more reproducible in both sensitivity and impedance, and is less sluggish (has smaller time constants) than a thermistor. There are good and bad points for the barretter and the thermistor which must be carefully weighed, and tradeoffs made, when choosing the type for your application.

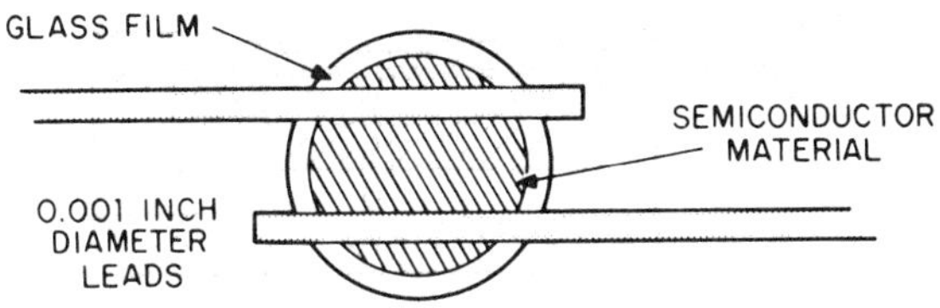

Fig. 7.9 Microwave thermistor.

7.2.3 Calorimeter

The calorimetric technique is the most fundamental method of measuring microwave power. Power is dissipated in a load and its temperature rise, or that of a cooling fluid flowing in or around it, is measured. Calibration of a calorimeter includes correction for heat capacity, heat losses, operating frequency, and mismatch. The procedure has been used for years in chemical and physical measurements. Basically, the calorimetric power meter dissipates power (a continuous flow of energy), controls the resulting flow of heat through some thermal path, and measures the temperature rise thereby set up.

Calorimetric power meters used for microwave work are divided into two categories: *flow* (liquid) and *static* (dry). The flow type uses a liquid to carry the heat away from the load in a controlled manner. The dry, or static, calorimeter measures the EMF generated in a thermopile placed between a reference load and an active one dissipating input power. Figure 7.10 shows both types of systems. The calorimetric method is used extensively in high power measurements.

The basic microwave power measurement setup is similar to that illustrated in Figure 7.11; the components shown are typical. The source, of course, depends on the power level needed, and whether you want CW, swept, or pulse measurements. The protection component could be a circulator or isolator, a step attenuator, or a fixed value of attenuation; its main function is to provide a good match for the source used. The directional coupler would be employed at power levels above approximately 10 mW (+10 dBm) since most sensing devices are designed to operate at

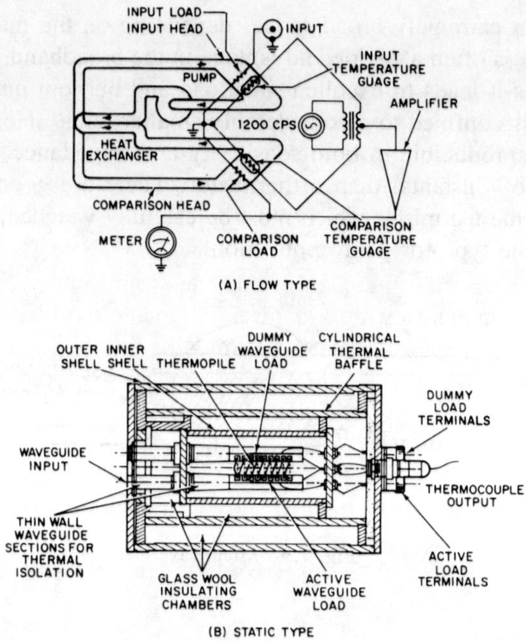

Fig. 7.10 Calorimetric power meters.

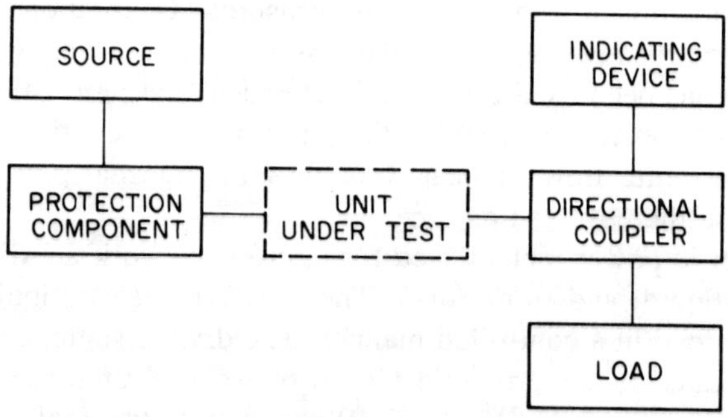

Fig. 7.11 Basic power test setup.

approximately this level. The load, or termination, is used to provide a good output match for the system, and its power handling capability must be adequate for the level you are to be reading.

7.2.4 Power Meter

The real heart of the power measuring setup is the indicating device. The three general types of thermally sensitive devices produce a reading on a meter or other display device. Figure 7.12 is an indicating unit utilizing a thermistor, the HP432A power meter (a fairly familiar piece of test equipment to most people in microwaves). Figure 7.12(A) is the actual component, Figure 7.12(B) is a block diagram. From (B) you can tell just what circuits you are adjusting when you move a control on the front panel. Figures 7.13 and 7.14 show actual thermistor mounts and their schematics. Many power meters today have digital displays.

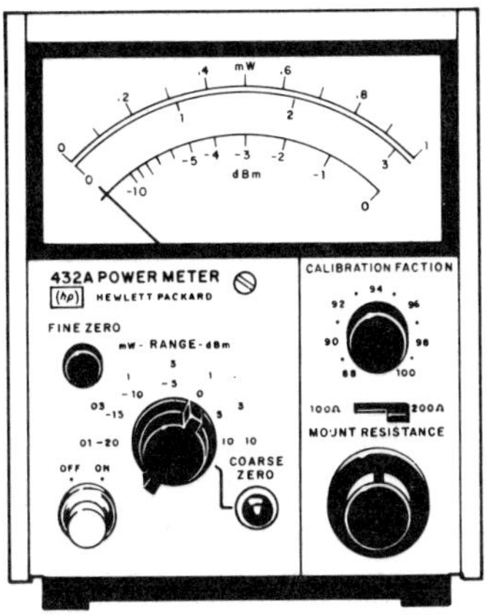

(A)

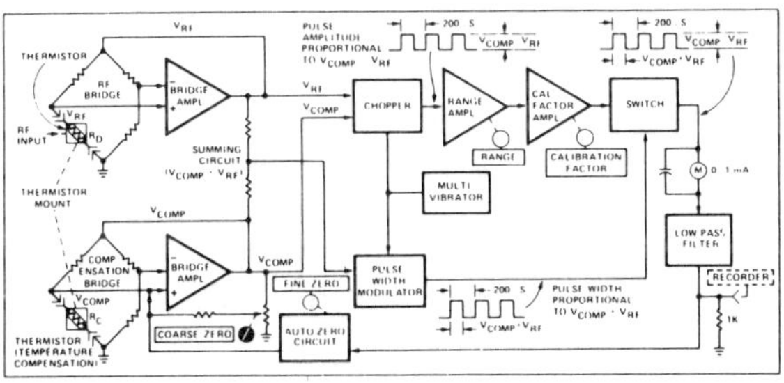

(B)

Fig. 7.12 HP432A power meter.

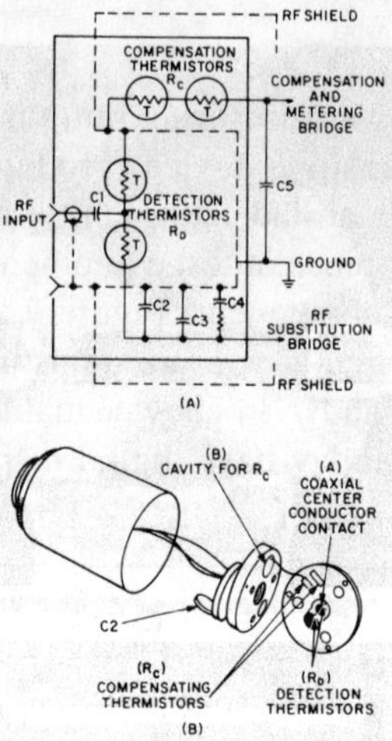

Fig. 7.13 Coaxial thermistor mount.

7.2.5 High Power Testing

Figure 7.15 shows a test setup to check the output power of a 100 watt power amplifier. This particular arrangement is one that actually was constructed when a 100 watt amplifier had been built and there were limited components available to check its output power. The components were an HP432A power meter with a 478A thermistor mount, a series of two watt attenuators ranging from 6 to 20 dB, type N one-watt terminations, and 20 dB coaxial couplers.

The first consideration is the power handling capabilities of each of the components. As mentioned above, the attenuators can handle two watts; the terminations, one. The thermistor mount has a maximum rating (printed in large red letters) of 30 mW (+15 dBm). A catalog is consulted regarding the coupler—it can handle 200 W of CW power.

The 3 dB attenuator can be used either with or without the circulator. With, it ensures that there is a good generator match for the unit under test by attenuating any reflections caused by either the circulator or the device; without, the attenuator has the complete responsibility of matching the generator and the device. Thus, we have shown both components in our setups, so that each does a good job and neither is left on its own. The attenuator is also used when a higher power generator is available (a 10 W maximum) and it is necessary to operate on the low end of its power range (a 200 mW input). Generators of this sort tend to be rather unstable, non linear, and inaccurate when used at their very low extreme. Therefore, if a pad of the proper power capability is inserted, the generator can put out a higher power while our device receives its required input level. In our example we needed 200 mW input. If we put a 6 dB pad (two W capability) at the input, the generator would

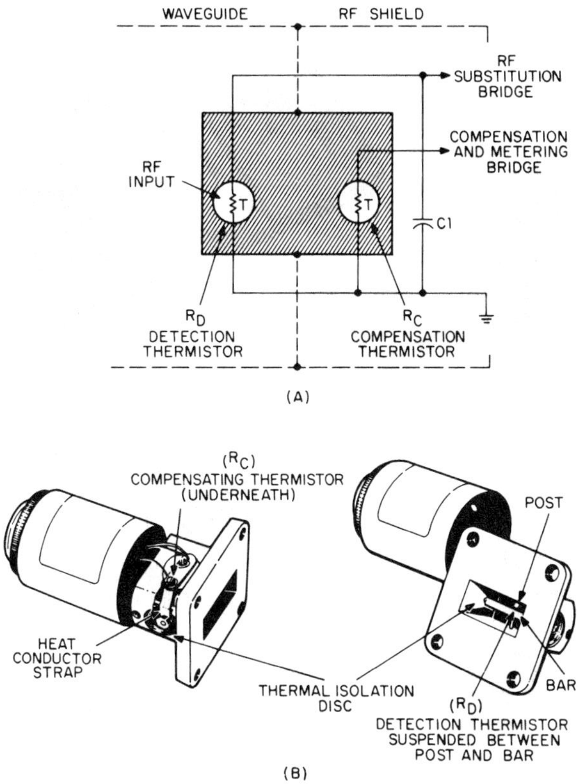

Fig. 7.14 Waveguide thermistor mount.

actually be putting out 800 mW—much more stable and accurate than a 200 mW signal. So, in addition to knowing the ratings of the components you use, also know the ratings of your equipment. If you need a low power but have a high power generator, arrange to operate it at a reasonable level, if at all possible. Your measurements will be much more meaningful.

Using the above ratings as our guide we set about the task of deriving a test setup that can pass 100 W without burning up any of the components. One fact is very evident—we cannot use the basic power test setup shown in Figure 7.11 for the simple reason that we do not have a 100 watt termination to put on the output of the directional coupler. Thus, we cannot use port one, in Figure 7.15, as our input port, and port two as the coupled one. This problem, however, does not take too long to solve. From our earlier discussions, coupling takes place from one to two and the insertion loss is from one to three. But, coupling also occurs from two to one; port two is isolated from port three due to the basic definitions and properties of a directional coupler. Therefore, one of our problems is resolved. Port two would

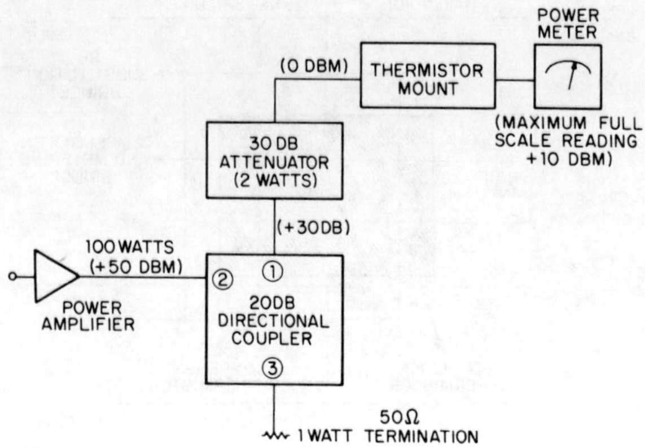

Fig. 7.15 100 watt power amplifier test setup.

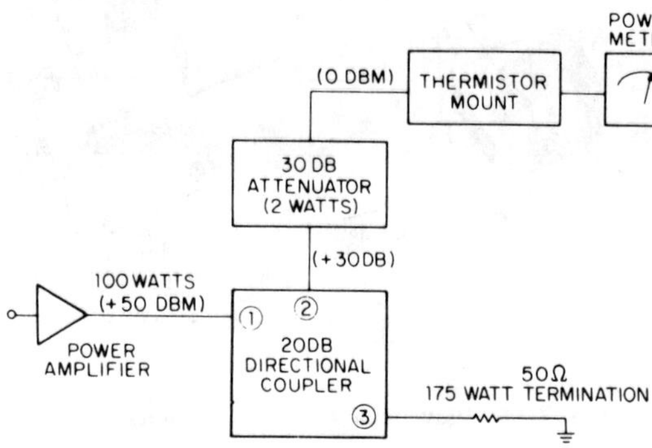

Fig. 7.16 Alternate setup for 100 watt test.

be our input; port one, the coupled port; and, because our coupler has a directivity of 25 dB, port three would be our terminated port (the one watt termination is an adequate safety margin).

At this point we have a 100 watt (+50 dBm) signal going through a 20 dB coupler and emerging at the coupled port at a one watt level.

Our next chore is to get this signal to our indicating device, the power meter. The thermistor mount can take only 30 mW (+15 dBm) before it burns up; therefore, we need a minimum of 15 dB of attenuation between the coupler and the thermistor mount. As an extra safety margin, 30 dB of attenuation is added, so that our thermistor mount has an input level of 0 dBm, well below its maximum rating. The power meter, when put on the 0 dBm scale, has a full scale reading of 100 watts—just what we needed to read. However, a much more accurate reading is obtained close to center scale. Therefore, we move the meter one scale higher to the +5 dBm scale—full scale is now 300 watts. Our desired reading is in the first one-third of the scale and our reading accuracy is increased.

An alternative approach to measuring the 100 watts, other than that described above, would be to have a high power termination. For example, you can buy a coaxial termination off the shelf that can handle 175 watts of power. With this termination, your test setup would be as shown in Figure 7.16; all of the other individual components are the same as those in Figure 7.15. Utilizing this high power load, the coupler now assumes a more familiar orientation—the input at port one, the coupled port at port two, and the straight through (or insertion) port at port three.

7.2.6 Medium Power Testing

Test setups for the medium power range are similar to that shown in Figure 7.16. The major variables are the termination (lower power) and the attenuator off the coupled port (lower value). The first is chosen to handle more power than you expect to be measuring; a good margin is 10 to 20 percent higher rating. The attenuator is adjusted to maintain the input rating of the thermistor mount, so that the power meter can be used on the −5 dBm to +10 dBm range. Operation on a lower scale usually involves checking meter drift—the thermistor mount is much more sensitive and can drift off its zero settings. Two setups for testing a one watt amplifier (medium power) are shown in Figure 7.17; they are identical except for the amplifier outputs. In Figure 7.17 (A) we use the directional coupler in much the same manner as we did in our 100 watt setup. The differences are that the termination needs to have only a two W power handling capability, and the attenuator off the coupled port is reduced to a 10 dB value. The result is a reading of one W full scale when the meter is in the 0 dBm range. If the amplifier is found to put out more than one watt, the range can be switched to the +5 dBm scale for a full scale reading of three W.

Figure 7.17(B) is a setup that is widely used, particularly when a directional coupler is not available. The output is simply attenuated by a 30 dB pad to produce a 0 dBm level at the input to the thermistor mount. The value of the pad, of course, can be almost anything, as long as you can conveniently translate what it does to your readings on the power meter, and on which scale you should read. The primary concern in such a setup is that it can safely handle the power pumped into it. Slightly above the one watt level is the maximum for this arrangement. Higher powers will rapidly approach the rating of the pad; it could suddenly start to smell quite burned. At this point, your power meter reads zero and your boss reads you the riot act. So, about one to one and one-half watts should be a good point to switch to a directional coupler for this type of power measurement.

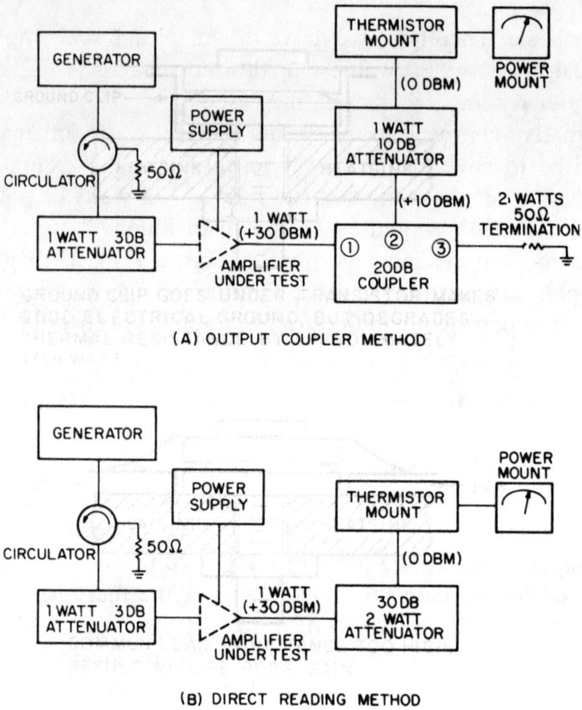

Fig. 7.17 Medium power test setups.

7.2.7 Low Power Testing

When we talk of low power devices in microwaves, we are usually talking about those used in a receiver. The tests of prime importance are noise figure, gain, modulation products, gain compression, and, generally, those other than of power. However, there are occasions when a power reading is needed; sometimes it is relative (compared to some other device), sometimes it must be absolute (a definite dBm value). Although there are power meters that can indicate to −50 or −60 dBm, the most useful tool for measuring these low power values is the spectrum analyzer.

Figure 7.18 is a typical test setup that can employ a spectrum analyzer for a low level power reading. The analyzer, placed directly at the output of the device under test, can read either a relative or absolute power. Figure 7.19 illustrates an analyzer display and how the absolute power reading is made; the test setup is shown above the display. The controls are set at −20 dBm on the log reference level and −3 dB on the display position. Therefore, the top of the display, at the 0 line, is −23 dBm. Our signal is on the −10 line; thus, the level of the signal at the input to the analyzer is −33 dBm. If there were no padding at the output of the test device we would have a −33 dBm signal and that would be it. However, as you can see from the setup, we have a 3 dB pad at the output—effectively making our readings 3 dB worse than they actually are, and so, the real output power from this particular device is −30 dBm. If you were to connect a direct reading power meter, (with adequate filtering to remove harmonics) to the output of this device, you would find a value of −30 dBm.

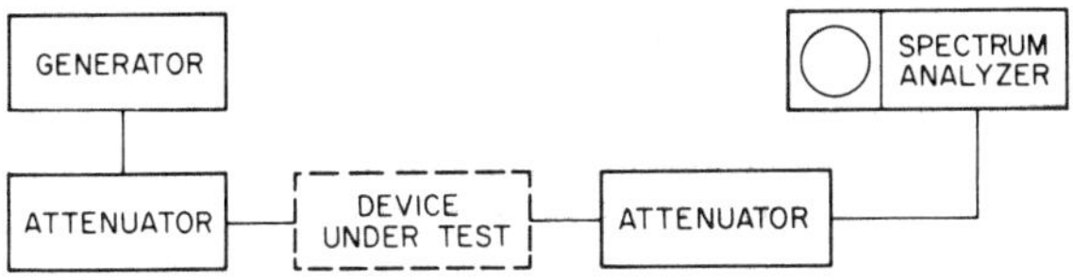

Fig. 7.18 Low power test setup.

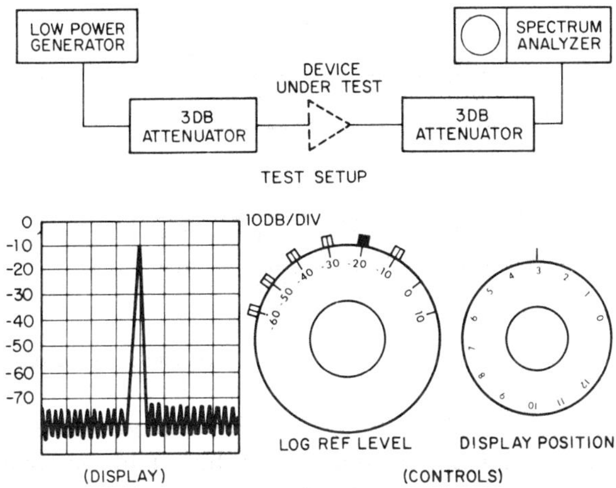

Fig. 7.19 Absolute power measurement using a spectrum analyzer.

One phrase in the statement above should be explained—". . . with adequate filtering to remove harmonics." A power meter's sensing system is completely independent of frequency. It does not care if it sees a 1 GHz, 2GHz, or a 3 GHz signal; it only knows that there is microwave power present which produces heat that it can pick up and that results in a meter reading. This reason is why we refer to having adequate filtering, usually a bandpass or low-pass filter, to ensure that only the desired fundamental frequency is being indicated.

This point is one great advantage of using a spectrum analyzer. You are able to see the fundamental frequency, the second harmonic, and any other signals generated along with their associated absolute level. Each can be very clearly distinguished from one another.

7.2.8 Peak Power Testing

The power measurements taken to this point have been cw readings at various levels. Another important type is the *peak pulse power reading*. We cover methods used in waveguide and coaxial setups; an additional procedure is also covered—that of a system called the *Notch wattmeter*.

Average Power-Duty Cycle Method

The first type of waveguide test is the *average power-duty cycle method*, as shown in Figure 7.20. It uses the idea that the peak power of a rectangular RF pulse may be determined by measuring average power, pulse width, and repetition rate.

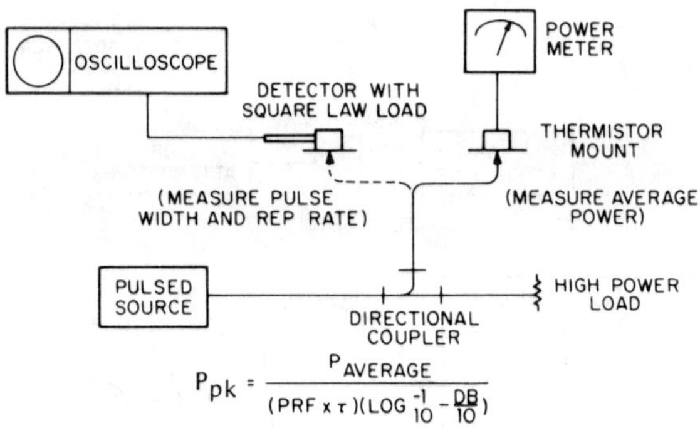

Fig. 7.20 Average power-duty cycle method of measuring peak power.

The product of pulse width and repetition rate is called the duty cycle of the pulsed source; it relates the duration of the pulse to the period of the pulse rate. Thus, the name *average power-duty cycle* specifies what actually is measured to obtain the peak power reading. Referring to Figure 7.20, it is very important that the maximum peak power and energy-per-pulse ratings of the particular thermistor being used are not exceeded—they are printed on the mount as watt-microsecond figures. If for any reason they are not obvious, you should consult a manual. Another point that should be made is that, when measuring peak power in a waveguide system, you should be sure that all flanges are carefully aligned and that the waveguide is clean and dry to avoid arc-over. The procedure is as follows:

1. Connect the crystal detector and its square law load to the coupled arm of the directional coupler. The square law load is used to improve the rise and fall times—which are very important in pulse measurements—of the detector.
2. Measure with the oscilloscope and record the pulse repetition frequency (PRF) and the pulse width (τ).

3. Replace the detector with the thermistor mount and the power meter.
4. Note the average power in the meter.
5. Calculate the peak pulse power as follows:

$$P_{peak} = \frac{P_{average}}{(\text{PRF} \times \tau)\left[\log_{10}^{-1}\left(\frac{\text{dB}}{10}\right)\right]}$$

when

P_{peak} = peak power of the source;
$P_{average}$ = average power read on the power meter;
dB = coupling factor of the directional coupler.

The term, $\log_{10}^{-1}$, indicates the anti-log of a number. To illustrate, let us take the log of a number and then reverse the procedure. If we want to determine $\log_{10}$ 200, we can review the table in Appendix R and find a value of 2.303 (because 200 is 2×10^2, which gives the first significant number, and the log of 2 is 0.303).

Now, if we were given a ratio of two numbers equal to 2.303 and wanted to find its anti-log, we would reverse the order. We could look in tables or use a slide rule. For the latter, first find 0.303 on the log scale; looking on the C or D scale would give us 2, meaning that it is some number times 10. That number we have already determined to be 2. Therefore, the anti-log of 2.303 is 2×10^2, or 200. Written mathematically:
$\log_{10}^{-1}$ 2.303

$$\log_{10} 200 = 2.303$$
$$\log_{10}^{-1} 2.303 = 2 \times 10^2 = 200$$

In the case above there is another (−) sign in front of the dB/10 term, meaning that the power of 10 is negative. Thus, $\log_{10}^{-1}(-2.303) = 2 \times 10^{-2} = 1/200$.

To illustrate this method, let us show an example. Suppose we took the following readings:

PRF = 1 kHz;
$\tau = 1.5\ \mu s\ (1.5 \times 10^{-6})$;
$P_{average} = 7.5$ mW (7.5×10^{-3});
Coupling factor = 40 dB.

Peak power would be calculated as

$$P_{peak} = \frac{7.5 \times 10^{-3}\ \text{W}}{(1 \times 10^{3})\,(1.5 \times 10^{-6})\left[\log_{10}^{-1}\left(-\frac{40}{10}\right)\right]}$$

$$= 50\ \text{kW}$$

Direct Reading Method

The peak power range of the system shown in Figure 7.20 is around 100 kW at X-band, limited by the maximum power ratings of the termination, thermistor mount, detector, and coupler.

An improved waveguide setup for peak power measurement of non-rectangular pulses or multipulse systems is shown in Figure 7.21. In this arrangement, the pulse amplitude is first detected by a fast-response crystal detector with a square law load. The detected pulse amplitude is noted on the scope and then accurately duplicated with a known CW power. The procedure is as follows:

1. Connect the equipment as shown in Figure 7.21, Step 1. Set the attenuator at a convenient reference to limit the expected peak power to about 1 mW.
2. Adjust the oscilloscope trigger and vertical sensitivity controls for a stable pulse display. *Use dc coupling* at the oscilloscope vertical input, and *make no further adjustments to the oscilloscope sensitivity once the reference pulse amplitude is set.*

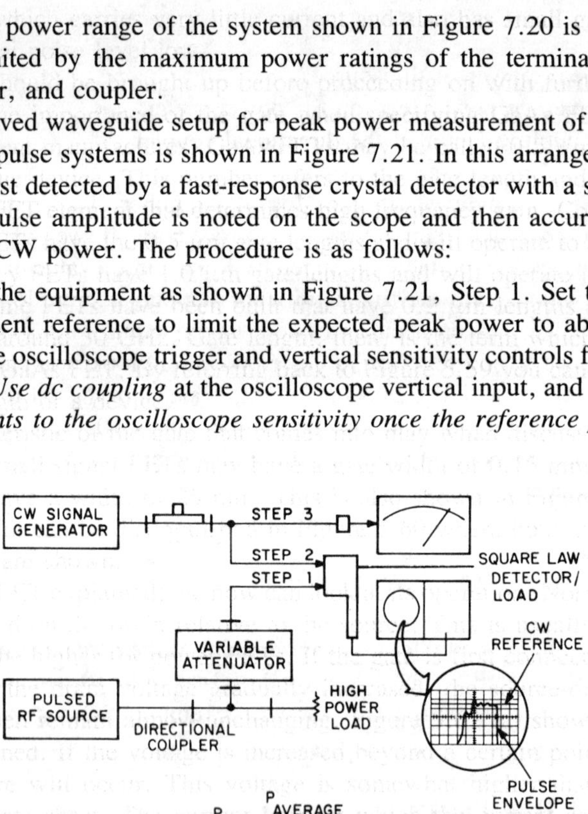

Fig. 7.21 Direct pulse power method.

3. Move the detector/square law load combination to the output of the slide screw tuner and signal generator, as shown for Step 2.
4. Adjust the signal generator for a CW output frequency equal to the pulsed source's RF frequency.

5. Adjust the slide screw tuner for maximum deflection of the oscilloscope trace. *Do not adjust the vertical sensitivity*. Use the output attenuator on the generator to keep the trace on the screen during these adjustments.
6. Set the generator for a trace deflection on the oscilloscope equal to the pulse amplitude noted in Step 1. Make no further adjustments to the output attenuator.
7. Connect the thermistor mount and power meter to the generator as shown for Step 3.
8. Re-adjust the slide screw tuner for a maximum reading on the power meter and note the average power reading.
9. Calculate the peak power from the equation

$$P_{\text{peak(source)}} = \frac{P_{\text{average}}}{\left[\log_{10}^{-1}\left(-\frac{\text{dB}}{10}\right)\right]}$$

where

dB = coupling factor of the directional coupler (including the variable attenuator setting) in dB;
P_{average} = average power measured on the power meter.

An example would be given

$$P_{\text{average}} = 0.5 \text{ mW}$$

$$\text{dB} = 80 \text{ dB}$$

$$P_{\text{peak}} = \frac{5 \times 10^{-4} \text{ watts}}{\left[\log_{10}^{-1}\left(-\frac{80}{10}\right)\right]} = 50 \text{ kW}$$

Coaxial Setup

Test setups talked about so far have been for waveguide setups, but peak pulse measurements are also made in coaxial systems by using a peak power calibrator. This device is one whose operation is based on a dc-to-pulse-power comparison. The calibrator is direct reading and does not rely on pulse shape for its accuracy. An example of such a unit is the BOONTON 8900B. The test setup for peak pulse power is shown in Figure 7.22. The directional coupler is a 50 dB coupler to ensure that the calibrator input is no greater: than 200 mW (+23 dBm) peak. The test procedure used is as follows:

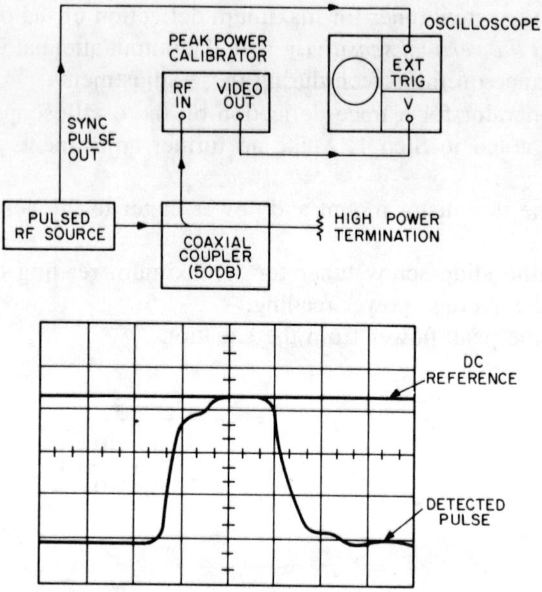

Fig. 7.22 Coaxial peak power measurement.

1. Calibrate the peak power calibration as outlined in its operating manual. Leave the pulsed source power *off*.
2. Set the oscilloscope *vertical sensitivity* to about a 50 mV/cm range and use a dc input setting.
3. Set the oscilloscope sweep time and trigger for a free running trace. The display should appear as two horizontal lines on the CRT.
4. Adjust the *null* on the calibrator until the two horizontal lines are on top of one another (balancing out the detector diode bias from the video output waveform which might otherwise be read as RF power).
5. Turn on the pulsed RF source.
6. Synchronize the oscilloscope for a stable display.
7. Using the *coarse* and *fine* controls on the calibrator, adjust the dc reference voltage (which now appears as a straight line on the screen) until it coincides with the peak of the displayed pulse. The scope display should be as in Figure 7.22.
8. Read the peak power on the calibrator and use the formula below to find the actual power.

$$P_{\text{peak(source)}} = \frac{P_{\text{peak(calibrator)}}}{\left[\log_{10}^{-1}\left(-\frac{\text{dB}}{10}\right)\right]}$$

where

$P_{\text{peak(calibrator)}}$ = reading of peak power on the peak-power calibrator;

dB = coupling factor of the directional coupler.

As an example of the above procedure suppose we read 135 mW on the calibrator and the coupling factor was 50 dB. Then,

$$P_{\text{peak(source)}} = \frac{135 \times 10^{-3}}{\left[\log_{10}^{-1}\left(-\frac{50}{10}\right)\right]} = \frac{135 \times 10^{-3}\ \text{W}}{1 \times 10^{-5}}$$
$$= 13.5\ \text{kW}$$

Notch Wattmeter

The *notch wattmeter* system is used in applications where the average power in a pulsed RF wave is too low to be measured by a power meter. Various null, or notch, techniques are possible; one is shown in Figure 7.23. The RF pulse to be measured is compared with a CW RF signal which is pulsed off to form a notch. Waveform (A) in Figure 7.23 represents the detected CW signal from the signal generator. Waveform (B) shows the detected RF pulse and the detected notch when they are not coincident. At waveform (C), the two waves are coincident and the pulse to be measured is placed in the notch. Coincidence is obtained by adjusting the pulse width, delay, and repetition frequency at the pulse generator. The signal generator has been adjusted to be equal to the pulse power of the pulse under consideration. The peak power can be determined from the calibrated output of the signal generator or from the reading of the power meter to which the two signals are simultaneously applied. When the latter method is used, the following formula applies

$$P_{\text{peak}} = \frac{P_{\text{power meter}}}{\left[\log_{10}^{-1}\left(-\frac{\text{dB}}{10}\right)\right]}$$

If the power read on the meter were 1 mW and the total attenuation between the source and the thermistor mount were found to be 73 dB, the peak power would be

$$P_{\text{peak}} = \frac{1 \times 10^{-3}}{\left[\log_{10}^{-1}\left(-\frac{73}{10}\right)\right]} = 20\ \text{kW}$$

Direct Reading Method

There are a variety of peak power meters available today which will read peak power directly. These meters are used in the same way as the CW power meters are used, and can be inserted into a test setup either in series, or at the coupled port of a directional coupler (similar to the two setups shown in Figure 7.17). Each of these meters will have unique properties and settings and it is suggested that you consult the appropriate manual before using them.

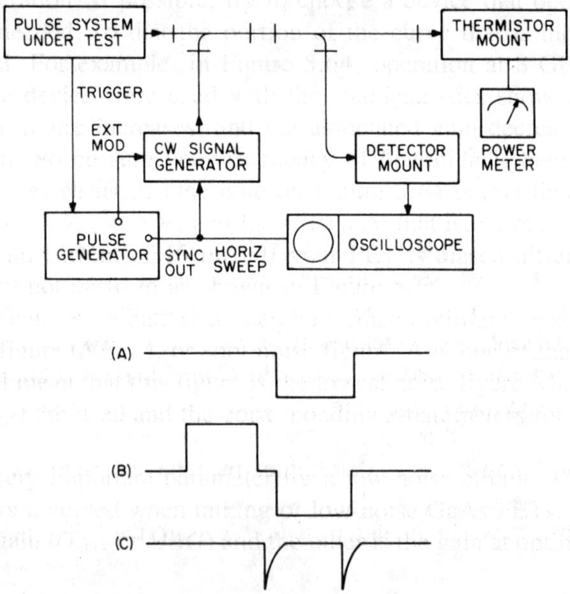

Fig. 7.23 Notch wattmeter setup.

7.2.9 Sources of Error

Up to this point we have not mentioned anything about possible errors in power measurements, instead concentrating on introducing the different methods available to obtain a reading of power for a particular application. There are, however, definite sources of error which must be considered, or compensated for, if you are to obtain accurate readings. The errors you encounter in microwave power measurements are due to *mismatch loss, RF loss, substitution loss,* and *error in substituted power measurement.*

Mismatch Loss—In basic electronics courses it is taught that a maximum transfer of power is achieved when the resistance of the load equals that of the generator supplying the load. This situation is true for dc and low frequency systems; it is also the case in microwaves. However, it is not quite so simple regarding microwaves. There is a generator resistance, or impedance, and a load impedance, but there is also the characteristic impedance of the transmission line or the components in between the generator and the load. The most frequent culprit for mismatches in a power measuring system is the thermistor mount. The power delivered to the mount, of course, is maximum when the mount impedance equals the characteristic impedance of the feeding transmission line. When the VSWR of the mount is not low (with respect to unity), power is reflected toward the source. Also, power may be reflected if the source is not matched. These mismatches can be reduced by the use of either a slide screw tuner to tune out the VSWR at the source of the load, or by insertion of additional attenuation to reduce the reflections and present a good match to both the load and the source. However, each of these add work to your measurement. First, the tuner has an RF pass to be considered as another source of error; secondly, the attenuation value inserted must be figured into your calibration so that the power meter reading is on the scale that you expect.

RF Losses—These losses are the ones in the RF components of the power measuring system. Two such components were mentioned above: the slide screw tuner and the attenuator. Other losses are the resistive ones in the walls of the thermistor mount and associated transmission line. If the mount has adjustable elements, there are resistive losses due to poor contacts. If there is a dielectric material present in the mount, it also can cause RF losses. The resistive losses in the system's transmission lines can be evaluated by an attenuation measurement. The losses in the thermistor mount are usually described in terms of mount efficiency, η.

$$\eta = \frac{\text{power absorbed by the thermistor}}{\text{power delivered to the thermistor mount}}$$

The RF losses and substitution error are regarded as one loss called the calibration *factor*, referred to earlier in this chapter in Section 7.1.2. The calibration factor of a good thermistor mount may be as high as 0.98. You should now be able to see why the calibration figure illustrated in Figure 7.1 varies with frequency—it is actually a plot of the RF losses with frequency.

Substitution Error—This error is due to the difference in heating effects of RF and low frequency or dc power, which results in an uneven current distribution along the wire because of standing waves. If the bolometer element (thermistor, for example) is small enough compared to a wavelength of RF, the error may be less than one percent.

Error in Substituted Power Measurement—The accuracy of substituted power measurements depends on the accuracy of the instrumentation equipment. As an example, if a 200Ω thermistor operating at 15 mA is used to measure 100 mW of power, an accuracy of current determination of almost 0.3 percent is necessary to obtain an accuracy of 10 percent in the power measurement. The more accurate the instrumentation, the lower the errors in your power calculation.

The first of our five basic microwave measurements has been covered. As we mentioned in the beginning of this section, there is no mention of voltage, current, or power factor. The terminology is strictly watts, regardless of the type of power being measured—low, medium, or high CW or peak pulse power in a waveguide or coaxial system. The method of determining power is by measuring the heat generated by that microwave power. This process emphasizes two very important points about microwave measurements. When you realize that microwave power does produce heat, you can understand why it is of prime importance to check the power handling capability of your components and equipment. We also have all learned early in life that something which produces heat can burn and hurt. That important lesson should not be forgotten when making power measurements. The circuits are hot, both in temperature and RF. They can burn a finger with their heat, or produce very painful RF burns, especially if they are concentrated at a small point such as the center conductor of a connector. If you have any doubts about microwave energy's ability to burn (or cook) refresh your memory with a device called a *microwave oven*.

7.3 TRANSMISSION MEASUREMENTS

The term *transmission measurement* is a very broad term that takes in such familiar areas as insertion loss, gain, conversion loss, and attenuation. What actually is being referred to is how much an input microwave signal has been increased or decreased when it is monitored at the output of a particular device or system. In other words, how much power is gained or lost through a particular device in our system.

In order to simplify the portion on transmission measurements, we concentrate on one particular area: attenuation. We investigate the different types of attenuation measurement, and then relate them to such factors as insertion loss (attenuation), gain (the opposite of attenuation), and conversion loss (attenuation through a mixer). In all of these note one key difference from any of the measurement setups in the preceding section. In that portion of the book we were concentrating on one particular measurement, that being at the output of some device to determine the power at that point. In our transmission measurements we basically are concentrating on a comparison—the power at the input of a device or system *versus* how much is at the output. The ratio of these two amounts gives us either our attenuation, gain, or conversion loss.

In the section on attenuators in Chapter Four, we define "to attenuate" as "to lessen the value of something;" in microwaves, that "something" is the microwave power through a device. The attenuation measurement setups we describe here are designed to tell us just how much of this power we are absorbing or reflecting—in general, by how much we have lessened the value of the input power applied to a device or system by the time it appears at the output.

Although several procedures for making attenuation measurements exist, the three used most generally are:

1. *RF substitution method*;
2. *Audio substitution method*;
3. *IF substitution method*.

For each of these methods, we describe the setup, the components involved, and the rationale behind the use of each particular component. We also discuss the procedure for obtaining attenuation measurements by each process. Methods for both CW and swept techniques are covered where applicable, and we conclude the section by pointing out general precautions and potential sources of error.

7.3.1 RF Substitution Method

When we refer to RF substitution, we mean just that—the test involves the inserting of a component into the RF line after a standard attenuator has been used to calibrate the system. Figure 7.24 illustrates an RF substitution setup for a cw measurement. The procedure for obtaining an attenuation reading in this case is as follows:

1. Choose the frequency in which you are interested, and set up your CW source to that frequency according to its specific instructions.
2. Short terminals A-A together.
3. Set a convenient level on whichever indicating device you choose. Be careful to note and observe the maximum input level to the detector or thermistor

mount. The level is set by the calibration attenuator and the output level control on the source.

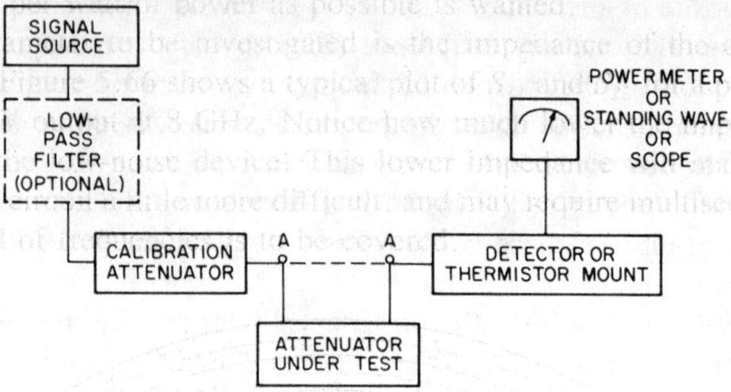

Fig. 7.24 CW measurement using RF substitution.

4. Record this level as your reference.
5. Insert the device to be tested at point A-A and note the new reading on the indicating device.
6. The difference between the reference and the figure in Step five is the attenuation through the particular device under test.

In an RF substitution test, the signal source may be a signal generator, a frequency synthesizer, or a sweep generator in the CW mode. The basic requirements are that it can produce a CW signal at the proper frequency, has reasonable stability (does not drift significantly), and is capable of providing a sufficient power level to make the test meaningful; it also may be necessary for it to have modulation capabilities, usually 1 kHz square wave. The choice of a source depends on any or all of the above qualifications that apply to your particular needs. For example, your situation may demand a source that has good stability at 6 GHz and a 10 mW output level. With the advances in the equipment available today, any one of the three source types could be fit into your setup and the required power, frequency, and stability be obtained. A couple of years ago you would have been limited to a cw generator or a sweeper; however, improvements using YIG-tuned oscillators and Gunn oscillators have made synthesizers available that can operate as high as 26.5 GHz.

The calibration attenuator can be either fixed or variable. Factors to consider are its attenuation accuracy, minimum insertion loss (if a variable attenuator), and VSWR characteristics (which is of prime importance). Whenever you insert a component into a system you should know what kind of reflections will be caused. A great effort should be made to use attenuators with a low VSWR to minimize reflections, and thereby enable your readings and overall setup to exhibit a high degree of accuracy.

The minimum insertion loss of any variable attenuator is another factor to consider seriously. Certain types have a loss of 15 to 20 dB, such as the waveguides below cut-off variety. This high a figure limits both the dynamic range of the overall setup and the devices that can be tested.

The indicating unit in your setup may be any of several types; the specific one used dictates the sort of detector that should be employed. If a power meter is used as an indicating device, a thermistor mount would be selected. When a scope is utilized for indication, the detector is used. The precedure when using the scope is similar to that for a meter. A reference level is set by the calibration attenuator. When the component is inserted, the attenuator is adjusted so that the deflection on the scope is equal to the original deflection. The value of attenuation is then read, either off the calibration attenuator, or off the face of the scope using the graticules on the face and the vertical settings.

Note that there is one area in Figure 7.24 we have not covered—the optional low-pass filter. It is utilized to ensure that what you are reading on your meter or scope is a true reading. The detectors and thermistor mounts used in microwave measurements are not frequency selective; that is, they pass and respond to frequencies from 100 MHz to 12.4 GHz and above. Most signal sources internally generate harmonics that can be mistaken very easily on the output meters as part of the power level. Therefore, to be sure that you are reading only the desired frequency, insert a low-pass filter. One way to determine the size of the harmonics is to use a spectrum analyzer as an indicating device; such a setup is shown in Figure 7.25. There are no detectors or thermistor mounts, but there is a matching attenuator to reduce any mismatch losses between the analyzer and the device under test. The analyzer level is set with the terminals A-A shorted and recorded as a reference. When a unit is inserted, the new level is the attenuation and may be read directly off the face of the CRT; alternately, the RF attenuator may be used to return the signal to its original level, with the attenuation value directly readable off the attenuator.

Though the discussion is concerned solely with CW RF substitution measurements, there are times when you need to know a particular device's capabilities over a band of frequencies. When this type of measurement is desired, a swept setup is required; Figure 7.26 shows two applicable arrangements. Figure 7.26 (A) is a setup using the HP415E standing wave meter with an X–Y recorder; Figure 7.26(B) uses a scope as an indicating device. An alternate arrangement to Figure 7.26(A) employs a power meter instead of standing wave meter; in the process, detector number 2 becomes unnecessary. A thermistor mount is used instead at the output of the calibration attenuator, and the "recorder" output on the power meter goes to the Y-terminal of the X–Y recorder.

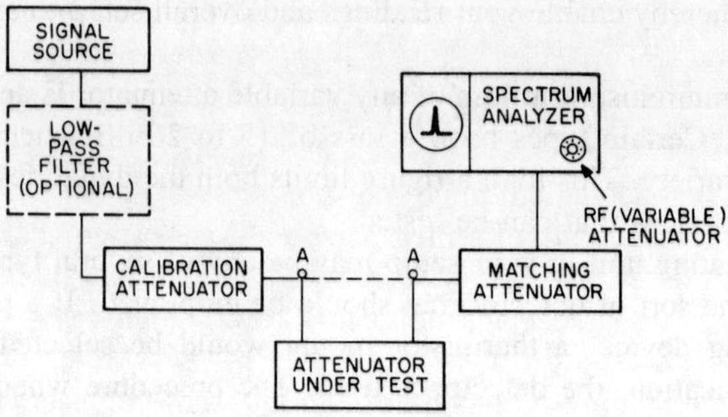

Fig. 7.25 RF substitution using a spectrum analyzer.

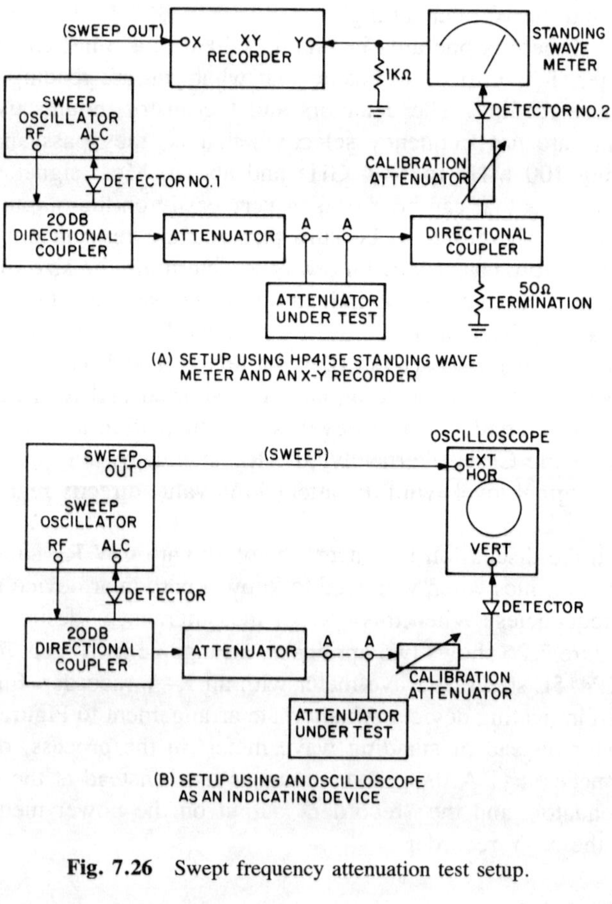

Fig. 7.26 Swept frequency attenuation test setup.

Before getting into any explanation of these test setups, one important area should be isolated and discussed—the *automatic leveling circuit* (ALC). Figure 7.27 shows only the ALC portion of the setups and is used to explain what is happening. Consider this example:

Suppose you were asked to hold a five pound weight in front of you at arm's length at a certain height for a period of, say, one-half hour. You begin by keeping the weight very close to the required level. As time goes by, your arms start to drop. Your eyes sense that the weight is no longer at the necessary height and tell your brain that some correction must be made; your brain signals the muscles in your arms to raise the weight. When the load is at its specified mark, your eyes sense that everything is once again all right and signal your brain to stop correcting. The brain in turn tells the muscle to stop—everything is again stationary. If your arms overcorrect and the weight goes above the mark, the process is reversed and the weight is lowered.

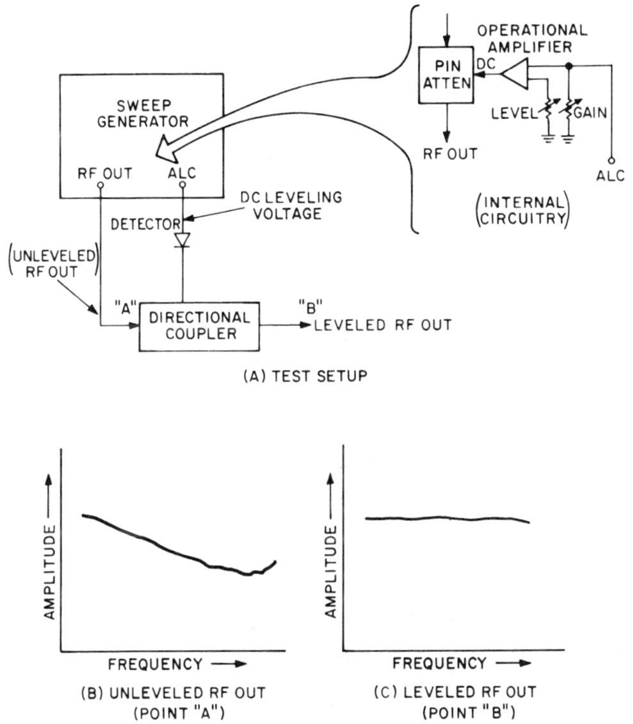

Fig. 7.27 Automatic leveling circuit (ALC).

The above example is a description of one of the most basic and simplest feedback systems known—your own body. The basic elements are: an original force (your arms holding the weight), a sensing device (your eyes), a command control device (your brain), and a corrective force (your muscles).

Now examine Figure 7.27(A). The RF output from the sweep generator is an unleveled signal similar to that in 7.27(B), assuming a frequency range wider than just a few MHz. This unleveled RF power is sampled by a directional coupler. The detector at the coupled port senses any change in RF power level and connects this fluctuation to a corresponding dc voltage. This dc voltage is fed to one input of an operational amplifier where the overall gain of the leveling "loop" can be set. The other input has a reference level established to determine the required operating point. The dc output current is used to control a *pin* attenuator which increases or decreases the level of the RF output. This process explains how we get an output as shown in Figure 7.27 (C). If the RF output decreases (as it does at higher frequencies), the power coupled to the detector also is decreased. The lower power means a lower dc voltage to the operational amplifier; in turn, the current to the *pin* diodes in the attenuator is decreased. With less current through the diodes there is less attenuation in the attenuator. Thus, RF output is increased to compensate for the sag in power (B), and the leveled output (C) results.

The four basic ingredients for a successful feedback system should be obvious in the explanation above.

1. The original force—*the unleveled RF output.*
2. The sensing device—*the coupler and detector.*
3. The command control device—*the operational amplifier.*
4. The correction force—*the PIN attenuator.*

Obviously, there are very good reasons why a microwave engineer uses an ALC loop to obtain a good swept measurement. It should be clear why a directional coupler and detector (or combined as a directional detector) are put at the sweeper output and an RG-58 cable with BNC connectors is run from the detector to a jack called "ALC" or "crystal." When the ALC button is pushed you are applying operating voltage to the amplifier so that the loop can be closed and leveling can take place.

With a leveled signal at our output (In Figure 7.27(A), point B) and a working knowledge of automatic leveling circuits, let us proceed to a discussion of swept attenuation measurements.

At this point we should backtrack and again refer to Figure 7.26. Figure 7.26(A) is a basic setup that is very useful since it results in a plot of your data; the entire left side is the ALC loop discussed above. In addition the signal is modulated with a 1 kHz square wave—a necessary requirement to have the standing wave meter work properly—and the leveled output you desire is produced. The first attenuator (on the order of 3 to 6 dB) cuts down reflection between the source and the device under test. The directional coupler may or may not be used, depending on the amount of power. For higher levels it must be inserted to prevent the standing wave meter from being overdriven. The termination should have an adequate power handling capability for the levels at which you are testing. The calibration attenuator is used to adjust for a proper level on the indicating device. The detector (number 2) removes the RF component and allows the 1 kHz signal to be used to operate the meter.

The SWR meter is a tuned low-noise amplifier-voltmeter calibrated in dB and SWR for use with square law detectors. It is an extremely useful instrument for directly measuring SWR, attenuation, and gain from metered scales; or as a tuned amplifier for driving a recorder when making RF substitution measurements. It responds to a standard tuned frequency of 1,000 Hz, which is front panel adjustable over a range of 7 percent for exact matching to the internal 1 kHz modulation of the signal source. Amplifier bandwidth is also adjustable from 15 to 130 Hz. The narrow extreme allows maximum sensitivity at CW frequencies, while the wider bandwidths enable swept tests to be displayed on an oscilloscope or X–Y recorder.

A 60 dB attenuator with an accuracy of 0.05 dB/10 dB ensures high precision in substitution measurements. An expand–offset feature allows any 2 dB range to be enlarged to full scale for maximum resolution. Linearity is ±0.02 dB on expanded ranges and is limited only by meter resolution on normal scales. This performance, together with the inherently low noise figure, permits maximum measurement range with exceptional resolution and linearity.

The 1kΩ resistor at the recorder Y–input is to match the meter output at the jack to the input of the X–Y recorder itself. The vertical information (Y–axis) is obtained from the standing wave meter; horizontal (X–axis), from the sweep generator. The speed of the pen on the recorder is determined by the sweep voltage from the sweeper and should be set to provide maximum resolution on your plots (that is, no peaks or valleys in your data are missed because the recorder swept past them too quickly).

With the individual pieces defined, we can go through a test procedure with the setup.

1. Connect the equipment as shown in Figure 7.26(A).
2. Adjust the calibration attenuator for maximum attenuation and set the *range* switch on the standing wave meter to *norm* and 0 dB.
3. Connect points A–A together with an adapter or a cable.
4. Turn on the sweep generator, X–Y recorder, and SWR meter power. (The sweep generator has a 30 second time delay if it uses a *backward wave oscillator* (BWO); if solid-state, it has RF available right away.)
5. When RF power is available, set the sweeper on a CW position at the lowest frequency of interest. Turn on the sweeper's square wave modulation.
6. Decrease the attenuation in the calibration attenuator until an indication of power is seen on the SWR meter. Adjust the square wave modulation on the

sweeper for a maximum reading on the SWR meter. (It may be necessary to alternate among the calibration attenuator, the *range* switch, and the *gain* control on the meter to obtain a 0 dB reading.)

7. Adjust the recorder's Y–axis for a point at the top of the paper. Insert the required number of dB; for example, you may need 1 dB steps up to 5 dB, then go in 10 dB steps up to 30. You thus would have 0, 1, 2, 3, 4, 5, 10, 20, and 30 dB points on your paper, with the 30 dB mark at the bottom of the page. A plot of this type is shown in Figure 7.28(A).
8. Set the Y–axis back to the 0 dB point.
9. Set the low and high frequency limits and set the sweeper to a *manual* mode. Run the frequency from low to high and set the X–axis of the recorder so that it covers both sides of paper from left to right.
10. Set the sweeper to a *trigger* mode and set the sweep speed to a 15 to 20 second full sweep time.
11. Drop the pen on the paper and hit the *trigger* button. The recorder traces the power variations across the paper, as shown in Figure 7.28(A), for a 0 dB setting. Lift the pen when it gets to the right side of the recorder, and the arm swings back to the left.
12. Repeat the above steps for all of the settings you read, using the *range* switch and the *calibration attenuator. Do not touch the gain control once you have made your initial setting.*
13. Set the attenuator and *range* switches back to their original settings and check to see that the pen in the recorder returns to its original zero setting.
14. Insert the device to be tested at A–A. The pen on the recorder should now indicate the value of attenuation present through the device.
15. Hit the *trigger* switch and drop the pen. The recorder plots the response as a function of the calibration lines originally plotted on the paper, and you have a record of your device and its actual variations with frequency. Figure 7.28(A) shows a 3 dB coupler that has been swept. It can be seen that at the center of the band it goes to a value of 3.2, while it is about 2.8 dB at each end. If we did not have the calibration marks on the paper to tell us what the generator did as a function of frequency, the readings on the meter and the recorder would be meaningless.

You may have noticed one potentially disturbing item in Figure 7.28(A)—the way the calibration lines go together as the attenuation is increased. This convergence is due to the fact that the output of the SWR meter is a logrithmetric voltage; as such, the spacing between readings decreases as the attenuation is increased. To counteract this crowding at higher attenuation values and thereby obtain an accurate reading, you can calibrate the X–Y recorder for 10 dB full scale. At higher figures, you can add 10 dB steps with the *range* switch. To illustrate this point, suppose we have a statement on the plot in Figure 7.28(A) that says +10 dB. We would have a 13 dB measurement with the accuracy of a 3 dB reading. The value and versatility of such a setup cannot be overestimated.

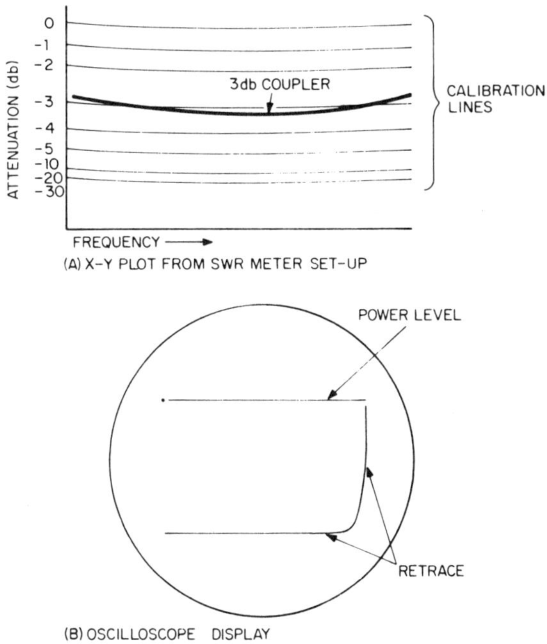

Fig. 7.28 Swept measurement displays.

A second method of RF substitution when a standing wave meter is not available is shown in Figure 7.26(B), using an oscilloscope as an indicating device. The scope actually takes the place of the SWR meter and the X–Y recorder. The calibration attenuator must do all of the setup in this case because there is no calibrated attenuation on the scope. The precedure for testing is as follows:

1. Connect the setup as shown in Figure 7.26(B). Join points A–A together by means of an adapter or cable. The cables carrying the sweep voltage and going to the scope and the ALC jack all are carrying dc—their RF characteristics are of no importance. All other cables should be chosen to give the lowest possible insertion loss at the frequencies of operation.
2. Set the calibration attenuator to its maximum attenuation.
3. Turn on the power to the sweep generator and the scope. Allow for a time delay if applicable.
4. Set the sweeper start and stop frequencies to the desired settings and put the sweeper in an *automatic* mode. Set the sweep time to its maximum setting.

(The scope should be in an *external horizontal* mode.) You should now have a line on the scope.

5. Adjust the scope for a full scale sweep and adjust the calibration attenuator until you have a presentation similar to that in Figure 7.28(B). Note the value of the calibration attenuation. The scope vertical should be around 0.1 V/cm to obtain a good display, which is your 0 dB setting and your reference. Mark this line and shape on the scope with a grease pencil.
6. Insert the device to be tested between points A–A. The presentation on the scope should be as shown in Figure 7.28(B), but at a reduced level. Take out attenuation from the calibration attenuation until the image is at its original reference level. Subtract this reading on the attenuator from that in 7.28; the result is the value of attenuation of your device.

One disadvantage of using a scope should be readily apparent. Although it provides a nice visual presentation, no permanent record of your data is created, unlike with the recorder. It is possible, however, to document your data by means of a scope camera. It supplies a photograph which can be marked with all necessary settings and scales. There are two very easy ways to mark your photos. Before laminating them with their protective coating, you can either scratch your information on them with a scribe, or use a fine soldering iron. Either method produces a professional looking picture that carries permanent and concise data for future reference.

7.3.2 Audio Substitution Method

Audio substitution (sometimes called dc substitution) is a method that uses basically the same setup as RF substitution, except that the standards used (calibration attenuator) for substitution are at low frequencies or dc (where calibration can be to accuracies better by several orders of magnitude). One audio setup is shown in Figure 7.29. Because of the nature of this measurement procedure, the setup is good only for single frequency readings; a second method for swept measurements, is treated following our initial explanation.

The test setup in Figure 7.29 appears to be very similar to previous setups. The source is a cw microwave generator modulated with an audio signal (usually 1 kHz). The isolator-attenuator combination is inserted to provide a good impedance match between the source and the attenuator to be tested. The bolometer mount, which may contain a barretter or a thermistor, is used to detect the RF power and connect it to a useful form. The frequency range of the mount should be checked to ensure that you have the proper one for your application. The audio attenuator, output amplifier, and indicating meter are all combined within a commercial device termed an *attenuation calibrator*.

Several commercial attenuators can be used as calibration standards. In order for the settings of the attenuator to hold, the input impedance to the following amplifier must be adjusted to the attenuator's characteristic, and the impedance looking back from the standard attenuator must be matched. In case of mismatches, appropriate corrections have to be made (within the instrument itself). The calibrator also provides bias current for the bolometer if a barretter is used.

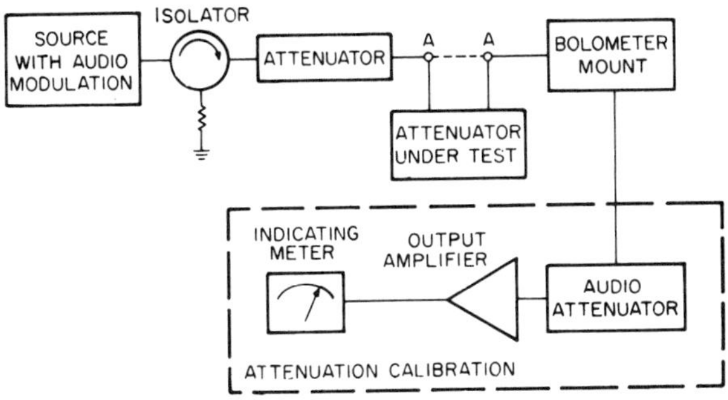

Fig. 7.29 Audio substitution setup.

In the single frequency audio substitution method of measurement, a calibrated audio attenuation standard is used to maintain a constant level at the output indicator with the unknown in and out of the line. The output level is obtained at the indicator with the unknown out of the system; the unknown then is placed in the system, and the necessary attenuation is removed from the audio standard attenuator to obtain the original reference level. Since the input power is changed when the unknown is placed in the line, the accuracy of measurement depends upon the response of the bolometer to the audio-modulated RF signal. The square-law response characteristics of bolometers are such that the decibel change on a linear output voltmeter is divided by two in order to obtain the RF power level change (in decibels) when using a square-law detector. In the present measurement the value of the unknown (RF attenuator) is obtained by taking one-half the value of the attenuation change of the standard. This adjustment is necessary because the input to the standard attenuator is ac voltage, which is proportional to the RF power level.

The general procedure for performing attenuation measurements with the calibrator setup is as follows:

1. The source is adjusted to the desired frequency, the modulation frequency is adjusted to the proper value, and the necessary source and load tuning is accomplished.
2. The RF power level at the barretter is set with the unit under test out of the system. This value is obtained by setting all attenuators on the calibrator to zero, lowering the gain controls for minimum gain, and then adjusting the RF power with the level set attenuator to obtain full-scale reading on the output meter.

3. The precision attenuators are then set to a value greater than that anticipated for the unknown. The gain on the attenuation calibrator is then increased to obtain a convenient reference level.
4. The unit under test is now placed in the system, causing a decrease in the output reference level. The precision attenuators are adjusted to return the output level to the reference point, and the value of attenuation is recorded.

This type of measurement procedure is used over a 20 to 30 dB attenuation range. The limiting factors to its range are the square-law response to the bolometer (barretter) and the internal noise level of the calibrator. This range can be extended by using an attenuator substituted in the RF line as a fixed value, with the readings obtained by the procedure above added to this figure. For example, if you have a fixed 20 dB attenuation in the line during initial calibration and the calibrator equipment indicates a value of 11.2 dB, the total attenuation of your device is 31.2 dB.

An alternate setup that can be used to obtain swept attenuation measurements is shown in Figure 7.30. The ALC loop described earlier is used again to maintain a leveled output; the attenuator ensures a good source match for the detector and the attenuator under test. The detector itself could be crystal or a barretter; its main job is to detect the RF signal so that its only component remaining is the audio modulation which had been applied at the source. The standard audio attenuator is where the method gets its name. It is an attenuator, fixed or variable, calibrated at an audio frequency to a very high degree of accuracy. Attenuators for use in the RF substitution method, discussed previously, are calibrated to hold ± 0.3 to ± 0.5 dB; attenuators in the audio range have accuracies on the order of $\pm .005$ to $\pm .01$ dB. The overall improvement in the accuracy of your setup is obvious. The indicating device in Figure 7.30 is shown only as a block, because many combinations could be used to provide a display of the attenuation information. One possibility is a standing wave meter by itself or with an X–Y recorder; another is an oscilloscope, with a camera to record the data. The type of display depends on the application and the available equipment. You may end up with a meter to provide the readings, and a pen to jot down the data directly as observed, in a lab notebook for future use.

The test procedure is similar to that of the RF substitution method.

1. Connect the test setup as shown in Figure 7.30 (Connect points A–A together.)
2. Set either a standard fixed audio attenuation in the line, or set in a high value of attenuation if it is a variable attenuator.
3. Turn on the power to the sweep generator and indicating device. (Allow for any built-in time delays.)
4. Set and record a convenient level on the indicating device by adjusting the RF level on the sweeper, the audio attenuation, and any gain controls. Be sure that the RF level at the in-line detector (prior to the audio attenuator) does not exceed its maximum power rating. If an X–Y recorder is used, run a series of calibration lines to give you an accurate reading of your device relative to the generator output.

5. Insert the device to be tested at points A–A.
6. Record the new power level reading on the indicating device, or adjust the standard audio attenuator until the original calibration level is obtained and read the attenuation off the attenuator. If an X–Y recorder is used, be sure to return all settings so that the recorder reads 0 dB before inserting the device to be tested. In this way, the recorder goes directly to the actual attenuation on your X–Y plot when the device is inserted, thereby providing an accurate reading.

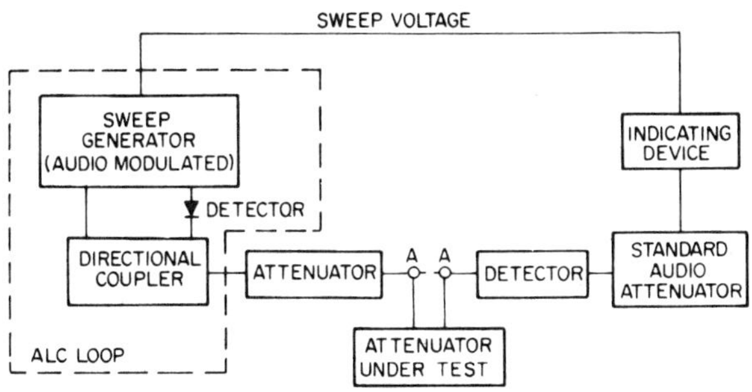

Fig. 7.30 Swept measurement using audio substitution.

7.3.3 IF Substitution Method

To this point we have presented a test method which substitutes a standard in the RF or microwave region, and one that substitutes a standard in the audio or dc region. To present a complete spectrum of measurements we need a method that bridges the gap in between the two. IF substitution is such a process. Once again we can come up with setups for both single frequency and swept measurements. The first uses a simple mixer to down-convert the RF signal being tested. The second, although more complex (you need to lock the RF and LO frequencies with a set increment between them), is one of the most accurate methods of attenuation measurement over a band of frequencies.

Figure 7.31 is a block diagram of a single frequency IF substitution scheme. The process called *superheterodyning* is employed—the mixing of two RF signals to produce an intermediate frequency which is either the sum or difference of the two signals; the resulting output frequency depends on whether the local oscillator is higher or lower than the RF input. The level of the RF input can vary over a large range since the local oscillator level usually saturates the mixer, allowing the mixing process to take place with either a low or high RF level. Thus, superheterodyning achieves high sensitivity; it is particularly applicable to large dynamic range attenuation measurements. The two methods discussed previously employ square-law detection in systems that are capable of measuring medium-range attenuation up to about 50 dB. With linear detection heterodyning-type measurements, dynamic ranges in excess of 150 dB can be achieved. In the IF substitution method, a precisely calibrated IF attenuator is used. The procedure is identical to the RF and audio methods in that a level is set up, the component to be tested is inserted, and the disparity between output levels is the unknown attenuation. One difference is that the output of the standard attenuator is fed to either an amplifier (where it is amplified to an appropriate level for indication) or a receiver (where it is detected and used to give a meter indication).

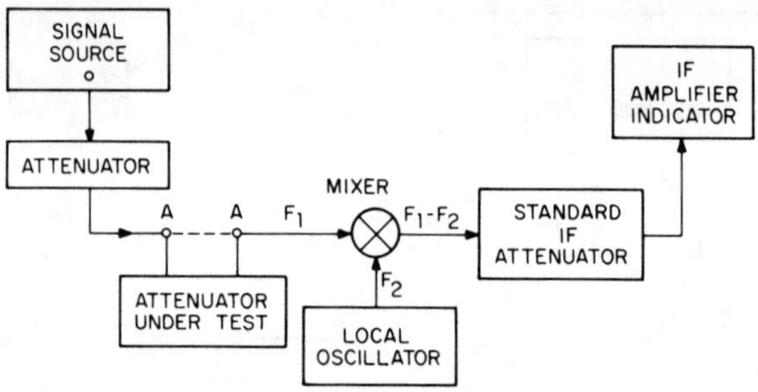

Fig. 7.31 Single frequency IF substitution method.

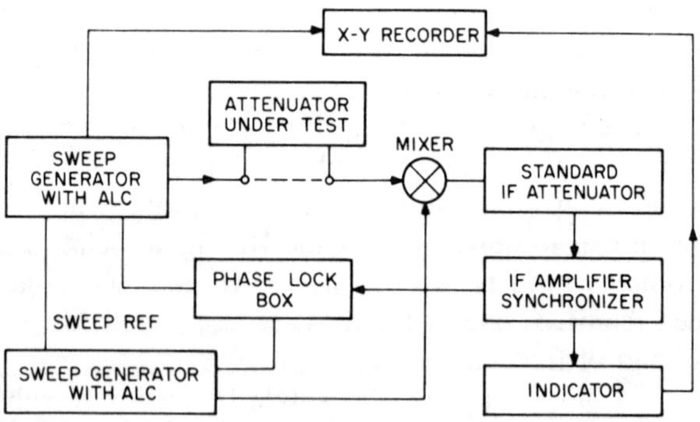

Fig. 7.32 Swept-frequency IF substitution test setup.

Several IF attenuation measurement systems are available for use in making swept-frequency measurements. One common setup is shown in Figure 7.32.

The two sweep generators are offset by the desired IF frequency and the sweep reference of one sweeper drives the other, to ensure synchronization in time. The output of the synchronous detector provides the reference by which the phase-lock box develops an error signal to keep both the sweeper frequencies as initially set and the spacing constant. This error signal is the same type as that we discussed regarding ALC. The only difference is that, with the ALC circuit, any change in amplitude creates an error signal; in our phase-lock system, a change in the IF frequency beyond a set tolerance causes an error. This scheme achieves a constant IF frequency that allows the use of narrow bandwidths in the IF amplifier; the result is a very wide dynamic range in the system.

With the two sweepers locked to a set IF frequency over your required frequency band, the procedure would be the same as if we were using only a single frequency band rather than a mixer. The major difference in performance is the greatly increased dynamic range of the IF setup that we mentioned before.

The test arrangements for swept measurements shown so far have been broken down, with each component's use explained. To this point one area has been taken pretty much for granted—the back side of the sweep generators where a line comes out at a point marked "sweep voltage." What exactly is this voltage that causes an X–Y recorder arm to move across the paper, or a dot to move across the face of an oscilloscope and, if it moves fast enough, form a line? The answer to these questions is shown in Figure 7.33. These three illustrations are actual reproductions of the oscilloscope displays of the sweep output voltages of three randomly chosen sweep generators. The voltage is either a ramp, as in Figures 7.33(A) and (C), or a triangular shape, as in Figure 7.33(B). The one characteristic common to all is that the voltage starts at a point, rises linearly to a maximum value, and then falls, more rapidly than it rose, back to its origin. The ramp functions stay at the maximum value for anywhere from 1 to 2 milliseconds before they fall. This pause produces a slight delay at the end of both the X–Y recorder trace and a scope sweep (to allow for any hysteresis in the system before the device is reset, such as the X–Y recorder arm returning to its starting point). The amplitude of the sweep voltage in the illustrated three cases ranged from 8 to 15 V. The time required to reach these values is determined by the speed setting on the sweep generator. The times in Figure 7.33 range from 9 to 10 milliseconds—set for convenience to get only one cycle of the voltage on the face of the scope.

You now should have some idea of what is coming out of the BNC jack on the front of the sweep generator and what is being sent to the X-axis of an X–Y recorder on the *ext horizontal* input of a scope. Also, you now can see why it makes no difference whether a piece of RG-58 cable or a piece of RG-62 is used to connect the jack to the recorder. The RF impedance of the cable is of no concern since the *sweep voltage* on the sweeper is a dc voltage.

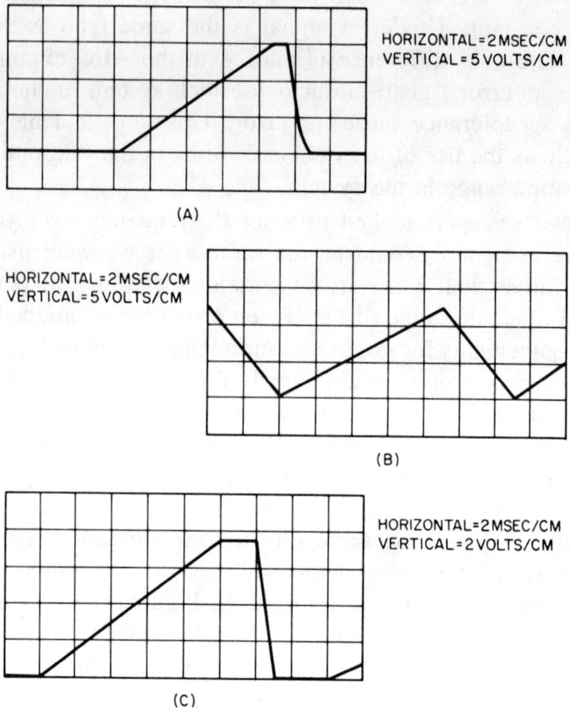

Fig. 7.33 Sweep voltage illustrations.

7.3.4 Precautions and Errors

These terms are considered together because there is a distinct relationship between the two. Taking certain basic precautions in a test setup can eliminate many of the errors that cause inaccuracies. Similarly, an error analysis of a setup often can reveal someone having been careless in the choice of a component.

Attenuation and other types of measurements are generally comparisons of an unknown value to a known standard. The sources of error can be classified as three basic types: *scalar, uncertainty,* and *vectorial.*

Scalar errors are those that usually can be either removed by calibration or reduced in size so as to minimize their effect. One such error is that of the calibration of the standard (IF, RF, or audio) attenuator. (More accurate calibration of the standard results in a higher degree of precision). Drift of the signal source and indicator are also scalar errors, but they can be easily checked and compensated for by constant recalibration. We can classify *scalar errors* as those that are readily visible and concerned mainly with setup calibration.

Uncertainty errors are not reduced or eliminated by calibration because they are random. They are caused by such things as connector repeatability, readability, and resetability of the standards dial, and by readability of the indicator.

Vectorial errors are such that they can be represented as vectors rotating with frequency. If determined individually at each frequency and summed vectorially, corrections can be made to cancel the effect of these errors. The cause of such multiple mismatch errors is the interaction of two or more reflections offset from each other along a transmission line. In other words, these errors are the vector sum of the mismatch present at the source, connectors, junction of components, or any place where two pieces are joined. Such mismatches have different phases; some of them add directly, some subtract, and some produce amplitude at a new phase angle. Many of these errors can be reduced greatly, or eliminated entirely, by a careful choice of components and equipment and by proper consideration as to compatible connectors.

As an example of how each of these small errors add up in a system, the following list was compiled to illustrate just the scalar and uncertainty errors when measuring a 20 dB pad.

Standard attenuator calibration accuracy	±2% of dB
Indicator readability	±0.05 dB
Connector repeatability	±0.05 dB
System drift	−0.05 dB
Maximum possible error due to uncertainty and Scalar errors if 20 dB attenuation is measured	±0.55 dB

Without considering any reflections, there is the possibility of a ±2.75 percent error in our measurement, much of which cannot be eliminated due to its nature (such as indicator readability). However, there are precautions that can be taken to improve your readings; they must be observed if a high degree of accuracy in your attenuation measurements is to be attained. Important factors to consider are signal *leakage*, *impedance match*, *generator protection*, and *equipment stability*. Let us examine each to understand their nature and effect, and see what can be done about them.

Leakage

The word *leakage* is used to describe both loss of internal energy from the microwave circuit and the entrance of energy into the microwave circuit from external sources. Leakage errors can easily be as high as 1 dB, sometimes considerably greater. Usually, they are due to poor mechanical conditions involving bent or dirty flanges on waveguides, or too poor assembly of the microwave setup. Leakage can also result from improper design or from faulty parts and components, such as detector mounts and tube mounts. If a leakage path from the signal source to the detector is present in a setup, the resulting error depends upon the amount of leakage relative to the attenuation being measured. To prevent significant errors, the total attenuation required in the leakage path may range from 40 dB to more than 150 dB.

The presence and, to some degree, effect of leakage in a particular setup, may be detected by covering various points in the setup with the hand or shielding material, while watching the output indicator for changes. In some setups, leakage can be so severe that the movement of the operator in the vicinity of the test bench may cause variations. Under such conditions measurements are almost meaningless, even if the attenuation involved in the component under test is quite small.

The leakage problem becomes especially acute during production testing when many test setups are in operation in the same general area. Under these conditions, not only the individual setups, but also the relation of each to surrounding equipment must be considered.

No extreme measures are required with the majority of setups. However, when accuracies of approximately 0.25 dB are involved, the attenuation in the leakage path should exceed the loss to be measured by 30 dB or more. In practice, flanges and other connections should be clean, undamaged, and properly assembled; correctly designed components must be used; and all directional couplers or signal sampling devices should be terminated. When even lower leakage levels are required, joints can be wrapped with foil or copper wool, slots and openings in slotted lines or other devices can be covered with polyiron, and physical location of components can be selected by trial and error to determine the best layout.

There is no substitute, of course, for well-designed and well-constructed components. Poor design in detector mounts, slotted lines, and other setup parts may make it impossible to reduce leakage to an acceptable level by any reasonable means.

Impedance Matching

Proper *impedance matching* is essential in attenuation measurements, particularly when low attenuation or insertion loss is involved. It is important that the setup be matched in both directions from the point of measurement.

In such setups, the load is usually a detector, and the detector is frequently a bolometer. The bolometer almost always has a better VSWR than the crystal detector, although commercial bolometer mounts are seldom better than 1.10 at their best frequency. In addition, there may be an impedance change with power level change, making it difficult to maintain a good match, especially with crystal detectors.

These errors may be reduced in several ways, the simplest of which is the insertion of a pad between the detector mount and the line. This action, though, has the obvious disadvantage of causing loss of signal and dynamic range. The amount of padding required depends upon how badly the detector is mismatched initially. A better solution is the use of a matching device such as a slide-screw or stub tuner.

This adjustment is for a single frequency, but VSWR can be reduced to very low values. Since the tuner is almost entirely reactive, it introduces very little power loss and is particularly valuable when power is limited. One other low loss method of correcting mismatch is through the use of an isolator.

Generator Protection

Generator protection, as used here, refers to steps required to minimize interaction of the microwave signal source with reflected power or other external factors which might cause changes in operating conditions during a measurement. A mismatch at a generator with high output VSWR can result in frequency pulling and instability. Changes in match as the device under test is inserted, or its attenuation is varied, may react on the source and affect its output level in the measurement. For general laboratory tests, a well-matched pad of approximately 10 dB is inserted in the line after the source. While this action helps to isolate the source from line variations, it does so at the expense of signal level. A better means is by using a ferrite isolator which virtually eliminates reflected power to the source. When the loss in signal caused by a pad cannot be tolerated and an isolator is unavailable, the only solution is to minimize mismatches and discontinuities in the setup wherever possible—eliminate unnecessary adapters, tighten all connectors to form a good ground, and even check all connectors to be sure that they are clean.

Equipment Stability

It is obvious that accurate measurements are difficult to make with power meters, SWR indicators, oscilloscopes, or power supplies that are unstable. This problem is particularly evident in attenuation measurements, where there is an appreciable time-lapse between an initial reading for a reference and a later reading for the actual determination. Sometimes two different channels are involved; their combined stability is a factor. With well-engineered equipment, errors in measurement due to drift and other variations can be reduced to the order of 0.05 to 0.1 dB. However, even with well designed instruments, accuracies can be improved by frequently checking balance and calibration adjustments. This process may seem to be a time-consuming bother, but it pays off in the long run, through greater accuracy and much more meaningful measurements.

General Considerations

Coaxial cables and other flexible sections are also frequent causes of error. Flexing a cable may change its insertion loss, thus altering a reference level; where coaxial cables must be used, every attempt should be made not to move them during a measurement. Further errors can be introduced when it is necessary to break and make a connection as the test piece is inserted in the line. Frequent unrecognized sources of error are adapters which are used in establishing a level or inserting the component, then left out of the line in subsequent measurements. It is usually possible to compensate for the insertion of the adapter, or else eliminate its use, but this point illustrates why we have been harping on employing the proper connectors on your components. Though the mismatch loss may seem small when taken at one single adapter, it becomes quite substantial when you total system losses.

The measurements covered in this section on transmission are all concerned with attenuation; they also apply to gain. When gain measurements are made, attenuation is usually added to the setup to bring the indicator back to its original calibration setting. The gain value is then read off the attenuator or indicating device. There is one important consideration—wherever a gain figure is involved, there is an active device (by active we mean that there is a dc voltage associated with it).

Great care should be taken to be sure that there is a dc block (a capacitor) at the output and/or input of the device; any dc voltage at either point not confined to the actual unit causes inaccurate and erroneous readings.

The conversion loss of a mixer is also an attenuation measurement—it is how much RF energy is lost through the device as a result of the conversion from RF to a lower IF frequency. This determination is usually made with a spectrum analyzer so that the individual frequencies can be distinguished. A level is set for the RF input to which the IF output is compared—the difference is the conversion loss.

The term "transmission" is broadly defined as the passing from one point to another, a description that certainly would characterize the measurements described above. RF energy is passed from one point to another, measured at each, and the two readings are compared to obtain the transmission measurement.

It should be pointed out here that many transmission measurements are conducted with network analyzers (covered in detail in Chapter Eight).

7.4 IMPEDANCE MEASUREMENTS

Thus far, we have referred many times to such terms as VSWR, reflection, and matching, all of which help to describe one of the most important specifications in microwaves—*microwave impedance*. The immediate connection with impedance is resistance, mainly because an RF impedance is expressed as 50Ω, for example. When we hear "ohms" we think of resistance: however, the 50Ω is, in actuality, 50 + *j*0 ohms—a real part (50) and an imaginary part (+*j*0). When we need a resistance in a dc circuit, we measure a voltage and a current, divide E by 1, and come up with a number. It is not that easy with RF impedance. Since there are two terms (50 + *j*0), each impedance has not only a magnitude (50Ω), but a phase angle (0° in this case) as well. Clearly, there is more to an RF impedance measurement than just reading a voltage and a current.

The parameter which characterizes microwave impedance is the VSWR (voltage standing wave ratio)—the ratio of the amplitude of the power transmitted in a forward direction and the power reflected back toward the source due to a mismatch at the load. A VSWR reading of unity indicates that the generator and its load are perfectly matched. Any difference in the load impedance causes reflections which, in turn, produce a standing wave ratio greater than unity that can be measured and interpreted. We cover three methods of measuring VSWR, and thus impedance, in a microwave system:

VSWR bridge method,
reflectometer method, and
slotted line method.

7.4.1 VSWR Bridge Method

We discussed bridge circuits in microwave applications regarding power measurements. In such an instance, a temperature sensing device causes an imbalance of the bridge that results in a dc voltage proportional to the initial power level. In the VSWR bridge, the imbalance is caused by the magnitude of reflection due to any mismatch present in the device under test. Figure 7.34 shows actual VSWR bridges that are commercially available, while Figure 7.35 is a schematic representation. For the bridge to have good accuracy, it is necessary that its components be matched throughout the designated operational frequency range. A particularly important factor is the impedance in the lower arm opposite the test port, the one marked "Rc = 50Ω *(reference)*."

Since this arm is the one to which the measured load is compared, it is essential that it have excellent characteristics over the entire frequency range of operation. This termination may be physically integrated into the bridge, both to achieve maximum bridge accuracy, and so that its performance is part of the bridge specifications.

Physical integration of the reference termination also guards against damage, poor contact, unknown performance, and a loss of the termination, all of which are far more likely if the arm is constructed as an accessory to the bridge.

The bridges in Figures 7.34 and 7.35 are just what they appear to be—resistive Wheatstone bridges which are used to measure reflections on an unknown load. The bridges operate to 12 GHz and exhibit less than a 1.02:1 residual VSWR over most of their range. Such a low VSWR is attained in slotted line sections only at great expense; here it is achieved in a unit that is inexpensive and small enough to fit in the palm of your hand. These bridges also exhibit the characteristic of being wide band devices which are suited to sweeping and reflectometer work—which slotted sections are not.

Compared to couplers, the bridges have wider frequency ranges, lower coupling loss, superior reflection performance, and lower cost. Another advantage is

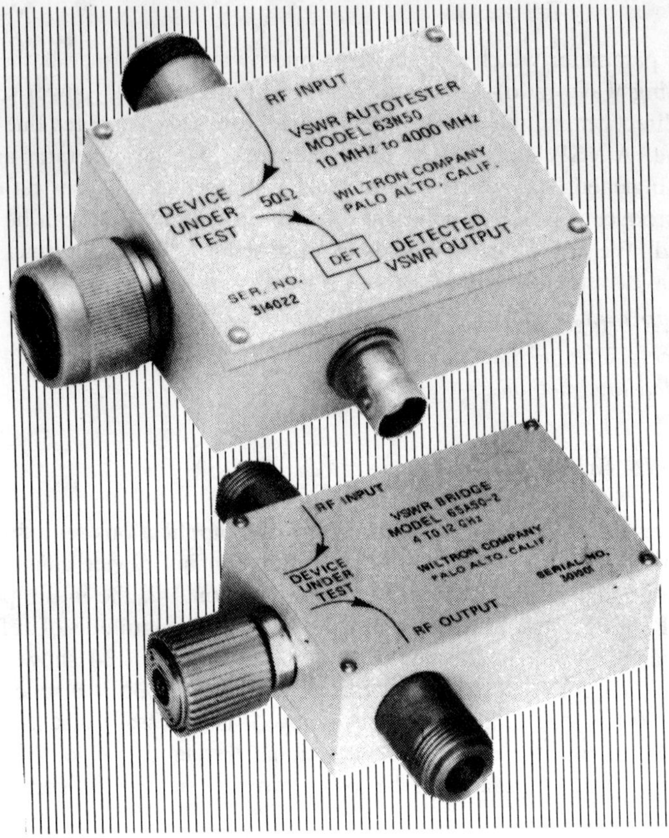

Fig. 7.34 VSWR bridges.

that they are available with a wide variety of connectors; their high performance includes the effects of a given connector. In other words, the bridge performance is not degraded by adapters needed to fit your particular system.

To understand the VSWR bridge and how it can be used, let us investigate a typical, appropriate test setup (Figure 7.36). This particular arrangement is for swept measurements; thus, the ALC loop is outlined and operates the same as described in the previous section. The calibration attenuator in the setup is used to set in specific values of attenuation as reference points. For example, if a VSWR of 1.15:1 is needed, 20 dB would be set in and unknown devices compared to it. The bridge, or course, is used to measure any reflections from the unknown. The oscilloscope displays the calibration line and the VSWR of the device under test. The *level meter* converts the dc voltage from an external detector (in the bridge, in this case) to an indication of RF power (calibrated in dB per division). The device is used to increase the sensitivity of the measuring system; it permits accurate low level RF measurements to −40 dBm.

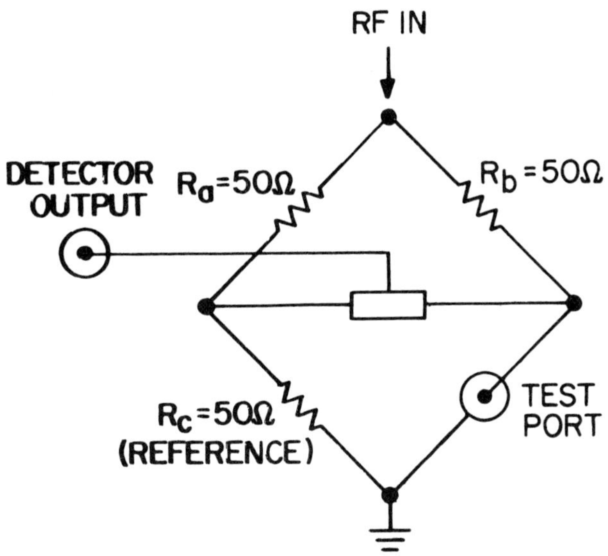

Fig. 7.35 VSWR bridge schematic.

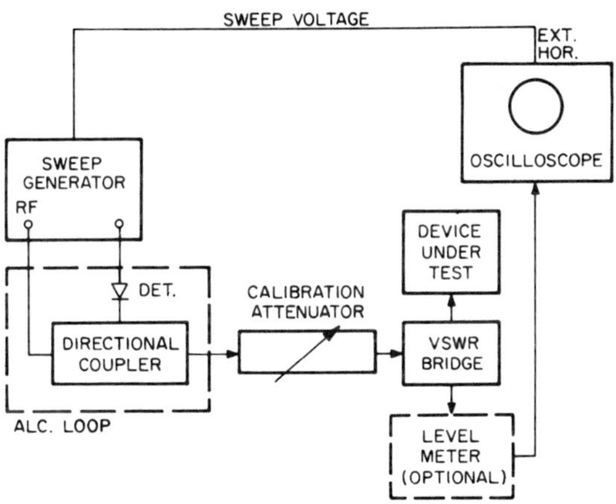

Fig. 7.36 VSWR bridge test setup.

With the test setup described and explained, the one remaining job is to outline the measurement procedure.

1. Connect the test setup as shown in Figure 7.36.
2. Adjust the calibration attenuator for minimum attenuation and turn on the power for the sweep generator and the oscilloscope.

3. When the sweeper and scope have power (after any time delay), be sure that the RF output is leveled. This point usually can be checked by observing the "RF unleveled" light on the front of most sweepers—when the light is on, the output is unleveled.
4. Leave the test port open; which corresponds to an infinite VSWR load or a 0 dB return loss. With the source now sweeping and the gain of the oscilloscope properly set, a trace corresponding to the 0 dB return loss (VSWR $= \infty$) appears on the scope.
5. The device to be tested is now connected to the test port of the bridge; the display moves to some level on the scope corresponding to the return loss, and thus VSWR, of the device. Overlays for the oscilloscope relating return loss to VSWR are available and are shown in Figure 7.37. They are transparent so that the value of the reflection can be read on the face of the scope directly, in either VSWR or dB of return loss. The overlays include four lower value VSWR scales, in addition to the one for infinity, so that intermediate and low VSWR values can be measured with higher resolution. The lowest scale is such, that a VSWR of as small as 1.066 occupies the full scale of the scope. To illustrate how that scale would work, consider an example. Suppose we calibrated the setup with 10 dB of attenuation set in at the calibration attenuator. When our device is attached, the trace on the scope sweeps from 5 dB at the low frequencies down to the 3 dB level at the high end. Thus, the VSWR is approximately 1.42 at the low end (15 dB return loss) and 1.58 at the high (13 dB return loss).

To provide a permanent record of the swept VSWR of a device, an X–Y recorder is used in place of the oscilloscope and overlay. Values of attenuation ranging from 0 to 30 dB are inserted by way of the calibration attenuator and a swept calibration line is recorded for each. The attenuator is then returned to 0 dB and the device is attached and swept over the calibrated frequency range. The values of return loss indicated can then be converted to VSWR and your measurement is complete. Table 7.1 shows the relationship between some typical VSWR values and their associated return losses.

Table 7.1 also verifies the fact that the lower the return loss (higher dB number), the lower the VSWR and the better the match. This relationship should be very easy to understand since the closer the load impedance is to the system characteristic impedance, the more power absorbed at the load, and the less there is to be reflected (lower value of return loss). So the value of return loss, and thus the VSWR, yields the impedance of a device.

Two important factors regarding the VSWR bridge are its directivity and the output VSWR of the test port. *Residual VSWR* is often used interchangeably with *directivity*; it refers to the ratio of the signal levels at the detector and test ports when the test port is operated into a matched load.

Table 7.1 VSWR Values and Associated Return Losses

VSWR	*Return Loss (dB)*	*VSWR*	*Return Loss (dB)*
1.01	46	1.30	18
1.02	40	1.50	14
1.04	34	1.80	11
1.066	30	1.93	10
1.10	27	3.00	6
1.15	23	5.00	3.5
1.22	20		

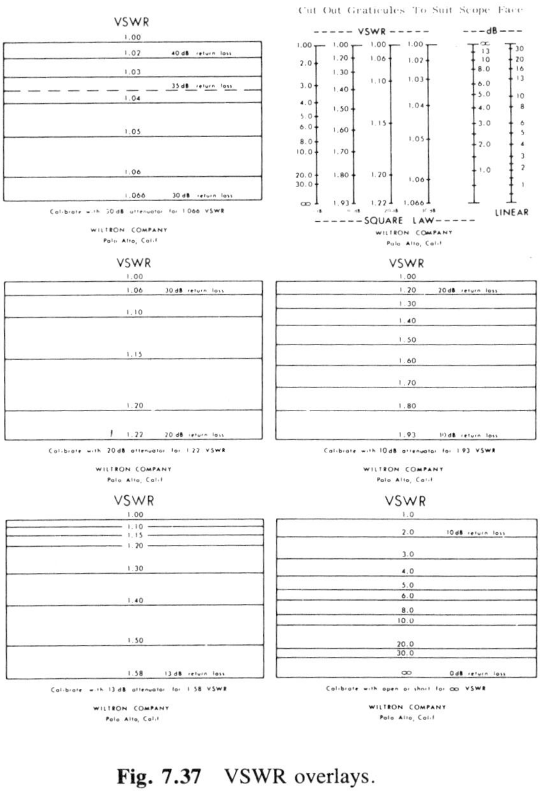

Fig. 7.37 VSWR overlays.

The *directivity* figure is the main factor that determines measurement accuracy for VSWR's of about 1.5:1 and below. Most bridges available today have directivities in excess of 40 dB or a residual VSWR equal to, or less than, 1.02:1. This VSWR value, as mentioned before, rivals the highly accurate and widely used slotted line.

The great care taken in designing a VSWR bridge for high directivity, a low test port VSWR, and the incorporation of a precision reference termination, bears fruit in the form of high measurement accuracy. And, as mentioned before, the proper connectors on your bridge eliminate adapters and any related additional mismatches which would decrease measurement accuracy.

7.4.2 Reflectometer Method

A thermometer would be used to measure the temperature of an object or an area. A speedometer would be used to determine the speed at which you are travelling in your car. It seems very reasonable that, if you need a reading of the reflections (and thus the impedances) of a device, you would use a *reflectometer*. This method of impedance measurement uses the reflection coefficient as the basis for its operation.

The reflection coefficient (ρ) of a device or system is another useful term in establishing the impedance match of microwave devices. The following relationship of ρ and SWR is frequently used in impedance work:

$$\rho = \frac{E_{\text{reflected}}}{E_{\text{incident}}} = \frac{\text{SWR} - 1}{\text{SWR} + 1}$$

The coefficient is a linear quantity varying from zero to one. The logarithmic expression of ρ is known as return loss and defined as

$$\text{dB}_{\text{return loss}} = -20 \log_{10} |\rho|$$

A reflection coefficient of 1.0 (total reflection), corresponds to a return loss of 0 dB; a zero coefficient equals an infinite dB return loss. An example of an intermediary value is a reflected signal from a test device at a level of 26 dB below the incident signal. In this case, ρ would be 0.05 (5 percent of the power is reflected back toward the source). The value of 0.05 is determined as follows:

$$\text{dB} = -20 \log_{20} |\rho|$$

$$26 = -20 \log_{10} |\rho|$$

$$\frac{-26}{20} = \log_{10} |\rho|$$

$$-1.3 = \log_{10} |\rho|$$

$$|\rho| = \log_{10}^{-1} (-1.3)$$

$$|\rho| = 0.05$$

The load reflection coefficient is measured by separating the incident and reflected waves propagated in the transmission line connecting the source and load. The reflectometer uses either coaxial or waveguide couplers to accomplish this division, and it permits oscilloscope displays or permanent X–Y recordings of reflection coefficient or return loss across entire operating bands. The incident wave referred to above is the power being transmitted in the desired, or forward, direction; the reflected wave is that power reflected in the reverse direction due to mismatches at the load.

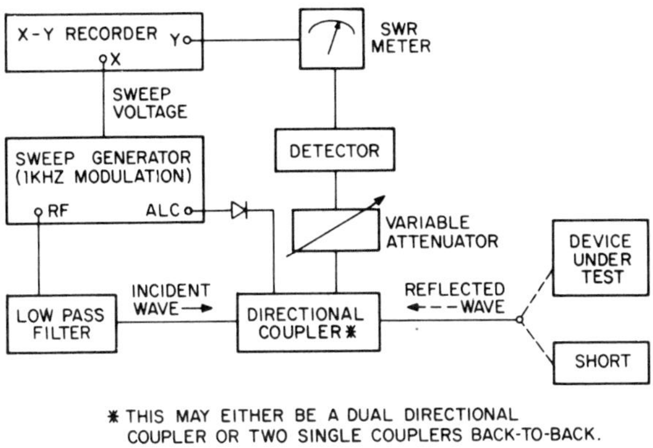

Fig. 7.38 Reflectometer setup.

The reflectometer setup shown in Figure 7.38 is designed to hold the incident power constant. With automatic leveling, only the relative amplitude of the reflected wave need be measured to determine the reflection coefficient.

The ALC circuit is identical to those described in previous sections. A low pass filter is inserted to eliminate any harmonics which could give false readings. The directional coupler, as indicated, can be either dual directional or two single ones back-to-back—the important property that must be present for good measurement is high directivity. The greater the coupler directivity, the more accurate the readings (the incident and reflected waves are isolated to such a degree that the only thing being read is the reflected wave). The SWR meter and X–Y recorder are standard devices described in detail in previous discussions.

A very important part of the reflectometer setup is one of the diagram's smallest blocks—the one entitled "short." This little device is used as a reference for all of our reflection measurements. What makes a microwave short suitable for use as a calibration standard? It certainly does not sound as though it would work very well; most people think of a short in an electrical circuit as something very dangerous and undesirable. However, to clarify matters, remember that it is called a *microwave short*; that is, at microwave frequencies it is a short circuit. (Granted, at dc you also read a short with an ohmmeter; this only affects you, though, if you have some active device in your setup that does not have a dc block at its input or output). The microwave short causes all RF energy from the incident wave to be reflected (100 percent reflection, $\rho = 1$). The value of a device with a known reflection coefficient is obvious, but since an open circuit also has a reflection coefficient of 1.0, would not it be easier to leave the coupler port open and eliminate the possibility of any mismatches between the coupler and the short? The answer is yes, it would be simpler, but it would not be as accurate. With an open circuit, a capacitance is set up that varies with frequency between the center conductor and the outside shield, making it difficult to tell just how good an open circuit you have. In order to obtain highly accurate readings, the capacitance must be calibrated out, but how much capacitance is there in a type N, SMA, or TNC connector at 1 GHz, 10 GHz, or 18 GHz? And, even if you knew the proportions of the capacitance, how would you calibrate it out of the setup in Figure 7.38? Thus, we use the short instead. There is no capacitance to consider, the reflection coefficient is 1.0, and you can be sure that it has a good ground (because it is a plane and not a point, a prerequisite in our introduction). By using computer-aided measurements, the open circuit capacitance *can* be calibrated out, but, for general test setups, the short is much easier to work with and it produces more accurate and reliable results. If an automatic network analyzer, for example, were available, the measurement would be much more accurate since variations in both an open and short would be calibrated out.

To calibrate the reflectometer setup in Figure 7.38, the short circuit is placed at the output port, thus reflecting all of the incident power (zero dB return loss). The detector in the reverse-arm coupler samples the reflected power and provides a proportional dc voltage for readout. By placing a calibrated attenuator ahead of the detector, specific amounts of return loss may be pre-inserted for calibrating the recorder gain. The attenuator is then returned to zero, the short removed, and the test device connected and measured on the pre-calibrated display. Measurements are also possible without the pre-insertion attenuator if the detector remains within its square-law region.

When we refer to a square-law region of a detector, what do we mean?

When we speak of a detector, we are talking about an RF input with a dc output; in other words, we are relating power to voltage or current. From basic electronics we know three relationships that exist between power and voltage or current:

$P = EI$;

$P = E^2/R$; and

$P = I^2R$.

The one that is of importance to us is the second one ($P = E^2/R$).

A power applied to the input of a detector produces a voltage proportional to the square root of the power times the system impedance (R). This fact can be seen by multiplying expression number 2 by R, and taking the square root ($E = \sqrt{PR}$). For another power level we would achieve a different voltage; and so on. One term has remained a constant over this range—the impedance (R). This value stays constant over a certain input power range, within which the relation of $P = E^2$ is valid; thus the term "square-law region." At either extreme of this area the R-term appears, eliminating the linear relationship of $P = E^2$. By keeping the power at the proper level, you can stay in the square-law region of the detector and obtain accurate readings.

What you have with the Figure 7.39 setup is an X–Y plot that has various levels which correspond to different return losses. When the device to be tested is connected and swept, the recorder traces the value of return loss as a function of frequency.

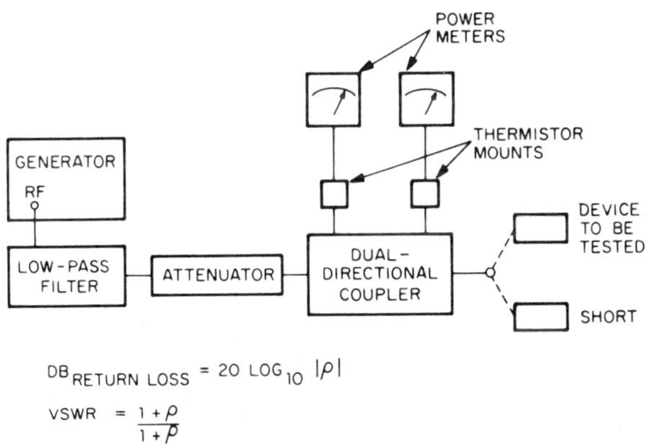

Fig. 7.39 CW reflector test setup.

When a swept response is not needed, the arrangement in Figure 7.39 is used; it is one, and only one, way of making a CW reflectometer setup. There are many changes that could be made to it, depending mainly on the available equipment.

The objective of the setup is to provide a reading of the power both incident (forward) and reflected (reverse) to the operator. From these values, it is possible to calculate the reflection coefficient (ρ) and VSWR at a single frequency. The procedure for testing is as follows:

1. Connect the setup as shown in Figure 7.39 with the short at the coupler output.
2. Turn on the generator and power meters.
3. Run up the input power level and check to see how closely the reflected power reads to this value, checking the calibration of the two coupler sections. It also gives you an indication of the magnitude of potential errors. Record the differences.
4. Remove the short and place the device you want to test at the output port.
5. Record the values of the power incident and reflected, as read on the dB scales of the power meter. The difference is the return loss of the device. For example, if you have a +10 dBm incident level and a −10 dBm reflected, the return loss is 20 dB. A reflection coefficient of 0.1 is achieved from the 20 dB figure by applying the equation: $dB_{\text{return loss}} = -20 \log_{10} |\rho|$. With a value of $\rho = 0.1$ the VSWR is 1.22:1, using the equation: $VSWR = 1 + \rho/1 - \rho$.

The procedure outline above is used for single frequency measurements. You can measure a band of frequencies by processing each frequency individually, then plotting your results.

One more swept measurement technique should be investigated before leaving the topic of reflectometry—one which employs an instrument called a *ratio meter*. This instrument does virtually all of the necessary calculations in the procedure above. The block diagram of a commercially available ratio meter is shown in Figure 7.40. It is basically a two channel device that takes the reference (incident wave) and test (reflected) signals, compares them in the log ratio detection, and produces a dc voltage proportional to their ratio. This voltage is displayed on the main meter and is available for use with an X–Y recorder or an oscilloscope.

A reflectometer setup using the RF ratio meter is shown in Figure 7.41; notice the similarity between this setup and that shown in Figure 7.39. In each case we are comparing an input level (reference) to that reflected (test).

The generator does not have an ALC loop. There is no need to level the generator—in the first case, we are taking only single frequency readings; in the second, the comparison takes place in the ratio meter so that there is no need to spend the extra time and money needed to provide a leveling loop.

They also both use either a dual-directional coupler, or two couplers placed back-to-back to sense the forward and reflected power, and a pair of bolometer mounts to correct this power to a useful form.

The sweep generator in the ratio meter setup is in the sweep mode and is unleveled. The attenuator at the output reduces reflection and so provides a better source match for the system. The coupler(s) samples both forward and reverse power and

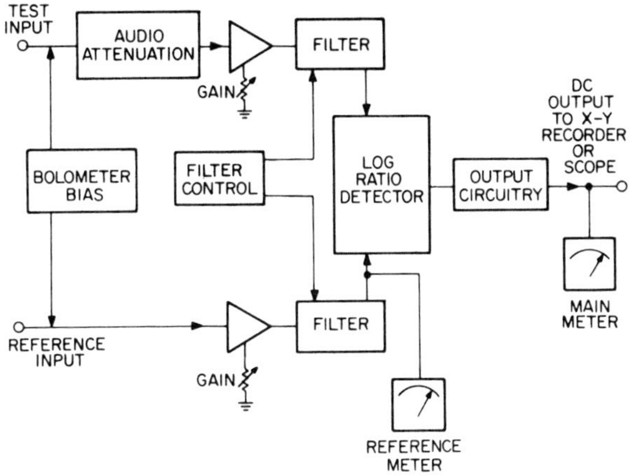

Fig. 7.40 Ratio meter block diagram.

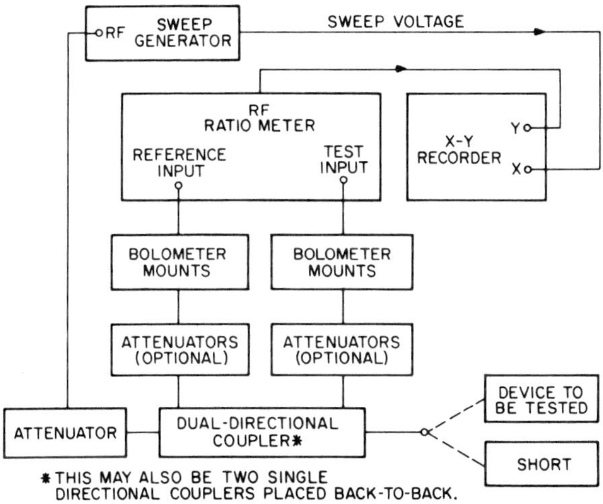

Fig. 7.41 Reflectometer setup using an RF ratio meter.

has a high value of directivity. The optional attenuators are used if there is a danger of either the input levels of the bolometer mounts being exceeded, or a mismatch existing between the coupler and the mount. The ratio meter itself is as described and shown in Figure 7.40. It provides a dc voltage output which is used at the Y-axis of the X–Y recorder.

The test procedure using the Ratio Meter would be as follows:

1. Connect the setup as shown in Figure 7.41. Turn on the sweep generator, ratio meter, and the X–Y recorder on the oscilloscope, whichever is used.
2. When all of the equipment has come up to power, attach the short at the coupler output as shown.
3. Set the RF level control on the sweeper until a convenient reference level is obtained on the *reference meter* on the ratio meter.
4. By using the audio attenuator, adjust for a 0 dB return loss in the system. If the X–Y recorder is used, run this on your paper as a 0 dB calibration line.
5. Repeat the above for various values of return loss that correspond to certain VSWRs. For example, 6 dB return loss = 3.0:1, 14 dB = 1.5:1, and 30 dB = 1.06:1. Use the audio attenuation to get these values.
6. When your whole calibration scale is run, return the attenuator to the original, 0 dB return loss value.
7. Attach the device to be tested in place of the short. The X–Y recorder should now move to the value of return loss that corresponds to the device's particular impedance.
8. Run the X–Y recorder—you now have a plot of return loss as a function of frequency.

Note—Different manufacturers of ratio meters have variations in their particular instrument. Each device has a specific way of being set up initially for your measurement. It is a good idea to read the instruction manual that comes with each and become familiar with the instrument before attempting a measurement. A little time spent in reading saves a lot of time during the actual measurement.

Sources of error from the reflectometer types of measurement are the directivity of the directional couplers, the main line mismatch of the directional couplers, the square-law response of the detectors, the match of any crystal detectors, the tracking of the couplers and detectors, and the calibration of the ratio meter, power meters (if used), or the SWR meter.

The reflectometer provides a rapid, accurate means of gathering information on microwave device performance over broad frequency ranges. The information would be utterly useless, however, if we could not display it properly. There are different methods for displaying swept-frequency measurements, the first of which is the X–Y recorder. It is a convenient means of obtaining a permanent record of swept-frequency measurements. The voltage to operate the Y-axis of the recorder is obtained from either the ratio meter, power meters, or SWR meter. The X-drive is obtained from the sweep generator, as shown in Figure 7.33.

The oscilloscope can also be used in place of an X–Y recorder. The oscilloscope display of swept-frequency measurements is particularly useful when the device under test is to be tuned or adjusted, such as filters, couplers, or amplifiers. Response limitations of a ratio meter require a direct-coupled oscilloscope and a long-persistence tube. Calibration lines can be drawn directly on the face of the tube or transparent overlays may be used.

A third method is to use the front panel meter on the ratio meter, power meter, or SWR meter you are using. These readings are a very convenient indication of performance at a single frequency. In swept-frequency applications where a high degree of accuracy is not necessary, limits can be marked on the meter and then you can watch the swing of the indicator to see if it stays within these boundaries.

It should be pointed out that various manufacturers of microwave devices and equipment provide reflectometer calculators to individuals who work in microwaves. These calculators are very handy, giving reflection coefficient (ρ), standing wave ratio, directivity, return loss in dB, and mismatch loss, all of which are derived from the basic equations we have been using. These devices save time when you are making measurements, but you should not forget how each specification is derived and what terms (such as ρ, VSWR, *et cetera*) are needed for that derivation.

7.4.3 Slotted Line Method

One of the most useful and most accurate instruments used in microwave measurements is the slotted line. A slotted line is a section of uniform, lossless (very low-loss) transmission line, with a longitudinally-oriented slot which provides access to a pickup probe that detects voltage variations on the line as it slides along it. Very simply put, it is a piece of low-loss line slotted in such a way to accommodate an inductive probe which measures the voltage along the line.

Slotted lines are made in coaxial and waveguide configurations; a typical one is shown in Figure 7.42. We cover two types of slotted line measurements in this section—single- and swept-frequency techniques.

Single-Frequency Measurements

Figure 7.43 shows a single-frequency slotted line test setup, from which the standing-wave ratio can be read directly. The easiest way to understand the single-frequency setup is to go through a measurement procedure.

1. Connect the test setup as shown in Figure 7.43. Turn on power to the signal generator and SWR meter. Allow for any warmup time and set the generator frequency.
2. Place a short on the output port of the slotted line (point A). Adjust the probe depth just deep enough to give a good indication of the meter.
3. Run the carriage of the slotted line down the line (in either direction) until you find a maximum reading on the meter. You may have to adjust the gain of the meter to keep its needle from being pegged. When the first maximum is found,

record the position of the carriage. (By a maximum reading we mean a maximum meter deflection to the right; it actually indicates a *voltage minimum*).
4. Move the carriage (in either direction) until a minimum is reached, then keep moving in the same direction until a second maximum is found. Record the position of the carriage.
5. Subtract the two readings. The resulting number is one-half the wavelength of the frequency at which you are operating times two; it is λ (one wavelength).
6. Remove the short and connect the device to be tested.
7. By moving the carriage and adjusting the gain of the SWR meter, adjust the meter for its maximum reading at one on the VSWR scale.
8. Move the carriage (either toward the load or toward the generator) until the meter reaches a minimum value. This figure is the VSWR of the device being tested at this particular frequency.

Fig. 7.42 Microwave slotted line.

This VSWR figure can be used to find a value of reflection coefficient or return loss; it can also be used to obtain a direct value of impedance, by using a *Smith chart*. This tool was covered extensively in Chapter Six, but how the slotted line/Smith chart combination gives an impedance reading is significant here. The easiest way of demonstrating this fact is with an example.

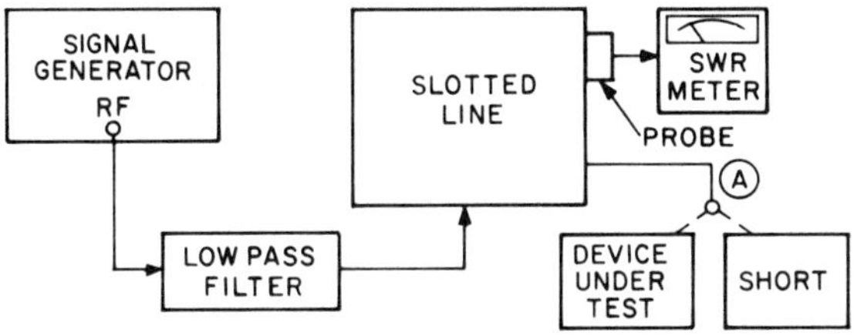

Fig. 7.43 Single frequency slotted line test setup.

Let us suppose that we have the following specifications:

(With short connected) Peak 1 = 12.21 cm, Peak 2 = 10.21 cm.
VSWR = 1.5:1

Distance to 1.5:1 point = 1.3 cm (toward the generator).
Our solution is outlined below:

Step 1 $\lambda/2$ at our frequency = 12.21 − 10.21 = 2.00 cm. Therefore λ = 4 cm.

Step 2 Our VSWR measured was 1.5:1. We now place a compass point at the center of the Smith chart in Figure 7.44, set the pencil on 1.5, and draw a circle. The impedance of our device lies somewhere on this circle.

Step 3 To obtain this 1.5 value, we found a meter reading corresponding to a minimum voltage, or 1.0:1 VSWR, and moved toward the generator until we found our 1.5 reading where the meter stopped moving to the left. This distance was measured as 1.3 cm. Therefore we must move 1.3/4.0, or 0.325λ, toward the generator to find our actual impedance.

Step 4 The next step is to determine where to start so that we can move 0.325λ toward the generator. We said that we started from a point where the meter read a VSWR = 1.0:1, a voltage minimum. Where there is a voltage minimum there must be a low resistance—at the left side of the chart where R = 0.

Step 5 Starting at R = 0, move along the outside edge of the chart in a clockwise direction using the top set of numbers—"Wavelengths toward the generator." When you reach a point where the numbers say .325 place a mark (I).

Step 6 Place a straight edge through I and the center of the chart (II); connect these points together with a line. Where the line intersects the 1.5:1 circle is the actual impedance of the device at the particular frequency (III). In our case it is 1.2-*j* 0.4Ω; this value is *normalized,* so that the actual impedance is 60-*j* 20Ω.

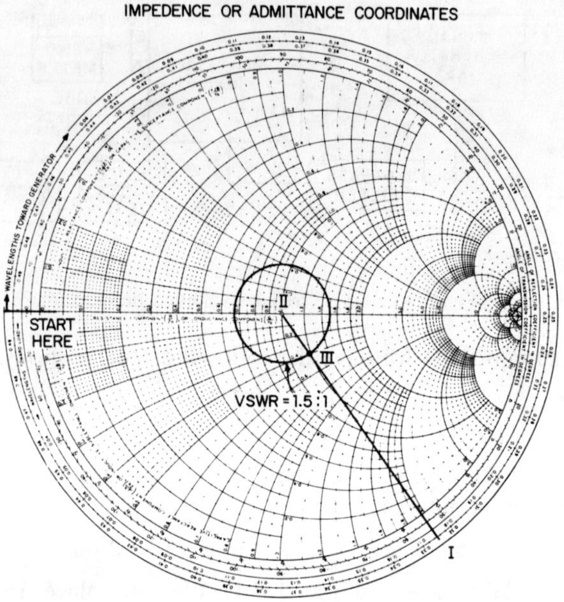

Fig. 7.44 Impedance example.

So you now can see how handy the slotted line setup is for single frequency measurements. The accuracy of such a system is very high, but it does have some obvious disadvantages.

1. *Time-consuming for multiple point measurements*
   Since the measurement described above must be repeated at every frequency of interest, this technique is seldom used for other than single frequency measurements.
2. *Loss of measurement continuity*
   When the single-frequency technique is used for broadband measurements, only a discrete number of points can be measured. For example, if measurements are being made in X-band, the following frequencies might typically be chosen: 8.0, 8.5, 9.0, 9.5, 10.0, 10.5, 11.0, 11.5, 12.0, and 12.4 GHz. It is not uncommon for an impedance mismatch of great magnitude to occur over a narrow frequency range, of 200 MHz or less. If this mismatch were centered about 11.7 GHz, for instance, it would go completely undetected.
3. *Residual SWR*
   Reflections due to the measurement port connector and discontinuity at the slot end are typically 1.02 SWR at 2 GHz (40 dB equivalent directivity), 1.03 at 10 GHz (36.5 dB), and 1.04 at 18 GHz (34 dB) for slotted lines with a pre-

cision 7mm connector. However, one precision line, made by Hewlett-Packard, offers only 1.015 SWR (42.5 dB) to 18 GHz.

4. *Error due to source residual FM*
   The minima of a standing wave are obscured in the presence of FM, thereby producing an inaccurate ratio of maxima and minima.
5. *Probe tuning errors*
   One of the major sources of error in standing wave measurements is excessive probe penetration. The presence of the probe affects the VSWR because it is essentially shunting the line. This fact is why we emphasized that the probe be set only deep enough to obtain an indication on the meter.
6. *Line Variations*
   Dimensional variations along the line introduce a measurement error.
7. *Errors due to deteector characteristics*
   Since SWR indicators are calibrated in terms of square-law response, errors are introduced if the detector departs from its square-law response.

Swept Slotted Line Measurements

There are various methods of sweeping a slotted line—one is to modify the setup in Figure 7.43 by replacing the generator with a sweep generator, and the SWR meter with an oscilloscope with storage capability. The sweep generator is set to sweep the frequency band of interest. Multiple sweeps are made while the carriage of the slotted line is moved manually. The measurement is displayed as an envelope of SWR (the scope calibrated in dB). At any given frequency, the ratio of the maximum and minimum amplitude of the envelope is related to SWR.

This swept slotted line technique eliminates the loss of measurement continuity described previously, and reduces measurement time (though it still requires considerably more time than reflectometer methods and requires an expensive storage scope). But the other disadvantages discussed in regard to the single frequency setup are still present, and, in addition, the display is not a direct readout.

A variation of the above procedure is available using the RF ratio meter described earlier. The test setup for such a measurement is shown in Figure 7.45(A). The 1 kHz modulated RF source is divided by a power divider to feed the reference and test channels equally. The attenuators ensure good matches among the power divider, slotted line, and reference detector. The reference detector is merely any device which converts the RF input to a dc or audio signal that is useful to the Ratio Meter.

VSWR measurements employing this technique are made in basically three steps:

1. The X–Y recorder is adjusted for the desired sensitivity over the selected frequency range (a sensitivity of 1 dB/inch is used in Figure 7.45(B)).

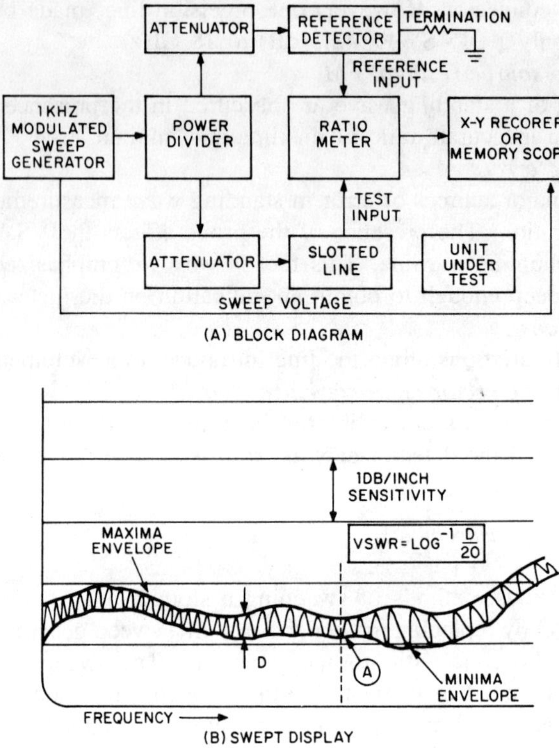

Fig. 7.45 Swept slotted line setup.

2. Repeated swept recordings are made to produce an envelope of the VSWR of the full frequency range.
3. VSWR is read in dB at any frequency in the swept range.

The audio attenuator of the ratio meter is used to calibrate the vertical displacement of the X–Y recorder to the desired resolution in dB. Then, with the slotted-line probe in a set position, the frequency band of interest is swept repeatedly, each time with the probe reset to a slightly different position. This action eventually results in clearly visible maxima and minima envelopes. The VSWR, in dB, at any given frequency is equal to the vertical distance between the maxima and minima envelopes, as illustrated in Figure 7.45(B). The envelopes are clearly visible, with a variety of plots between them.

To illustrate how you convert the display in (B) to a VSWR, consider an example. Suppose we have chosen the point in the figure marked (A). The distance, D, between the maxima and minima envelopes appears to be on the order of 0.5 dB. Our formula states that VSWR = $\log^{-1} d/20$. Therefore

$$\text{VSWR} = \log^{-1} d/20;$$

$$\text{VSWR} = \log^{-1} 0.5/20;$$

$$\text{VSWR} = \log^{-1} (.025).$$

The VSWR at A is equal to the number whose log is .025; if you consult a slide rule or a book of tables, that number is 1.05. Thus, the VSWR at point A is 1.05:1.

The actual measuring time can be greatly reduced by using a memory scope that can store each of the many sweeps necessary for a good measurement. This process saves the time otherwise taken as the X–Y recorder moves repeatedly from one side to the other.

Fixed frequency VSWR measurements using this technique can be made by moving the probe between maxima and minima along the slotted line at the selected frequency and reading directly from the VSWR scale of the ratio meter.

The swept slotted line method for measuring VSWR is a highly accurate and useful procedure; however, its usefulness and accuracy can be improved by the use of the HP8755 frequency response test set, or any of the network analyzer systems. This system operates throughout the frequency range from 1.8 to 18 GHz. Measurement results are displayed on a storage oscilloscope as an envelope of the SWR in dB, as before. At any given frequency, the ratio of the maximum and minimum amplitude of the envelope is the SWR. A plot of SWR can be generated in a few seconds and retained on the CRT for evaluation or photography. Accuracy of slotted-line measurements is limited primarily by the residual SWR of the line itself, 1.01 in waveguide and 1.02 to 1.06 in coax, depending upon the frequency and type of connector.

The 8757 system and others are covered thoroughly later in Chapter Eight.

And so we add one more important microwave parameter to our store of knowledge—*impedance*. Test procedures have ranged from a single bridge to a swept slotted line. Each is unique in its own way, but all have the same thing in common—providing an indication of how well-matched are the devices we use in our systems.

7.5 FREQUENCY MEASUREMENTS

We have mentioned the term "frequency" many times thus far, to characterize the area of operation of both components and test setups. Now we discuss how to measure this very important parameter.

To begin with, there are two general classes of frequency measuring devices: *active* and *passive*. Active devices are also referred to as *electronic* since they operate on a frequency-beating technique as a form of measurement. Similarly, passive devices are also called *electromechanical,* because they measure frequency by measuring its wavelength—f = v/λ (where f = frequency, v = velocity of propagation, and λ = wavelength).

7.5.1 Active (Electronic) Technique

Purely electronic frequency measurements are those in which a standard is compared to an unknown by a null-beating technique. This technique is used to calibrate wavemeters because of its high accuracy. Typically, the signal from a frequency standard is taken and amplified, then connected into a harmonic generator, to provide a so-called "comb" of standard frequencies. Ahead of the harmonic generator, this signal contains the frequency from the standard, for example, 10 MHz. When amplified and connected into the harmonic generator, its waveshape is altered, resulting in harmonic distortion. This distortion creates a spectrum out of the harmonic generator consisting of harmonics of the original frequency. Good generators provide harmonics up to the 100th or higher harmonic number. If one uses a mixer, blending the harmonics with a swept-frequency signal, and connects the result through a detector to an indicator such as an oscilloscope, beat frequency notes can be seen. Since a mixer passes the sum and the difference of the frequencies, a low-pass filter is usually connected to the end of the mixer to allow only the difference of the frequencies to go through. If the signal is connected to the vertical amplifier of an oscilloscope having a few MHz frequency responses, indication in the vertical direction occurs only when the beat frequency is in the passband of that amplifier. Consequently, when a difference frequency from the sweeper and the particular harmonic generator begins to approach a few MHz, indication is given. As the difference frequency decreases, the amplitude of the signal becomes larger and larger until zero beat occurs. There is no amplitude at zero beat, of course—as the sweeper frequency passes through zero beat, a mirror image occurs and the signal decreases from maximum amplitude to zero. These beat notes appear at every harmonic number on the spectrum being swept. If the fundamental frequency being multiplied is 100 MHz, then a beat note is obtained at every 100 MHz on the spectrum.

Figure 7.46 shows a typical frequency measuring arrangement that is adaptable for use as a wavemeter calibration setup, simply through the insertion of the wavemeter at point A-A. The accuracy of this type of setup can be quite high, because a very stable 1 MHz source can be used to initiate the comb generator. Thus the beat note accuracy depends on that of the standard being used.

Another method of measuring the frequency of a device is through the use of a frequency counter—a unit which counts the frequency, then displays it. Although it may not be as accurate as the frequency conversion technique discussed above, it is much more convenient and can be left in your circuit while all your measurements are being taken. Figure 7.47 shows two ways of connecting a test setup when using the frequency counter. In (A) a directional coupler is used; first to sample the RF energy going down the line, then to feed it to the counter. The coupler can be 10 or 20 dB to ensure the proper level going to the counter. There is, of course, a maximum level for the counter input, but there also is a minimum level required for counter operation. Some counters have a meter which signifies the proper level; others have an indication that says "lock" to tell you that the level is of a sufficient value to make your measurements valid. Figure 7.47(B) uses an attenuator to guarantee the proper level at the counter input.

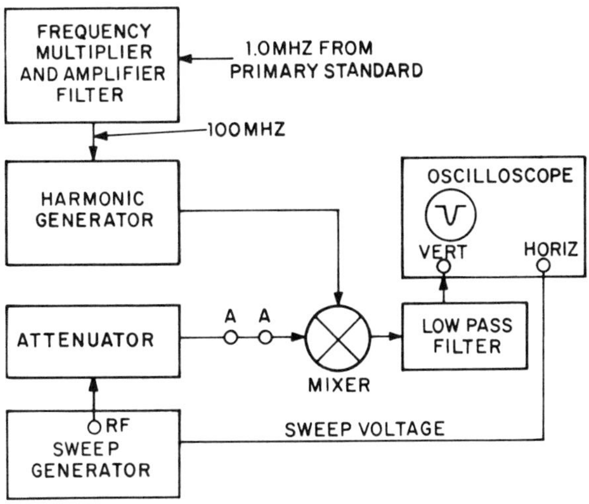

Fig. 7.46 Frequency measuring setup.

The counter is the most widely used form of electronic frequency measuring in everyday systems. If great accuracy were needed, the frequency-beating technique would be used, but the accuracies needed for most test setups are met very nicely by using the counter. Some counters give a direct reading on the display; with others you must go through the process of heterodyning, mentioned earlier. In the second case, you have to do a little figuring to find your frequency.

Frequency counters are available today that read directly up to 18 GHZ. These pieces of equipment are tremendously useful since, at one point in time, you could only read frequencies in that range on wavemeters or through an elaborate system of frequency conversion. And even with good conversion setups, the accuracy still left something to be desired.

A simple method of obtaining a relative frequency reading is to use a spectrum analyzer. We used the analyzer to obtain an absolute power reading in a previous section—that same setup (Figure 7.19) can be employed for a relative frequency reading. The spectrum analyzer can be tuned to the general frequency before any RF is applied to your device, or before your generator is turned on (if you are checking the frequency of a generator). The bandwidth switch should be in the widest position, to give you an indication on the screen of where your signal is. When the power is turned on and you see your desired signal, you should run through the identification process outlined for each specific analyzer. This procedure is an absolute necessity since an analyzer can generate signals internally that can show up close to your desired signal. Most analyzers that cover any frequency range at all use a heterodyning process, with the local oscillator generally operating in the 2 to 4 GHz range; many harmonics and sub-harmonics easily can result from such a situation. It would seem that you would have one big mess on your analyzer out of which you would have to pick the desired signal. However, this is not the case. There are instances when random signals are present, but you generally can pick out the desired signal quite easily. One very fast method of separating good signals from the bad is to note each level, and then either increase or decrease the input RF attenuator by 10 dB. The true signals will go up or down by 10 dB with the attenuator, the internal ones may go 20 dB or any random amount other than 10 dB. Thus, you can eliminate many signals quickly by operating only one switch on the analyzer.

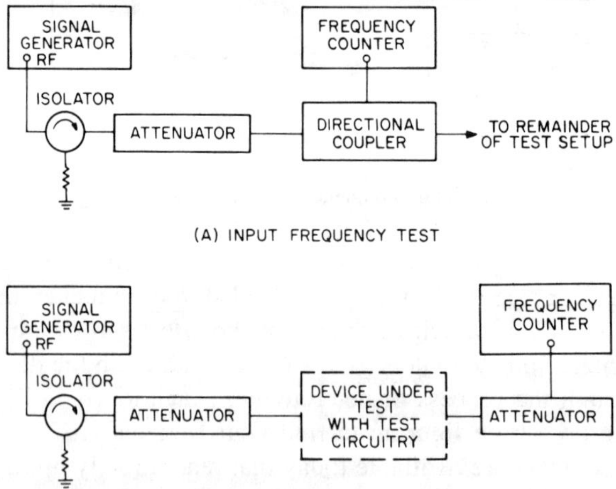

Fig. 7.47 Frequency testing using a frequency counter.

When you have determined which signal is the one you want, you can decrease the bandwidth on the analyzer end, and, at the same time, tune the signal to the center of the screen. When the bandwidth is decreased to where you have a good presentation, you can read an appropriate frequency on the analyzer scale.

We have listed three methods to measure frequency by *active,* or *electronic* means, in decreasing order of accuracy. Each of these methods finds extensive use in microwave test setups. There are areas where a high degree of accuracy is required, and there are times when only a general indication is needed. Whichever degree of accuracy is desired, there is an electronic method available to achieve it.

7.5.2 Passive (Electromechanical) Technique

Where the accuracy of active devices is not required, passive devices offer direct readout at a considerable saving in cost. Passive transmission-type frequency meters are two-port devices that absorb part of the input power in a tunable cavity. When the cavity is tuned to resonance, a dip occurs in the transmitted power level which can be observed on a meter or oscilloscope display of the detected RF voltage. Frequency is then read from a calibrated dial driven by the cavity tuning mechanism.

The accuracy of cavity frequency meters depends upon the cavity, *Q;* dial calibration; backlash; and the effects of temperature and humidity variations. Waveguide and coaxial passive frequency meters achieve accuracies of a few parts in 10^4.

We referred to the frequency meter (also called wavemeters) above as a passive transmission type, which is one of three classifications given to such devices. The three are: *transmission, reaction,* and *absorption*. Figure 7.48 shows diagrams of each of these, with power out *versus* frequency for two. The two-port cavity in (A) is the transmission type. Maximum energy passes through the cavity only when the frequency of the input signal is equal to the resonant frequency setting of the cavity. The cavity is effectively a short circuit, except at frequencies near resonance. The bandwidth characteristics of the cavity are also shown in (A); this response curve should not be mistaken as the response of the input signal *versus* output signal. The graph illustrates why we see a dip in our test setup level when we are tuned to a resonant frequency—a maximum amount of energy is present in the cavity itself at this frequency, F_o, causing energy to be taken from the main line of the test setup. Thus, there is a dip in the overall response. This wavemeter is one of the most widely used.

The *reaction* variety is shown in (B). The cavity absorbs energy from the transmission line at the resonant frequency as shown on the P_o *versus* frequency plot.

The *absorption* meter is represented by the one-entry coaxial wavemeter in (C). Viewed from the input, the cavity exhibits the same characteristics as that of the transmission type; that is, it causes a pronounced dip in the RF energy at the resonant frequency.

A typical frequency measuring setup is shown in Figure 7.49(A); though this particular one is a swept setup, a single frequency arrangement, with the sweeper in CW, can also be used. One important point should be emphasized—be sure to place the wavemeter after the ALC loop. If, for some reason, you want to measure the frequency right at the RF output port, disconnect the leveling loop. The ALC loop cannot level the wavemeter because it is a very sharply tuned resonant circuit; the loop cannot respond fast enough to level that resonance. So when using an ALC loop, place the wavemeter *after* the loop. (It is probably a good idea to think of the ALC loop as an actual part of your generator whenever it is used, so that everything always follows it.)

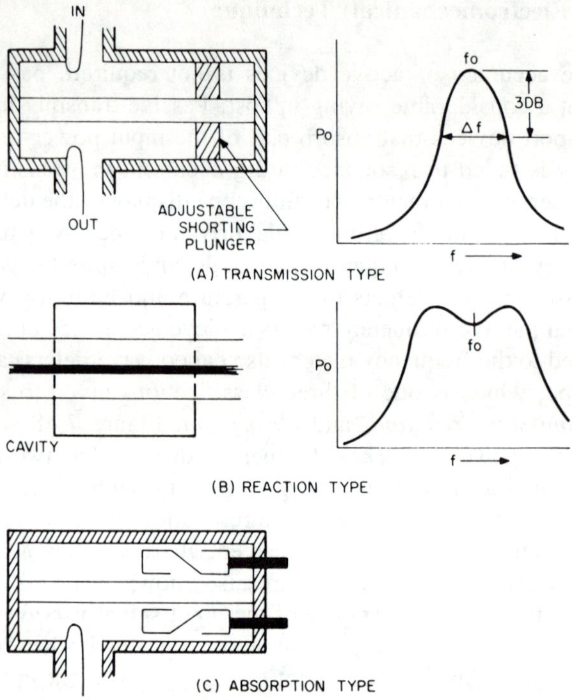

Fig. 7.48 Wavemeter diagrams.

The ALC loop is a standard setup in (A), with an attenuator immediately following it to provide a good match between the loop and the wavemeter. The resonant properties of the meter, which could cause mismatches as the device is tuned, make this inclusion necessary. The output indicators could be any combination of devices; in the figure, they are a power meter and either an X–Y recorder or oscilloscope. An SWR meter with a detector input could also be used with either the recorder or scope, but remember that the generator must then have a 1 kHZ modulation.

1. Connect the test setup as shown in Figure 7.49(A). Turn on the power to all equipment.
2. Set the wavemeter far out of the frequency range you are to sweep.

3. Run a set of calibration lines as shown in Figure 7.49(B).
4. Set the X–Y recorder level back to 0 dB and the wavemeter at your first frequency. Sweep the system.
5. Repeat the above (at the 0 dB line) for f_2, f_3, . . . *et cetera,* giving you a set of calibrated attenuation and frequency lines.

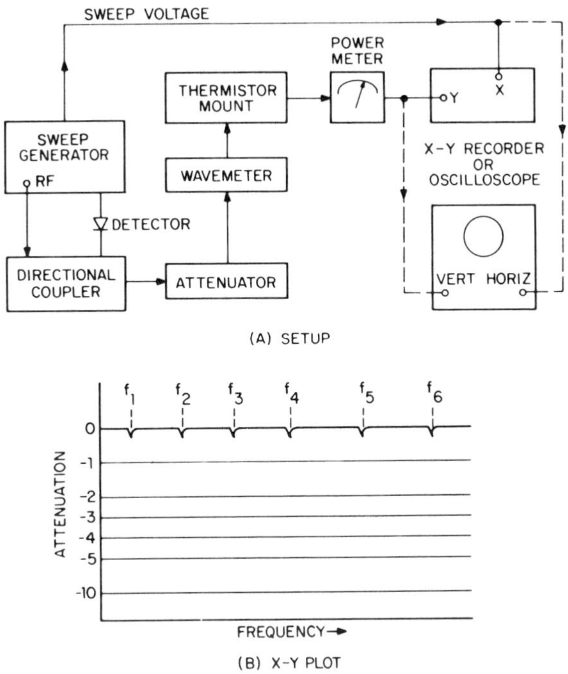

Fig. 7.49 Frequency measuring setup using a wavemeter.

Note—Sweep the system at a speed that is slow enough to indicate the dip in the recorder, but fast enough to give only a narrow dip.

If you want to check the generator for frequency calibration, this setup also can be uscd. Put the generator in a CW position, and set it on the frequency you want checked. Set an RF level for an indication on the output meter (the recorder or oscilloscope can be disconnected). Move the wavemeter very slowly until you observe a dip in the meter. When the initial dip is found, rock the tuning mechanism back and forth until you find the most pronounced dip. Read the frequency directly off the wavemeter scale. This procedure can be repeated as necessary to calibrate the whole scale of a sweep generator.

A typical coaxial wavemeter is shown in Figure 7.50. This instrument consists of a special transmission section with a high-Q resonant cavity which is tuned by a choke plunger. A 1 dB or greater dip in output indicates resonance; virtually full power is transmitted off resonance. Tuning is by a precision lead screw, spring loaded to eliminate backlash. Resolution is increased by a long, spiral scale calibrated in small frequency increments. Resettability is extremely good, and all frequency calibrations are clearly visible so you can tell at a glance the specific portion of the band you are measuring. Overall accuracy of this particular device is better than 2 percent.

A less often used mechanical frequency measuring technique is done with a slotted line. This technique was discussed earlier in reference to impedance determinations; we run through a typical measurement here as a quick review. The setup is a generator, slotted line, and SWR meter. The probe is set to a depth only deep enough to get an indication on the meter. A short is connected to the output of the slotted line. By moving the carriage and adjusting the gain of the SWR meter, we set up for a maximum deflection to the right on the meter (a minimum voltage). Record the position of this maximum (call this point x_1). Move the carriage either toward the load (short), or toward the generator until you go through a dip and come to another maximum. Record the location of this second maximum (x_2). You have now gone through one-half the wavelength of the frequency you are generating. To find this frequency, subtract the two readings ($x_1 - x_2$) to get $\lambda/2$. Then solve this equation

$$f = \frac{v}{\lambda} = \frac{v}{2|x_1 - x_2|}$$

As an example, suppose $x_1 = 12.2$ cm, $x_2 = 10.2$ cm, and $v = 3 \times 10^{10}$ cm/s (a standard value).

$$f = \frac{v}{2|x_1 - x_2|} = \frac{3 \times 10^{10}}{2(12.2 - 10.2)} = \frac{3 \times 10^{10}}{4}$$

$$f = .750 \times 10^{10} = 7.5 \times 10^{9} = 7.5 \text{ GHz}$$

A slotted line thus gives you a frequency reading. It is not a direct reading, because you must do some figuring, but it gives you a reading with an accuracy of less than 1 percent.

7.6 NOISE MEASUREMENTS

Every network, whether a mixer, an amplifier or complete receiving system, has a certain amount of internal noise associated with its operation. A comparison of this noise level to the level of the desired signal in the network is the signal-to-noise ratio. There are input and output signal-to-noise ratios for each network under

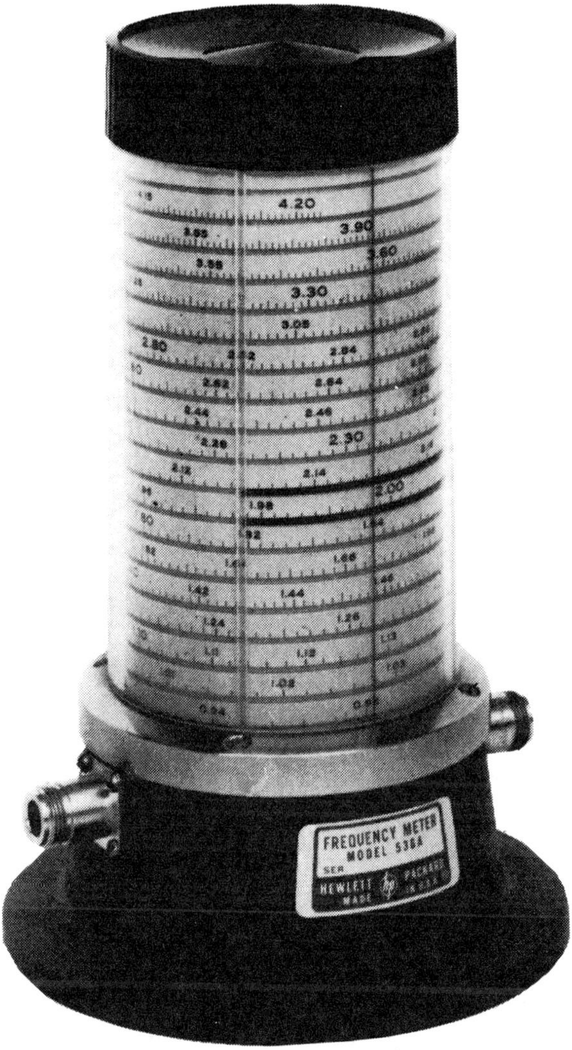

Fig. 7.50 Coaxial wavemeter.

consideration. The term *noise figure* is a measure of the reduction of a signal-to-noise ratio by the network we are using (a mixer, amplifier, receiver, *et cetera*). It is a term expressed in dB which compares the two ratios; expressed mathematically it is

$$NF = 10 \log \frac{S_i/N_i}{S_o/N_o}$$

where:

NF = noise figure in dB;
S_i/N_i = input signal-to-noise ratio;
S_o/N_o = output signal-to-noise ratio.

Noise figure is sometimes referred to as a "figure of merit" of a device or system, which is an accurate appraisal. The noise figure of a device is a measure of how the device handles its internal noise in order to produce an output that is as noise-free as possible. If the noise is dealt with effectively, it is a measure of merit of a device.

Though noise is usually expressed in dB, it can be termed as *noise temperature,* in °K. A formula describes the relationship between *noise figure* and *noise temperature*.

$$NF = 10 \log_{10} \left(1 + \frac{T_N}{290\text{K}} \right)$$

where:

NF = noise figure in dB
TN = noise temperature in K

A 40-inch spiral chart which lists an assortment of noise figure/noise temperature determinations found from the formula above is shown in Figure 7.51. The chart is rolled into a spiral to combine accuracy and convenience; it is most accurate at the low-noise end of the scale. The next time someone refers to a noise temperature of 1,100K, 300K, or 500K, check the spiral chart to convert that specification to the more familiar noise figure.

There are many different methods of measuring noise, such as the *twice factor, Y-factor* and *hot-cold noise source* processes. Though valid means of determining noise figures, they are time-consuming and, thus, not used extensively. Instead, they illustrate a variety of noise measurement procedures.

The method most widely used in microwaves is the *automatic noise figure measurement* (illustrated by the setup in Figure 7.52). Versatility and accuracy are its two greatest advantages. Figure 7.52 is a swept-frequency arrangement but, by eliminating the X–Y recorder and putting the sweeper in CW mode, the same setup can be used for single-frequency measurements.

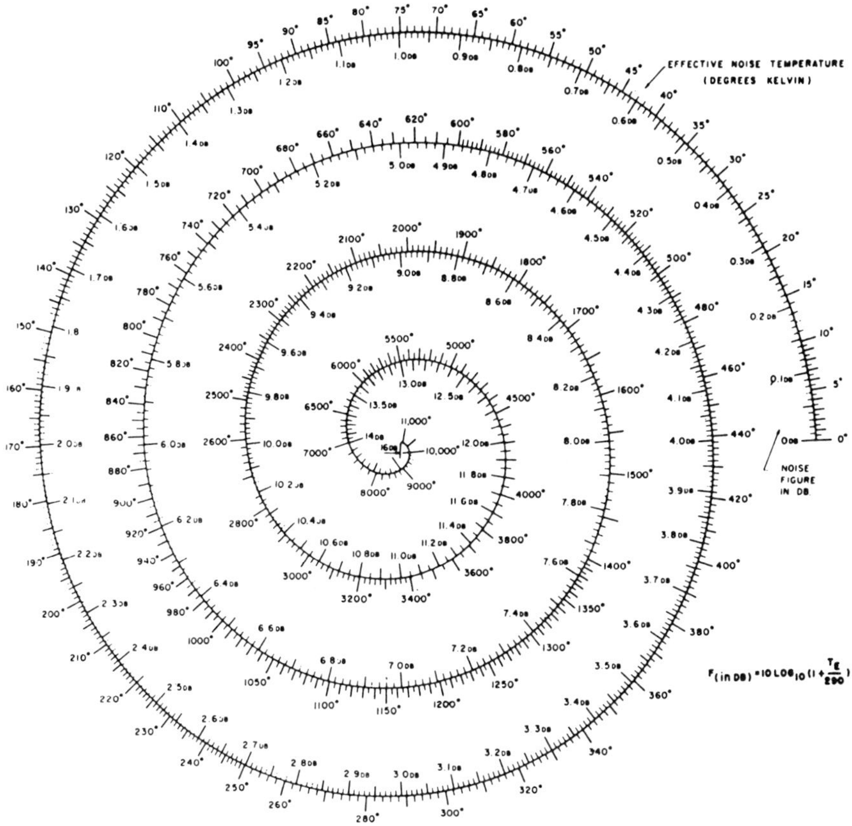

Fig. 7.51 Spiral chart relating noise figure and noise temperature.

The noise source shown in the test setup is usually a gas-discharge type for the microwave region, although temperature-limited diode sources are used at the lower end of the spectrum (under 1 GHz). The latter source consists of a diode mounted in parallel, with a resistive load equal to the source resistance for which the receiver, or device, to be tested was designed (for example, 50Ω). The diode is operated temperature-limited (effectively current-limited), and a meter, placed in the plate circuit of the diode, is calibrated directly in noise figure in terms of diode current.

Noise available from this type of source can be varied. The gas-discharge noise source operates at a very high temperature and has a high available noise power. The effective thermal agitation of an argon tube noise source, for example, represents an equivalent temperature of approximately 10,000K (about a 15.8 dB noise figure), compared with a normal room temperature of 290K. When used in waveguide systems, the gas tube (either argon or neon) is matched to the structure by mounting it

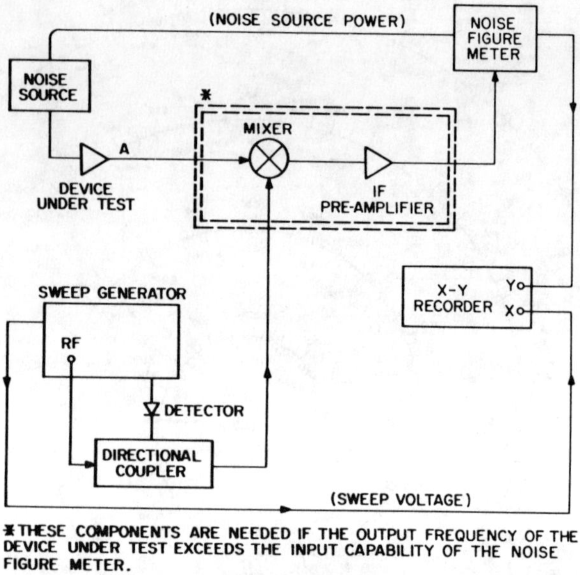

Fig. 7.52 Automatic noise figure setup.

so that it extends through the broad sides of the waveguide at an angle of approximately 10 degrees. Solid-state noise sources are also available to 18 GHz; they exhibit an excessive noise ratio of around 15 dB (9,000K). The mixer and IF preamplifier, as stated in Figure 7.52, are used when the output frequency of the device under test exceeds the input capability of the noise figure meter. The automatic noise figure meter has a tuned input; that is, it cannot operate at just any frequency. The input could be set steps (30, 60, 120, 200 MHz) or a band of acceptable frequencies (10 to 1,000 MHz). Regardless of how the input is tuned, however, the absolute maximum input frequency on currently available noise figure meters is 1 GHz. Therefore, the majority of the microwave spectrum has to undergo some sort of frequency conversion—the mixer, and its associated amplifier, provides that change. The amplifier also helps to increase the overall setup gain to the proper level, which may be indicated by a light or a meter reading of infinity (∞).

The sweep generator and its associated-leveling circuitry provide the second signal to the mixer. If you are taking a single-frequency measurement, it is simply a source for a second signal to be used for mixing; if you are taking swept-frequency noise-figure measurements, it also supplies the calibration lines on the X–Y recorder. To allow the system to have calibration lines, a step attenuator can be placed at point A to allow the insertion of various levels.

The noise figure meter has the ability to automatically display measured values while operating with either gas discharge, temperature-limited diode, or solid-state sources. Two important settings are to be made when using an automatic noise figure meter. The first is a zero setting, which sets a reference for all of your noise readings. The second is a check if there is a high enough level into the meter (a high enough gain in the system) to operate the meter properly, which may be a visual indication in the form of a light on the front panel or an "∞" setting on a meter. No meaningful data can be obtained unless the input level to the meter is sufficient.

The X-Y recorder, of course, is used to plot the noise figure as a function of frequency. Levels of "zero" and "∞" can be used for initial calibration.

We briefly outline a procedure for making a noise figure measurement; it is very general since each noise figure meter made by a different manufacturer is unique in the way it is set up. It would be wise to consult the manual for your specific meter and learn its operation.

1. Connect the setup as shown in Figure 7.52. Turn on all ac power.
2. Adjust the sweeper for its proper frequency range and the noise source for its proper current (which can be found in the manual). If the mixer and preamplifier are used, be sure that an output frequency that is compatible with the input of the meter is produced.
3. Set the meter "zero" and "∞" controls.
4. Turn the noise figure meter to "noise figure" and read the value. If the mixer and preamplifier are used, some calculation is necessary because you are not only looking at the device under test, but also at extra components (mixer and preamplifier) with specific noise figures and gains). The formula used to calculate noise figure is

$$NF_S = NF_1 + \frac{(NF_2 - 1)}{G_1} + \frac{(NF_3 - 1)}{G_2} + \ldots \textit{ et cetera}$$

where

NF_S = noise figure as read on the meter;
NF_1 = noise figure of the device under test;
NF_2 = noise figure of the mixer;
NF_3 = noise figure of the preamplifier;
G_1 = gain of device under test;
G_2 = gain of mixer;
G_3 = gain of preamplifier.

It can be seen that the overall noise figure of any cascaded amplifying system depends primarily on that of the first stage. The effects of subsequent stages are reduced by the gain up to that point. On the other hand, the use of a passive stage with gain less than one increases the importance of the subsequent stage's noise figure.

As in any system there are factors which decrease the overall accuracy of your measurements. Actual measurement techniques must consider the possibility of system errors caused by mismatch, temperature, and image and spurious responses. The instrument accuracy of the noise source and noise figure meter should also be taken into acount.

Mismatch

Noise power obeys all power transfer laws but, since it is random in phase, mismatches cause ambiguous errors rather than known amounts of power loss.

The critical matching situation, then, involves the excess noise source. Noise sources are rated in *available* excess noise power, thus mismatches cause an ambiguous amount of excess noise power to be coupled to the system. Figure 7.53 shows the effect of several possible mismatch conditions.

Note that the possible error is maximum at low noise figures where the greatest accuracy is usually desired. The importance of well-matched noise sources over the entire frequency range of interest is apparent.

Temperature

To obtain a value of excess noise, the standard temperature (T_o) is assumed to be 290K (62.6°F). Manufacturers know the value of the noise source and rate it in terms of this standard value. A variation of 20 degrees from the assumed 290K, for example, causes an error of about 0.3 dB in the measured noise figure. If you are trying to read a 3 dB noise figure, you have already introduced a 10 percent error.

Image and Spurious Response

When using a broadband excess noise source, an automatic noise figure meter measures the true noise figure of the total pass band of the device under test. If in its operation the device does not utilize the full pass band for signal information (as would be the case with a radar receiver with an image response), its operating noise figure is higher than that measured. This apparent value can be calculated by the equation:

$$\text{Operating } NF = NF \text{ (reading)} + 10 \log (B_t/B_u),$$

where B_t is the total bandpass of the device, and B_u is the operational bandpass.

This equation is a convenient simplification that assumes constant gain in the device under test. Therefore, great care should be taken in noise figure measurements to be sure that you are measuring the desired frequency and not an image or spur.

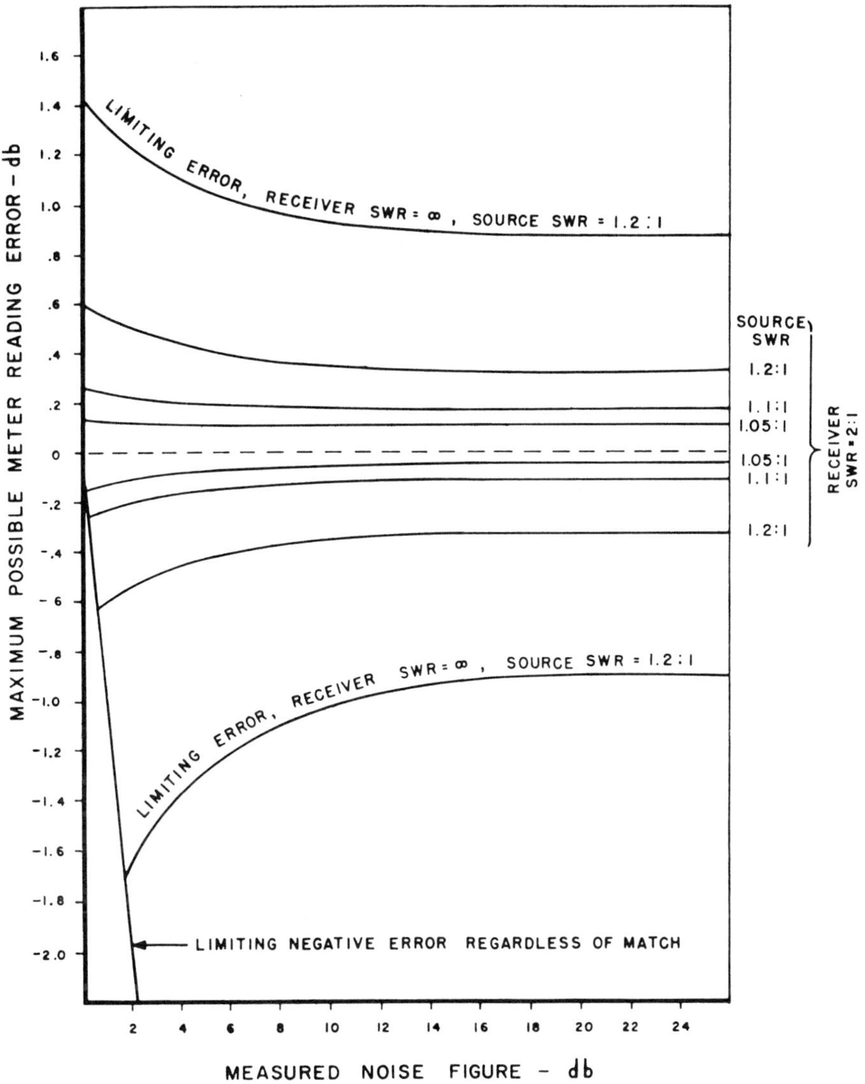

Fig. 7.53 Typical error effects for several mismatches between noise source and receiver.

Excess Noise Source Accuracy

Accuracy of pulse-type gas discharge noise source tubes is specified as ± 0.5 dB. Temperature-limited diode sources can be made to much closer specifications, but are limited to the VHF region and below.

Correction of the excess noise ratio of a noise source sometimes is necessary to take into account temperature variations from the standard 290K. This situation is particularly applicable to diode sources, where the heat of the temperature-limited diode tends to raise the ambient temperature of the output resistor.

Noise Figure Meter Accuracy

Automatic noise figure meters are specified as accurate within $\pm {}^{1}/_{2}$ dB over most of their range, ±1 dB over the remainder. This accuracy figure includes the effects of meter tracking, variation from square law, and aging effects. With the current specifications of noise sources, and the possible errors caused by mismatch and temperature, such meter accuracy would seem to be consistent with that of the overall system.

For measuring low-noise devices, modified noise figure meters which provide increased resolution and accuracy are available. The modification expands a 4 dB (up to 6 dB) portion of the scale over the full range of the meter scale. Noise figure meter accuracy in the expanded mode of operation is ±0.2 dB; either expanded or normal operation can be selected.

7.7 SUMMARY

This chapter has been on microwave measurements. Many setups have been shown while others were not, but the whole idea of this chapter was not to shower you with a great number of test setups that you were expected to memorize; rather, the purpose was to show you basic and useful setups. With these arrangements as a starting point, you could assemble your own setups as needed—not from memory, but from your knowledge of each individual piece in the basic setup and its applications. Very simply put, you should think out a test setup instead of memorizing it; that way, if something goes wrong, you know exactly where to look for the problem. There is nothing worse than being under a deadline to produce some accurate data and having your memory fail you. Learn about the setups, don't memorize them.

Chapter 8
Microwave Network Analyzer

The term *network analyzer* conjures up ideas of many high level mathematical formulas and relationships. This, however, is not the case with this chapter. We are going to present equipment used extensively throughout the microwave industry for network analysis.

What the network analyzer actually does is to combine many of the pieces of equipment (discussed in the previous chapter) to make a variety of measurements. Many of the analyzers have calibration procedures which are much easier than those discussed previously, and the speed and accuracy of the readings are greatly enhanced. Also, most of the analyzers have signal sources (synthesized) which are an integral part of the overall system.

There are two types of analyzers that will be discussed: *scalar* and *vector*. A scalar analyzer will measure and display only the amplitude of the signal being analyzed. This is displayed as a function of frequency. The vector analyzer measures and displays both the amplitude *and* the phase. These are also displayed as a function of frequency but are not only shown on linear scales but also on Smith chart displays. All four S-parameters may be measured and displayed with the vector analyzer. The analyzer which you should use depends on your particular application. Two models of scalar analyzers and two models of vector analyzers will be presented. After the discussions, the choice is yours. It should be pointed out that everything presented here is of a general nature to introduce you to the network analyzer. For specific details or tests you should consult the manuals or application notes for that particular piece of equipment.

8.1 SCALAR ANALYZERS

As we have stated above, the *scalar network analyzer* will measure and display only the amplitude characteristics of a device or system being tested. This is not to insinuate that this type of analyzer is inferior by any means. It actually is a very valuable piece of equipment when only the amplitude characteristics are required. This is the case for many applications. Therefore, the scalar network analyzer has many uses throughout the microwave industry and is used to perform many test requirements on components and systems.

8.1.1 Hewlett-Packard 8757A

Figure 8.1 shows the HP8757A scalar network analyzer. The HP8757A is a unit which measures insertion loss, gain, return loss, SWR, and power. It can operate from 10 MHz to 60 GHz with the proper options attached. The unit has four independent display channels and up to four detector inputs. This permits simultaneous viewing of up to four measurement parameters at independent resolutions. Other features of the HP8757A are

- −60 dBm sensitivity in all inputs
- Ac/dc measurement flexibility
- High (>40 dB) directivity bridges
- Low SWR detectors (10 Mhz to 60 GHz)
- Direct plotter and printer output
- Save/recall calibration and test set
- Calibration adapts to Δf
- Fully HB-IB programmable
- Built-in self test diagnostics

As mentioned above in the features, there are accessories used with the basic 8757A system. These are the high directivity bridges and the low SWR detectors. These components play a vital part in determining the overall 8757A capabilities.

The directional bridges available are the HP85027A, 85027B, 85027C, 85020A and the 85020B. The 85027A and 85027C operate from 10 MHz to 18 GHz. The "A" model has an APC-7 connector on the test port, and a type N at the input. The 85027B has APC-3.5 connectors at both the input and test ports. The unit also has a frequency range extending from 10 MHZ to 26.5 GHz. The HP 85020A bridge covers the range from 10 MHz to 4.3 GHz (type-N connectors), while the 85020B operates from 10 MHz to 2.4 GHz with a 75 Ω input and test port (type-N connectors also).

The low SWR detectors reduce mismatch errors during the measurement. The HP85025 and 85026 detectors allow the tester to choose either modulated (ac) or unmodulated (dc) detection. With ac detection, you obtain greater immunity to noise and drift with time and temperature. With dc detection, you have greater power measurement accuracy and it is generally used when measuring oscillators and modulation-sensitive devices. The HP85025A covers a range of 10 MHz to 18 GHz and uses either type-N or APC-7 connectors. The HP85025B has an SMA connector and goes from 10 MHz to 26.5 GHz.

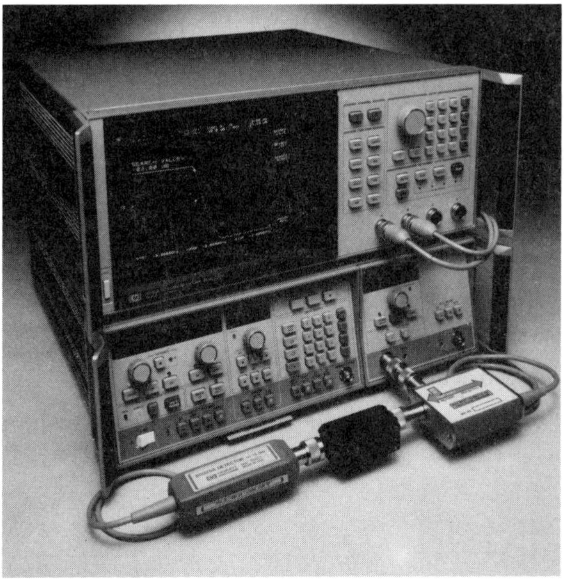

Fig. 8.1 HP8757A scalar network analyzer.

The HP 85026 series of detectors covers from 26.5 GHz to 60 GHz with a good match and at least 55 dB of dynamic range. The HP11664A and HP11664E detectors are also available, and are equivalent to the HP85025A and HP85025B, respectively, with the exception that the 11664 series operate in the ac mode only.

An additional accessory which may be used with the HP8757A is a power splitter. This can be used in ration applications such as swept power tests, and measurements where remote sources are required. The power splitters will be seen in test setups shown later in this section. The HP11667A and HP11667B splitters are available. The 11667A operates from dc to 18 GHz with type-N connectors. The 11667B operates from dc to 26.5 GHz with APC-3.5 connectors.

With the system presented and some accessories explained, we will now show how a test might be run with the HP8757A. The following are general steps which indicate how a test might be conducted.

1. First you need to calibrate the system. Prompts lead you through the calibration procedure. Once the calibration is complete you are allowed to narrow the frequency span (to check the ripple on a bandpass filter, for example) without needing to recalibrate. This speeds up the measurement time considerably.
2. Next you connect the device to be tested. When you press *autoscale,* the full response of the device is brought to the screen. This full response display will allow you to determine which specific area you need to examine in more detail.

3. You can now zero in on a specific region to 0.01 dB by means of the cursor functions. This may be a bandwidth (3 dB, 1 dB, ripple BW, *et cetera*), or it may search for maximum and minimum points to measure a change, or difference, in power or frequency.
4. Now it is possible to plot the test results on a plotter or printer. A description can be entered on the HP8757A front panel. This will give your plot the detailed information needed to end up with complete data.
5. You may save your complete test setup in one of the *save/recall* registers so that the same measurement may be duplicated later.
6. An additional test may be made which determines *pass/fail* characteristics of a particular device. This is done by comparing the device to on-screen limits which have been entered on the HP8757A front panel. Thus, a *go/no go* response is achieved with this method of testing.

To illustrate the capabilities of the HP8757A scalar network analyzer, we will present two test setups which can be achieved with the 8757A.

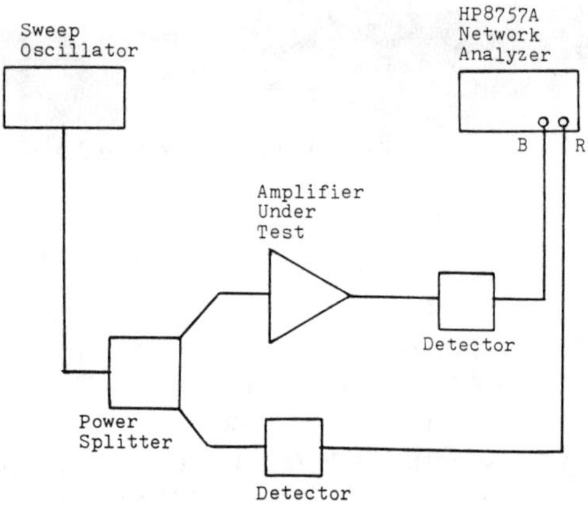

Fig. 8.2 Amplifier test setup.

The first setup is shown in Figure 8.2 and is designed to measure gain, gain flatness, and output power. Notice that there are no specific model numbers put on the sweep oscillator, power splitter, or detectors. This will enable the setup to be a general setup and allow you to insert the proper devices for your particular application.

In the setup of Figure 8.2, the R (reference) detector responds to the input power to the amplifier, and the B detector is the output power of the amplifier. The ratio of B/R is the amplifier gain. Output power can be read off the analyzer and

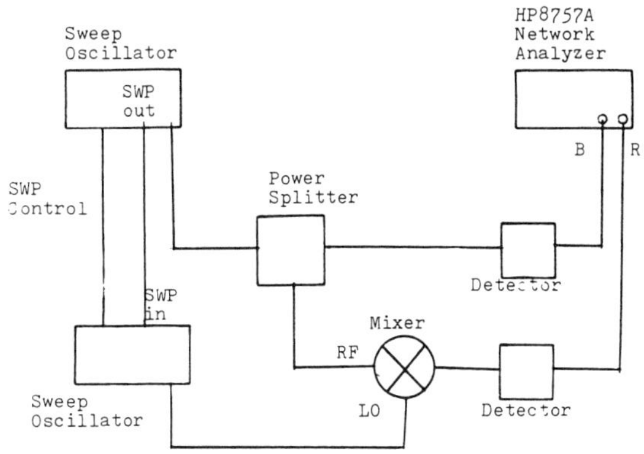

Fig. 8.3 Mixer test setup.

the gain flatness can be achieved by using the cursor maximum, minimum, and delta functions.

A second test setup is shown in Figure 8.3. This is a setup for testing mixers. It can be seen how a conversion loss measurement may be accomplished with this test setup. The RF signal is shown on the R (reference) input while the IF output is displayed on the B input. A B/R ratio will indicate the conversion loss. This method of testing may be enhanced by using ac detection at the mixer output. The setup would have the RF input modulated and the LO input unmodulated. This will result in the IF output modulated and displayed but there would be no LO feedthrough since the LO is not modulated. Thus there will be no LO feedthrough to cause inaccuracies in the conversion loss measurement.

Before leaving the HP8757A, we should look at one more area which has been in our test setups and is of vital importance: the sweep oscillator. Generally, two types are recommended for use with the HP8757A. These are the HP8350B sweep oscillator (with plug-ins), and the HP8341A or HP8340A synthesized sweeper (depending on the upper frequency needed). The HP8350B is an excellent source for broadband swept measurements from 10 MHz to 60 GHz. The oscillator is available with more than 30 different RF plug-ins, which allow it to be configured for a wide range of applications such as high output power, low harmonics, low cost, and wide frequency range.

When increased frequency and accuracy are required, the HP8341A or HP8340A synthesized sweepers should be used. The 8341A covers 10 MHz to 20 GHz, and the 8340A goes from 10 MHz to 26.5 GHz. Both sweepers have better than 4 Hz resolution.

8.1.2 Wiltron 561 Scalar Network Analyzer

The Wiltron 561 scalar network analyzer is shown in Figure 8.4. The 561 measures transmission, return loss, and power, without the need for an external controller. Hard copies of all displayed test data and test parameters can be produced on a dot-matrix printer or GPIB plotter. Features of the Wiltron 561 are:

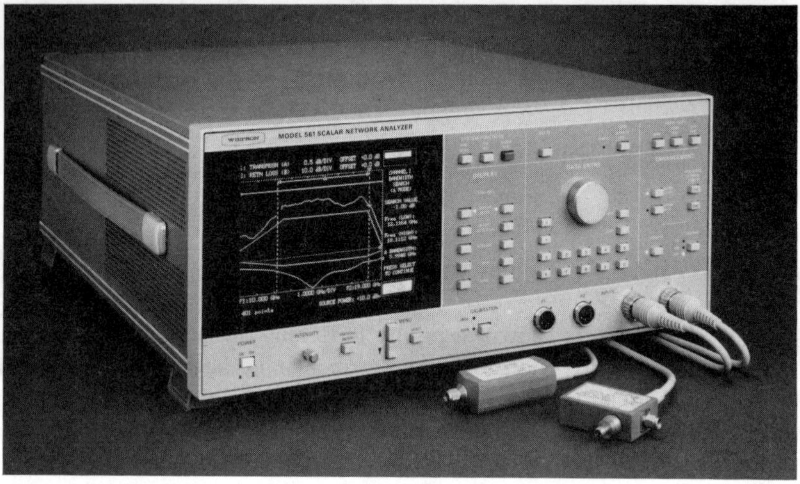

Fig. 8.4 Wiltron 561 scalar network analyzer.

- 10 MHz to 40 GHz continuous coverage with one set of components in coaxial;
- Coupled system outputs fully annotated display to printer or plotter;
- Simplified calibration and measurement;
- 2,000 point calibration allows narrowing of the sweep without recalibration;
- Interpolation allows precise, versatile cursor and marker functions;
- Automatic scaling of resolution and offset.

Accessories which accompany the 561 are the SWR autotester and the microprocessor compensated detector. The 560 series SWR autotesters integrate in one package a broadband high directivity bridge, a detector, a low reflection test port, a reference termination, and a connecting cable. The output of the autotester is a detected signal, varying in proportion to reflections from the test device connected to the test port. A variety of connectors are available depending on the range of frequencies needed. From 10 MHz to 18 GHz the test port may have a GPC-7 or type-*N* connector, while the input connector is a type-*N*. Connectors on both ports from 10 MHz to 26.5 GHz are SMA, and from 10 MHz to 40 GHz are K connectors.

The 560 series of detectors are used for coaxial transmission loss or gain and power measurements. The detectors are zero-biased with replaceable Schottky diodes and provide −60 dBm sensitivity. Once again, a variety of connectors are available. GPC-7 connectors or type-*N* (10 MHz to 18.5 GHz), SMA (10 MHz to 26.5 GHz) or K connectors (10 MHz to 40 GHz) can be specified.

The 561 has the most extensive set of cursor functions (seven) that are available on any scalar analyzer. The following is a list of these cursor functions:

Main cursor—Position of the cursor is continuously variable with the tuning knob. The frequency and amplitude of the test data at the cursor on both traces are displayed digitally.

Cursor delta—The difference in amplitude and frequency between the reference cursor and the main cursor positions on the test data are displayed for both traces. To establish a new reference, the position of the two cursors can be reversed by making a menu selection.

Cursor min/max—The 561 moves the cursor automatically to the minimum or maximum value of test data on either trace and displays the value in dB or dBm.

Cursor "X" dB—The cursor moves automatically to the amplitude on either trace where the test data is equal to the entered value of "X" dB or dBm.

Cursor "X" *bandwidth*—Cursors are displayed automatically above and below the cursor at the frequencies where the test data are equal to the entered value "X" dB. The frequencies of the low and high cursors and the bandwidth between them are displayed.

Cursor next marker—The cursor moves automatically to the next highest frequency marker.

Cursor alive marker—The cursor moves automatically to the frequency of the active marker.

With the addition of a sweep generator ranging from upper frequency limits of 8 GHz, 20 GHz, 26.5 GHz, and 40 GHz (1,5,10, or 40 mw choices), the 561 becomes an automated transmission, return loss (SWR), and power measuring system. Operating over the 10 MHz to 40 GHz range from a single coaxial test port, the system provides fully annotated displays of test data and measurement parameters.

Under internal microprocessor control, the 561 normalizes and simultaneously displays any two inputs on channels A, B, R1, and R2. The same inputs can be displayed as ratios A/R1, A/R2, B/R1, or B/R2. The dynamic range of every channel is 71 dB (−55 dBm to +16 dBm). Typically, the noise floor is less than −62 dBm, providing a greater than 76 dB dynamic range in almost all applications.

Figure 8.5 shows a typical 561 test setup. It can be seen that both reflected

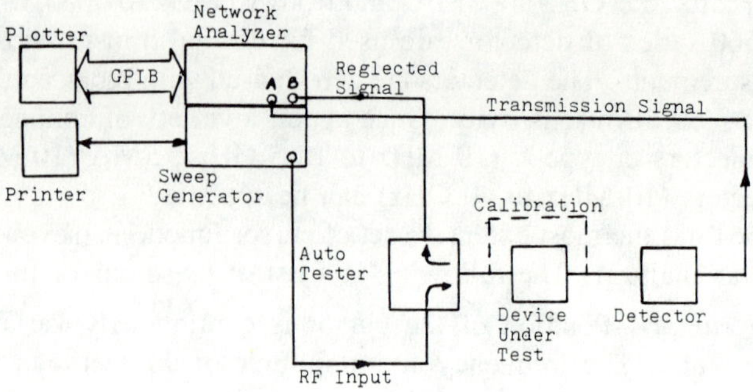

Fig. 8.5 Typical 561 test setup.

signals and transmission signals can be displayed and worked with using this system. Detected signals from the SWR autotester vary in proportion to the reflections, while the detector output varies in proportion to transmission loss or gain. The detector can be used to measure power in dBM.

During normalization, procedural guidance is provided automatically for transmission and return loss measurements. For return loss tests, a 0 dB reference is established by connecting an open, and then a short, to the SWR autotester test port. The normalization data are taken independently of sensitivity settings at 2,001 points with 0.002 dB resolution and stored in memory for correction of test data, or for recall. Furthermore, an algorithm interpolates between data points to hold interpolated test data accuracy to within +/− 0.1 dB. Therefore, once the 561 has been normalized across a user-selected frequency range, measurements can be made over any portion of the range without renormalization. Setup time is virtually eliminated by storing parameters for up to nine test setups.

During measurements, data are taken at 101, 201, or 401 points (user selected) with 0.005 dB vertical resolution on both channels. Updating occurs every 100 ms allowing "real time" adjustments of the test device.

8.2 VECTOR ANALYZERS

The *vector analyzer,* as the name implies, is an analyzer which will give you information in vector form, that includes both a magnitude *and* an angle. This means that the device must be able to not only measure the amplitude characteristics of a device, but also its phase characteristics.

In 1967 Hewlett-Packard brought the HP8410 system. This vector analyzer was uncontested by anyone in the test equipment industry for nearly 20 years. There are now, however, other analyzers which are capable of providing vector measurements at microwaves and we will present two of them. One is the next generation of the 8410—the HP8510, and the other is Wiltron's 360 system.

8.2.1 HP8510 Vector Analyzer

We characterized the HP8510 as the next generation of the HP8410, but it is much more than just the next step. It is an instrument which makes error-corrected measurements of magnitude, phase, and group delay from 45 MHz to 100 GHz (with the proper accessories). One way the HP8510 does resemble the HP8410 is that it is made up of different pieces of equipment. There is the HP8510B itself (the "B" version being the latest base unit at the time of this publication), an HP8511A frequency converter, S-parameter test sets ranging to 20 or 26.5 GHz, and signal sources appropriate for your application. Figure 8.6 shows a combined HP8510 system. Some key features of the 8510 are:

- Digital display;
- Two independent channels;
- Menus;
- Calibration-store-and-retrieve up to eight different calibrations;
- Frequency of time domain;
- Versatile displays possible;
- Five fully annotated markers;
- Save and recall eight instrument states;
- Auxiliary menus;
- Easy data entry;
- Tape or external HP-IB disk drive can be used for data entry;
- Principal analyzer functions are divided into four sections:
    - *Stimulus*–Controls sources
    - *Parameter*—Selects paramater to be measured
    - *Format*—Places measured data into required format
    - *Response*—Sets scale/division, reference level or position, or puts on AUTO

As we stated above, the complete 8510 system is made up of other pieces of equipment. One of these is the 8511A frequency converter. This unit converts the incoming RF signal down to a lower IF signal which can be more easily processed. When the 8511A is combined with the 8510B network analyzer, a four channel receiver/signal processor results that operates over the frequency range of 45 MHz to 26.5 GHz. The HP8511A contains four separate RAF-to-IF converters, all of which operate over the entire dynamic range of the system. This dynamic range is 105 dB (−10 to −115 dBm) from 45 MHz to 20 GHz, and 98 dB (−15 to −113 dBm) from 20 GHz to 26.5 GHz.

The S-parameter test sets are also an important part of the HP8510 system. There are two test sets available: the HP8514B and HP8515A. The HP8514B operates from 45 MHz to 20 GHz, has an input operating power level of 0 to −5 dBm, and has 3.5 mm connectors on it. The HP8515A operates from 45 MHz to 26.5 GHz, has an input operating power level of −5 to −25 dBm, and also has 3.5 mm connectors. When the HP8514B or HP8515A S-parameter test sets are combined

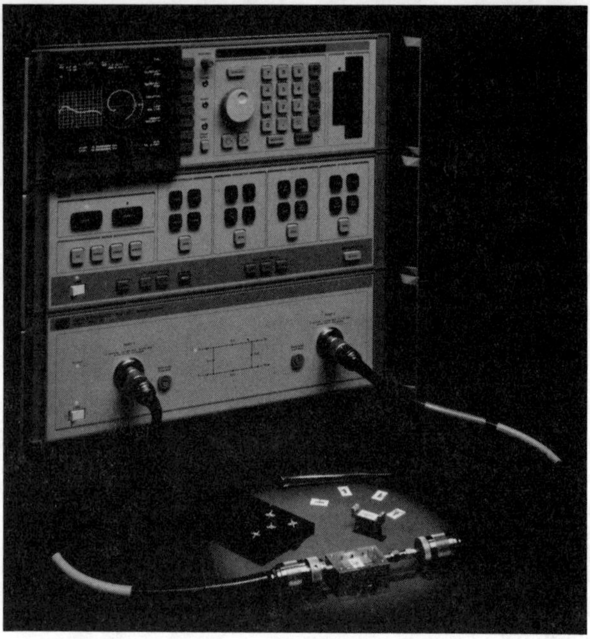

Fig. 8.6 HP8510 vector network analyzer.

with the 8510B system, the result is a system for making full S-parameter measurements. The dual port construction on the test sets develops a separate reference channel for each incident port. RF switching is done with a single built-in electronic switch. When taking active measurements, the test sets include a 90 dB step attenuator for changing the incident power level at either port, as well as the ability to apply dc bias (external) to the test port.

Calibration kits are available for a variety of connector types: 7 mm, 3.5 mm, and type-*N*. The kits contain open circuits, short circuits, fixed loads, and sliding loads. The kits provide very accurate calibration of your system, which is an absolute necessity for obtaining useable and repeatable measurements.

The number of test setups possible with the HP8510B system are too numerous to mention. Application notes are available for a variety of measurements. We will however, choose one which shows a typical setup to be used with the HP8510B system. It is shown in Figure 8.7 and is a swept mixer test setup. This, traditionally, has been a rather difficult and time consuming test to run, since locking of the two sources created difficulty. This, however, is now a rather straightforward measurement with the HP8510B. The setup in Figure 8.7 is for making conversion loss and relative phase measurements on a mixer. The measurements are made relative to a

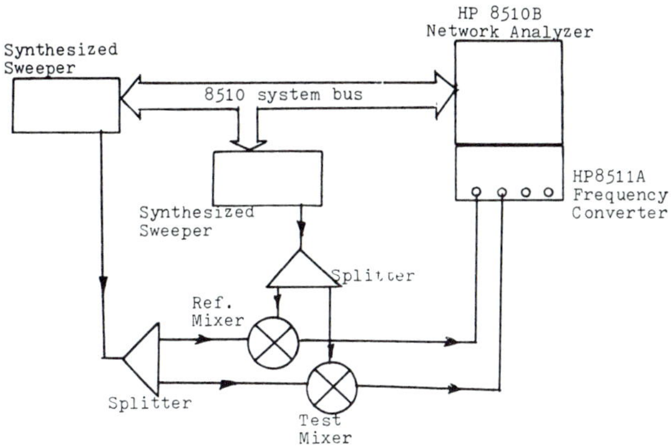

Fig. 8.7 Mixer test setup.

reference mixer, which is used to provide a reference signal at the same frequency as the test signal. The RF frequency ranges from 3 GHz to 7 GHz and the LO frequency ranges from 3.1 GHz to 7.1 GHz. The IF frequency is held constant at 100 MHz. In the multiple frequency mode, all frequency control is set up through the HP8510B front panel, and the HP8510 provides direct control of the entire system without the need for an external controller.

In conclusion, it can be said that the HP8510B is a very valuable tool for use in research and development as well as in production and manufacturing. For research it offers system accuracy, which gives confidence in design; its system speed helps in design evaluation; time domain gives powerful insight into design behavior; and the data obtained is easily transferred to CAE programs.

For production and manufacturing, its system speed permits "real time" device adjustments; the data received is a standard for comparison in final tests; disk drive interface allows unlimited storage of measurement steups and data; full HP-IB programmability makes automation easy; and verification kit data can be certified to meet US MIL-STD 45662.

8.2.2 Wiltron 360 Vector Network Analyzer

The Wiltron 360 vector network analyzer came into being in 1987 as the first 0.5 to 40 GHz coaxial vector network analyzer. This claim is valid since other vector analyzers switch to waveguide above 18 GHz. The 360, however, has the benefit of K connectors which are rated to greater than 40 GHZ and are, of course, coaxial connectors. A 360 system is shown in Figure 8.8. Features of the 360 are shown below.

- Frequency coverage to 40 GHz directly in coax;
- Fully reversing test sets including an active device test set with bias tees and attenuators;
- Large color display;
- Four simultaneous S-parameters displayed;
- Buffered output to printer/plotter;
- MS-DOS disk;
- Frequencies may be changed without recalibrating;
- Discrete frequencies may be entered;
- Precision calibration kits—SMA, 3.5mm, 7mm, K, N;
- Possible to select the data display you prefer: magnitude (real or imaginary), phase, rectangular, polar, Smith chart, distance domain, time domain, or group delay;
- Observe test data as single; dual; or four-channel display (large 190 mm diagonal color screen);
- Four independent channels permit simultaneous display of all four S-parameters and up to eight traces;
- Dot matrix printer provides a hard copy record in tabular or graphical format for up to four complex impedances or transmission measurements.

The 360 also has other features which are useful in microwave measurements. For example, on the 360 there is no need to waste time or degrade accuracy by having to physically reverse the test device. Every test set includes automatic signal reversing so measurement of all parameters proceeds without interruption or loss of accuracy. The active device test sets also have 70 dB level set attenuators in both port lines, with which the drive signal can be adjusted to the proper test level. In addition, a step attenuator within the forward transmission measurement path permits measurements on devices with an output power level of up to one watt. Full S-parameter measurements on active devices are possible.

Also, when isolation of components within a test device is required or the distance to an impedance discontinuity must be measured, the 360's time/distance domain software provides the solution.

The time domain displays are fully mixable with other display modes. Using four-channel display, one can observe a normal frequency domain display on channel one, time domain response on channel two, time domain with gate on channel three, and frequency domain with gate on channel four. Whichever display is chosen there is >100 dB dynamic range up to 18 GHz, 70 dB up to 40 GHz. At very low signal levels, test data accuracy may be improved by the use of two different methods of signal enhancement: averaging and IF bandwidth reduction. At all signal levels and at all measurement speeds, automatic 12-term vector error correction with a synthesized microwave test signal removes system residual errors to produce accurate phase and magnitude measurements.

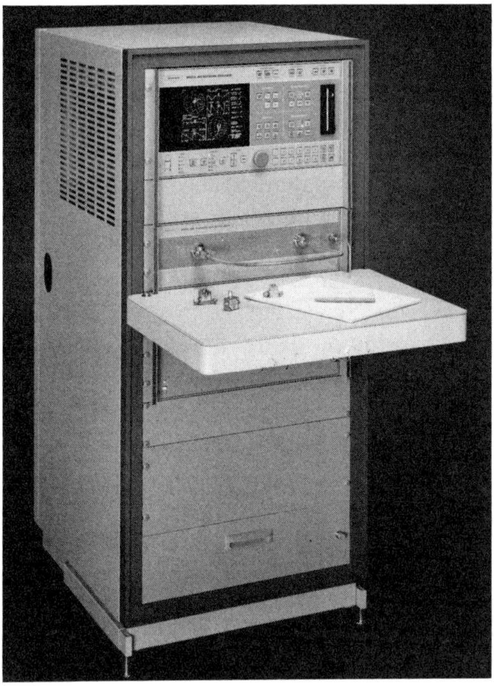

Fig. 8.8 Wiltron 360 vector network analyzer.

The tests that can be run on the 360 are many and varied. We have mentioned some of them in general terms previously, but now we will list six specific tests that can be performed.

1. *Complex impedance and transmission characteristics*—The 360 tests two-port devices in both directions without disconnecting. It then applies 12-term vector error correction to the test data and displays all four S-parameters simultaneously in "real time" with phase-locked accuracy. Such parameters as forward and reverse transfer characteristics, as well as input and output Smith chart impedances of FET amplifiers over ranges such as 1 to 8 GHz may be tested and displayed.
2. *Group delay*—With a "real time" display of signal transit time, it is easy to adjust the test device to optimize group delay. Variable frequency aperture settings can be made without recalibrating. The deviation from linear phase and group delay may both be displayed.
3. *Amplitude and phase matching*—To match the amplitude and phase of a test

device to a standard or to a known device, the characteristics of one can be placed in memory and compared automatically to that of the second, using the "trace memory" feature of the 360. Differences between the two units can be directly measured and compared.

4. *Insertion loss/gain and phase*—Because of its low noise floor, 0.001 dB resolution, IF bandwidth reduction, averaging, and automatic scaling, the system can measure attenuators, high-isolation switches, and high rejection filters. The 360 has two major advantages over scalar analyzers for magnitude measurements: first; with 12-term error correction, the test results are more accurate, and secondly, the dynamic range is >100 dB up to 18 GHz. This is a 30 to 40 dB improvement over the scalar analyzer.
5. *Time domain*—With the addition of the time or distance domain, the system displays discrete discontinuities as a function of time or distance. Unwanted reflections can be removed from the display by windowing and gating. As an example, a discontinuity can be located on a cable at a specific point from the input connector.
6. *Semiconductor characteristics*—There are three features of the 360 which enhance the method of testing semiconductors. The first active device test sets include bias tees and step attenuators for adjusting signal levels at the measurement ports. A unique attenuator allows direct measurement of devices up to one watt output power. The second is that all four S-parameters can be displayed simultaneously, and the third is that software is available that implements de-embedding and customizes output format. In a typical semiconductor test application, all four S-parameters on a GaAs FET over a broad frequency range can be tested and displayed.

The tests performed on the 360 system could not be run without specific test sets and the appropriate signal source. There are four test sets from which to choose the unit to use for your application. All models include automatic signal reversing with which full S-parameter tests can be made. Models 3620 and 3621 have 70 dB variable attenuators in both port lines for adjusting the test signal level. A step attenuator in the forward transmission line attenuates the test device output power, up to one watt.

There are two signal sources designed specifically for use with the 360: the 360SS45 covering 10 MHz to 18 GHz, and the 360SS69 operating from 10 MHz to 40 GHz. Both are under the control of the 360 analyzer and provide clean, phase-locked test signals for precise test data. Frequency resolution is 100 kHz.

8.3 SUMMARY

This chapter has presented the most revolutionary type of measurement system the microwave industry has seen—the *network analyzer*. We have presented both the scalar analyzer (HP8757 and the Wiltron 561) and the vector analyzer (HP8510 and Wiltron 360). The capability of each of these analyzers has been shown to a certain extent. Obviously, details of operation can be found in specific documents presented by HP and Wiltron. This chapter only provided an overview of each type. Regardless of which type or model you choose, you have a very powerful measurement tool.

Chapter 9
Microwave Tubes

In this day of fast moving modern technology with semiconductors being made to operate in increasingly higher frequencies and emphasis placed no miniaturization, it may seen strange to be talking about *microwave tubes*. Many people think of tubes as being used during the war or back in the "old days"; however, with microwaves, such is not the case. Tubes are very much in use today and are planned to be used very extensively in many systems in the future. Proof of this fact can be found in numerous articles that outline advances in tube technology to be compatible with microwave systems of the 1980s.

It may be true that tubes, as we normally think of them, are used very little anymore in today's radios, televisions, and computers. The type of tube used in microwave circuits however, is quite different from those employed in the applications mentioned above. Such terms as velocity modulation, buncher, catcher, helix, split-anode, reentrant-stream, and drift space are not everyday terms in the world of ordinary "tubes," but they are household items in microwave tube technology. These terms are used to describe the action or operation of such tubes as *traveling wave tubes (TWTs), klystrons, magnetrons, and cross-field amplifiers (CFAs)*. These are tubes that have been developed in the 1980s, and are discussed individually in this chapter.

9.1 TRAVELING WAVE TUBES (TWT)

Many electron devices have been developed to extend operation into the microwave region. These devices either eliminate or capitalize on the transit time phenomenon that limits operation of conventional tubes at these frequencies. Of all microwave tubes, only the *traveling wave tube (TWT)* offers extremely broad bandwidth operation.

Obtaining broad bandwidth in a microwave device had been investigated as early as the late thirties. Several schemes were developed and patented involving the interaction of electrons with distributed circuits. The use of the latter was studied since it was recognized that the bandwidth of resonators (whether LC-circuit or resonant cavity) is inherently limited.

The first practical utilization of the traveling wave concept was demonstrated by R. Kompfner at Oxford in 1944. Working at S-band, Kompfner reported a power gain of 14 (11.5 dB). Later, work by J. R. Pierce (in association with R. Kompfner) led to devices that became the forerunners of the tubes available today. In 1946, a device was described in the *Proceedings of the IRE* that featured 23 dB gain with 800 MHz bandwidth at S-band and an output power of 200 milliwatts. The weighting factor (the product of gain times bandwidth times power) of this tube would be 3.2 $\times 10^{10}$. By comparison, today one may obtain tubes with a weighting factor, or gain-bandwidth product, in excess of 10^{19}.

As discussed above, the paramount feature of the TWT is its extremely broad bandwidth. Bandwidths of one octave are usually offered in production tubes, however, even greater values have been achieved. (A bandwidth of one octave is one in which the upper frequency is twice the lower frequency.) As an indication of what can be achieved, test results from a one octave TWT have been recorded over a range of 2.25 octaves. Power in excess of one watt can be obtained from 2.2 GHz to 11.0 GHz. Significantly, these tubes are production units and not just laboratory models.

The TWT is an electronic amplifying device, accepting a weak RF input signal and amplifying it many thousands of times. It performs the same functions as its principal predecessors, the triode and klystron, with one uncommon characteristic—extremely wide bandwidth.

Figure 9.1 is a simplified sketch of a helix-type TWT structure. The major elements are the electron gun assembly, the RF interaction circuit, the electron beam focusing magnets, and the collector. Before examining each of these elements let us take a quick look at the overall TWT and its operation. In the electron gun assembly, when the cathode is heated it emits a continuous stream of electrons, which are drawn through the anode, focused into a tight, narrow beam by a magnetic field, and made to travel the length of the tube (eventually to dissipate in the collector in the form of heat). At the same time the electron beam is fed into a slow-wave structure, consisting, in this case, of a tightly wound helix. The RF energy travels along the helix at the velocity of light. However, because of the helical path, the energy progresses along the length of the tube at a considerably lower velocity, determined primarily by the pitch of the helix. (Refer to Section 4.9.2 on helical antennas for a clearer understanding of the pitch of a helix.) The speed at which the energy slides forward is made synchronous with the velocity of the electron beam, resulting in an interaction between the beam and the RF signal—some electrons are slowed by the RF field; others are accelerated. As these "velocity-modulated" electrons move down through the helix they form bunches, which, in turn, interact with the helix RF wave and surrender energy to it. This activity results in a great amplification of the RF

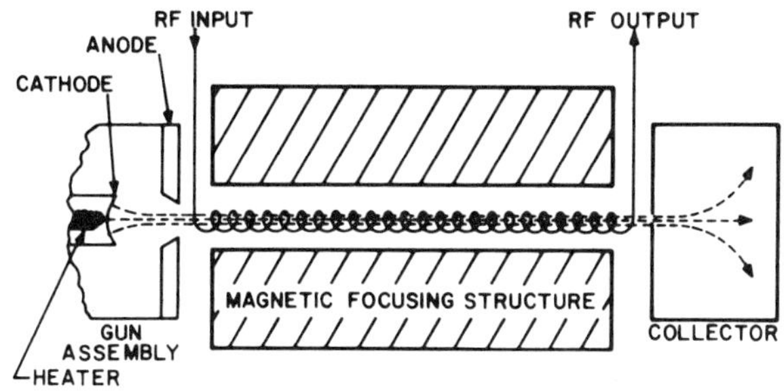

Fig. 9.1 Helix-type TWT.

signal by the time it reaches the output. Single TWTs have been built with power gains of more than 10,000,000 (70 dB).

9.1.1 The Electron Gun Assembly

The electron gun is similar to the lens in a projector—the objective is to get as much electron current flowing into as small a region as possible without distortion or fuzzy edges. As the gun is the source of electrons for the beam, its design is extremely important. A wide variety have been developed in an effort to provide better electron beams.

Most TWTs are equipped with control grids which allow the beam to be turned on and off rapidly, with a much smaller voltage swing than is required when the cathode is modulated.

The typical grid controlled gun has six main elements: the gun shells or support structures (usually ceramic), the heater, the cathode or electron emitter, a control grid, a focus electrode to aid in proper formation of the electron beam, and an anode which effectively provides the accelerating field for the electrons. Figure 9.2 shows a typical gun with the various elements listed above.

9.1.2 RF Interaction Circuit

Many different interaction structures are used in traveling wave tubes such as the helix, ring bar, the ring loop, or the coupled cavity, illustrated in Figure 9.3. However, selection of the RF circuit is a complex trade-off analysis, based on many interlocking parameters such as output power, beam voltge, bandwidth, and frequency. Starting at lower power levels and octave bandwidths, helics are suitable; as power and beam voltage levels are increased the ring bar and ring loop structures offer advantages, particularly in pulsed radar applications. Good thermal conductivity is a requirement for all RF structures. As power level and frequency are increased,

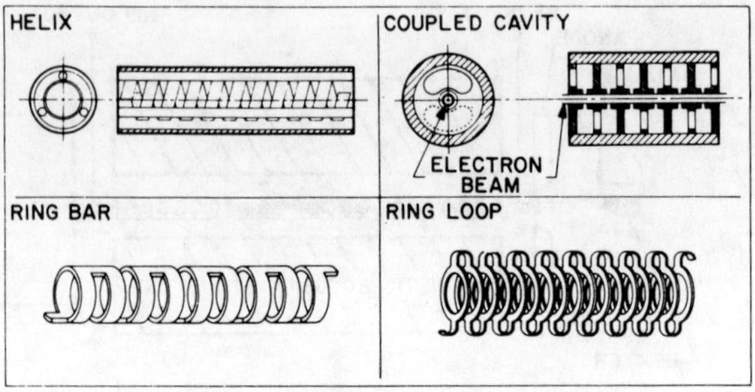

Fig. 9.2 TWT electron gun assembly.

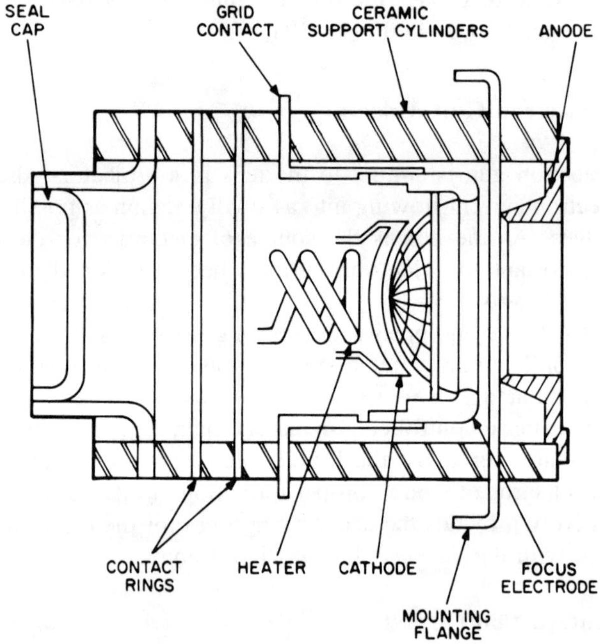

Fig. 9.3 TWT RF structures.

RF losses on the circuit become appreciable, and either massive stub-supported ring bar circuits or coupled cavities must be used to dissipate the heat.

The helix circuit is regarded as the most efficient TWT form. As mentioned above, it is capable of producing bandwidths wider than those of the more recently developed structures. In the basic form, its design is restricted to tubes having peak-power outputs of less than 3,000 W. A number of configurations utilizing the basic helix structure have been explored in an attempt to extend its many fine properties to provide higher output powers. Early among these efforts was a scheme of using two helixes wound in opposite directions. This device, known as a contra-wound or folded helix, extends the useful range of operating voltages up to the 20 to 70 kV range and allows the use of larger transverse dimensions for a given frequency range.

The couple cavity, as mentioned previously, is used for high-power applications. It is also known as a folded-waveguide circuit since its structure resembles a waveguide folded in accordion fashion. This configuration effectively slows down the RF energy, allowing synchronization with the electron beam; the process is the same as that which the helix accomplishes. Originally these structures provided frequency bandwidths in the order of 10 to 15 percent. Recently, however, means have been developed for increasing the bandwidths of these circuits to 40 percent and more; tubes utilizing this circuit have been built which produce several hundred kilowatts of peak power at S- through Ku-band with up to 60 dB gain.

9.1.3 Electron Beam Focusing

All TWTs require some means of holding the electron beam together as it travels through the interaction structure of the tube—the beam tends to disperse or spread out as a result of the natural repulsive forces between electrons ("like" charges repel). A magnetic field is used for this purpose. Such a field of proper magnitude confines the electron beam to the pencil-like cylindrical shape it must maintain. The four principal types of magnetic focusing are shown in Fig. 9.4.

The solenoid is still regarded as one of the best magnetic focusing structures. Its magnetic lines are parallel to the direction of travel of the electrons, and it can be accurately aligned with the beam. It is used whenever high powers are needed and tube size and weight are not critical. Most of the very high power TWT's today use solenoids as a focusing structure.

In certain structures, where the interaction structure is short enough, permanent magnets are used in place of solenoids. Because of the length limitations, this type of focusing is generally restricted to low gain or lower power tubes.

Perhaps the most profound development in beam control has been the evolution of periodic-permanent-magnet (PPM) focusing, particularly its adaptation to high power TWTs. Previously, its fundamental limitations were thought to be so restrictive, it was used only in low power TWTs where the beam density is quite low.

One of the greatest needs for this lightweight focusing method is found at the high power levels where TWTs (with focusing solenoids) have been too large and heavy for many airborne and space applications. PPM focusing has been successfully used to achieve 10 kW of average beam power in a 125 kW peak power pulse tube at S-band frequencies.

There are two types of PPM focusing structures, named for the direction of the magnetic lines with respect to the tube beam axis. The first is the *radial PPM* type, meaning that the magnetic segments have a field that is at right angles to the tube axis. The second, the *axial PPM* type, has magnetic discs with a field parallel to the tube axis. Both structures produce similar periodically reversing magnetic fields along the electron beam axis.

Electrostatic focusing techniques have been tried experimentally in low and medium power TWTs with some success. They require no magnetic field and, in principle, could provide even greater compactness and lighter weight. However, such techniques, abundant though they are, have not yet become practical enough to compete successfully with PPM focusing techniques.

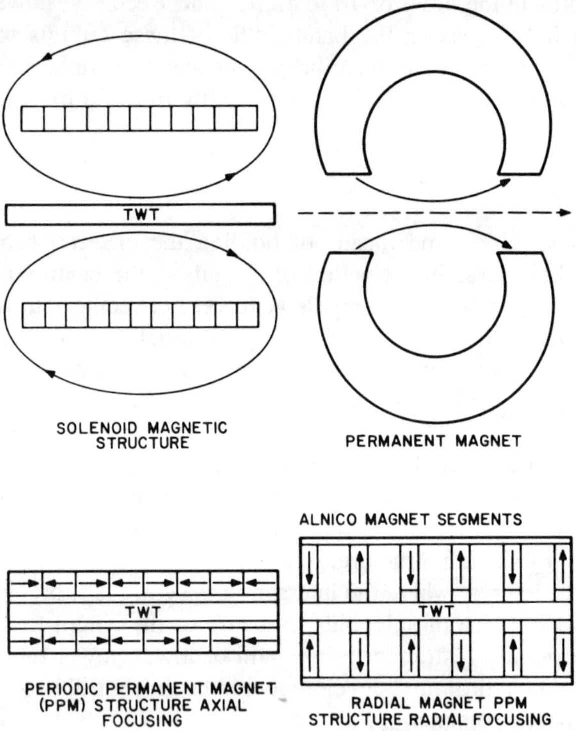

Fig. 9.4 Methods of magnetic focusing.

9.1.4 The Collector

The collector performs no other function than to dissipate the electrons in the form of heat as they emerge from the slow-wave structure, usually accomplished by thermal conduction to a colder outside surface where the heat is absorbed by circulated air or a liquid. Its size is determined by the method of cooling used and the amount of energy that must be dissipated.

In order to improve efficiency, the collector of some tubes is operated at a voltage between ground and that of the cathode. The voltage of the collector is brought closer to that of the cathode by placing a power supply between the two structures. Under these conditions the tube is said to be operating with a "depressed collector."

A typical collector may be "depressed" as much as 50 percent; that is, having one-half the voltage of the cathode. Since the electron beam is collected at some value less than the full cathode voltage, less energy is wasted, resulting in higher efficiency.

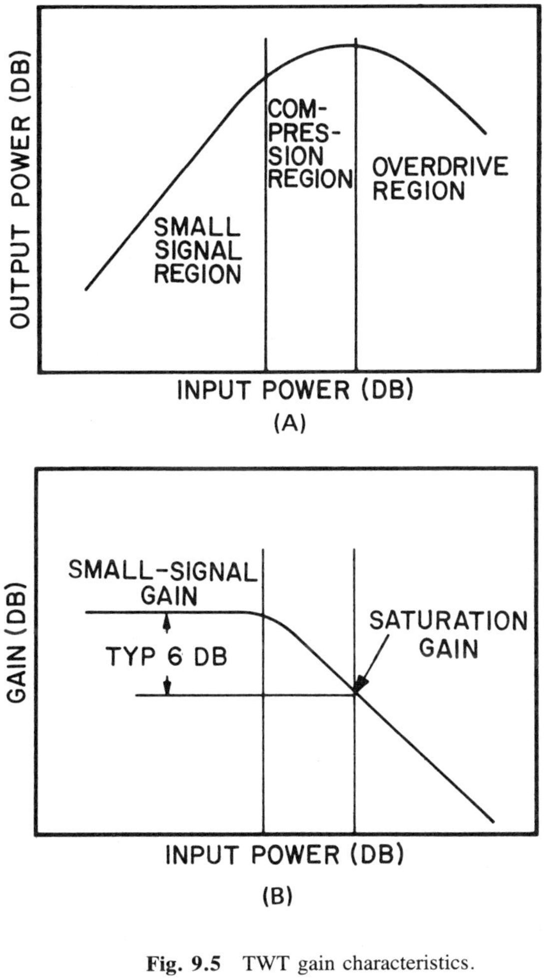

Fig. 9.5 TWT gain characteristics.

9.1.5 TWT Terms

In order to grasp any concept, the language in which it is conveyed must be initially understood. You certainly would have a difficult time in Paris or Rome if you did not speak at least a little bit of French or Italian, respectively. The same qualification holds true for a subject like TWTs. If you do not understand the terminology (language), comprehending the device itself is very difficult. For this reason, the following glossary of terms is presented. Once these words are mastered, TWTs become much easier to use and understand.

Gain Compression—With reference to Figures 9.5(A) and (B), as input or drive power is increased in the linear region of operation, the small signal is approximately constant and the output power rises in a linear fashion. As the maximum power level is approached, the gain starts to decrease, because the beam slows down as energy is extracted from it. Gain compression is the amount of gain decrease from the small-signal condition. In a traveling wave tube, the decrease in gain from small-signal to the saturation point is typically 6 dB.

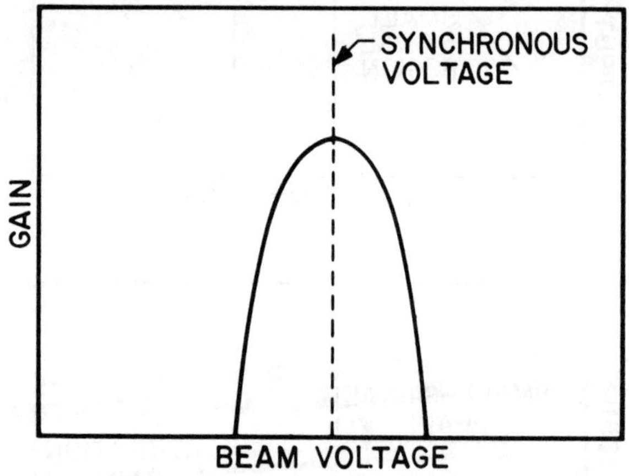

Fig. 9.6 Gain *versus* beam voltage.

Beam Voltage—The voltage between the cathode and the RF structure. As can be seen from Figure 9.6, gain is somewhat sensitive to beam voltage in a traveling wave tube because of the requirement for the electron beam and the RF wave to travel at near synchronous velocities.

Synchronous Voltage—As shown in Figure 9.6, the synchronous voltage is the beam voltage necessary to obtain the greatest interaction between the electrons in the electron beam and the RF wave on the circuit.

Gain—Gain in a traveling wave tube is usually defined as the ratio of RF output power to RF input power. For convenience, it is almost always expressed in decibels using the formula

$$G_{dB} = 10 \log_{10} \frac{P_o}{P_i}$$

In a traveling wave tube, gain is a fuction of beam voltage, current, and frequency. To obtain maximum gain, electron and phase velocities must be nearly synchronous. Conversely, varying the beam voltage to change the output power reduces the gain at a given frequency, as shown in Figure 9.6.

Gain variation over an octave band can vary as much as 10 dB, as shown in Figure 9.7, although gain equalizers or feedback techniques can reduce the span to 2 or 3 dB. Fine grain variations result primarily from internal reflections, so rigid control of construction and assembly techniques are used to minimize such effects. Under these conditions, fine grain variations about the gross curve can often be held to within ±1 dB.

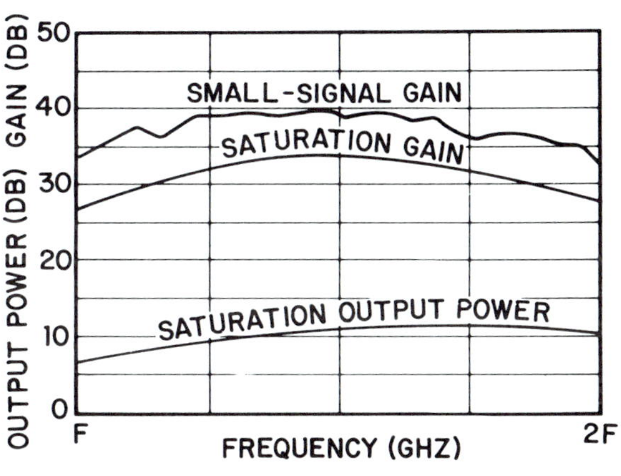

Fig. 9.7 Gain *versus* frequency.

Efficiency—The efficiency of a traveling wave tube is defined as

$$\text{Eff} = \frac{\text{Microwave Output Power}}{\text{Total Input Power}} \times 100\ (\%)$$

The total input power consists of beam power, heated power, RF input signal power, and, if used, solenoid power. Since the gain of a traveling wave tube is usually 30 dB or more, RF drive power can be normally neglected.

There is a difference between the efficiency at rated power of a traveling wave tube and the actual efficiency which is almost always higher. The rated power and efficiency of a tube are selected to give maximum assurance that these ratings can be met even under the worst conditions and at the worst point in the band. As can be seen from Figure 9.8, the difference between this conservative rating and a delivered tube can be considerable.

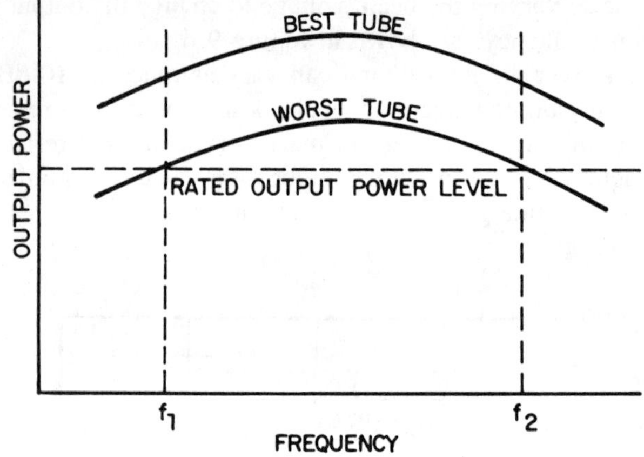

Fig. 9.8 Tube rating variations.

Phase Characteristics—Phase shift in a traveling wave tube is defined as the phase of output signal relative to the input signal. A typical helix traveling wave tube has a total phase shift of approximately 10,000 degrees. The relationship is sensitive to many parameters and is dependent on the operationg conditions of the tube. A small change in many of these conditions can change phase a large fraction of a cycle.

Phase sensitivity is defined as the rate of change of phase with a specific operating parameter. The most critical factor is that of beam voltage, in which a 0.1 percent change causes a 3 degree phase shift in a typical traveling wave tube. Other factors that change phase include beam current, solenoid current, and heater voltage. Phase is not particularly sensitive to collector voltage, but does change with drive level at a rate of 5°/dB—a variation known as AM/PM conversion.

The linearity of the phase shift through a traveling wave tube is determined by the effects of mismatches that cause variations in phase with all paramters held constant except the frequency of the input signal. Mismatches are found either outside the tube, or within the circuit terminations.

Depressed Collector Operation—The efficiency of a traveling wave tube usually may be increased by operation with a depressed collector voltage. Separate power supplies are used for the helix and the collector. Since the RF energy has already been extracted from the beam by the time it reaches the collector, the regulation requirements of the latter are not severe. In a typical case, the helix voltage might require regulation to ± 1 percent, and the collector voltage to ± 10 percent. Depressed collector operation increases efficiency when the same amount of energy in the electron beam can be collected at a lower voltage, thereby decreasing the total input power at the same output level. With lower voltage, less energy is dissipated in the collector, making it easier to cool. A variety of collector designs have been developed to operate at high voltage, as power densities have increased. Special insulation techniques, including gas filled collectors, are available for this purpose. Multi-state depressed collectors are also provided for applications where maximization of efficiency is the major criterion.

AM to PM Conversion—Amplitude modulation/phase modulation conversion, or AM/PM conversion, is defined as the change in phase angle of the output RF voltage produced by variations in input signal level, and is usually expressed in degrees/dB. Dynamic measurements are usually made of this parameter and the results given at a specified output power level. The slope of the curve relating output phase in degrees to drive level in dB is nearly zero for very small signals, but begins to increase when the input signal increases to about 18 to 20 dB below the level necessary to saturate the tube. Traveling wave tube non-linearities increase with output power level, and at output powers that are about 6 to 10 dB below saturation, this relationship becomes more pronounced. These features are illustrated in Figure 9.9.

The principal reason that AM/PM conversion occurs in a traveling wave tube is that the average beam velocity decreases as the input signal is amplified, due to the energy exchange between the beam and the growing RF wave. This decrease creates the same effect as if the beam were constant and the electrical length of the tube were increased. This phenomenon makes parameters like phase variable functions of RF input signal level.

Noise Figure—A traveling wave tube, like any other electron tube, generates some noise, primarily due to variations in the current density of the electron beam, electron velocity fluctuations, and electron interception on tube elements not associated with the interaction process.

Noise figure, usually expressed in dB, is defined as the ratio of the signal to noise ratio on the input signal to the signal-to-noise ratio on the output signal.

Dynamic Range—Dynamic range is an important factor in the selection of low noise traveling wave tubes that can be defined for both linear and overdriven operation conditions. Referring to Figure 9.5(A), the limit of linear operation is defined as the point where an increase in the drive signal causes a gain compression of 1 dB. Dynamic range for linear operation can then be defined as the ratio of this maximum signal level to the reference noise level of the tube expressed in dB. A 2 to 4 GHz traveling wave tube with a 5 dB noise figure has a typical linear dynamic, range of 85 dB for a bandwidth of 1 MHz.

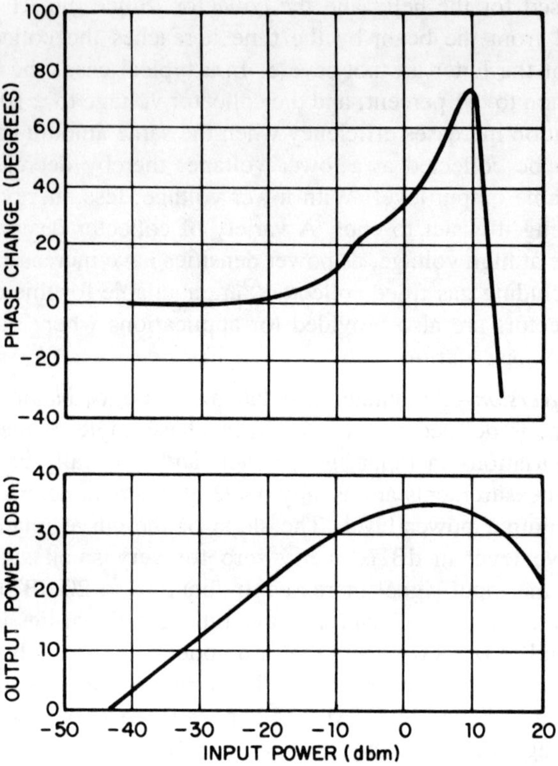

Fig. 9.9 Phase and output power *versus* input power.

The dynamic range from the linear region to full saturation for a low noise tube is typically 7 dB. Low noise traveling wave tubes are also used above saturation by taking advantage of their limiting action under overdriven conditions. They can be designed for such applications with input drive levels that vary as much as 10 to 25 dB above saturation with almost no increase in power being applied to the following stage.

9.1.6 TWT Applications

As seen in Chapter Four, a device (or component) is utterly useless unless specific applications can be listed for it.

One wide application of TWTs is in *electronic countermeasures (ECM)* systems. Active ECM systems are designed to receive a wideband of input radar frequencies and retransmit erroneous range information. They must perform effectively over a broad range of frequencies, input signal levels, and pulse waveforms. The traveling wave tube is ideally tailored to such microwave system designs by its inherent ability to provide the highest currently available gain-bandwidth product over octave bandwidths.

Other military applications of TWTs include phased-array radars, drone vehicle guidance and control, and fire control. Its low-noise capability makes it attractive for long-range communications systems where optimum weak signal amplification is critical. Additional benefits include high-power handling capability, high efficiency, and linear increase in phase shift with frequency or constant time delay.

Traveling wave tubes can also perform certain information processing functions—amplitude, pulse, and phase modulation—and certain signal processing functions—serrodyning, frequency multiplication, limiting, automatic gain control, *et cetera*. Also, certain specialized techniques are possible, such as very short pulse generation and wide-band oscillators using feedback circuits of high Q and wide tuning ability.

Serrodyning has not yet been covered and may be causing some concern. A basic explanation of its operation is as follows:

> The electron velocity in the beam may be varied by making small changes in the helix voltage. This effect disturbs the RF wave and the RF circuit and changes the phase of the output signal with respect to the input signal. Many low power helix traveling wave tubes are furnished with isolated helix circuits that allow sawtooth voltage modulation that effects a 360 degree phase shift in the RF signal. This produces what is the equivalent of a single sideband or Doppler pulse modulation.

Serrodyning finds its principal use in simulating a shift in frequency, as occurs in Doppler radar systems. It also permits a received signal entering a system to be retransmitted at a different frequency in relay stations and electronic scanning of an antenna.

9.1.7 TWT Data Sheet

In Chapter Four each component is explained by the use of a typical data sheet. The same method is now used to tie all of the information on TWTs together. A typical data sheet supplied by a vendor for a C-Band (4 to 8 GHz) TWT is shown in Figure 9.10, while Figure 9.11 shows two plots that apply to the data offered.

To understand what the figures mean, consider the fact that the tube is specified as follows:

Frequency	4–8 GHz
Power output	2 W (+33 dBm)
Minimum small signal gain	30 dB
Gain variation	8 dB (maximum)

TEST DATA

TUBE TYPE TWT-1 SERIAL NUMBER 101
HEATER VOLTAGE 6.3 (VOLTS) HEATER CURRENT 0.8 (AMPS)
GRID VOLTAGE 34.0 (VOLTS) HELIX CURRENT 1.8 (MA)
CATHODE VOLTAGE 1900 (VOLTS) CATHODE CURRENT 24 (MA)

RF PERFORMANCE	MIN	MAX	
SMALL SIGNAL GAIN	30DB		SEE FIG. 9.11
GAIN VARIATION		8DB	SEE FIG. 9.11
GAIN DEPRESSION		1DB	ΔDB .40 ΔDB .10 ΔDB .80
POWER OUT	+33DBM		SEE FIG. 9.11
VSWR (INPUT)		2.5:1	<2.5:1
VSWR (OUTPUT)		2.5:1	<2.5:1
GAIN STABILITY		NO OSCILLATION	F_1 OK F_2 OK F_3 OK
INSERTION LOSS	50DB		>50DB
GRID CUT OFF	80DB		>80DB

WIRING COLOR CODE

BLACK - ANODE
YELLOW - CATHODE HEATER
BROWN - HEATER
GREEN - GRID
WHITE - (COAX) HELIX

Fig. 9.10 TWT data sheet.

These are the important specifications for the TWT and the figures prove that it meets every one quite easily. This data sheet typifies manufacturers' specifications on TWT's. Usually offered are a printed form (Figure 9.10) and actual X-Y plots of the small signal gain and power output (Figure 9.11). By using this sheet and the terms defined in Section 9.1.5, a good working knowledge of the TWT in the real and practical world can be obtained.

Figure 9.12 shows some TWTs that are available today.

9.2 KLYSTRONS

The *klystron,* in the microwave region, performs the same functions that ordinary vacuum tubes do at conventional radio frequencies; that is, it generates, receives, and amplifies radio (microwave) signals. The reason that klystrons were in-

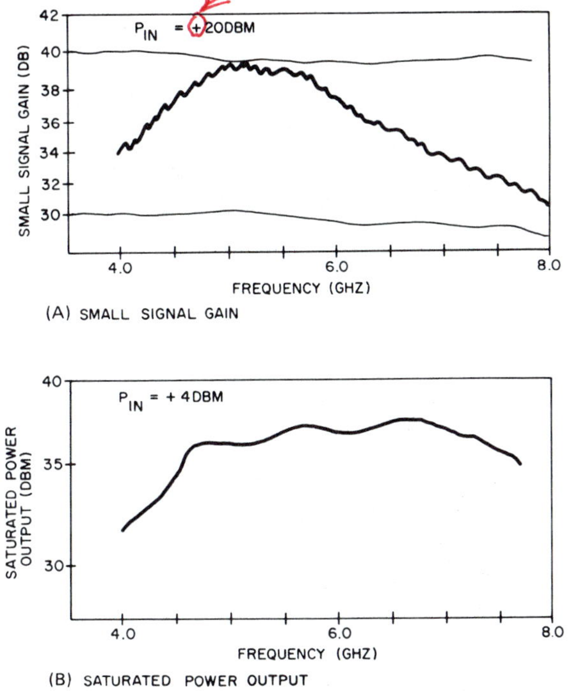

Fig. 9.11 Data plots of a TWT.

vented is that the ordinary triode tube cannot handle microwaves effectively. The following is a basic review of how a triode operates.

The triode tube consists of three elements: a filament that emits a stream of electronics, a grid that stands in the path of the stream, and a plate that attracts the electrons and catches them after they pass through the grid. The grid acts like a valve, opening or closing to the passage of electrons according to the voltage found on it. (The vacuum tube itself is actually referred to as a *valve* in England.) Energy intercepted by an antenna, for example, comes to the grid as a weak alternating current oscillating at the frequency of the RF energy, modulating the flow of the electrons crossing the tube. The electron stream then delivers at the plate an alternating current which reproduces, with great amplification, the weak input signal on the grid.

It takes an electron about a billionth of a second to cross a tube. This transit time is short compared to a cycle of a long RF wave (around one millionth of a second); hence the electron is slowed or speeded up by the voltage on the grid, and the flow of electrons therefore can follow the voltage fluctuations. In the case of microwaves, however, the oscillations are so rapid (that is, the cycle is so short) that the voltage on the grid may go through several complete oscillations during an

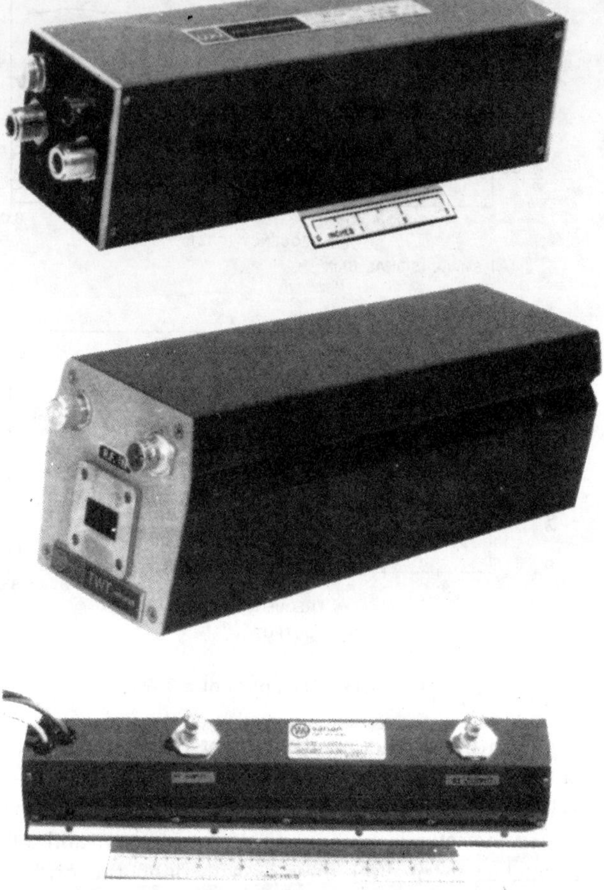

Fig. 9.12 Traveling wave tube.

electron's travel cross the tube. In other words, the grid voltage changes too fast and produces complete chaos among the electrons. Several types of tubes are available which overcome this transit time difficulty, such as the TWT and BWO, as well as magnetrons (covered in the next section). Yet, of those tube types, none is quite as simple or versatile as the klystron, with the possible exception of the TWT.

The klystron tube makes a virtue of the very factor which defeats the triode—the transit time of the electrons. It "modulates" the velocity of electrons so that as they travel through the tube, they sort themselves into groups and arrive at their destination in bunches, thus delivering an oscillating current of controlled frequency.

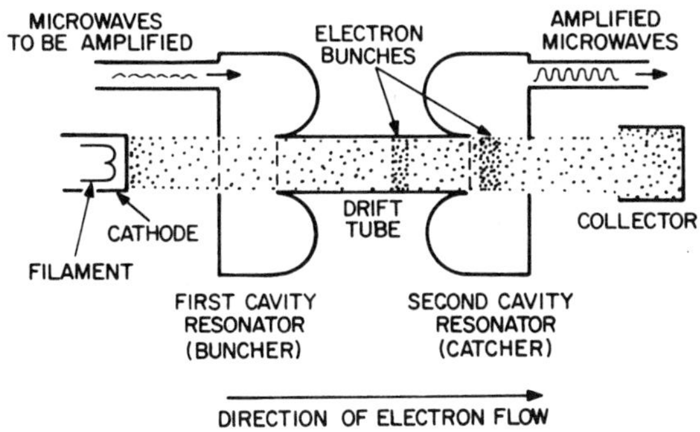

Fig. 9.13 Two-cavity klystron.

Two basic types of klystrons are *multi-cavity* and *reflex*. As the name implies, the first variety has more than one cavity, while the latter has only one. The multi-cavity klystron can be best understood in its simplest form—with only two cavities.

9.2.1 Two-Cavity Klystron

The two-cavity klystron (Figure 9.13) is the oldest form of what are termed "velocity variation" devices. Its operation is outlined below:

1. The filament is heated by an electric current, causing the cathode to emit electrons (shown as dots).
2. The electrons are pulled away from the cathode toward the other end of the tube by the positive charge on the two cavity resonators, drift tube, and collector.
3. The electrons pass through the first cavity resonator, called the buncher—a metal-enclosed chamber or cavity. Here the electrons encounter the microwaves to be amplified.
4. These microwaves in the buncher interact with the electrons, speeding some of them up while slowing others down.
5. In the drift tube, the faster electrons catch up with the slower ones, and bunches of electrons are formed.
6. Where the bunching becomes greatest, there is placed a second cavity resonator, called the catcher—a metal-enclosed chamber like the buncher.
7. The catcher "catches" some of the energy in the electron bunches; that is, the bunches give up energy to the catcher by exciting microwaves there. The oscillations of these microwaves are stronger (have larger amplitude) than those of the microwaves in the buncher. Thus, amplification has taken place.
8. The amplified microwaves excited in the catcher are removed and sent to wherever they are to be used.

9. In the meantime, the electrons continue to travel through the catcher, are caught by the collector, where their remaining energy is dissipated in heat. The same tube may also be used to generate microwaves (oscillations), by feeding in the proper phase part of the power from the second cavity, back to the first, using sufficient amplitude to overcome the losses in the system. The initial start of oscillation is caused by random fluctuations of the electron beam current and by noise voltages which generate fields in the cavity.

In the above explanation the cavities are the crux of the device. For present purposes, each can be regarded as a tank (parallel resonant) circuit. A cavity of given size resonates to waves of a given frequency. Thus, in the klystron the cavity is made small enough to resonate at microwave frequencies.

One might wonder why the electrons traveling from the cathode to the first cavity resonator (buncher) are not absorbed by the cavity since the latter has a positive charge which attracts the electrons. To understand why, refer to Figure 9.14.

The resonator has a hole all the way through it, through which the electrons can travel. They enter and leave the cavity resonator by passing through a fine wire screen on each side of the hole, known as a grid (shown as vertical dotted lines in the diagram in Figure 9.13.) The grids are very finely meshed and do not appreciably hinder the flow of electrons, which are so tiny that most pass through the grid just as very fine-grained sand passes through a coarse screen.

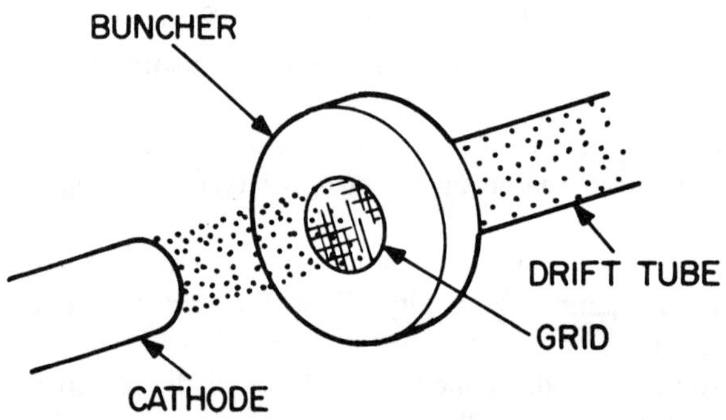

Fig. 9.14 Buncher and grid construction of a klystron.

Now the positive charge on the grid of the first resonator attracts the electrons; by the time they reach the first resonator they are traveling so fast that most of them pass through without being caught.

The two-cavity klystron, as mentioned before, can be used as either an amplifier (its most common use) or as an oscillator (with limited applications). Although most multi-cavity (two-cavity) klystrons are used as high power amplifiers, there are still a large number employed in the medium power range as both amplifiers and oscillators. Almost all of these medium power tubes are at X- or Ku-band. To illustrate this point, here are some ranges of characteristics for medium power klystrons.

Amplifiers (CW)	
Power	2–250 W
Bandwidth	0.1–0.5%
Gain	5–50 dB
Beam voltage	750–4500 V
Amplifiers (Pulse)	
Peak power	500–1500 W
Bandwidth	0.1–0.5%
Gain	5–55 dB
Beam voltage	5–10 kV
Oscillators	
Power	1–250 W
Tuning range	0–250 MHz
Beam voltage	750–14,000 V

9.2.2 Reflex Klystron

Unlike the two-cavity type, the reflex klystron can be used only as an oscillator. It has been both the most widely used source of microwave power in laboratory applications for years, and an ideal microwave local oscillator for very sensitive receivers (especially in ECM work).

The reflex klystron (Figure 9.15) differs from the two-cavity klystron in that is has a reflector and only *one* cavity. The single cavity resonator performs the bunching and catching process; feedback is developed by reflecting the electron beam so that is passes through the resonator a second time.

The electrons pass through the resonant cavity and interact with the very weak microwaves created by the random motion of the electrons already present, causing them to bunch. From this point they are turned around and returned to the cavity by the reflector, with its negative voltage.

In the cavity, the bunched electrons strengthen the very weak microwaves. After giving the latter a "boost," the electrons are caught by the cavity, return to the cathode (through an external path) and repeat the entire process. Each time they return to the cavity they make the microwave oscillations larger until, as with the oscillations of a swing, it can no longer grow in size. In a very short time, the sufficient amplitude is reached, and the microwave energy can be taken at the cavity output port.

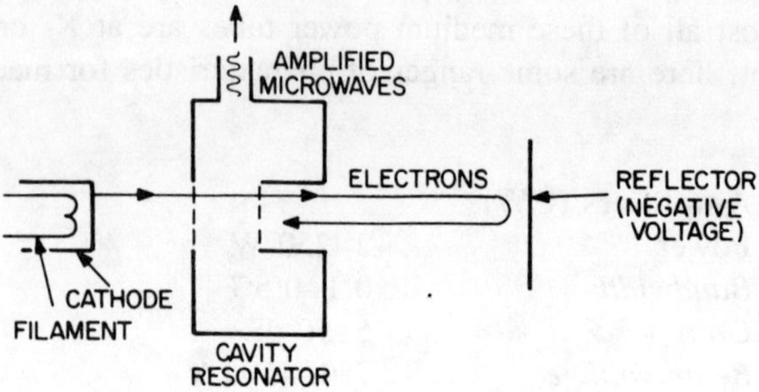

Fig. 9.15 The reflex klystron.

An area of prime importance in the operation of the klystrons is that of *bunching*.

To understand why bunching occurs in a reflex klystron, two facts must be remembered:

1. An electron that has been speeded up by the action of the RF field travels farther into the reflecting field and takes longer than the average time to return to the resonator.
2. An electron that has been slowed down by the action of the RF field travels less distance into the reflecting field and takes less than the average time to return to the resonator.

This phenomenon can be clearly seen in Figure 9.16—a velocity, or Applegate, diagram.

The electron that passes the resonator gap at time t_a is accelerated and takes longer to return than an electron leaving at time t_b, when the RF field is zero. Electrons that leave at time t_c, later in the cycle, require less than the average transit time. In consequence, all these electrons return to the resonator in a bunch at time t_r.

The arrival time, t_r, of a group of electrons returning to the resonator depends upon the physical dimensions of the klystron and both the acceleration and reflector voltages. (The lower the accelerator voltage, the shorter the distance traveled by the electrons into the reflecting field, and the shorter the arrival time. The lower the reflector voltage, the greater the distance and the longer the arrival time.)

The transit time of the electron with average velocity, leaving at time t_b, may correspond to any number of cycles of the radio frequency. But to sustain oscillations, the electron bunch must arrive back at the resonator during the time that the RF field is retarding the returning electrons. In this way, the electron velocity is reduced and some of the kinetic energy of the electrons is transferred into electromagnetic energy in the cavity resonator field.

To attain maximum energy transfer from the bunched electrons, maximum retardation is needed; it is achieved when the RF field is at the positive peak of the

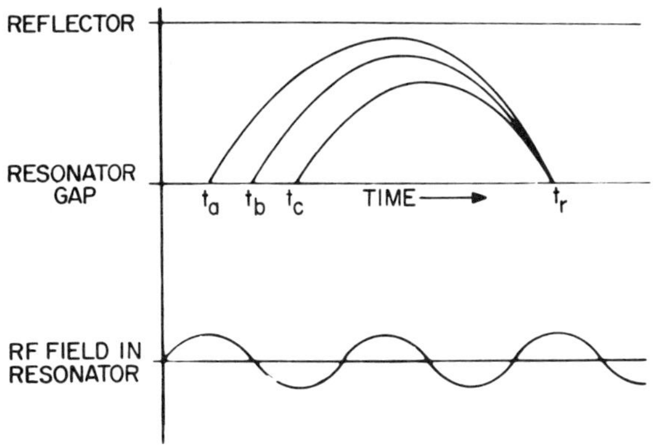

Fig. 9.16 Velocity (Applegate) diagram.

sine wave shown in Figure 9.16. At this time the electron going from the cathode encounters the maximum accelerating value in the gap while one returning from the reflector meets the maximum retarding value in the gap. Therefore, for maximum energy transfer, the transit time for an electron that enters the reflecting field with average velocity must correspond to one quarter cycle less than an integral number of cycles.

An alternate way of stating this condition is noted in Figure 9.16. The unaccelerated electrons at "b" must remain in the reflecting field for three-quarters of an RF cycle, or for an integral number of cycles plus three-quarters of one. This condition is designated one and three-quarters cycle distance on the diagram. The term "mode" is again used to describe a certain set of conditions, in this case refering to the repeller or reflector voltage settings required to keep the electrons in the reflector field for the designated ($n - {}^1/_4$) cycles. The term therefore defines the number of cycles of electron transit time N and is given by:

$$\left(N = n - \frac{1}{4}\right)$$

where n is a positive integer greater than zero. Klystron reflector voltage modes are designated three-quarter mode, one and three-quarter mode, *et cetera*.

The basic frequency of oscillation of the klystron is determined by its cavity dimensions, which can be varied by flexing a portion of the cavity wall or changing the spacing of the cavity grids. In addition, the oscillation frequency can be adjusted by the reflector voltage. The maximum power output point is referred to as the "top" of the mode. If the reflector voltage is set to any other voltage, the transit time is changed, and the klystron oscillates at a different frequency with decreased power output. Therefore, in order to obtain maximum power output at a specific frequency, it is necessary to adjust the cavity resonant frequency and the reflector voltage to obtain maximum power output for any fixed setting of other electrode potentials.

The variations of frequency and power output are plotted in Figure 9.17 as a function of reflector voltage. These curves can be obtained by plotting the power output as the reflector voltage is varied to successive values, or by sweeping the klystron reflector voltage with a sawtooth waveform and observing the rectified power output as displayed on an oscilloscope. Oscillations occur over the range of reflector voltage values, causing the electron bunches to return through the cavity grids in the proper phase to deliver energy to the cavity.

The points "a," "b," and "c" at the top of each mode represent the maximum power output point where the transit time is $(n - {}^{1}/_{4})$ cycles. Therefore, each of these points also represents the same frequency.

Consider a klystron oscillating with a reflector voltage indicated at "A" in Figure 9.17. If this value is made more negative, the electrons are retarded more and the transit time is made shorter. Therefore, the bunches are not properly formed and returned through the cavity too soon. Since bunching is not complete, the power delivered to the cavity is less, the resulting RF signal has less amplitude, and the power output is lowered. The klystron now oscillates at a higher frequency with less output power. Alternately, if the reflector voltage is made less negative than "A," the electrons travel farther into the reflector region and become separated by the time they arrive back at the cavity grids. The "debunching" again causes a decrease in RF amplitude, power output, and, in this case, frequency. The latter two stabilize at this new value. The power supplied to the cavity in the bunching process increases as the oscillation grows and continues to increase until the RF current reaches its maximum possible value, or optimum bunching point, where maximum power is supplied to the cavity. If the RF voltage increases to a larger value, the electron bunches are formed too soon, less power is delivered to the cavity, the RF current decreases, and the amplitude of the RF voltage is automatically decreased to the value where maximum power is again supplied.

The frequency sensitivity characteristics, shown in Figure 9.17, indicate that tuning sensitivity and tuning range are higher for the higher order modes. Frequency tuning sensitivity is described as the number of MHz/volt of reflection voltage change. The klystron data sheet should give information on the tuning range of the device and the range of the reflector voltage.

One term that should be covered before the topic of klystrons is completed is *hysteresis*. As applied to the klystron, the term refers to the dependence of power output and frequency upon the direction from which the reflector potential is made to approach a given voltage. Recall from the discussion on YIG filters how hysteresis is applied to them. This process is the same as in the klystron tuning procedure. When increasing the reflector voltage through a mode, the amplitude and frequency may exhibit smooth tuning characteristics up to a certain critical value of reflector

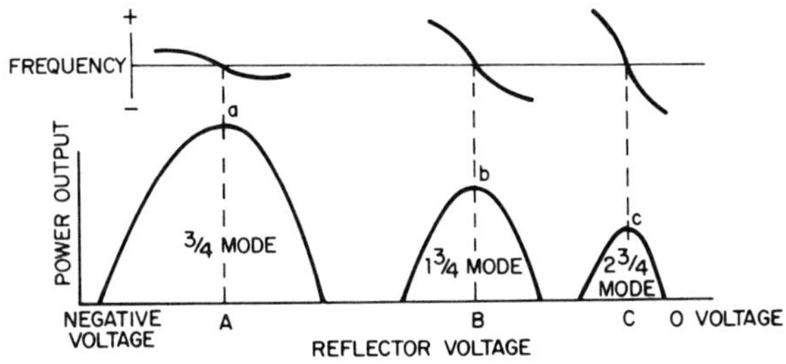

Fig. 9.17 Power and frequency *versus* reflex voltage.

voltage when a sudden change may be encountered. Several such changes may occur on one side, or even on both sides, of the mode. If the reflector voltage is now varied in a decreasing direction, some of the discontinuities may not occur at all, or amplitude and frequency changes may occur at different points. Also, the maximum power output may occur at a different reflector voltage. Thus, great care should be taken when tuning a klystron, making sure that both the mode and frequency are correct. A glance at the manufacturer's instructions saves much time later.

A few applications of reflex klystrons are listed below. The list is not complete by any means, but illustrates general applications.

1. *Doppler navigation* (missiles, aircraft, *et cetera*)—reflex klystron used as a low power transmitter and receiver.
2. *Police doppler radar*—reflex klystron used as a low power transmitter.
3. *Radar installations*—reflex klystron used as a local oscillator in transmitters and receivers.
4. *Communications relay systems*—reflex klystron used as a low power transmitter.
5. *Laboratory signal generator.*
6. *Laboratory test equipment.*
7. *Pump tubes for parametric amplifiers.*

Below is a glossary of klystron terms that may prove to be very useful.

Buncher—The first cavity resonator in a two-cavity klystron; a metal-enclosed chamber or cavity where the microwaves interact with the electrons, causing the latter to bunch further down the tube.

Catcher—The second cavity resonator in a two-cavity klystron; a metal-enclosed chamber or cavity where the electron bunches give up energy.

Cathode—A specially-prepared metallic surface, which when heated, gives off electrons.

Cavity resonator—A metal-enclosed chamber or cavity; a part of every klystron.

Collector—In a two-cavity klystron, a piece of metal that catches the electrons after they have traveled the length of the tube.

Drift tube—In a two-cavity klystron, the space between the two cavity resonators where the electrons form bunches.

Multi-cavity klystron—A klystron having more than one cavity resonator.

Reflector—In a reflex klystron, a piece of metal with a negative voltage, which repels or reflects the electrons back into the cavity.

Reflex klystron—A one-cavity klystron, widely used as an oscillator.

Two-cavity klystron—A klystron having two cavity resonators. (Figure 9.18 shows klystrons that are available today.)

9.3 MAGNETRONS

The conventional magnetron was developed during World War II. It is the cheapest, smallest, and lightest source of RF power than can be used in high power radar. Radar designers have been using this tube for the past thirty years during which time the basic design has changed very little.

The limitations of the conventional magnetron were known early in its application; while much effort went into improving its performance and capability, research continued to find a device capable of longer life, improved efficiency, and superior stability. The development of the *coaxial magnetron* is the result of this research.

In order to understand the capabilities and limitations of magnetrons, both the *conventional* and *coaxial* types are covered.

9.3.1 The Conventional Magnetron

The magnetron is a type of diode in which a magnetic field is set up perpendicular to the electric field existing between the cathode and the plate. This field is provided by an electromagnet in some instances, and by a permanent magnet in others. The three basic types of magnetron oscillators are identified on the basis of the means by which energy is transferred to the RF field.

1. The *cyclotron-frequency* magnetron operates by virtue of resonance between the period of RF oscillation and the rotational motion of the electron.
2. The *negative-resistance* or *split-anode* magnetron operates on the principle of a static negative resistance characteristic between the anode sections.
3. The *traveling wave* or *resonant-cavity* magnetron depends upon the mean velocity of the electron being synchronized with the velocity of the traveling-wave component of the RF interaction field between the cathode and anode.

Fig. 9.18 Klystrons.

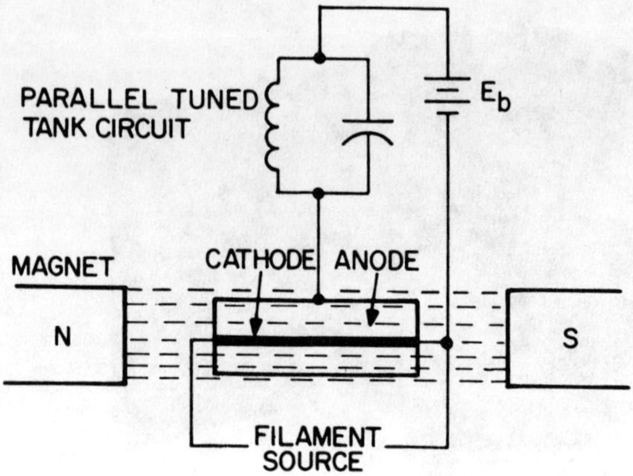

Fig. 9.19 Cyclotron-frequency magnetron.

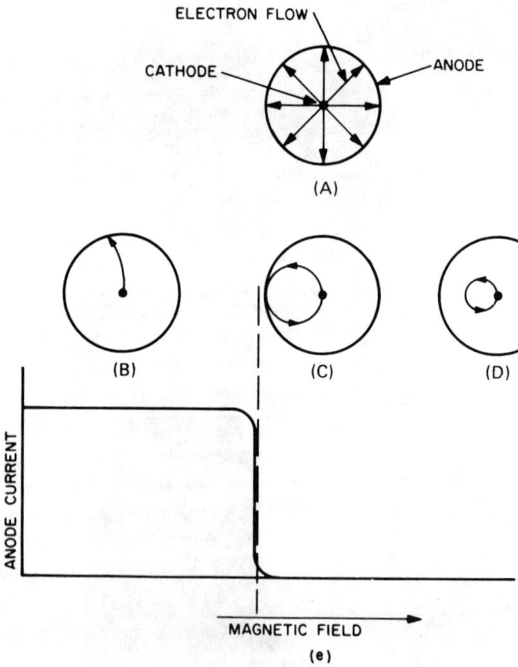

Fig. 9.20 Effect of the magnetic field on electrons.

9.3.1.1 The Cyclotron-Frequency Magnetron

The simplest form of cyclotron-frequency magnetron is shown in Figure 9.19. The diode, formed by the cathode cylinder inside the anode cylinder, is positioned between the poles of the magnet (permanent or electromagnet) so that the magnetic lines are parallel to the axis of the anode-cathode combination.

With the basic construction of the tube in mind, the effect of the magnet on overall operation is considered. Figure 9.20 shows how the magnetic field effects the electrons traveling from the cathode to the anode. When no magnetic field is present, heating the cathode results in a uniform and direct flow of electrons to the anode (plate), as shown in Figure 9.20(A). However, as a magnetic field is applied and is continually increased, the results are as shown in Figure 9.20(B), (C), and (D). In Figure 9.20(B) the magnetic field has increased to a point where the electron shown proceeds to the plate in a curved rather than in a direct path. In Figure 9.20(C), the magnetic field has reached a value great enough to cause the electron to barely graze the anode, known as the *critical value* of the magnetic field. Any increase in the field from this value results in electrons completely missing the anode as shown in Figure 9.20(D). At the *critical* or *cut-off* value of the magnetic field, there is an abrupt decrease in anode current (Figure 9.20(E)). When the magnetron is adjusted to the cut-off or critical field value and the electrons fail to reach the plate, the magnetrons can produce oscillations by virtue of the currents induced electrostatically by the moving electrons. The frequency of these oscillations is determined by the time it takes the electrons to travel from the cathode toward the plate and back again.

9.3.1.2 Negative-Resistance Magnetron

The *negative-resistance magnetron* has the anode split parallel to the axis in two halves, and for this reason is sometimes referred to as the *split-anode magnetron,* a variation of the basic type, but operating at a higher frequency and capable of greater output. Its general construction, except for the split anode, is basically the same, as well. The half plates are operated at different potentials to provide an electron motion as shown in Figure 9.21. The electron leaving the cathode and moving toward the high potential plate is deflected by the magnetic field at a certain radius of curvature. After passing the split between the two plates, the electron enters the electrostatic field set up by the lower potential plate. Here the magnetic field has more effect on the electron, which is deflected at a smaller radius of curvature. The electron then continues to make a series of loops through the magnetic and electric fields until it finally falls on the low potential plate. During the time that the electrons

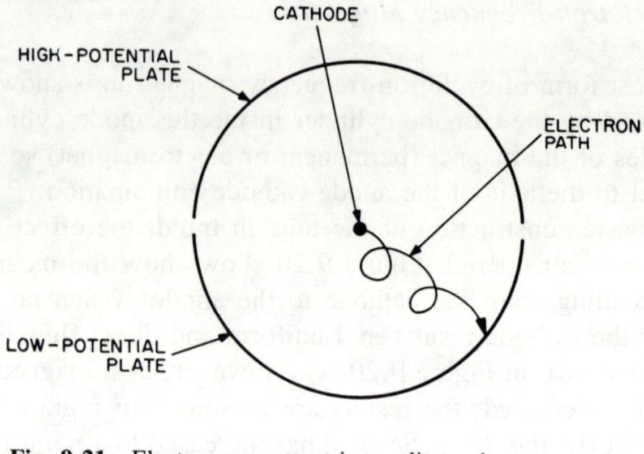

Fig. 9.21 Electron movement in a split-anode magnetron.

complete the several loops, the alternating RF field does not change significantly. As mentioned, the electrons terminate on the low potential plate and thus the negative resistance characteristic is produced.

Oscillations can be started by applying the proper value of magnetic field to the tube. The required figure is somewhat beyond the critical value which, for the split-anode tube, is the field required to cause all electrons to miss the plate when its halves are operating at the same potential.

Since a very concentrated magnetic field is required for the negative-resistance magnetron oscillator, the length of the tube plate is limited to a few centimeters for a magnet of reasonable dimensions. In addition, a small-diameter tube is required to make the magnetron operate efficiently at the higher frequencies. Both of these facts greatly limit the plate size and therefore impose serious limitation on the permissible plate-heat dissipation. For this reason, a heavy-walled plate is used to increase the radiating properties of the tube. To obtain still greater dissipation, tubes with high outputs use an artificial cooling method such as forced air or water cooling.

9.3.1.3 Traveling-Wave Magnetron

The *traveling-wave* or *multi-cavity* magnetron consists of a cylindrical anode structure containing a number of equally spaced cavity resonators. In this type of magnetron a permanent magnet is usually used. The cathode anode arrangement of an eight-cavity magnetron is shown in Figure 9.22. An electron traveling between the cathode and anode of such a tube is subject to accelerations and decelerations, resulting in a spiral pattern going completely around the anode, with the electron effectd by all eight cavities.

The definite phase relationship of the fields appearing across adjacent slots is used as a means of defining different *modes* of oscillation. the total phase shift in

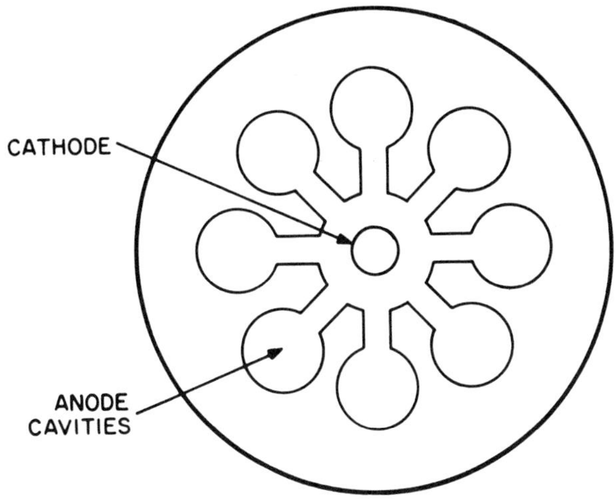

Fig. 9.22 Eight-cavity magnetron.

such a system of resonators must be a multiple of 2π radians (one complete revolution or 360 degrees). The phase difference between adjacent resonators for a system of N resonators (eight in this case) may assume values of $2\pi\ n/N$, where $n = 0, 1, 2, \ldots N/z$. The phase angle is π for the N/z mode, called the *pi mode*.

Two identical resonant cavities resonate at two frequencies when the cavities are tightly coupled together. The two resonant frequencies lie above and below that of the individual resonators. As in the case of two coupled resonators, N resonators have N modes of oscillation because of the effect of mutual coupling. A common means of separating the π mode from adjacent modes is by a method called strapping. The straps consist of wires of either circular or rectangular cross section connected to alternate segments of the anode block, usually recessed into the resonator structure, as shown in Figure 9.23. In the π mode, each strap joins anode segments at the same voltage, and is capacitively coupled to the segment it passes over. This capacitance, effectively in parallel with the resonator, reduces the frequency of the π mode. For lower modes, the strap causes a decrease in the capacitance because of the different voltages existing at the end of the straps. There is also inductance associated with the strap. The combined effects increase the resonant frequency of the lower modes, thus increasing the separation between them.

The rising-sun type resonator of Figure 9.24 obtains division of modes into two groups by employing an equal number of small and large cavities.

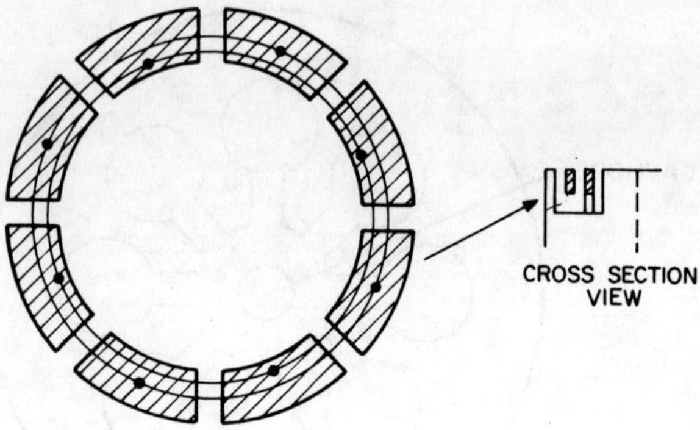

Fig. 9.23 Cavity strapping.

Fig. 9.24 Rising-sun type of resonator system.

9.3.2 The Coaxial Magnetron

What are the differences between the conventional magnetron (discussed above) and the coaxial magnetron? At first glance similarities abound—both have a similar slow wave structure, a cylindrical cathode structure (usually inside the anode), and a magnetic circuit producing an axial magnetic field. Furthermore, they are almost identical in external appearance.

The reasons for the improved performance of the coaxial magnetron become apparent upon examination of its internal design. The coaxial magnetron is so named because of the large coaxial cavity coupled to the inner resonators of the anode block. Such a design concept leads to a high Q RF oscillator with improved performance in areas such as frequency stability, freedom from arcing, lifetime and better reliability (lower "mean time between failure)". Tuning can be accomplished easily over wide ranges with no performance degradation. With these, and other qualities, the coaxial magnetron is able to overcome most of the difficulties encountered in the conventional tubes. It is actually an extension of the conventional magnetron structure (cathode surrounded by an anode) by the addition of a third element, a coaxial cavity that surrounds the anode forming its inner wall. Its structure is shown in Figure 9.25.

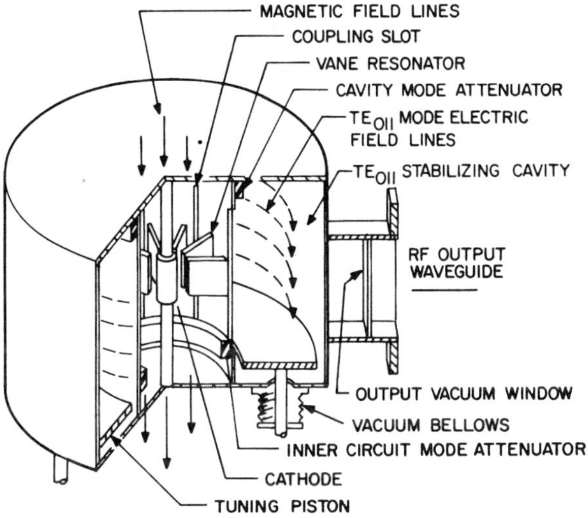

Fig. 9.25 Coaxial magnetron.

Just as the original concept of strapping the anode is considered a method of tying together points of the structure of the same potential for the π mode, the coaxial magnetron concept ensures that the RF currents at the backs of the cavities all are phase correctly for π mode operation.

In contrast to the conventional magnetron, where the anode wave structure is the frequency determining element, the frequency of operation of the coaxial magnetron is determined by the resonant frequency of the surrounding coaxial cavity, with the resonators being quarter wave transformers changing the low impedances at the cylinder walls to a higher impedance necessary for good interaction efficiency.

With the above type of structure π mode operation can be maintained with many more resonators than is possible with either strap or rising-sun anodes because, by the very nature of the cavity field patterns, highly selective damping of unwanted modes is feasible. Up to 144 cavities have been used in the coaxial magnetron; in the normal case, though, this figure usually does not exceed 26. Thus, the small resonators remain the same size as the anode diameter, and hence the cathode diameter and anode-cathode spacing can be increased to very large values, resulting in lower cathode current density, lower anode power densities, longer life, and less arcing.

In a conventional magnetron, the output is coupled to one resonator only so that the extraction of power from the other cavities depends on their mutual coupling and the capacitive coupling between the vane tips and the cathode. In a coaxial magnetron the loading is much more symmetrical as every cavity is coupled to the coaxial cavity from which the power is taken.

Since the frequency of the magnetron is determined by the coaxial cavity, the simplicity of the device becomes very apparent. Thus, a tunable coaxial magnetron can be built, the performance of which equals that of a fixed frequency version—a condition difficult to achieve with a conventional tunable magnetron because the mechanism for tuning is much more complex, possibly leading to greater RF losses and arcing.

Thus far, advantages of the coaxial magnetron over the conventional tube have been discussed. These traits are neatly compiled in the following list:

1. Positive control of operating mode.
2. Other modes are absorbed internally, thus no dependence on load characteristics for mode control.
3. Very high Q, giving low pulling, low pushing, a good spectrum and high efficiency.
4. Absence of straps in the structure—hence, lower fields, less loss and less chance of arcing.
5. They can be designed with a large number of resonators—hence a large cathode giving lower current density, lower cathode temperature and longer life.
6. Any evaporated barium has very small effect on tube frequency.
7. Large internal volume giving good storage life.
8. Ease of tuning with no degradation of magnetron performance.

The only real disadvantage of the tube is its size—it is physically larger and heavier than conventional magnetrons. If space and weight are no problem for the application, coaxial magnetrons are desirable.

Two terms listed in the third advantage above, *pushing* and *pulling* factors, may not be familiar at this point. Thus, individual explanations are offered below.

Pushing Factor

The spokes of the electron cloud in the interaction region appear as a current modulated beam reacting with multiple gaps of a resonant structure. Such a beam appears as a reactive load across the resonator, changing its resonant frequency by a small amount, as a function of the fields in the interaction region and of the beam current density.

The variation of frequency with the change of beam current is called the pushing factor of the tube. Both the reduced interaction fields and the stabilizing influence of the high Q cavity reduce the pushing factor of coaxial magnetrons as compared with conventional ones. For example, the specification maximums for the conventional S-band device is 200 kHz/A whereas the coaxial device is limited to 15 kHz/A maximum.

Improved spectrum, use of narrowband receivers, and range enhancement (increased operating range) are all derived from the great improvement in the pushing factor shown above. The low pushing factor also makes the coaxial magnetron 5 to 10 times more tolerant of power supply ripple.

Pulling Factor

The pulling factor of a coaxial magnetron is the change of frequency of a magnetron caused by a mismatch in the output load which reflects back a reactive element loading the resonant cavity. The phase of the reflected power changes the value of this reactive element, and the pulling factor is defined as the total oscillating frequency excursion when a mismatch, usually of VSWR 1.5, is varied in phase through 180 degrees.

The high Q of the resonant cavity in a coaxial magnetron reduces the optimum coupling to the load and hence reduces the reactive effects of the mismatch. Typical reductions of pulling factor for a VSWR 1.5 at S-band are from 20 MHz for a conventional magnetron, to 2.5 MHz for a coaxial one.

As mentioned before, the coaxial magnetron can be tuned easily; in fact, the method of tuning used gives the device its particular name. *Frequency agile magnetrons* (or *dither-tuned magnetrons*) and *accu-tune magnetrons* are coaxial types with different methods of tuning.

A frequency agile magnetron differs from the basic tunable form only in terms of tuning speed. This type of tube requires a velocity fast enough to produce a required pulse-to-pulse frequency difference, but not so fast as to produce a significant frequency change during the time of a single pulse. The proper speed of tuning is achieved by attaching the tuning rod from a standard tunable magnetron to a motor through a crankshaft-type coupling. Rotational motion of the motor shaft is thus translated into a linear motion of the tuning rod and produces sinusoidal variations in the RF output frequency.

The accu-tune magnetron utilizes the identical agile tuner configuration discussed above, except the tuner motor is replaced with a servo motor with two modes of operation. The first is as a typical induction motor (with the control voltage input disconnected); in the second, the motor shaft holds a particular angular position proportional to the magnitude of the control voltage applied. Therefore, the tube can be operated either as a dithered frequency magnetron with a continuously varying frequency, or any specific operating frequency within the band can be electrically selected. The advantage offered by this type of magnetron over the older tunable tubes is that frequency can be changed very rapidly from one fixed point to another by simply adjusting the control input of the servo motor. As an example, the total time required to tune from F_1 to F_1 + 400 MHz is less than 0.06 seconds.

Fig. 9.26 Magnetrons.

The great value of magnetrons in microwave systems is now realized, particularly in their application to radar systems. With the widespread use of the new coaxial types, it will be a long time before high power radar systems do not use magnetrons.

(Figure 9.26 shows magnetrons that are available.)

9.4 CROSSED-FIELD AMPLIFIERS

The *crossed-field amplifier (CFA)* is a microwave power amplifier which, for a given peak output power, is small, light, efficient, and operates at relatively low voltage. These features make it particularly suited for applications in which the transmitter must be either airborne (or spaceborne), transportable, mobile, or portable. The CFA, like the magnetron, is capable of generating very high peak powers. It operates, however, with a power gain of 10 to 20 dB, which is modest when compared with some traveling wave tubes (TWTs) and klystrons. Its average power handling capability is also less than that of many of the TWTs and klystrons, but is generally adequate for those classes of systems in which the CFAs advantages of mobility are important.

Conceived in the early 1950s, the CFA was first applied late in that decade and has since been developed at a rapid rate, primarily for military systems. However, CFAs are used in non-military applications, such as air route surveillance radar. Because of its relatively recent origin, the CFA has not yet been widely deployed in operating systems. The large number of benefits it offers to the system design and performance, together with the growing complexity of missions which must be performed by new radars, assures the CFA of a bright future.

The CFA is a tube which might best be described as part magnetron and part traveling wave tube. It is like a magnetron because it utilizes the same type of electronic interaction (crossed electric and magnetic fields); as a result, it has many of the magnetron's characteristics. It is like a TWT because electronic interaction occurs with a traveling wave and, of course, it is an amplifier. (In a magnetron the electronic interaction is with a standing wave.) Power can be generated with high efficiency (for the same reasons a magnetron can do so) and at comparable voltage levels.

In fact, many CFAs look quite like a magnetron, except for the addition of an input port. Figure 9.27 shows actual CFAs.

The device derives its name from the fact that electronic interaction occurs in a region of crossed electric and magnetic fields, distinguishing the CFA from the conventional traveling wave tube. Otherwise they are made up of the same basic elements, each having a slow wave circuit, an input-output system, and an electronic system.

The slow wave circuit, or delay line, is a periodic structure which has the circuit characteristics of a bandpass filter. It must be capable of propagating RF

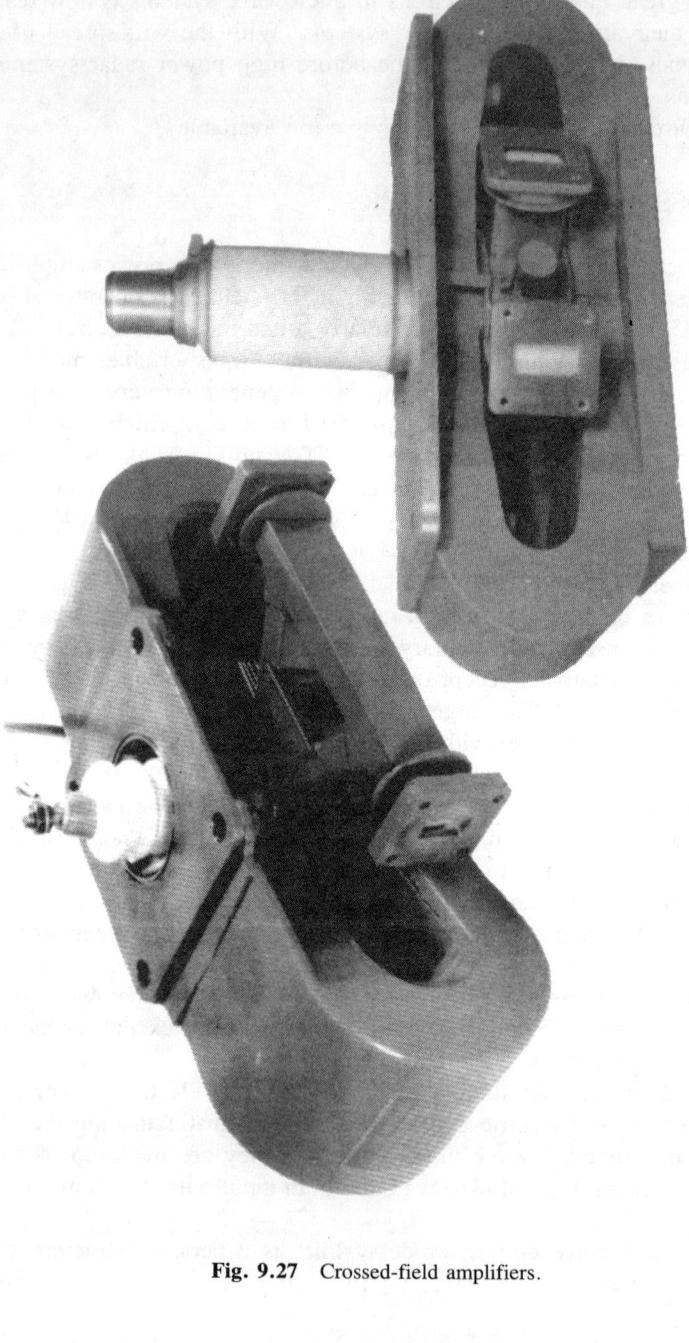

Fig. 9.27 Crossed-field amplifiers.

energy over the frequency range of interest, while simultaneously providing fringing electric field lines to interact with electrons. These fields must have a phase velocity approximately equal to the velocity of the electronic stream.

The input-output system provides impedance transformation between the RF transmission line system external to the amplifier and the slow wave circuit itself. These transformations are called circuit matches and the bandwidth necessary for a low VSWR often determines the useful bandwidth of the CFA itself.

The electronic system generates electrons, confines them to the interaction area, and collects them when they are spent. Some crossed-field, or *injection beam,* amplifiers have electron guns which inject a beam of electrons into the interaction area. Others, called *distributed emission* amplifiers, have a relatively large cathode which extends the entire length of the slow wave circuit and along which electrons are generated. Our discussion is confined to the latter variety, represented schematically in Figure 9.28(A). The cathode is also called the sole; distributed emission amplifiers are sometimes referred to as *emitting sole* amplifiers. The words "cathode" and "sole" are used interchangeably. The magnetic field is shown normal to the plane in the sketch of Figure 9.28(A), and the cathode during operation is negative with respect to the slow wave circuit. The magnetic field in this type of amplifier provides focusing forces for the electron stream and establishes the crossed-fields which give the amplifier its desirable characteristics. In a TWT, the magnetic field is used only to focus the electron beam.

The distributed emission amplifier can be arranged in a number of ways, in either a linear or circular format. In the former, the slow wave circuit is set in a straight line; in the latter, the circuit is bent into a circle, as in Figure 9.28(B). Amplifiers made with the circular format may collect electrons at one end of the circuit, or the input and output sections may be brought close enough so that electrons from the output are permitted to reenter the interaction area at the input. Reentrancy is employed in many amplifiers to enhance efficiency; however, when it is employed, the reentering electrons may be modulated with information which could subsequently be amplified. This process is equivalent to providing an RF feedback which must be considered in determining the behavior of the amplifier. It is possible, however, to obtain reentrancy after demodulation, thereby eliminating feedback.

A schematic of the interaction region of a crossed-field amplifier is shown in Figure 9.29(A). An emitting cathode (or sole) is placed opposite an anode slow wave structure which propagates an RF wave at velocities considerably less than the speed of light. A magnetic field is supplied perpendicular to the plane of the paper, and a dc electric field is supplied between anode and cathode. Electrons entering the illustration from the left can have a drift velocity equal to the ratio of dc electric and magnetic fields. The drift velocity is adjusted to be approximately equal to the velocity of the wave along the circuit. In the presence of an RF drive wave on the circuit, electrons can either gain or lose energy, depending upon the direction of the

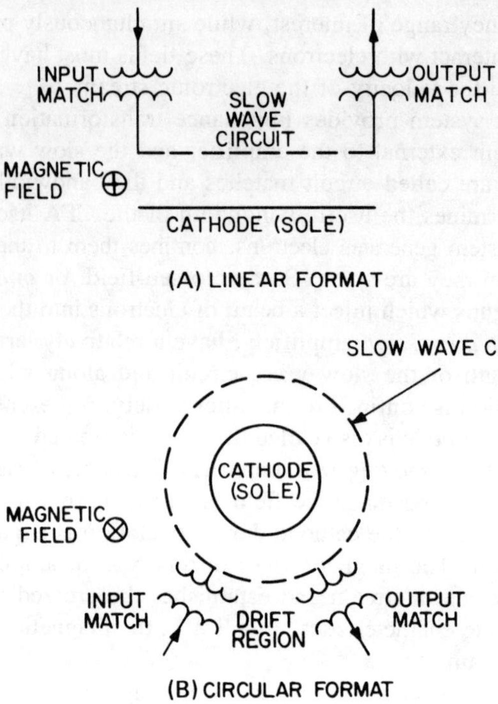

Fig. 9.28 Distributed emission amplifier.

RF field in which they are found. The result of this energy exchange is the formation of space charge bunches, or spokes, and the amplification of the RF wave as it travels along the circuit. The RF fields therefore focus the electrons into spokes which drift along the circuit in phase with the wave. Similar formations occur in magnetrons.

The spokes of space charge emanate from a cloud of electrons near the cathode surface, or hub. Figure 9.29(B) shows a sketch of the spoke pattern relative to the RF field pattern on the circuit. The orthogonal RF electric field and the magnetic field within the spoke produce a force on the electrons directed toward the anode. In reaching this destination, the electrons follow cycloidal trajectories. When they finally strike the anode, they give up essentially the same kinetic energy they had when they left the hub. Therefore, potential energy equivalent to the dc potential difference between anode and hub surface is converted into RF energy. The potential difference between the hub surface and the cathode is essentially the dissipation on the anode and can be small when compared with the power output.

To obtain higher power output from the crossed-field amplifier, it is necessary to apply more voltage between anode and cathode than is necessary to give a drift velocity equal to the wave velocity on the circuit. When this procedure is finished,

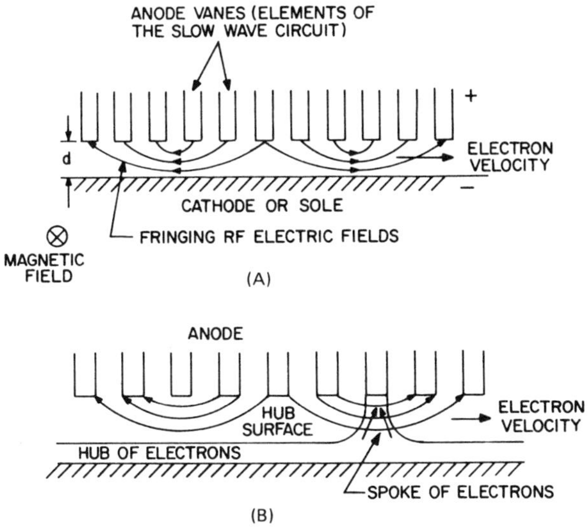

Fig. 9.29 Interaction region of a CFA.

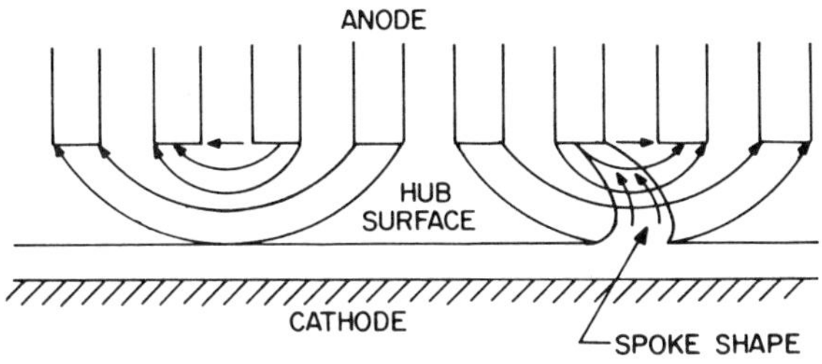

Fig. 9.30 Hub and spoke pattern with slightly higher anode voltage.

more anode current is drawn and the shape of the spoke changes so that it appears as in Figure 9.30. The spoke remains in synchronism with the wave although the effects of this overvoltage, as it is called, are discernible as phase pushing. The synchronous behavior of the spoke is the reason the CFA has an inherently higher phase stability than linear beam tubes.

There appears to be no ideal way of presenting CFA performance because of the number of variables involved. For example, one might plot such dependent characteristics as power output, gain, efficiency, phase characteristics, spurious output, and so on, as a function of frequency over the rated band of operation for the amplifier. Each one of these displays, however, might be made for only one specified operating current and for only one specified RF input power at a time. Since it is of interest to see how these dependent chararacteristics do change as a function of peak current and RF input, other types of presentations are made. Obviously the amount of data required is large and the method of presenting changes can be somewhat complex. To simplify the performance information, four primary characteristics that are important in any microwave amplifier are discussed: *compression, spurious output, bandwidth,* and *jitter*.

Compression Characteristics

The compression curve of an amplifier shows power output *versus* power input with all other parameters held constant. The data can be displayed either in linear or log form. In either, display lines of constant gain can be drawn to indicate the departure of the amplifier from linearity. The compression curve is also a useful display on which may be indicated a lower power limit of operation for a particular amplifier. The signal-to-noise ratio of a crossed-field amplifier typically decreases as the RF drive power is decreased. The lower power limit for a particular application, of course, is governed by the sensitivity of that application to a lower signal-to-noise ratio.

For each amplifier, an RF drive power level even lower than the minimum acceptable for systems use is present, at which the starting process for the amplifier via secondary emission becomes erratic.

The compression characteristics of crossed-field amplifiers are such that the curve is monotonic, or does not peak at a particular level of RF drive. One reason for this fact is that the amplifiers are transparent, and additional elements of drive power which are applied appear in the output of the amplifier, even if they do not increase the energy conversion in the amplifier itself. The compression curve is therefore expected to retain some positive slope for very high levels of drive power, differing from the curves of some types of linear beam amplifiers, such as traveling wave tubes and klystrons (which have an optimum RF drive level for obtaining maximum output power).

Spurious Output Characteristics

The spurious output characteristics of an amplifier, or its signal-to-noise ratio performance, varies primarily with two parameters—the RF drive power and the peak current through the amplifier. The assessment of an amplifier's performance, with regard to spurious output, requires essentially that a measuring system be set up in the laboratory which to some extent duplicates the receiver to be used in the system.

The measurements of spurious outputs at extraspectral frequencies can be made with a number of spectrum analyzer techniques. In presenting the data it is desirable to specify the spurious output, either in absolute levels or in dB below the signal, with an indication given of the analyzer bandwidth over which the measured power is integrated. An example typical of crossed-field amplifiers is that spurious output is 32 to 40 dB below the signal using an analyzer bandwidth of 6 MHz. Measurements of this type have been made, for example, for X-band amplifiers over the frequency range from 7 to 10 GHz. The frequency band over which an examination is made for spurious output should, if possible, include both upper and lower band edges of the amplifier under test. Since the RF drive source cannot be made free of all spurious outputs, compensation must be made for its contribution.

Bandwidth

The bandwidth of a crossed-field amplifier is most often considered as the frequency range over which its output is constant within 1 dB. The actual levels applied to this definition depend upon the application. Variations of up to 2 dB have on occasion been found permissible. Because of the interaction between the power supply or modulator and the amplifier, it is generally advisable to specify whether the bandwidth desired is a constant current bandwidth or constant voltage bandwidth, or if it is to be realized with a fixed setting of a line type modulator.

The bandwidth of an amplifier is also often referred to as "instantaneous," a term which warrants some interpretation. One might define instantaneous bandwidth as the bandwidth required to accommodate the simultaneous amplification of a certain band of frequencies. Crossed-field amplifiers have not often been used in this way. Because they are not linear, there has been a question about whether undesirable intermodulation effects are produced. This concern is probably of more importance in communications applications, which require CW rather than pulsed operation. CW CFAs, however, have shown non-linearities of the same type as pulsed CFAs, since the electronics of both classes of amplifier are the same.

Another interpretation of instantaneous bandwidth might be that the amplifier should be capable of delivering successive pulses at different frequencies anywhere in its operating band, requiring that it be capable of being "tuned" during the interpulse period. This requirement may occur often in practical systems, and amplifiers are typically capable of this type of instantaneous bandwidth.

A third interpretation of instantaneous bandwidth revolves around the fact that the amplifier should be capable of being driven at different frequencies on an intrapulse basis, as might be required in pulse compression systems where frequency modulation is applied during the pulse. Likewise, amplifiers have been demonstrated to have this type of bandwidth.

Perhaps a further interpretation might even be, that the amplifier should have instantaneous bandwidth in order to be capable of transmitting extremely short pulses whose equivalent RF bandwidth is large.

Generally, it is desirable to indicate the type of bandwidth required in the application for which the amplifier is intended.

Jitter

Jitter in the output of a transmitter is of concern in a large variety of applications; it falls into two major categories: amplitude and leading edge. *Amplitude jitter* is produced almost entirely by modulator or power supply effects; it is predictable from the performance chart characteristics for the amplifier.

Leading edge jitter is also of interest, and is examined particularly with regard to the reliability of cold cathode starting. Unlike the magnetron (with the leading edge jitter sometimes observed on its output), the crossed-field amplifier does not build up its power from noise on each pulse. The CFA is driven by a relatively large drive signal and the secondary emission process has been found to be quite rapid. The CFA likewise has been found to be insensitive to rate of rise of applied voltage and, as yet, no limits have been found for this quantity.

The distributed emission crossed-field amplifier is a compact and lightweight microwave device which can provide bandwidth comparable with the TWT at low to moderate gain levels. It generates large amounts of peak power with relatively low applied voltages; it operates with a high efficiency (40 to 75 percent typically) that is generally not offset by the need for additional focusing power, since it is usually equipped with an integral permanent magnet.

The CFA is capable of low to moderate average power levels compared with klystrons and TWTs. It has good phase stability and can be used over a wide range of pulse lengths. Its electrical characterization is made more difficult by the fact that it can operate with a wide range of RF drive levels and modulator input levels, and generally performs over a wide range of frequencies. The amplifiers are novel in that they can be made to operate with cold secondary emitting cathodes requiring no heater or warmup time. Some types can be operated in a way analogous to grid modulated TWTs and klystrons, through the use of RF turn on and control electrode turn off techniques.

Crossed-field amplifiers have been built at frequencies ranging from UHF to Ka-band, and at peak power levels of several megawatts. Bandwidths of more than 25 percent have been demonstrated. One may expect the CFA to find growing application in pulsed radar systems.

9.5 SUMMARY

Four very important devices that make the field of microwaves what it is today have been covered. They range from tubes developed in the early 1940s to the more modern crossed-field amplifier that combines good qualities of many devices. You should be aware by now that microwave tubes are here to stay, by no means having gone the way of the conventional vacuum tubes that now have been almost completely replaced by transistors in radio and television. However, a transistor capable of producing a megawatt of power at X-band is still a long way off.

Each of the tube types discussed has been touched on lightly. If more information is wanted on any one or all of these tubes, manufacturers are happy to supply application notes and data sheets that do an outstanding job of explanation.

The next time someone says that tubes are dead in electronics, show him or her a klystron, TWT, magnetron, or a crossed-field amplifier. One person's mind undoubtedly will have changed. *Tubes are very much alive and living in the world of microwaves!*

Appendices A through S

APPENDIX A
MICROWAVE BANDS

		Official JCS Band Designation				
A	Band	0	MHz	to	250	MHz
B	Band	250	MHz	to	500	MHz
C	Band	500	MHz	to	1000	MHz
D	Band	1.0	GHz	to	2.0	GHz
E	Band	2.0	GHz	to	3.0	GHz
F	Band	3.0	GHz	to	4.0	GHz
G	Band	4.0	GHz	to	6.0	GHz
H	Band	6.0	GHz	to	8.0	GHz
I	Band	8.0	GHz	to	10.0	GHz
J	Band	10.0	GHz	to	20.0	GHz
K	Band	20.0	GHz	to	40.0	GHz
L	Band	40.0	GHz	to	60.0	GHz
M	Band	60.0	GHz	to	100.0	GHz
		US Microwave Band				
P	Band	225	MHz	to	390	MHz
L	Band	390	MHz	to	1.55	GHz
S	Band	1.55	GHz	to	5.20	GHz
C	Band	3.90	GHz	to	6.20	GHz
X	Band	5.20	GHz	to	10.9	GHz
K	Band	10.9	GHz	to	36.0	GHz
Ku	Band	15.25	GHz	to	17.25	GHz
K_1	Band	15.35	GHz	to	24.5	GHz

APPENDIX A continued

Q	Band	36.0	GHz	to	46.0	GHz
V	Band	46.0	GHz	to	56.0	GHz
W	Band	56.0	GHz	to	100.0	GHz

Old JCS Band Designation

VHF	Band	100	MHz	to	300	MHz
UHF	Band	300	MHz	to	1.0	GHz
L	Band	1.0	GHz	to	2.0	GHz
S	Band	2.0	GHz	to	4.0	GHz
C	Band	4.0	GHz	to	8.0	GHz
X	Band	8.0	GHz	to	12.4	GHz
Ku	Band	12.4	GHz	to	18.0	GHz
K	Band	18.0	GHz	to	26.0	GHz
Ka	Band	26.0	GHz	to	40.0	GHz
Millimeter		40.0	GHz	to	100.0	GHz

APPENDIX B dBm TO WATTS

dBm	Milli-Watt	dBm	Milli-Watt	dBm	Milli-Watt	dBm	Milli-Watt	dBm	Milli-Watt	dBm	Milli-Watt
-18.0	.0158	-9.1	.123	-0.2	.955	8.7	7.41	17.6	57.5	26.5	447
-17.9	.0162	-9.0	.126	-0.1	.977	8.8	7.59	17.7	58.9	26.6	457
-17.8	.0166	-8.9	.129	0.0	1.00	8.9	7.76	17.8	60.3	26.7	468
-17.7	.0170	-8.8	.132	0.1	1.02	9.0	7.94	17.9	61.7	26.8	479
-17.6	.0174	-8.7	.135	0.2	1.05	9.1	8.13	18.0	63.1	26.9	490
-17.5	.0178	-8.6	.138	0.3	1.07	9.2	8.32	18.1	64.6	27.0	501
-17.4	.0182	-8.5	.141	0.4	1.10	9.3	8.51	18.2	66.1	27.1	513
-17.3	.0186	-8.4	.145	0.5	1.12	9.4	8.71	18.3	67.6	27.2	525
-17.2	.0191	-8.3	.148	0.6	1.15	9.5	8.91	18.4	69.2	27.3	537
-17.1	.0195	-8.2	.151	0.7	1.17	9.6	9.12	18.5	70.8	27.4	550
-17.0	.0200	-8.1	.155	0.8	1.20	9.7	9.33	18.6	72.4	27.5	562
-16.9	.0204	-8.0	.158	0.9	1.23	9.8	9.55	18.7	74.1	27.6	575
-16.8	.0209	-7.9	.162	1.0	1.26	9.9	9.77	18.8	75.9	27.7	589
-16.7	.0214	-7.8	.166	1.1	1.29	10.0	10.0	18.9	77.6	27.8	603
-16.6	.0219	-7.7	.170	1.2	1.32	10.1	10.2	19.0	79.4	27.9	617
-16.5	.0224	-7.6	.174	1.3	1.35	10.2	10.5	19.1	81.3	28.0	631
-16.4	.0229	-7.5	.178	1.4	1.38	10.3	10.7	19.2	83.2	28.1	646
-16.3	.0234	-7.4	.182	1.5	1.41	10.4	11.0	19.3	85.1	28.2	661
-16.2	.0240	-7.3	.186	1.6	1.45	10.5	11.2	19.4	87.1	28.3	676
-16.1	.0245	-7.2	.191	1.7	1.48	10.6	11.5	19.5	89.1	28.4	692
-16.0	.0251	-7.1	.195	1.8	1.51	10.7	11.7	19.6	91.2	28.5	708
-15.9	.0257	-7.0	.200	1.9	1.55	10.8	12.0	19.7	93.3	28.6	724
-15.8	.0263	-6.9	.204	2.0	1.58	10.9	12.3	19.8	95.5	28.7	741
-15.7	.0269	-6.8	.209	2.1	1.62	11.0	12.6	19.9	97.7	28.8	759
-15.6	.0275	-6.7	.214	2.2	1.66	11.1	12.9	20.0	100	28.9	776
-15.5	.0282	-6.6	.219	2.3	1.70	11.2	13.2	20.1	102	29.0	794
-15.4	.0288	-6.5	.224	2.4	1.74	11.3	13.5	20.2	105	29.1	813
-15.3	.0295	-6.4	.229	2.5	1.78	11.4	13.8	20.3	107	29.2	832
-15.2	.0302	-6.3	.234	2.6	1.82	11.5	14.1	20.4	110	29.3	851
-15.1	.0309	-6.2	.240	2.7	1.86	11.6	14.5	20.5	112	29.4	871
-15.0	.0316	-6.1	.245	2.8	1.91	11.7	14.8	20.6	115	29.5	891
-14.9	.0324	-6.0	.251	2.9	1.95	11.8	15.1	20.7	117	29.6	912
-14.8	.0331	-5.9	.257	3.0	2.00	11.9	15.5	20.8	120	29.7	933
-14.7	.0339	-5.8	.263	3.1	2.04	12.0	15.8	20.9	123	29.8	955
-14.6	.0347	-5.7	.269	3.2	2.09	12.1	16.2	21.0	126	29.9	977
-14.5	.0355	-5.6	.275	3.3	2.14	12.2	16.6	21.1	129	30.0	1000
-14.4	.0363	-5.5	.282	3.4	2.19	12.3	17.0	21.2	132		
-14.3	.0372	-5.4	.288	3.5	2.24	12.4	17.4	21.3	135		
-14.2	.0380	-5.3	.295	3.6	2.29	12.5	17.8	21.4	138		
-14.1	.0389	-5.2	.302	3.7	2.34	12.6	18.2	21.5	141		
-14.0	.0398	-5.1	.309	3.8	2.40	12.7	18.6	21.6	145		
-13.9	.0407	-5.0	.316	3.9	2.45	12.8	19.1	21.7	148		
-13.8	.0417	-4.9	.324	4.0	2.51	12.9	19.5	21.8	151		
-13.7	.0427	-4.8	.331	4.1	2.57	13.0	20.0	21.9	155		
-13.6	.0437	-4.7	.339	4.2	2.63	13.1	20.4	22.0	158		
-13.5	.0447	-4.6	.347	4.3	2.69	13.2	20.9	22.1	162		
-13.4	.0457	-4.5	.355	4.4	2.75	13.3	21.4	22.2	166		
-13.3	.0468	-4.4	.363	4.5	2.82	13.4	21.9	22.3	170		
-13.2	.0479	-4.3	.372	4.6	2.88	13.5	22.4	22.4	174		
-13.1	.0490	-4.2	.380	4.7	2.95	13.6	22.9	22.5	178		
-13.0	.0501	-4.1	.389	4.8	3.02	13.7	23.4	22.6	182		
-12.9	.0513	-4.0	.398	4.9	3.09	13.8	24.0	22.7	186		
-12.8	.0525	-3.9	.407	5.0	3.16	13.9	24.5	22.8	191		
-12.7	.0537	-3.8	.417	5.1	3.24	14.0	25.1	22.9	195		
-12.6	.0550	-3.7	.427	5.2	3.31	14.1	25.7	23.0	200		
-12.5	.0562	-3.6	.437	5.3	3.39	14.2	26.3	23.1	204		
-12.4	.0575	-3.5	.447	5.4	3.47	14.3	26.9	23.2	209		
-12.3	.0589	-3.4	.457	5.5	3.55	14.4	27.5	23.3	214		
-12.2	.0603	-3.3	.468	5.6	3.63	14.5	28.2	23.4	219		
-12.1	.0617	-3.2	.479	5.7	3.72	14.6	28.8	23.5	224		
-12.0	.0631	-3.1	.490	5.8	3.80	14.7	29.5	23.6	229		
-11.9	.0646	-3.0	.501	5.9	3.89	14.8	30.2	23.7	234		
-11.8	.0661	-2.9	.513	6.0	3.98	14.9	30.9	23.8	240		
-11.7	.0676	-2.8	.525	6.1	4.07	15.0	31.6	23.9	245		
-11.6	.0692	-2.7	.537	6.2	4.17	15.1	32.4	24.0	251		
-11.5	.0708	-2.6	.550	6.3	4.27	15.2	33.1	24.1	257		
-11.4	.0724	-2.5	.562	6.4	4.37	15.3	33.9	24.2	263		
-11.3	.0741	-2.4	.575	6.5	4.47	15.4	34.7	24.3	269		
-11.2	.0759	-2.3	.589	6.6	4.57	15.5	35.5	24.4	275		
-11.1	.0776	-2.2	.603	6.7	4.68	15.6	36.3	24.5	282		
-11.0	.0794	-2.1	.617	6.8	4.79	15.7	37.2	24.6	288		
-10.9	.0813	-2.0	.631	6.9	4.90	15.8	38.0	24.7	295		
-10.8	.0832	-1.9	.646	7.0	5.01	15.9	38.9	24.8	302		
-10.7	.0851	-1.8	.661	7.1	5.13	16.0	39.8	24.9	309		
-10.6	.0871	-1.7	.676	7.2	5.25	16.1	40.7	25.0	316		
-10.5	.0891	-1.6	.692	7.3	5.37	16.2	41.7	25.1	324		
-10.4	.0912	-1.5	.708	7.4	5.50	16.3	42.7	25.2	331		
-10.3	.0933	-1.4	.724	7.5	5.62	16.4	43.7	25.3	339		
-10.2	.0955	-1.3	.741	7.6	5.75	16.5	44.7	25.4	347		
-10.1	.0977	-1.2	.759	7.7	5.89	16.6	45.7	25.5	355		
-10.0	.100	-1.1	.776	7.8	6.03	16.7	46.8	25.6	363		
-9.9	.102	-1.0	.794	7.9	6.17	16.8	47.9	25.7	372		
-9.8	.105	-0.9	.813	8.0	6.31	16.9	49.0	25.8	380		
-9.7	.107	-0.8	.832	8.1	6.46	17.0	50.1	25.9	389		
-9.6	.110	-0.7	.851	8.2	6.61	17.1	51.3	26.0	398		
-9.5	.112	-0.6	.871	8.3	6.76	17.2	52.5	26.1	407		
-9.4	.115	-0.5	.891	8.4	6.92	17.3	53.7	26.2	417		
-9.3	.117	-0.4	.912	8.5	7.08	17.4	55.0	26.3	427		
-9.2	.120	-0.3	.933	8.6	7.24	17.5	56.2	26.4	437		

APPENDIX B continued

dBm	Watts	dBm	Watts	dBm	Watts	dBm	Watts	dBm	Watts
30.1	1.02	39.4	8.71	48.7	74.10	58.0	631.00	67.3	5370.00
30.2	1.05	39.5	8.91	48.8	75.90	58.1	646.00	67.4	5500.00
30.3	1.07	39.6	9.12	48.9	77.60	58.2	661.00	67.5	5620.00
30.4	1.10	39.7	9.33	49.0	79.40	58.3	676.00	67.6	5750.00
30.5	1.12	39.8	9.55	49.1	81.30	58.4	692.00	67.7	5890.00
30.6	1.15	39.9	9.77	49.2	83.20	58.5	708.00	67.8	6030.00
30.7	1.17	40.0	10.00	49.3	85.10	58.6	724.00	67.9	6170.00
30.8	1.20	40.1	10.20	49.4	87.10	58.7	741.00	68.0	6310.00
30.9	1.23	40.2	10.50	49.5	89.10	58.8	759.00	68.1	6460.00
31.0	1.26	40.3	10.70	49.6	91.20	58.9	776.00	68.2	6610.00
31.1	1.29	40.4	11.00	49.7	93.30	59.0	794.00	68.3	6760.00
31.2	1.32	40.5	11.20	49.8	95.50	59.1	813.00	68.4	6920.00
31.3	1.35	40.6	11.50	49.9	97.70	59.2	832.00	68.5	7080.00
31.4	1.38	40.7	11.70	50.0	100.00	59.3	851.00	68.6	7240.00
31.5	1.41	40.8	12.00	50.1	102.00	59.4	871.00	68.7	7410.00
31.6	1.45	40.9	12.30	50.2	105.00	59.5	891.00	68.8	7590.00
31.7	1.48	41.0	12.60	50.3	107.00	59.6	912.00	68.9	7760.00
31.8	1.51	41.1	12.90	50.4	110.00	59.7	933.00	69.0	7940.00
31.9	1.55	41.2	13.20	50.5	112.00	59.8	955.00	69.1	8130.00
32.0	1.58	41.3	13.50	50.6	115.00	59.9	977.00	69.2	8320.00
32.1	1.62	41.4	13.80	50.7	117.00	60.0	1000.00	69.3	8510.00
32.2	1.66	41.5	14.10	50.8	120.00	60.1	1020.00	69.4	8710.00
32.3	1.70	41.6	14.50	50.9	123.00	60.2	1050.00	69.5	8910.00
32.4	1.74	41.7	14.80	51.0	126.00	60.3	1070.00	69.6	9120.00
32.5	1.78	41.8	15.10	51.1	129.00	60.4	1100.00	69.7	9330.00
32.6	1.82	41.9	15.50	51.2	132.00	60.5	1120.00	69.8	9550.00
32.7	1.86	42.0	15.80	51.3	135.00	60.6	1150.00	69.9	9770.00
32.8	1.91	42.1	16.20	51.4	138.00	60.7	1170.00	70.0	10000.00
32.9	1.95	42.2	16.60	51.5	141.00	60.8	1200.00		
33.0	2.00	42.3	17.00	51.6	145.00	60.9	1230.00		
33.1	2.04	42.4	17.40	51.7	148.00	61.0	1260.00		
33.2	2.09	42.5	17.80	51.8	151.00	61.1	1290.00		
33.3	2.14	42.6	18.20	51.9	155.00	61.2	1320.00		
33.4	2.19	42.7	18.60	52.0	158.00	61.3	1350.00		
33.5	2.24	42.8	19.10	52.1	162.00	61.4	1380.00		
33.6	2.29	42.9	19.50	52.2	166.00	61.5	1410.00		
33.7	2.34	43.0	20.00	52.3	170.00	61.6	1450.00		
33.8	2.40	43.1	20.40	52.4	174.00	61.7	1480.00		
33.9	2.45	43.2	20.90	52.5	178.00	61.8	1510.00		
34.0	2.51	43.3	21.40	52.6	182.00	61.9	1550.00		
34.1	2.57	43.4	21.90	52.7	186.00	62.0	1580.00		
34.2	2.63	43.5	22.40	52.8	191.00	62.1	1620.00		
34.3	2.69	43.6	22.90	52.9	195.00	62.2	1660.00		
34.4	2.75	43.7	23.40	53.0	200.00	62.3	1700.00		
34.5	2.82	43.8	24.00	53.1	204.00	62.4	1740.00		
34.6	2.88	43.9	24.50	53.2	209.00	62.5	1780.00		
34.7	2.95	44.0	25.10	53.3	214.00	62.6	1820.00		
34.8	3.02	44.1	25.70	53.4	219.00	62.7	1860.00		
34.9	3.09	44.2	26.30	53.5	224.00	62.8	1910.00		
35.0	3.16	44.3	26.90	53.6	229.00	62.9	1950.00		
35.1	3.24	44.4	27.50	53.7	234.00	63.0	2000.00		
35.2	3.31	44.5	28.20	53.8	240.00	63.1	2040.00		
35.3	3.39	44.6	28.80	53.9	245.00	63.2	2090.00		
35.4	3.47	44.7	29.50	54.0	251.00	63.3	2140.00		
35.5	3.55	44.8	30.20	54.1	257.00	63.4	2190.00		
35.6	3.63	44.9	30.90	54.2	263.00	63.5	2240.00		
35.7	3.72	45.0	31.60	54.3	269.00	63.6	2290.00		
35.8	3.80	45.1	32.40	54.4	275.00	63.7	2340.00		
35.9	3.89	45.2	33.10	54.5	282.00	63.8	2400.00		
36.0	3.98	45.3	33.90	54.6	288.00	63.9	2450.00		
36.1	4.07	45.4	34.70	54.7	295.00	64.0	2510.00		
36.2	4.17	45.5	35.50	54.8	302.00	64.1	2570.00		
36.3	4.27	45.6	36.30	54.9	309.00	64.2	2630.00		
36.4	4.37	45.7	37.20	55.0	316.00	64.3	2690.00		
36.5	4.47	45.8	38.00	55.1	324.00	64.4	2750.00		
36.6	4.57	45.9	38.90	55.2	331.00	64.5	2820.00		
36.7	4.68	46.0	39.80	55.3	339.00	64.6	2880.00		
36.8	4.79	46.1	40.70	55.4	347.00	64.7	2950.00		
36.9	4.90	46.2	41.70	55.5	355.00	64.8	3020.00		
37.0	5.01	46.3	42.70	55.6	363.00	64.9	3090.00		
37.1	5.13	46.4	43.70	55.7	372.00	65.0	3160.00		
37.2	5.25	46.5	44.70	55.8	380.00	65.1	3240.00		
37.3	5.37	46.6	45.70	55.9	389.00	65.2	3310.00		
37.4	5.50	46.7	46.80	56.0	398.00	65.3	3390.00		
37.5	5.62	46.8	47.90	56.1	407.00	65.4	3470.00		
37.6	5.75	46.9	49.00	56.2	417.00	65.5	3550.00		
37.7	5.89	47.0	51.10	56.3	427.00	65.6	3630.00		
37.8	6.03	47.1	51.30	56.4	437.00	65.7	3720.00		
37.9	6.17	47.2	52.50	56.5	447.00	65.8	3800.00		
38.0	6.31	47.3	53.70	56.6	457.00	65.9	3890.00		
38.1	6.46	47.4	55.00	56.7	468.00	66.0	3980.00		
38.2	6.61	47.5	56.20	56.8	479.00	66.1	4070.00		
38.3	6.76	47.6	57.50	56.9	490.00	66.2	4170.00		
38.4	6.92	47.7	58.90	57.0	501.00	66.3	4270.00		
38.5	7.08	47.8	60.30	57.1	513.00	66.4	4370.00		
38.6	7.24	47.9	61.70	57.2	525.00	66.5	4470.00		
38.7	7.41	48.0	63.10	57.3	537.00	66.6	4570.00		
38.8	7.59	48.1	64.60	57.4	550.00	66.7	4680.00		
38.9	7.76	48.2	66.10	57.5	562.00	66.8	4790.00		
39.0	7.94	48.3	67.60	57.6	575.00	66.9	4900.00		
39.1	8.13	48.4	69.20	57.7	589.00	67.0	5010.00		
39.2	8.32	48.5	70.80	57.8	603.00	67.1	5130.00		
39.3	8.51	48.6	72.40	57.9	617.00	67.2	5250.00		

APPENDIX C DECIBELS VERSUS VOLTAGE AND POWER

Voltage Ratio	Power Ratio	− db +	Voltage Ratio	Power Ratio	Voltage Ratio	Power Ratio	− db +	Voltage Ratio	Power Ratio
1.0000	**1.0000**	**0**	**1.000**	**1.000**	**.4467**	**.1995**	**7.0**	**2.239**	**5.012**
.9886	.9772	.1	1.012	1.023	.4416	.1950	7.1	2.265	5.129
.9772	.9550	.2	1.023	1.047	.4365	.1905	7.2	2.291	5.248
.9661	.9333	.3	1.035	1.072	.4315	.1862	7.3	2.317	5.370
.9550	.9120	.4	1.047	1.096	.4266	.1820	7.4	2.344	5.495
.9441	.8913	.5	1.059	1.122	.4217	.1778	7.5	2.371	5.623
.9333	.8710	.6	1.072	1.148	.4169	.1738	7.6	2.399	5.754
.9226	.8511	.7	1.084	1.175	.4121	.1698	7.7	2.427	5.888
.9120	.8318	.8	1.096	1.202	.4074	.1660	7.8	2.455	6.026
.9016	.8128	.9	1.109	1.230	.4027	.1622	7.9	2.483	6.166
.8913	**.7943**	**1.0**	**1.122**	**1.259**	**.3981**	**.1585**	**8.0**	**2.512**	**6.310**
.881C	.7762	1.1	1.135	1.288	.3936	.1549	8.1	2.541	6.457
.8710	.7586	1.2	1.148	1.318	.3890	.1514	8.2	2.570	6.607
.8610	.7413	1.3	1.161	1.349	.3846	.1479	8.3	2.600	6.761
.8511	.7244	1.4	1.175	1.380	.3802	.1445	8.4	2.630	6.918
.8414	.7079	1.5	1.189	1.413	.3758	.1413	8.5	2.661	7.079
.8318	.6918	1.6	1.202	1.445	.3715	.1380	8.6	2.692	7.244
.8222	.6761	1.7	1.216	1.479	.3673	.1349	8.7	2.723	7.413
.8128	.6607	1.8	1.230	1.514	.3631	.1318	8.8	2.754	7.586
.8035	.6457	1.9	1.245	1.549	.3589	.1288	8.9	2.786	7.762
.7943	**.6310**	**2.0**	**1.259**	**1.585**	**.3548**	**.1259**	**9.0**	**2.818**	**7.943**
.7852	.6166	2.1	1.274	1.622	.3508	.1230	9.1	2.851	8.128
.7762	.6026	2.2	1.288	1.660	.3467	.1202	9.2	2.884	8.318
.7674	.5888	2.3	1.303	1.698	.3428	.1175	9.3	2.917	8.511
.7586	.5754	2.4	1.318	1.738	.3388	.1148	9.4	2.951	8.710
.7499	.5623	2.5	1.334	1.778	.3350	.1122	9.5	2.985	8.913
.7413	.5495	2.6	1.349	1.820	.3311	.1096	9.6	3.020	9.120
.7328	.5370	2.7	1.365	1.862	.3273	.1072	9.7	3.055	9.333
.7244	.5248	2.8	1.380	1.905	.3236	.1047	9.8	3.090	9.550
.7161	.5129	2.9	1.396	1.950	.3199	.1023	9.9	3.126	9.772
.7079	**.5012**	**3.0**	**1.413**	**1.995**	**.3162**	**.1000**	**10.0**	**3.162**	**10.000**
.6998	.4898	3.1	1.429	2.042	.3126	.09772	10.1	3.199	10.23
.6918	.4786	3.2	1.445	2.089	.3090	.09550	10.2	3.236	10.47
.6839	.4677	3.3	1.462	2.138	.3055	.09333	10.3	3.273	10.72
.6761	.4571	3.4	1.479	2.188	.3020	.09120	10.4	3.311	10.96
.6683	.4467	3.5	1.496	2.239	.2985	.08913	10.5	3.350	11.22
.6607	.4365	3.6	1.514	2.291	.2951	.08710	10.6	3.388	11.48
.6531	.4266	3.7	1.531	2.344	.2917	.08511	10.7	3.428	11.75
.6457	.4169	3.8	1.549	2.399	.2884	.08318	10.8	3.467	12.02
.6383	.4074	3.9	1.567	2.455	.2851	.08128	10.9	3.508	12.30
.6310	**.3981**	**4.0**	**1.585**	**2.512**	**.2818**	**.07943**	**11.0**	**3.548**	**12.59**
.6237	.3890	4.1	1.603	2.570	.2786	.07762	11.1	3.589	12.88
.6166	.3802	4.2	1.622	2.630	.2754	.07586	11.2	3.631	13.18
.6095	.3715	4.3	1.641	2.692	.2723	.07413	11.3	3.673	13.49
.6026	.3631	4.4	1.660	2.754	.2692	.07244	11.4	3.715	13.80
.5957	.3548	4.5	1.679	2.818	.2661	.07079	11.5	3.758	14.13
.5888	.3467	4.6	1.698	2.884	.2630	.06918	11.6	3.802	14.45
.5821	.3388	4.7	1.718	2.951	.2600	.06761	11.7	3.846	14.79
.5754	.3311	4.8	1.738	3.020	.2570	.06607	11.8	3.890	15.14
.5689	.3236	4.9	1.758	3.090	.2541	.06457	11.9	3.936	15.49
.5623	**.3162**	**5.0**	**1.776**	**3.162**	**.2512**	**06310**	**12.0**	**3.981**	**15.85**
.5559	.3090	5.1	1.799	3.236	.2483	.06166	12.1	4.027	16.22
.5495	.3020	5.2	1.820	3.311	.2455	.06026	12.2	4.074	16.60
.5433	.2951	5.3	1.841	3.388	.2427	.05888	12.3	4.121	16.98
.5370	.2884	5.4	1.862	3.467	.2399	.05754	12.4	4.169	17.38
.5309	.2818	5.5	1.884	3.548	.2371	.05623	12.5	4.217	17.78
.5248	.2754	5.6	1.905	3.631	.2344	.05495	12.6	4.266	18.20
.5188	.2692	5.7	1.928	3.715	.2317	.05370	12.7	4.315	18.62
.5129	.2630	5.8	1.950	3.802	.2291	.05248	12.8	4.365	19.05
.5070	.2570	5.9	1.972	3.890	.2265	.05129	12.9	4.416	19.50
.5012	**.2512**	**6.0**	**1.995**	**3.981**	**.2239**	**.05012**	**13.0**	**4.467**	**19.95**
.4955	.2455	6.1	2.018	4.074	.2213	.04898	13.1	4.519	20.42
.4898	.2399	6.2	2.042	4.169	.2188	.04786	13.2	4.571	20.89
.4842	.2344	6.3	2.065	4.266	.2163	.04677	13.3	4.624	21.38
.4786	.2291	6.4	2.089	4.365	.2138	.04571	13.4	4.677	21.88
.4732	.2239	6.5	2.113	4.467	.2113	.04467	13.5	4.732	22.39
.4677	.2188	6.6	2.138	4.571	.2089	.04365	13.6	4.786	22.91
.4624	.2138	6.7	2.163	4.677	.2065	.04266	13.7	4.842	23.44
.4571	.2089	6.8	2.188	4.786	.2042	.04169	13.8	4.898	23.99
.4519	.2042	6.9	2.213	4.898	.2018	.04074	13.9	4.955	24.55

APPENDIX C continued

Voltage Ratio	*Power Ratio*	*− db +*	*Voltage Ratio*	*Power Ratio*
.1995	**.03981**	**14.0**	**5.012**	**25.12**
.1972	.03890	14.1	5.070	25.70
.1950	.03802	14.2	5.129	26.30
.1928	.03715	14.3	5.188	26.92
.1905	.03631	14.4	5.248	27.54
.1884	.03548	14.5	5.309	28.18
.1862	.03467	14.6	5.370	28.84
.1841	.03388	14.7	5.433	29.51
.1820	.03311	14.8	5.495	30.20
.1799	.03236	14.9	5.559	30.90
.1778	**.03162**	**15.0**	**5.623**	**31.62**
.1758	.03090	15.1	5.689	32.36
.1738	.03020	15.2	5.754	33.11
.1718	.02951	15.3	5.821	33.88
.1698	.02884	15.4	5.888	34.67
.1679	.02818	15.5	5.957	35.48
.1660	.02754	15.6	6.026	36.31
.1641	.02692	15.7	6.095	37.15
.1622	.02630	15.8	6.166	38.02
.1603	.02570	15.9	6.237	38.90
.1585	**.02512**	**16.0**	**6.310**	**39.81**
.1567	.02455	16.1	6.383	40.74
.1549	.02399	16.2	6.457	41.69
.1531	.02344	16.3	6.531	42.66
.1514	.02291	16.4	6.607	43.65
.1496	02239	16.5	6.683	44.67
.1479	.02188	16.6	6.761	45.71
.1462	02138	16.7	6.839	46.77
.1445	.02089	16.8	6.918	47.86
.1429	.02042	16.9	6.998	48.98
.1413	**.01995**	**17.0**	**7.079**	**50.12**
.1396	.01950	17.1	7.161	51.29
.1380	.01905	17.2	7.244	52.48
.1365	.01862	17.3	7.328	53.70
.1349	.01820	17.4	7.413	54.95
.1334	.01778	17.5	7.499	56.23
.1318	.01738	17.6	7.586	57.54
.1303	.01698	17.7	7.674	58.88
.1288	.01660	17.8	7.762	60.26
.1274	.01622	17.9	7.852	61.66
.1259	**.01585**	**18.0**	**7.943**	**63.10**
.1245	.01549	18.1	8.035	64.57
.1230	.01514	18.2	8.128	66.07
.1216	.01479	18.3	8.222	67.61
.1202	.01445	18.4	8.318	69.18
.1189	.01413	18.5	8.414	70.79
.1175	.01380	18.6	8.511	72.44
.1161	.01349	18.7	8.610	74.13
.1148	.01318	18.8	8.710	75.86
.1135	.01288	18.9	8.811	77.62
.1122	**.01259**	**19.0**	**8.913**	**79.43**
.1109	.01230	19.1	9.016	81.28
.1096	.01202	19.2	9.120	83.18
.1084	.01175	19.3	9.226	85.11
.1072	.01148	19.4	9.333	87.10
.1059	.01122	19.5	9.441	89.13
.1047	.01096	19.6	9.550	91.20
.1035	.01072	19.7	9.661	93.33
.1023	.01047	19.8	9.772	95.50
.1012	.01023	19.9	9.886	97.72
.1000	**.01000**	**20.0**	**10.000**	**100.00**
	10^{-3}	30		10^{3}
10^{-2}	10^{-4}	40	10^{2}	10^{4}
		50		10^{5}
	10^{-5}	60		10^{6}
10^{-3}	10^{-6}	70	10^{3}	10^{7}
	10^{-7}	80		10^{8}
10^{-4}	10^{-8}	90	10^{4}	10^{9}
10^{-5}	10^{-9}			
	10^{-10}	100	10^{5}	10^{10}

APPENDIX D MICROWAVE FORMULAS

Inductive Reactance (X_L)

$$X_L = 2\pi fL$$

where

f = frequency (hertz)
L = inductance (Henries)

Capacitive Reactance (X_c)

$$X_c = \frac{1}{2\pi fc}$$

where
f = frequency (hertz)
c = capacitance (farads)

Wavelength (λ)

$$\lambda(\text{centimeters}) = \frac{3 \times 10^{10}}{f}$$

$$\lambda(\text{meters}) = \frac{3 \times 10^{8}}{f}$$

where

f = frequency (hertz)

dB *(Power and Voltage)*

$$\text{dB}_{(\text{power})} = 10 \log_{10} \frac{P1}{P2}$$

$$\text{dB}_{(\text{voltage})} = 20 \log_{10} \frac{E1}{E2}$$

APPENDIX D continued

where

$P1$ and $P2$ = system powers
$E1$ and $E2$ = system voltages

Characteristic Impedance (Z_0) of RF Cable

$$Z_0 = \frac{138}{\sqrt{\varepsilon}} \log_{10} \frac{D}{d}$$

where

ε = dielectric constant
D = inside diameter of outer conductor
d = outside diameter of inner conductor

% Velocity of Propagation (v)

$$v = \frac{1}{\sqrt{\varepsilon}} \times 100$$

where

ε = dielectric constant

Noise Figure (NF_{dB})

$$NF_{dB} = 10 \log_{10} \frac{S_i/N_i}{S_o/N_o}$$

where

NF_{dB} = noise figure (dB)
S_i/N_i = input signal-to-noise ratio
S_o/N_o = output signal-to-noise ratio

Reflection Coefficient (p)

$$\rho = \frac{\text{SWR} - 1}{\text{SWR} + 1}$$

APPENDIX D continued

where

$$\text{SWR} = \text{standing wave ratio}$$

Return Loss in dB

$$\text{dB} = -20 \log_{10} |\rho|$$

where

$$\rho = \text{reflection coefficient}$$

VSWR

$$\text{VSWR} = \frac{1 + \rho}{1 - \rho}$$

where

$$\rho = \text{reflection coefficient}$$

APPENDIX E S-PARAMETERS

What are *S*-parameters?

S-parameters are reflection and transmission coefficients—familiar concepts to RF and microwave designers. Transmission coefficients are commonly called gains or attenuations; reflection coefficients are directly related to VSWR's and impedances.

Conceptually, *s*-parameters are like *h*, *y*, or *z*-parameters because they describe the inputs and outputs of a black box. The inputs and outputs are in terms of power for *s*-parameters; for *h*, *y*, and *z*-parameters, they are voltages and currents. Using the convention that "a" is a signal into a port and "*b*" is a signal out, the figure below helps to explain *S*-parameters.

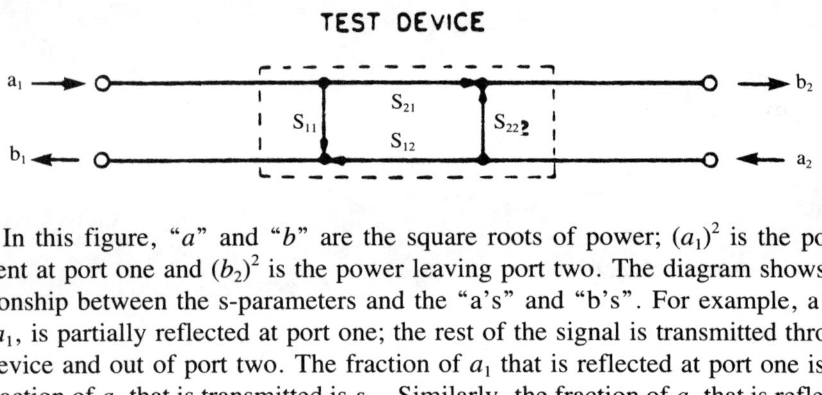

In this figure, "*a*" and "*b*" are the square roots of power; $(a_1)^2$ is the power incident at port one and $(b_2)^2$ is the power leaving port two. The diagram shows the relationship between the s-parameters and the "a's" and "b's". For example, a signal, a_1, is partially reflected at port one; the rest of the signal is transmitted through the device and out of port two. The fraction of a_1 that is reflected at port one is s_{11}; the fraction of a_1 that is transmitted is s_{21}. Similarly, the fraction of a_2 that is reflected at port two is s_{22}, and the fraction s_{12} is transmitted.

The signal, b_1, leaving port one is the sum of the fraction of a_1 that is reflected at port one and the fraction of a_2 that is transmitted from port two.

Thus, the outputs can be related to the inputs by the equations:

$$b_1 = s_{11}a_1 + s_{12}a_2$$

$$b_2 = s_{21}a_1 + s_{22}a_2$$

when

$$a_2 = 0 \qquad \text{and when } a_1 = 0$$

$$s_{11} = \frac{b_1}{a_1}, \; s_{21} = \frac{b_2}{a_1} \qquad s_{12} = \frac{b_1}{a_2}, \; s_{22} = \frac{b_2}{a_2}$$

APPENDIX E continued

The setup below shows how s_{11} and s_{22} are measured.

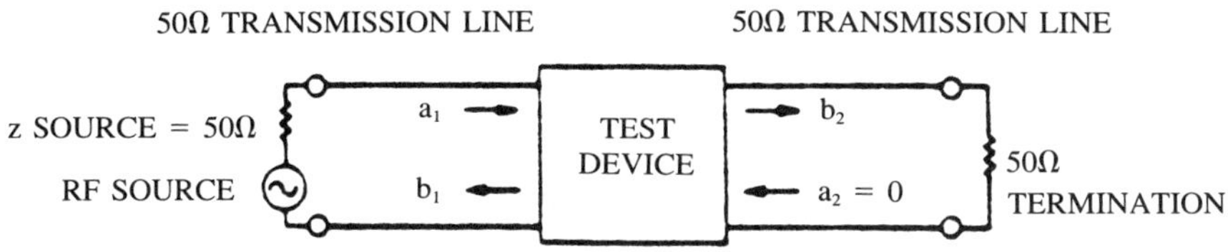

Port one is driven and a_2 is made zero by terminating the 50Ω transmission line coming out of port two in its characteristic 50Ω impedance. This termination ensures that none of the transmitted signal, b_2, is reflected toward the test device.

Similarly, the setup for measuring s_{12} and s_{22} is:

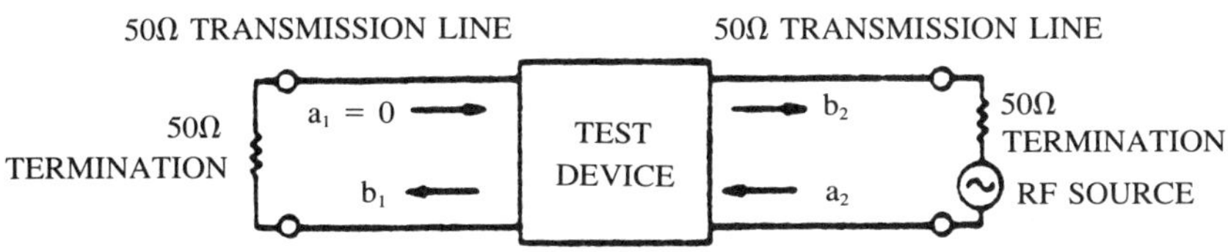

S-parameters are vector quantities—they give magnitude and phase information. Most measurements of microwave components, like attenuation, gain, and VSWR, have been measured historically only in terms of magnitude. Why? Mainly because it was too difficult to obtain both phase and magnitude information for those specifications.

S-parameters are measured so easily that the determination of accurate phase information is no longer a problem. Measurements like electrical length or dielectric coefficient can be determined readily from the phase of a transmission coefficient. Phase is the difference between knowing only a VSWR and knowing the exact impedance. VSWR's have been useful in calculating mismatch uncertainty, but when components are characterized with *S*-parameters, there is no mismatch uncertainty. The mismatch error can be calculated precisely.

Two-port *S*-parameters are easy to measure at high frequencies, because the device under test is terminated in the characteristic impedance of the measuring system. The characteristic impedance termination has the following advantages:

1. The termination is accurate at high frequencies—it is possible to build an accurate characteristic impedance load. "Open" or "short" terminations are required to determine *h*-, *y*-, or *z*-parameters, but lead inductance and capacitance make these terminations unrealistic at high frequencies.
2. No tuning is required to terminate a device in the characteristic impedance, but positioning an "open" or "short" at the terminals of a test device requires pre-

cision tuning. A "short" is placed at the end of a transmission line; the line length is varied precisely until an "open" or "short" is reflected to the device terminals. On the other hand, if a characteristic impedance load is placed at the end of the line, the device sees the characteristic impedance regardless of line length.

3. Broadband swept-frequency measurements are possible because the device remains terminated in the characteristic impedance as frequency changes. However, a carefully reflected "open" or "short" moves away from the device terminals as frequency is changed; it needs to be "tuned-in" at each frequency.
4. The termination enhances stability by providing a resistive termination that stabilizes many negative resistance devices which otherwise might oscillate.

An advantage due to the inherent nature of S-parameters is:

Different devices can be measured with one setup—probes do not have to be located right at the test device. Requiring probes to be located at the test device imposes severe limitations on the setup's ability to adapt to different types of devices.

Faster amd more accurate microwave design is possible with S-parameters. When a Smith chart is laid over a polar display of s_{11} and s_{22}, the input or output impedance is read directly. If a swept-frequency source is used, the display becomes a graph of input or output impedance versus frequency. Likewise, CW or swept-frequency displays of gain or attenuation can be made.

S-parameter design techniques have been used for some time; in combination with the Smith chart, they can be used to optimize matching networks and to design transistor amplifiers. Amplifiers can be designed for maximum gain or for a specific gain over a given frequency range; amplifier stability can be investigated, and oscillators can be designed.

APPENDIX F METRIC MULTIPLIER PREFIXES

Multiples and submultiples of fundamental units such as meter, gram, liter, second, ohm, farad, henry, volt, ampere, and watt may be indicated by the following prefixes.

Prefix	*Abbreviation*	*Multiplier*
tera	T	10^{12}
giga	G	10^{9}
mega	M	10^{6}
myria	ma	10^{4}
kilo	k	10^{3}
hecto	h	10^{2}
deca	da	10
deci	d	10^{-1}
centi	c	10^{-2}
milli	m	10^{-3}
micro	μ	10^{-6}
nano	n	10^{-9}
pico	p	10^{-12}

APPENDIX G DIELECTRIC CONSTANTS

Material	*Dielectric Constant*
Air	1.00
Teflon® (PTFE)	2.10
Polyethylene	2.25
Rexolite #1422	2.55
Neoprene Rubber	4.00
Balsawood	1.30
Snow (freshly fallen)	1.20
Snow (hard-packed)	1.50
Water (distilled)	76.70

APPENDIX H COEFFICIENTS OF EXPANSION

Element	Expansion (PPM)
Aluminum	24.3
Carbon	0.6–4.3
Chromium	6.2
Cobalt	12.3
Copper	16.56
Gold	14.2
Indium	33
Iron	11.7
Lead	28.7
Magnesium	25.2
Nickel	13.3
Silver	18.9
Tin	23
Titanium	8.5

(PPM = parts per million)

APPENDIX I CHEMICAL SYMBOLS

Element	Symbol
Aluminum	Al
Beryllium	Be
Boron	B
Carbon	C
Chromium	Cr
Cobalt	Co
Copper	Cu
Gallium	Ga
Germanium	Ge
Gold	Au
Indium	In
Iron	Fe
Lead	Pb
Magnesium	Mg
Nickel	Ni
Platinum	Pt
Silicon	Si
Silver	Ag
Tin	Sn
Titanium	Ti
Zinc	Zn

APPENDIX J MELTING POINTS

Material	*Melting Point (°C)*
Aluminum	660
Cobalt	1495
Copper	1083
Gold	1063
Indium	156
Iron	1535
Lead	327
Nickel	1453
Silver	960
Tin	232

APPENDIX K GREEK ALPHABET

Name	*Capital*	*Small*	*Common Designation*
alpha	A	α	Angles, coefficients, attenuation constant, absorption factor, area.
beta	B	β	Angles, coefficients, phase constant.
gamma	Γ	γ	Complex propagation constant (capital), specific gravity, angles, electrical conductivity, propagation constant.
delta	Δ	δ	Increment or decrement (capital or small), determinant (capital), permittivity (capital), density, angles.
epsilon	E	ε	Dielectric constant, permittivity, electric intensity.
zeta	Z	ζ	Coordinates, coefficients.
eta	H	η	Intrinsic impedance, efficiency, surface charge density, hysteresis, coordinates.
theta	Θ	$\vartheta\theta$	Angular phase displacement, time constant, reluctance, angles.
iota	I	ι	Unit vector.
kappa	K	κ	Susceptibility, coupling coefficient.
lambda	Λ	λ	Permeance (capital), wavelength, attenuation constant.
mu	M	μ	Permeability, amplification factor, prefix *micro* (millionth).
nu	N	ν	Reluctivity, frequency.
xi	Ξ	ξ	Coordinates.
omicron	O	o	
pi	Π	π	Radius of a circle = 3.1416.
rho	P	ρ	Resistivity, volume charge density, coordinates.
sigma	Σ	σ	Summation (capital), surface charge density, complex propagation constant, electrical conductivity, leakage coefficient.
tau	T	τ	Time constant, volume resistivity, time-phase displacement, transmission factor, density.
upsilon	Υ	υ	
phi	Φ	$\phi\varphi$	Scalar potential (capital), magnetic flux, angles.
chi	X	χ	Electric susceptibility, angles.
psi	Ψ	ψ	Dielectric flux, phase difference, coordinates, angles.
omega	Ω	ω	Resistance in ohms (capital), solid angle (capital), angular velocity.

Small letter is used except where capital is indicated.

APPENDIX L TEMPERATURE CONVERSION

Use the tables below as follows:

- Find your temperature in the center column.
- If the given temperature is °C, the value of °F can be found in the right side column.
- If the given temperature is °F, the value of °C can be found in the left side column.

Example 1:
Given 650°C, the converted value is 1202°F.

Example 2.
Given 230°F, the converted value is 110°C.

APPENDIX L continued

EXAMPLE 1

−459.4 to 0		
C		F
−273	−459.4	
−268	−450	
−262	−440	
−257	−430	
−251	−420	
−246	−410	
−240	−400	
−234	−390	
−229	−380	
−223	−370	
−218	−360	
−212	−350	
−207	−340	
−201	−330	
−196	−320	
−190	−310	
−184	−300	
−179	−290	
−173	−280	
−169	−273	−459.4
−168	−270	−454
−162	−260	−436
−157	−250	−418
−151	−240	−400
−146	−230	−382
−140	−220	−364
−134	−210	−346
−129	−200	−328
−123	−190	−310
−118	−180	−292
−112	−170	−274
−107	−160	−256
−101	−150	−238
− 96	−140	−220
− 90	−130	−202
− 84	−120	−184
− 79	−110	−166
− 73	−100	−148
− 68	− 90	−130
− 62	− 80	−112
− 57	− 70	− 94
− 51	− 60	− 76
− 46	− 50	− 58
− 40	− 40	− 40
− 34	− 30	− 22
− 29	− 20	− 4
− 23	− 10	+ 14
− 17.8	− 0	+ 32

0 to 100					
C		F	C		F
−17.8	0	32	10.0	50	122.0
−17.2	1	33.8	10.6	51	123.8
−16.7	2	35.6	11.1	52	125.6
−16.1	3	37.4	11.7	53	127.4
−15.6	4	39.2	12.2	54	129.2
−15.0	5	41.0	12.8	55	131.0
−14.4	6	42.8	13.3	56	132.8
−13.9	7	44.6	13.9	57	134.6
−13.3	8	46.4	14.4	58	136.4
−12.8	9	48.2	15.0	59	138.2
−12.2	10	50.0	15.6	60	140.0
−11.7	11	51.8	16.1	61	141.8
−11.1	12	53.6	16.7	62	143.6
−10.6	13	55.4	17.2	63	145.4
−10.0	14	57.2	17.8	64	147.2
− 9.4	15	59.0	18.3	65	149.0
− 8.9	16	60.8	18.9	66	150.8
− 8.3	17	62.6	19.4	67	152.6
− 7.8	18	64.4	20.0	68	154.4
− 7.2	19	66.2	20.6	69	156.2
− 6.7	20	68.0	21.1	70	158.0
− 6.1	21	69.8	21.7	71	159.8
− 5.6	22	71.6	22.2	72	161.6
− 5.0	23	73.4	22.8	73	163.4
− 4.4	24	75.2	23.3	74	165.2
− 3.9	25	77.0	23.9	75	167.0
− 3.3	26	78.8	24.4	76	168.8
− 2.8	27	80.6	25.0	77	170.6
− 2.2	28	82.4	25.6	78	172.4
− 1.7	29	84.2	26.1	79	174.2
− 1.1	30	86.0	26.7	80	176.0
− 0.6	31	87.8	27.2	81	177.8
0.0	32	89.6	27.8	82	179.6
0.6	33	91.4	28.3	83	181.4
1.1	34	93.2	28.9	84	183.2
1.7	35	95.0	29.4	85	185.0
2.2	36	96.8	30.0	86	186.8
2.8	37	98.6	30.6	87	188.6
3.3	38	100.4	31.1	88	190.4
3.9	39	102.2	31.7	89	192.2
4.4	40	104.0	32.2	90	194.0
5.0	41	105.8	32.8	91	195.8
5.6	42	107.6	33.3	92	197.6
6.1	43	109.4	33.9	93	199.4
6.7	44	111.2	34.4	94	201.2
7.2	45	113.0	35.0	95	203.0
7.8	46	114.8	35.6	96	204.8
8.3	47	116.6	36.1	97	206.6
8.9	48	118.4	36.7	98	208.4
9.4	49	120.2	37.2	99	210.2
			37.8	100	212.0

APPENDIX L continued

EXAMPLE 2

100 to 1000					
C		F	C		F
38	100	212	260	500	932
43	110	230	266	510	950
49	120	248	271	520	968
54	130	266	277	530	986
60	140	284	282	540	1004
66	150	302	288	550	1022
71	160	320	293	560	1040
77	170	338	299	570	1058
82	180	356	304	580	1076
88	190	374	310	590	1094
93	200	392	316	600	1112
99	210	410	321	610	1130
100	212	413.6	327	620	1148
104	220	428	332	630	1166
110	230	446	338	640	1184
116	240	464	343	650	1202
121	250	482	349	660	1220
127	260	500	354	670	1238
132	270	518	360	680	1256
138	280	536	366	690	1274
143	290	554	371	700	1292
149	300	572	377	710	1310
154	310	590	382	720	1328
160	320	608	388	730	1346
166	330	626	393	740	1364
171	340	644	399	750	1382
177	350	662	404	760	1400
182	360	680	410	770	1418
188	370	698	416	780	1436
193	380	716	421	790	1454
199	390	734	427	800	1472
204	400	752	432	810	1490
210	410	770	438	820	1508
216	420	788	443	830	1526
221	430	806	449	840	1544
227	440	824	454	850	1562
232	450	842	460	860	1580
238	460	860	466	870	1598
243	470	878	471	880	1616
249	480	896	477	890	1634
254	490	914	482	900	1652
			488	910	1670
			493	920	1688
			499	930	1706
			504	940	1724
			510	950	1742
			516	960	1760
			521	970	1778
			527	980	1796
			532	990	1814
			538	1000	1832

1000 to 2000					
C		F	C		F
538	1000	1832	816	1500	2732
543	1010	1850	821	1510	2750
549	1020	1868	827	1520	2768
554	1030	1886	832	1530	2786
560	1040	1904	838	1540	2804
566	1050	1922	843	1550	2822
571	1060	1940	849	1560	2840
577	1070	1958	854	1570	2858
582	1080	1976	860	1580	2876
588	1090	1994	866	1590	2894
593	1100	2012	871	1600	2912
599	1110	2030	877	1610	2930
604	1120	2048	882	1620	2948
610	1130	2066	888	1630	2966
616	1140	2084	893	1640	2984
621	1150	2102	899	1650	3002
627	1160	2120	904	1660	3020
632	1170	2138	910	1670	3038
638	1180	2156	916	1680	3056
643	1190	2174	921	1690	3074
649	1200	2192	927	1700	3092
654	1210	2210	932	1710	3110
660	1220	2228	938	1720	3128
666	1230	2246	943	1730	3146
671	1240	2264	949	1740	3164
677	1250	2282	954	1750	3182
682	1260	2300	960	1760	3200
688	1270	2318	966	1770	3218
693	1280	2336	971	1780	3236
699	1290	2354	977	1790	3254
704	1300	2372	982	1800	3272
710	1310	2390	988	1810	3290
716	1320	2408	993	1820	3308
721	1330	2426	999	1830	3326
727	1340	2444	1004	1840	3344
732	1350	2462	1010	1850	3362
738	1360	2480	1016	1860	3380
743	1370	2498	1021	1870	3398
749	1380	2516	1027	1880	3416
754	1390	2534	1032	1890	3434
760	1400	2552	1038	1900	3452
766	1410	2570	1043	1910	3470
771	1420	2588	1049	1920	3488
777	1430	2606	1054	1930	3506
782	1440	2624	1060	1940	3524
788	1450	2642	1066	1950	3542
793	1460	2660	1071	1960	3560
799	1470	2678	1077	1970	3578
804	1480	2696	1082	1980	3596
810	1490	2714	1088	1990	3614
			1093	2000	3632

APPENDIX M GLOSSARY OF MICROWAVE TRANSISTOR TERMINOLOGY

$V_{(BR)CBO}$	Breakdown voltage of a reverse biased collector base junction (measured with the emitter open).
$V_{(BR)EBO}$	Breakdown voltage of a reversed biased emitter base junction (measured with the collector open).
$V_{(BR)CEO}$	Breakdown voltage between the collector and emitter terminals (measured with the base open).
I_{CBO}	Leakage current of a reverse biased collector base junction (measured with the emitter open).
I_{EBO}	Leakage current of a reverse biased emitter base junction (measured with the collector open).
h_{fe}	dc common emitter current gain
C_{cb}	Collector base junction capacitance (measured with the emitter connected to the grounded terminal of a three terminal measurement system).
f_T	Transition Frequency—The frequency at which the magnitude of the small signal common emitter short circuit current gain approximates unity.
r'_bC_C	The collector base time constant.
$P_{T(\max)}$	Maximum continuous power dissipation below a reference temperature (usually 25°C).
$T_{J(\max)}$	Maximum allowable transistor junction temperature.
$I_{C(\max)}$	Maximum allowable collector current without destruction or degradation of the transistor.
NF	A measure of the noise generated by the transistor.
$G_{\max}$	The maximum available power gain (MAG) when the transistor is unconditionally stable and input and output ports are simultaneously conjugately matched.
$f_{\max}$	Maximum frequency of oscillation—the frequency at which $G_{\max}$ approximates unity.
s_{11}	Input reflection coefficient.
s_{12}	Reverse transfer coefficient.
s_{21}	Forward transfer coefficient.
s_{22}	Output reflection coefficient.
P_0	The power output at the one (1) dB gain compression point (amplifier). A measure of the RF power output (oscillator).

APPENDIX N MICROWAVE TERMS

Attenuation—The decrease in amplitude of a signal during its transmission from one point to another. It may be expressed as a ratio or in dB.

Characteristic Impedance—The driving-point impedance (or resistance) of a line, if the line were of infinite length. Microwave systems are usually 50Ω.

Coaxial Cable—Also called coaxial line, coaxial transmission line, and concentric line. A transmission line in which one conductor completely surrounds the other, the two being coaxial and separated by a continuous solid dielectric or by dielectric spacers.

Cut-Off Frequency—The frequency that marks the edge of the passband of a filter and the beginning of the transition to the stopband. Also the upper frequency limit (usually of a loaded transmission circuit) beyond which the attenuation rises very rapidly. In waveguide, this figure is the lowest frequency at which the waveguide propagates energy in some particular mode without attenuation.

dc Return—A path back to ground for a microwave semiconductor which usually consists of a low value of resistance or a coil; it can be used as an RF choke also.

Directivity—The property that causes an antenna to radiate or receive more energy in some directions than others. Also, the property that causes energy to be coupled in one direction and not in another.

Gigahertz (GHz)—A term for 10^9 cycles per second, which replaces the more cumbersome and obsolete term kilomegacycle.

Harmonic—A wave having a frequency that is an integral multiple of the fundamental frequency. For example, a wave with twice the frequency of the fundamental is called the *second harmonic*.

Iris (Diaphragm)—In a waveguide, a conducting plate (or plates) which is very thin (compared to a wavelength) and occupies part of the cross section of a waveguide.

Microstrip—A microwave transmission component in which a single conductor is supported above a ground plane.

Noice—Any unwanted electrical disturbance or spurious signal which modifies the transmitting, indicating, or recording of desired data. Also, random electrical variation generated internally in electronic components.

Noise Figure—The ratio of the signal-to-noise at the input of a device to the signal-to-noise ratio at the output of a device. If the log is taken of this ratio and multiplied by ten, you have the quantity in decibels.

Octave—A band of frequencies, the limits of which have a ratio of two to one. For example, 1 to 2 GHz, 2 to 4 GHz, 4 to 8 GHz, *et cetera*.

Passband—The band of frequencies which can pass through a filter with essentially no attenuation.

Reflection Coefficient—At any specified plane in a uniform transmission medium, the vector ratios between the electrical fields associated with the reflected and incident waves.

Return Loss—The difference between the power incident upon, and the power reflected from, a load.

Ripple—The wave-like variation in the amplitude response of a filter.

Stripline (Sandwich)—A transmission component in which a single conductor is surrounded by ground plane; the term "sandwich" is derived from this physical arrangement.

Subharmonic—A sinusoidal quantity, the frequency of which is an integral submultiple of the fundamental frequency. A wave with half the frequency of the fundamental is called the second subharmonic, one with a third of the fundamental is the third subharmonic, *et cetera*.

VSWR (Voltage Standing Wave Ratio)—The ratio of the amplitude of the electric field or voltage at a voltage maximum to the amplitude at an adjacent voltage minimum.

APPENDIX O REFERENCE TABLE OF RIGID RECTANGULAR WAVE GUIDE DATA AND FITTING

Waveguide										
EIA designation WR( )	MDL designation ( ) band	JAN designation RG( )/U	Material alloy	Dimensions (in.)					Recommended operating range for TE_{10} mode	
				Inside	Tol.	Outside	Tol.	Wall thickness nominal	Frequency (kmc/sec)	Wavelength (cm)
2300	2300	...	Alum.	23.000–11.500	±0.020	23.250–11.750	±0.020	0.125	0.32–0.49	93.68–61.18
2100	2100	...	Alum.	21.000–10.500	±0.020	21.250–10.750	±0.020	0.125	0.35–0.53	85.65–56.56
1800	1800	201	Alum.	18.000–9.000	±0.020	18.250–9.250	±0.020	0.125	0.41–0.625	73.11–47.96
1500	1500	202	Alum.	15.000–7.500	±0.015	15.250–7.750	±0.015	0.125	0.49–0.75	61.18–39.97
1150	1150	203	Alum.	11.500–5.750	±0.015	11.750–6.000	±0.015	0.125	0.64–0.96	46.84–31.23
975	975	204	Alum.	9.750–4.875	±0.010	10.000–5.125	±0.010	0.125	0.75–1.12	39.95–26.76
770	770	205	Alum.	7.700–3.850	±0.005	7.950–4.100	±0.005	0.125	0.96–1.45	31.23–20.67
650	*L*	69 103	Copper Aluminum	6.500–3.250	±0.005	6.660–3.410	±0.005	0.080	1.12–1.70	26.76–17.63
510	510	...	...	5.100–2.550	±0.005	5.260–2.710	±0.005	0.080	1.45–2.20	20.67–13.62
430	*W*	104 105	Copper Aluminum	4.300–2.150	±0.005	4.460–2.310	±0.005	0.080	1.70–2.60	17.63–11.53
340	340	112 113	Copper Aluminum	3.400–1.700	±0.005	3.560–1.860	±0.005	0.080	2.20–3.30	13.63–9.08
284	*S*	48 75	Copper Aluminum	2.840–1.340	±0.005	3.000–1.500	±0.005	0.080	2.60–3.95	11.53–7.59
229	229	...	...	2.290–1.145	±0.005	2.418–1.273	±0.005	0.064	3.30–4.90	9.08–6.12
187	*C*	49 95	Copper Aluminum	1.872–0.872	±0.005	2.000–1.000	±0.005	0.064	3.95–5.85	7.59–5.12
159	159	...	...	1.590–0.795	±0.004	1.718–0.923	±0.004	0.064	4.90–7.05	6.12–4.25
137	X_B	50 106	Copper Aluminum	1.372–0.622	±0.004	1.500–0.750	±0.004	0.064	5.85–8.20	5.12–3.66
112	X_L	51 68	Copper Aluminum	1.122–0.497	±0.004	1.250–0.625	±0.004	0.064	7.05–10.00	4.25–2.99
90	*X*	52 67	Copper Aluminum	0.900–0.400	±0.003	1.000–0.500	±0.003	0.050	8.20–12.40	3.66–2.42
75	75	...	...	0.750–0.375	±0.003	0.850–0.475	±0.003	0.050	10.00–15.00	2.99–2.00
62	K_U*	91 — 107	Copper Aluminum Silver	0.622–0.311	±0.0025	0.702–0.391	±0.003	0.040	12.4–18.00	2.42–1.66
51	51	...	...	0.510–0.255	±0.0025	0.590–0.335	±0.003	0.040	15.00–22.00	2.00–1.36
42	*K*	53 121 66	Copper Aluminum Silver	0.420–0.170	±0.0020	0.500–0.250	±0.003	0.040	18.00–26.50	1.66–1.13
34	34	...	...	0.340–0.170	±0.0020	0.420–0.250	±0.003	0.040	22.00–33.00	1.36–0.91
28	K_A	— — 96	Copper Aluminum Silver	0.280–0.140	±0.0015	0.360–0.220	±0.002	0.040	26.50–40.00	1.13–0.75
22	*Q*	— 97	Copper Silver	0.224–0.112	±0.0010	0.304–0.192	±0.002	0.040	33.00–50.00	0.91–0.60
19	19	...	...	0.188–0.094	±0.0010	0.268–0.174	±0.002	0.040	40.00–60.00	0.75–0.50
15	*V*	— 98	Copper Silver	0.148–0.074	±0.0010	0.228–0.154	±0.002	0.040	50.00–75.00	0.60–0.40
12	12	— 99	Copper Silver	0.122–0.061	±0.0005	0.202–0.141	±0.002	0.040	60.00–90.00	0.50–0.33
10	10	...	...	0.100–0.050	±0.0005	0.180–0.130	±0.002	0.040	75.00–110.00	0.40–0.27

* This is an MDL flange number.
SOURCE: Microwave Development Laboratories, Inc.

Cutoff for TE_{10} mode		Range in $\frac{2\lambda}{\lambda c}$	Range in $\frac{\lambda g}{\lambda}$	Theoretical attenuation lowest to highest frequency (db/100 ft)	Theoretical c-w power rating lowest to highest frequency (mw)	Fittings		
						Flange		
Frequency (kmc/sec)	Wavelength (cm)					Choke UG( )/U	Cover UG( )/U	EIA designation WR ( )
0.256	116.84	1.60–1.05	1.68–1.17	0.051–0.031	153.0–212.0	...		2300
0.281	106.68	1.62–1.06	1.68–1.18	0.054–0.034	120.0–173.0	...	FA168A*	2100
0.328	91.44	1.60–1.05	1.67–1.18	0.056–0.038	93.4–131.9	...		1800
0.393	76.20	1.61–1.05	1.62–1.17	0.069–0.050	67.6–93.3	...		1500
0.513	58.42	1.60–1.07	1.82–1.18	0.128–0.075	35.0–53.8	...		1150
0.605	49.53	1.61–1.08	1.70–1.19	0.137–0.095	27.0–38.5	...		975
0.766	39.12	1.60–1.06	1.66–1.18	0.201–0.136	17.2–24.1	...		770
0.908	33.02	1.62–1.07	1.70–1.18	0.317–0.212 0.269–0.178	11.9–17.2	... ...	417A 418A	650
1.157	25.91	1.60–1.05	1.67–1.18	...	...	...		510
1.372	21.84	1.61–1.06	1.70–1.18	0.588–0.385 0.501–0.330	5.2–7.5	... ...	435A 437A	430
1.736	17.27	1.58–1.05	1.78–1.22	0.877–0.572 0.751–0.492	3.1–4.5	... ...	553 554	340
2.078	14.43	1.60–1.05	1.67–1.17	1.102–0.752 0.940–0.641	2.2–3.2	54A 585	53 584	284
2.577	11.63	1.56–1.05	1.62–1.17	...	...	...		229
3.152	9.510	1.60–1.08	1.67–1.19	2.08–1.44 1.77–1.12	1.4–2.0	148B 406A	149A 407	187
3.711	8.078	1.51–1.05	1.52–1.19	...	...	...		159
4.301	6.970	1.47–1.05	1.48–1.17	2.87–2.30 2.45–1.94	0.56–0.71	343A 440A	344 441	137
5.259	5.700	1.49–1.05	1.51–1.17	4.12–3.21 3.50–2.74	0.35–0.46	52A 137A	51 138	112
6.557	4.572	1.60–1.06	1.68–1.18	6.45–4.48 5.49–3.83	0.20–0.29	40A 136A	39 135	90
7.868	3.810	1.57–1.05	1.64–1.17	...	...	...		75
9.486	3.160	1.53–1.05	1.55–1.18	9.51–8.31 ... 6.14–5.36	0.12–0.16	541 FA190A* ...	419 FA191A* ...	62
11.574	2.590	1.54–1.05	1.58–1.18	...	...	...		51
14.047	2.134	1.56–1.06	1.60–1.18	20.7–14.8 17.6–12.6 13.3–9.5	0.043–0.058	596 598 ...	595 597 ...	42
17.328	1.730	1.57–1.05	1.62–1.18	...	...	...		34
21.081	1.422	1.59–1.05	1.65–1.17	... ... 21.9–15.0	0.022–0.031	600 FA1241A* ...	599 FA1242A* ...	28
26.342	1.138	1.60–1.05	1.67–1.17	... 31.0–20.9	0.014–0.020	... ...	383 ...	22
31.357	0.956	1.57–1.05	1.63–1.16	...	...	...		19
39.863	0.752	1.60–1.06	1.67–1.17	... 52.9–39.1	0.0063–0.0090	... ...	385 ...	15
48.350	0.620	1.61–1.06	1.68–1.18	... 93.3–52.2	0.0042–0.060	... ...	387 ...	12
59.010	0.508	1.57–1.06	1.61–1.18	...	...	...		10

APPENDIX P TRANSMISSION LINES BY RG/U NUMBER

Abbreviation

Dielectric

PE	Solid Polyethylene
PTFE	Solid Polytetrafluoroethylene
PIB	Polyisobulyene, Type B, per MIL-C-17
Rubber	per MIL-C-17D
Sil	Silicone Rubber
PS	Polystyrene
AL	Aluminum

Conductors and Braid Materials

SCCA1	Silver covered copper covered aluminum
BC	Bare copper
SC	Silver covered copper
CCS	Copper covered steel
TC	Tinned copper
SCCS	Silver covered copper covered steel
SCCad Br	Silver covered cadmium bronze
GS	Galvanized steel
TCCS	Tinned copper covered steel
SSC	Silver covered strip
HR	High resistance wire
SA	Silver covered Alloy 135

Jacket Material

PVC-1	Black Polyvinylchloride, contaminating, Type 1, per MIL-C-17D
PVC-11	Grey polyvinylchloride, non-contaminating, Type II, per MIL-C-17D
PVC-11A	Black polyvinylchloride, non-contaminating, Type IIA, per MIL-C-17D
PE-111	Clear polyethylene
PE-111A	High molecular weight, black polyethylene, Type IIIA, per MIL-C-17D
FG Braid V.	Fiberglass, impregnated, Type V, per MIL-C-17D
FEP-IX	Fluorinated ethylene propylene, Type IX, per MIL-C-17D
PUR	Polyurethane, Black specific compounds

APPENDIX P continued

SIL/DAC-VI Dacron braid over silicon rubber, Type VI, per MIL-C-17D
Rubber Per MIL-C-17D

RG U Type	Inner Conductor	Dielectric Material	DOD (Inch)	Number & Type of Shielding Braids	Jacket Material	O.D. (Inch)	Weight (lbs/ft)	Nominal Imped. (ohms)	Nominal Capac-itance (pf/ft)	Max. Oper. Temp. Range (°C)	Max. Oper. Voltage (Volts RMS)	Comments
1-3	WAVEGUIDE											Waveguide Not manufac
4	0 0320" BC	PE	0 116	2 BC	PVC-I	0 226	0 025	50 0	30 8	40 +80	1 900	Use RG58C
5	0 0508" BC	PE	0 185	2 BC	PVC-I	0 332	0 088	52 5	28 5	40 +80	3 000	Use up to 100 MHz
5A	0 0508" SC	PE	0 181	2 SC	PVC-II	0 328	0 088	50 0	30 8	40 +80	3 000	Use RG212
5B	0 0508" SC	PE	0 181	2 SC	PVC-IIA	0 328	0 087	50 0	30 8	40 +80	3 000	Use RG212
6	0 0285" CCS	PE	0 185	2 Inner SC Outer BC	PVC-II	0 332	0 081	76 0	20 0	40 +80	2 700	Use RG6A
6A	0 0285" CCS	PE	0 185	2 Inner SC Outer BC	PVC-IIA	0 332	0 082	75 0	20 6	40 +80	2 700	Good attenuation stability
7	0 0359" BC	Air-space PE	0 250	1 BC	PVC-I	0 370	0 080	95 0	12 5	-40 +80	1 000	Use RG63B
8	0 0855" 7 0 0285" BC	PE	0 285	1 BC	PVC-I	0 405	0 106	52 0	29 5	-40 +80	4 000	Use RG213
8A	0 0855" 7 0 0285" BC	PE	0 285	1 BC	PVC-IIA	0 405	0 106	52 0	29 5	-40 +80	5 000	Use RG213
9	0 0855" 7 0 0285" SC	PE	0 280	2 Inner SC Outer BC	PVC-II	0 420	0 140	51 0	30 0	-40 +80	4,000	Use RG214
9A	0 0855" 7 0 0285" SC	PE	0 280	2 SC	PVC-II	0 420	0 140	51 0	30 0	40 +80	4,000	Use RG214
9B	0 0855" 7 0 0285" SC	PE	0 280	2 SC	PVC-IIA	0 420	0 150	50 0	30 8	-40 +80	5 000	Use RG214
10	0 0855" 7 0 0285" BC	PE	0 285	1 BC	PVC-II w Armor	0 475 max	0 146	52 0	29 5	-40 +80	4 000	Use RG215
10A	0 0855" 7 0 0285" BC	PE	0 285	1 BC	PVC-IIA w Armor	0 475 max	0 146	52 0	29 5	-40 +80	5 000	Use RG215
11	0 0477" 7 0 0159 TC	PE	0 285	1 BC	PVC-I	0 405	0 096	75 0	20 6	-40 +80	4 000	Use up to 100 MHz
11A	0 0477" 7 0 0159" TC	PE	0 285	1 BC	PVC-IIA	0 405	0 096	75 0	20 6	-40 +80	5 000	Use up to 1000 MHZ
12	0 0477" 7 0 0159" TC	PE	0 285	1 BC	PVC-II w Armor	0 475 max	0 141	75 0	20 6	-40 +80	4 000	Use RG12A
12A	0 0477" 7 0 0159" TC	PE	0 285	1 BC	PVC-IIA w Armor	0 475 max	0 141	75 0	20 6	-40 +80	5 000	Use up to 1000 MHz
13	0 0477" 7 0 0159" TC	PE	0 280	2 BC	PVC-I	0 420	0 126	74 0	20 8	-40 +80	4.000	Use RG216
13A	0 0477" 7 0 0159" TC	PE	0 280	2 BC	PVC-IIA	0 420	0 126	74 0	20 8	40 +80	5 000	Use RG216
14	0 0102" BC	PE	0 370	2 BC	PVC-II	0 545	0 216	52 0	29 5	-40 +80	5.500	Use RG217 [3]
14A	0 1020" BC	PE	0 370	2 BC	PVC-IIA	0 545	0 216	52 0	29 5	-40 +80	7 000	Use RG217
15	0 0571" CCS	PE	0 370	2 BC	PVC-I	0 545	0 197	76 0	20 0	-40 +80	5.000	Use up to 1000 MHz
16	0 1250" BC tube	PE	0 460	1 BC	PVC-I	0 630	0 254	52 0	29 5	-40 +80	6.000	Use up to 1000 MHz
17	0 1880" BC	PE	0 680	1 BC	PVC-II	0 870	0 460	52 0	29 5	-40 +80	11.000	Use up to 1000 MHz
17A	0 1880" BC	PE	0 680	1 BC	PVC-IIA	0 870	0 460	52 0	29 5	-40 +80	11.000	Use RG218
17B	CANCELLED REASSIGNED NEW NOMENCLATURE, RG177											
18	0 1880" BC	PE	0 680	1 BC	PVC-II w Armor	0 945 max	0 585	52 0	29 5	-40 +80	11 000	Use RG219
18A	0 1880" BC	PE	0 680	1 BC	PVC-IIA w Armor	0 945 max	0 585	52 0	29 5	-40 +80	11 000	Use RG219

APPENDIX P continued

RG U Type	Inner Conductor	Dielectric Material	DOD (Inch)	Number & Type of Shielding Braids	Jacket Material	O.D. (Inch)	Weight (lbs ft)	Nominal Imped. (ohms)	Nominal Capac-itance (pf ft)	Max. Oper Temp. Range (C)	Max. Oper. Voltage (Volts RMS)	Comments
19	0.2500" BC	PE	0.910	1 BC	PVC-II	1.120	0.740	52.0	29.5	-40 +80	14,000	Use RG220
19A	0.2500" BC	PE	0.910	1 BC	PVC-IIA	1.120	0.740	52.0	29.5	-40 +80	14,000	Use RG220
20	0.2500" BC	PE	0.910	1 BC	PVC-II w Armor	1.195 max	0.925	52.0	29.5	-40 +80	14,000	Use RG221
20A	0.2500" BC	PE	0.910	1 BC	PVC-IIA w Armor	1.195 max	0.925	52.0	29.5	-40 +80	14,000	Use RG221
21	0.0508" high resistance wire	PE	0.185	2 SC	PVC-II	0.332	0.087	53.0	29.0	-40 +80	2,700	Use RG222
21A	0.0508" high resistance wire	PE	0.185	2 SC	PVC-IIA	0.332	0.087	53.0	29.0	-40 +80	2,700	Use RG222
22	2 cond 0.0456" 7/0.0152" BC	PE	0.285	1 TC	PVC-I	0.405	0.105	95.0	16.0	-40 +80	1,000	Balanced line w twisted cond
22A	2 cond 0.0456" 7/0.0152" BC	PE	0.285	2 TC	PVC-II	0.420	0.151	95.0	16.0	-40 +80	1,000	Balanced line w twisted cond
22B	2 cond 0.0456" 7/0.0152" BC	PE	0.285	2 TC	PVC-IIA	0.420	0.151	95.0	16.0	-40 +80	1,000	Balanced line w twisted cond
23	2 cond 0.0855" 7/0.0285" BC	PE 2 cores	0.380	2 Individual inner Common outer BC	PVC-I	0.650x 0.945	0.490	125.0	12.0	-40 +80	3,000	Use RG23A
23A	2 cond 0.0855" 7/0.0285" BC	PE 2 cores	0.380	2 Individual inner Common outer BC	PVC-IIA	0.650x 0.945	0.490	125.0	12.0	-40 +80	3,000	Dual coaxial balanced line
24	2 cond 0.0855" 7/0.0285" BC	PE 2 cores	0.380	2 Individual inner Common outer BC	PVC-I w Armor	1.034x 0.735	0.670	125.0	12.0	-40 +80	3,000	Use RG24A (max)
24A	2 cond 0.0855" 7/0.0285" BC	PE 2 cores	0.380	2 Individual inner Common outer BC	PVC-IIA w Armor	1.034x 0.735	0.670	125.0	12.0	-40 +80	3,000	Armored RG23A (max)
25A	0.0585" 19/0.0117" TC	Rubber-E	0.288	2 TC	Rubber-IV	0.505	0.2[illegible]5	48.0	50.0	-40 +80	10,000	
26A	0.0585" 19/0.0117" TC	Rubber-E	0.288	1 TC	Rubber-IV w Armor	0.505 max	0.189	48.0	50.0	-40 +80	10,000	
27A	0.0925" 19/0.0185" TC	Rubber-D	0.455	1 TC	Rubber-IV w Armor	0.570 max	0.304	48.0	50.0	-40 +80	15,000	
28B	0.0925" 19/0.0185" TC	Rubber-D	0.455	2 TC GS	Rubber-IV	0.750	0.370	48.0	50.0	-40 +80	15,000	
29	0.0320" BC	PE	0.116	1 TC	PE-III	0.184 max	0.021	53.5	28.5	-55 +80	1,900	Use RG58
30	0.0477" 7/0.0159" BC	PIB	0.185	1 BC	PVC-I	0.250	0.044	50.0	27.0	-40 +80	1,500	Use RG58
31	0.0855" 7/0.0285" BC	PIB	0.285	1 BC	PVC-I	0.405	0.106	51.0	31.0	-40 +80	2,000	Use RG213
32	0.0855" 7/0.0285" BC	PIB	0.285	1 BC	PVC-I w Armor	0.465	0.141	51.0	29.0	-40 +80	2,000	Use RG215
33	0.1019" BC	PE	0.370	None	Lead	0.470	0.390	51.0	30.0	-55 +80	6,000	
34	0.0855" 7/0.0285" BC	PE	0.455	1 BC	PVC-I	0.625	0.224	71.0	21.5	-40 +80	5,200	Use RG31B
34A	0.0747" 7/0.0249" BC	PE	0.460	1 BC	PVC-IIA	0.630	0.224	75.0	20.6	-40 +80	6,500	Use RG34B
34B	0.0747" 7/0.0249" BC	PE	0.460	1 BC	PVC-IIA	0.630	0.224	75.0	20.6	-40 +80	6,500	Use up to 1000 MHz
35	0.1144" BC	PE	0.680	1 BC	PVC-II w Armor	0.945 max	0.525	71.0	21.5	-40 +80	10,000	Use RG35B
35A	0.1045" BC	PE	0.680	1 BC	PVC-IIA w Armor	0.945 max	0.525	75.0	20.6	-40 +80	10,000	Use RG35B
35B	0.1045" BC	PE	0.680	1 BC	PVC-11A w Armor	0.945 max	0.525	75.0	20.6	-40 +80	10,000	Unarmored see RG164
36	0.1620" PC	PE	0.910	1 BC	PVC-I	1.180	0.805	69.0	22.0	-40 +80	13,000	Use up to 1000 MHz
37	0.0320" TC	Rubber-C	0.140	1 TC	PE-III	0.210	0.040	52.5	38.0	-55 +80	750	

APPENDIX P continued

RG U Type	Inner Conductor	Dielectric Material	DOD (Inch)	Number & Type of Shielding Braids	Jacket Material	O.D. (Inch)	Weight (lbs/ft)	Nominal Imped. (ohms)	Nominal Capac-itance (pf/ft)	Max. Oper. Temp. Range (°C)	Max. Oper. Voltage (Volts RMS)	Comments
38	0 0453" TC	Rubber-C	0 196	2 TC	PE-III	0 312	0 110	52 5	38 0	−55 +80	1.000	
39	0 0253" CCS	Rubber-C	0 196	2 TC	PE-III	0 312	0 100	72 5	28 6	−55 +80	1.000	
40	0 0253" CCS	Rubber-C	0 196	2 TC	Rubber-IV	0 420	0 150	72 5	28 0	−40 +80	1 000	
41	0 0490" 16/0 0100" TC	Rubber-C	0 250	1 TC	Rubber-IV	0 425	0 150	67 5	27 6	−40 +80	3.000	
42	0 0285" Resistance wire	PE	0 196	2 SC	PVC-II	0 342	0 050	78 0	20 0	−40 +80	2.700	Use RG222
43	2 cond 0 0855" 7/0 0285" BC	Rubber-B	0 472	1 BC	PVC-I	0 617		95 0	17 6	−40 +80	1 500	Use RG57
44-47	STUD SUPPORTED RIGID LINES See Mil-HDBK 216, Para 5 5											
48-53	RECTANGULAR WAVEGUIDES COVERED BY MIL-W-85 See Mil HDBK 216 Para 6 23											
54	0 0477" 7 0 0159" BC	PE	0 185	1 BC	PVC-I	0 275	0 045	58 0	27 0	−40 +80	2 500	Use RG54A
54A	0 0456" 7 0 0152" BC	PE	0 178	1 TC	PE-III	0 250 (max)	0 041	58 0	26 5	−55 +80	3.000	Use up to 1000 MHz
55	0 0320" BC	PE	0 116	2 TC	PE-III	0 206 max	0 032	53 5	28 5	−55 +80	1.900	Use RG55B
55A	0 0350" SC	PE	0 116	2 SC	PVC-IIA	0 216 max	0 034	50 0	30 8	−40 +80	1.900	Use RG223
55B	0 0320" SC	PE	0 116	2 TC	PE-IIIA	0 206 max	0 033	53 5	28 5	−55 +80	1.900	Use up to 1000 MHz
56	0 0585" 19 0 0117" BC	Rubber-D	0 308	2 BC	PVC-I	0 535	0 243	48 0	50 0	−40 +80	8.000	
57	2 cond 0 0855" 7 0 0285" BC	PE	0 472	1 TC	PVC-I	0 625	0 225	95 0	17 0	−40 +80	3.000	Balanced line Parallel
57A	2 cond 0 0855" 7 0 0285" BC	PE	0 472	1 TC	PVC-IIA	0 625	0 225	95 0	17 0	−40 +80	3,000	Balanced line Parallel
58	0 0320" BC	PE	0 116	1 TC	PVC-I	0 195	0 029	53 5	28 5	−40 +80	1.900	Use RG58B
58A	0 0355" 19 0 0071" TC	PE	0 116	1 TC	PVC-I	0 195	0 029	52 0	28 5	−40 +80	1.900	Use RG58C
58B	0 0320" BC	PE	0 116	1 TC	PVC-IIA	0 195	0 029	53 5	28 5	−40 +80	1.900	Use up to 1000 MHz
58C	0 0355" 19 0 0071" TC	PE	0 116	1 TC	PVC-IIA	0 195	0 029	50 0	30 8	−40 +80	1.900	Extra flexible version RG58B
59	0 0253" CCS	PE	0 146	1 BC	PVC-I	0 242	0 032	73 0	21 0	−40 +80	2.300	Use RG59B
59A	0 0230" CCS	PE	0 146	1 BC	PVC-IIA	0 242	0 032	75 0	20 6	−40 +80	2.300	Use RG59B
59B	0 0230" CCS	PE	0 146	1 BC	PVC-IIA	0 242	0 032	75 0	20 6	−40 +80	2.300	Use up to 1000 MHz
60	0 0508" Str C	Rubber-C	0 250	1 BC	Rubber-IV	0 425	0 150	50 0	39 0	-40 +80	1.100	
61	SPECIAL 500 OHM LINE											
62	0 0253" CCS	Air-space PE	0 146	1 BC	PVC-I	0 242	0 038	93 0	13 5	−40 +80	750	Use RG62A
62A	0 0253" CCS	Air-space PE	0 146	1 BC	PVC-IIA	0 242	0 038	93 0	13 5	−40 +80	750	Low capacitance
62B	0 0240" 7/0 0080" CCS	Air-space PE	0 146	1 BC	PVC-IIA	0 242	0 038	93 0	13 5	−40 +80	750	Extra flexible RG62A
63	0 0253" CCS	Air-space PE	0 285	1 BC	PVC-I	0 405	0 083	125 0	10 0	−40 +80	1,000	Use RG63B
63A	0 0253" BC	Air-space PE	0 285	1 BC	PVC-I	0 405	0 083	125 0	10 0	−40 +80	1,000	Use RG63B

APPENDIX P continued

RG U Type	Inner Conductor	Dielectric Material	DOD (Inch)	Number & Type of Shielding Braids	Jacket Material	O.D. (Inch)	Weight (lbs/ft)	Nominal Imped. (ohms)	Nominal Capac-itance (pf/ft)	Max. Oper. Temp. Range (°C)	Max. Oper. Voltage (Volts RMS)	Comments
63B	0.0253" CCS	Air space PE	0.285	1 BC	PVC-IIA	0.405	0.083	125.0	10.0	−40 +80	1 000	Low capacitance
64	0.0585" 19/0.0117 TC	Rubber D	0.308	2 TC	Rubber IV	0.495	0.225	48.0	60.0	−40 +80	[illegible]	
64A	0.0585" 19/0.0117" TC	Rubber E	0.288	2 TC	Rubber IV	0.475 max	0.205	48.0	50.0	−40 +80	[illegible]	
65	0.0080" Formex-F 0.1280" dia. Helix	PE	0.285	1 BC	PVC-I	0.405	0.096	950.0	44.0	−40 +80	1 000	High impedance video delay line
65A	0.0080" Formex F 0.1280" dia. Helix	PE	0.285	1 BC	PVC-IIA	0.405	0.096	950.0	44.0	−40 +80	1 000	High impedance video delay line
66-69	RECTANGULAR WAVEGUIDE COVERED BY MIL-W-25. See Mil HDBK 216 Para 6.17-6.23											
71	0.0253" CCS	Air space PE	0.146	2 TC	PVC-I	0.250 max	0.046	93.0	13.5	−40 +80	750	Use RG71B
71A	0.0253" CCS	Air space PE	0.146	2 TC	PE III	0.250 max	0.046	93.0	13.5	−55 +80	750	Use RG71B
71B	0.0253" CCS	Air space PE	0.146	2 TC	PE IIIA	0.250 max	0.046	93.0	13.5	−55 +80	750	Low capacitance
72	0.0253" CCS	Air space PE	0.460	1 BC	PVC-I	0.630	0.169	150.0	7.8	−40 +80	750	Low capacitance
73	0.0320" BC	PE	0.116	2 BC	Copper braid	0.175	0.031	25.0	61.8	−55 +80	1 000	Low impedance
74	0.1020" BC	PE	0.370	2 BC	PVC-II w. Armor	0.615 max	0.310	52.0	29.5	−40 +80	5 500	Use RG224
74A	0.1020" BC	PE	0.370	2 BC	PVC-IIA w. Armor	0.615 max	0.310	52.0	29.5	−40 +80	7 000	Use RG224
75	RECTANGULAR WAVEGUIDE COVERED BY MIL-W-85. See Mil HDBK 216 Para 6.17-6.21											
76	STUD SUPPORTED RIGID LINE. See Mil HDBK 216 Para 5.5											
77A	0.0585" 19/0.0117" TC	Rubber E	0.288	2 TC	PVC-IIA	0.450	0.195	48.0	50.0	−40 +80	8 000 Peak	
78A	0.0585" 19/0.0117" TC	Rubber E	0.288	1 TC	PVC-IIA	0.420	0.149	48.0	50.0	−40 +80	8 000 Peak	
79	0.0253" CCS	Air-space PE	0.285	1 BC	PVC-I w. Armor	0.475 max	0.136		10.0	−40 +80	1 000	Low capacitance
79A	0.0253" CCS	Air-space PE	0.285	1 BC	PVC-I w. Armor	0.475 max	0.130	125.0	10.0	−40 +80	1 000	Low capacitance
79B	0.0253" CCS	Air-space PE	0.285	1 BC	PVC-IIA w. Armor	0.475 max	0.136	125.0	10.0	−40 +80	1 000	Low capacitance
80	RIGID LINE. See Mil HDBK 216 Para 5.2											
81	0.0525" BC	Magnesium Oxide G	0.321	None	Copper tube	0.325	0.172	50.0	37.0	> 250	3 000	
82	0.1250" BC	Magnesium Oxide G	0.650	None	Copper tube	0.750	0.698	50.0	36.0	> 250	5 000	
83	0.1020" BC	PE	0.240	1 BC	PVC-I	0.405	0.120	35.0	44.0	−40 +80	2 000	Low impedance
84A	0.1045" BC	PE	0.680	1 BC	PVC IIA w. lead sheath	1.000	1.325	75.0	20.6	−40 +80	10 000	RG35B with
85A	0.1045" BC	PE	0.680	1 BC	PVC IIA w. lead Armor	1.565 max	2.910	75.0	20.6	−40 +80	10 000	RG84A with special Armor
86	2 cond. 0.0855" 7/0.0285" BC	PE	0.300 x 0.650	None	None	0.300 x 0.650	0.100	200.0	7.8	−55 +80	10 000	Twin lead
87A	0.0960" 7/0.0320" SC	PTFE	0.280	2 SC	FG Braid V	0.425	0.180	50.0	29.4	−55 +250	5 000	Use RG225
88	0.0585" 19/0.0117" TC	Rubber E	0.288	4 TC	PVC-I	0.515 max	[illegible]	48.0	50.0	−40 +80	10 000	
88A	0.0585" 19/0.0117" TC	Rubber E	0.288	4 TC	PVC IIA	0.515 max	0.211	48.0	50.0	−40 +80	10 000	

APPENDIX P continued

RG U Type	Inner Conductor	Dielectric Material	DOD (Inch)	Number & Type of Shielding Braids	Jacket Material	O.D. (Inch)	Weight (lbs/ft)	Nominal Imped. (ohms)	Nominal Capac-itance (pf ft)	Max. Oper. Temp. Range (C)	Max. Oper. Voltage (Volts RMS)	Comments
88B	0 0585″ 19 0 0117″ TC	Rubber-E	0 288	4 TC	Rubber-IV	0 565	0 238	48 0	50 0	40 +80	10 000	
89	0 0253″ CCS	Air-space PE	0 285	1 BC	PVC-I	0 632	0 195	125 0	10 0	- 40 +80	1 000	Low capacitance
90	0 0603″ 7 0 0201″SC	PE	0 195	3 SC GS SC	PVC-IIA	0 425		50 0	30 8	- 40 +80	3 000	Excellent shielding
91	RECTANGULAR WAVEGUIDE COVERED BY MIL-W-85 See Mil HDBK 216 Para 6 17-6 23											
92	RIGID COAXIAL LINE See Mil HDBK 216 Para 5 2											
93	0 2000″ 19 0 0400″BC	Taped PTFE	0 573	1 BC	FG Braid-V	0 710	0 475	50 0	29 4	- 55 +250	10 000	Use RG211A
94	0 1125″ 19 0 0225″ SC	Taped PTFE	0 292	2 BC	FG Braid-V	0 445	0 270	50 0	29 4	- 55 +250	7 000	Use RG226
94A	0 1270″ 19 0 0254	Taped PTFE	0 370	2 BC	FG Braid-V	0 500	0 445	50 0	29 4	55 +250	7 000	Use RG226
95-99	RECTANGULAR WAVEGUIDES COVERED BY MIL W-85 See Mil HDBK 216 Para 6 17-6 23											
100	0 0735″ 19 0 0147″ BC	PE	0 146	1 BC	PVC-I	0 242	0 046	35 0	44 0	- 40 +80	2 000	Use up to 1000 MHz
101	0 0641″ BC	Rubber		1 TC		0 588		75 0				
102	2 cond 0 0808″ BC	Rubber		1 TC		1 088		140 0				
103-107	RECTANGULAR WAVEGUIDES COVERED BY MIL-W-85 See Mil HDBK 216 Para 6 17-6 23											
108	2 cond 0 0378″ 7 0 0126″ TC	PE	0 079 each	1 TC	PVC-II	0 235	0 032	78 0	24 5	- 40 +80	1 000	Use RG108A
108A	2 cond 0 0378″ 7 0 0126″ TC	PE	0 079 each	1 TC	PVC-IIA	0 235	0 032	78 0	24 5	- 40 +80	1 000	Balanced line
109-110	RECTANGULAR WAVEGUIDE COVERED BY MIL-W-85 See Mil HDBK 216 Para 6 17-6 23											
111	2 cond 0 0456″ 7 0 0152″ BC	PE	0 285	2 TC	PVC-II w Armor	0 490 max	0 146	95 0	16 0	- 40 +80	1 000	Use RG111A
111A	2 cond 0 0456″ 7 0 0152″ BC		0 285	2 TC	PVC-IIA w Armor	0 490 max	0 146	95 0	16 0	- 40 +80	1 000	Use RG22B w Armor
112-113	RECTANGULAR WAVEGUIDE COVERED BY MIL-W-85 See Mil HDBK 216 Para 6 17-6 23											
114	0 0070″ CCS	Air-space PE	0 285	1 BC	PVC-I	0 405	0 087	185 0	6 5	- 40 +80	1,000	Use RG114A
114A	0 0070″ CCS	Air-space PE	0 285	1 BC	PVC-IIA	0 405	0 087	185 0	6 5	40 +80	1 000	Low capacitance
115	0 0840″ 7 0 0280″ SC	Taped PTFE	0 250	2 SC	FG Braid-V	0 375	0 148	50 0	29 4	- 55 +250	5,000	Use RG115A
115A	0 0840″ 7 0 0280″ SC	Taped PTFE	0 255	2 SC	FG Braid-V	0 415	0 180	50 0	29 4	- 55 +250	5 000	Extra flexible RG225
116	0 0960″ 7 0 0320″ SC	PTFE	0 280	2 SC	FG Braid-V w Armor	0 475	0 198	50 0	29 4	- 55 +250	5 000	Use RG227
117	0 1880″ BC	PTFE	0 620	1 BC	FG Braid V	0 730	0 641	50 0	29 4	- 55 +250	7 000	Use RG211A
117A	0 1880″ BC	PTFE	0 620	1 BC	FG Braid-V	0 730	0 641	50 0	29 4	- 55 +250	7 000	Use RG211A
118	0 1880″ BC	PTFE	0 620	1 BC	FG Braid-V w Armor	0 780	0 682	50 0	29 4	- 55 +250	7,000	Use RG228A
118A	0 1880″ BC	PTFE	0 620	1 BC	FG Braid-V w Armor	0 780	0 682	50 0	29 4	- 55 +250	7 000	Use RG228A
119	0 1020″ BC	PTFE	0 332	2 BC	FG Braid-V	0 465	0 225	50 0	29 4	- 55 +250	6 000	Use up to 1000 MHz
120	0 1020″ BC	PTFE	0 332	2 BC	FG Braid-V w Armor	0 525 max	0 282	50 0	29 4	- 55 +250	6 000	RG119 w Armor

APPENDIX P continued

RG U Type	Inner Conductor	Dielectric Material	DOD (Inch)	Number & Type of Shielding Braids	Jacket Material	O.D. (Inch)	Weight (lbs/ft)	Nominal Imped. (ohms)	Nominal Capac-itance (pf/ft)	Max. Oper. Temp. Range (°C)	Max. Oper. Voltage (Volts RMS)	Comments
121	RECTANGULAR WAVEGUIDE COVERED BY MIL-W-85 See Mil HDBK 216 Para 17-6 23											
122	0 0300" 27 0 0050" TC	PE	0 096	1 TC	PVC-IIA	0 160	0 016	50 0	29 4	40 +80	1 900	Use up to 1000 MHz
124	0 0253" TCCS	Taped PTFE	0 135	1 TC	FG Braid-V	0 240	0 210	73 0	20 3	-55 +250	2 300	Use RG140
125	0 0159" CCS	Air-space PE	0 460	1 BC	PVC-IIA	0 600	0 180	150 0	7 8	40 +80	2 000	Low capacitance
126	0 0609" 7 0 0203" HR	PTFE	0 185	1 HR	FG Braid-V	0 280	0 070	50 0	29 4	-55 +250	3 000	High loss cable
127	RECTANGULAR WAVEGUIDE COVERED BY MIL-W-85 See Mil HDBK 216 Para 6 17-6 23											
128	RIGID LINE See Mil HDBK 216 Para 5 2											
129	RECTANGULAR WAVEGUIDE COVERED BY MIL-W-85 See Mil HDBK Para 6 17-6 23											
130	2 cond 0 0855" 7 0 0285" BC	PE	0 472	1 TC	PVC-I	0 625	0 220	95 0	17 0	40 +80	3 000	RG57 with twisted cond
131	2 cond 0 0855" 7 0 0285" BC	PE	0 472	1 TC	PVC-I w Armor	0 710 max	0 290	95 0	17 0	40 +80	3 000	Armored RG130
132	RECTANGULAR WAVEGUIDE COVERED BY MIL W-85 See Mil HDBK 216 Para 6 17-6 23											
133	0 0285" BC	PE	0 285	1 BC	PVC-I	0 405	0 094	95 0	16 2	-40 +80	4 000	Use RG133A
133A	0 0253" BC	PE	0 285	1 BC	PVC-IIA	0 405	0 094	95 0	16 2	-40 +80	4 000	95 ohm version RG8
134	RIGID LINE See Mil HDBK 216 Para 5 2											
135 139	RECTANGULAR WAVEGUIDES COVERED BY MIL-W-85 See Mil HDBK 216 Para 6 17-6 23											
140	0 0250" SCCS	PTFE	0 146	1 SC	FG Braid-V	0 233	0 056	75 0	19 5	-55 +250	2 300	See RG302 for FEP jacket
141	0 0359" SCCS	PTFE	0 116	1 SC	FG Braid-V	0 190	0 036	50 0	29 4	-55 +250	1 900	Use RG141A
141A	0 0390" SCCS	PTFE	0 116	1 SC	FG Braid-V	0 190	0 036	50 0	29 4	-55 +250	1 900	See RG303 for FEP jacket
142	0 0359" SCCS	PTFE	0 116	2 SC	FG Braid-V	0 206 max	0 047	50 0	29 4	-55 +250	1 900	Use RG142A
142A	0 0390" SCCS	PTFE	0 116	2 SC	FG Braid-V	0 206 max	0 047	50 0	29 4	-55 +250	1 900	See RG142B for FEP jacket
142B	0 0390" SCCS	PTFE	0 116	2 SC	FEP	0 195	0 050	50 0	29 4	-55 +200	1 900	Stranded center cond available
143	0 0570" SCCS	PTFE	0 185	2 SC	FG Braid-V	0 325	0 114	50 0	29 4	-55 +250	3 000	Use RG143A
143A	0 0590" SCCS	PTFE	0 185	2 SC	FG Braid-V	0 325	0 109	50 0	29 4	-55 +250	3 000	See RG304 for FEP jacket
144	0 0537" 7/0 0179" SCCS	PTFE	0 285	1 SC	FG Braid-V	0 410	0 137	75 0	19 5	-55 +250	5 000	High temp RG11A
145	2 cond 0 0720" BC	Air-space PE		Copper tube	Lead/tar			75 0	14 6			
146	0 0070" CCS	Air-space PTFE	0 285	1 BC	FG Braid-V	0 375	0 108	190 0	6 0	-55 +200	1 000	Low capacitance
147	0 2500" BC	PE	0 910	1 BC	PVC-I w Armor	1 937		52 0	29 5	-40 +80	14 000	RG19U with Armor
148	0 0855" 7/0 0285" BC	PE	0 285	1 BC	PVC-I w Armor	0 800		52 0	29 5	-40 +80	4 000	
149	0 048" 7/0 0159" TC	PE	0 285	1 BC	PVC-IIA	0 405	0 105	75 0	20 6	-40 +80	5 000	Use RG391
150	0 048" 7/0 0159" TC	PE	0 285	1 BC	PVC-IIA w Armor	0 475 max	0 112	75 0	20 6	-40 +80	5 000	Use RG392

APPENDIX P continued

RG U Type	Inner Conductor	Dielectric Material	DOD (Inch)	Number & Type of Shielding Braids	Jacket Material	O.D. (Inch)	Weight (lbs/ft)	Nominal Imped. (ohms)	Nominal Capac-itance (pf/ft)	Max. Oper. Temp. Range (°C)	Max. Oper. Voltage (Volts RMS)	Comments
151 155	RIGID LINES COVERED BY MIL-L-3890 See Mil HDBK 216. Para 5 4											
156	0 0855" 7 0 0285" TC	PE and cond PE	0 285	3 TC. GS TC	PVC-IIA	0 540	0 211	50 0	32 0	-40 +80	10.000	Triaxial pulse cable
157	0 1005" 19 0 0201" TC	PE and cond PE	0 455	3 TC. GS TC	PVC-IIA	0 725	0 317	50 0	38 0	-40 +80	15 000	Triaxial pulse cable
158	0 1988" 37 0 0284" TC	PE and cond PE	0 455	3 TC GS TC	PVC-IIA	0 725	0 380	25 0	78 0	-40 +80	15 000	Triaxial pulse cable
159	0 0320" SC	Taped PTFE	0 116	1 SC	FG Braid-V	0 195	0 035	50 0	29 4	-55 +250	1.900	Use RG141
160	4 cond 0 071" 19 0142" 2TC. 2BC	PE	0 322	1 BC	PVC-I	1 055		125 0	12 0	-40 +80	3.000	4 cond Balanced line
161	0 012" 7 0 004" S Cad BR	PTFE	0 057	1 SC	Nylon	0 082	0 015	70 0	20 0	-60 +120	1 000	Miniature
162	RIGID LINE See Mil HDBK 216. Para 5 2											
163	RECTANGULAR WAVEGUIDE COVERED BY MIL-W-85 See Mil HDBK 216. Para 6 17-6 23											
164	0 1045" BC	PE	0 680	1 BC	PVC-IIA	0 870	0 490	75 0	20 6	-40 +80	10.000	RG35B without Armor
165	0 0960" 7 0 0320" SC	PTFE	0 285	1 SC	FG Braid-V	0 410	0 121	50 0	29 4	-55 +250	5 000	RG225 with one braid
166	0 096" 7 0 0320" SC	PTFE	0 285	1 SC	FG Braid-V w Armor	0 460	0 144	50 0	29 4	-55 +250	5 000	RG165 with Armor
167 173	RECTANGULAR WAVEGUIDE COVERED BY MIL-W-85 See Mil HDBK 216. Para 6 17-6 23											
174	0 0189" 7 0 0063" CCS	PE	0 060	1 TC	PVC-I	0 100	0 008	50 0	30 8	-40 +80	1.500	Miniature data transmission
175	RIGID LINE											
176	0 135" Helix over Magnetic core	PE	0 285	1 Magnet Wire	PVC-I	0 405	0 120	2.240	49 0	-40 +80	5 000	
177	0 195" BC	PE	0 680	2 SC	PVC-IIA	0 895	0 470	50 0	30 8	-40 +80	11.000	High frequency RG218
178	0 0120" 7 0 0040" SCCS	PTFE	0 036	1 SC	KEL-F	0 079 max	0 0054	50 0	29 4	-40 +150	1 000	Use RG178B
178A	0 0120" 7 0 0040" SCCS	PTFE	0 034	1 SC	KEL-F	0 075 max	0 005	50 0	29 4	-40 +150	1.000	Use RG178B
178B	0 0120" 7 0 0040" SCCS	PTFE	0 034	1 SC	FEP-IX	0 075 max	0 0054	50 0	29 4	-55 +200	1 000	High strength conductors avail
179	0 0120" 7 0 0040" SCCS	PTFE	0.057	1 SC	KEL-F	0 094 max	0 010	70 0	20 4	-55 +150	1 200	Use RG179B
179A	0 0120" 7 0 0040" SCCS	PTFE	0.063	1 SC	KEL-F	0 105 max	0 010	75 0	19 5	-40 +150	1.200	Use RG179B
179B	0 0120" 7 0 0040" SCCS	PTFE	0 063	1 SC	FEP-IX	0 100	0 010	75 0	19 5	-55 +200	1.200	High strength conductors avail
180	0 0120" 7/0 0040" SCCS	PTFE	0 103	1 SC	KEL-F	0 141 max	0 019	93 0	15 4	-40 +150	1 500	Use RG180B
180A	0 0120" 7/0 0040" SCCS	PTFE	0 102	1 SC	KEL-F	0 145 max	0 019	95 0	15 4	-40 +150	1.500	Use RG180B
180B	0 0120" 7/0 0040" SCCS	PTFE	0 102	1 SC	FEP IX	0 145 max	0 019	95 0	15 4	-55 +200	1 500	High strength conductors avail
181	2 cond 0 0477" 7/0 0159" BC	PE	0 210	2 Individual inner Common Outer BC	PVC-IIA	0 640	0 198	125 0	12 0	-40 +80	3.500	Balanced line
182	2 of 19/0 0142" BC 2 of 19/0 0066" TC	4 cores PE	2/ 332 2/ 146	Each core. 1 BC overall shield. 1 BC	PVC-IIA each PVC-I overall	1 055		125 0 Each	12 0 Each	-40 +80	2.300 3.000	Special 4 coax
183	0 2510" BC	PS Helix	0 632	Al tube	None	0 750	0 380	50 0	23 0	-40 +80	1.800	
184	RECTANTULAR WAVEGUIDE											

RG U Type	Inner Conductor	Dielectric Material	DOD (Inch)	Number & Type of Shielding Braids	Jacket Material	O.D. (Inch)	Weight (lbs/ft)	Nominal Imped. (ohms)	Nominal Capac-itance (pf/ft)	Max. Oper. Temp. Range (°C)	Max. Oper. Voltage (Volts RMS)	Comments
185	0 0031" Mag wire Helix on PE core	Air-space PE	0 188	1 Magnet wire	PVC-IIA	0 282		2,000		−40 +80		Delay line cable
186	0 008" TFE Helix over core	Air-space PE	0 292	1 Magnet wire	PVC-IIA	0 405		1,000		−40 +80		Delay line cable
187	0 0120" 7/0 0040" SCCS	PTFE	060	1 SC	PTFE	0 110 max	0 010	75 0	19 5	−55 +250	1,200	Use RG179B
187A	0 0120" 7/0 0040" SCCS	PTFE	060	1 SC	PTFE	0 110 max	0 010	75 0	19 5	−55 +250	1,200	Use RG179B
188	0201 7 0067" SCCS	PTFE	0 060	1 SC	PTFE	0 110 max	0 011	50 0	29 4	−55 +250	1,200	Use RG316
188A	0201" 7 0067" SCCS	PTFE	0 060	1 SC	PTFE	0 110 max	0 011	50 0	29 4	−55 +250	1,200	Use RG316
189	0 2510" BC	PS Helix	0 632	2 SC	PE-IIIA	0 875	0 570	50 0	23 0	−55 +80	3,500	Use RG389
190	0 0585" 19 0 0117" TC	Rubber H J	0 380	3 TC. GS. TC	Neoprene VIII	0 700	0 353	50 0	50 0	−55 +80	15,000	
191	0 485" TC Braid	Rubber H J H	1 065	3 TC. GS. GS	Neoprene VIII	1 460	1 469	25 0	85 0	−55 +80	15,000	
192	1 055" GS Tube TC Braid	Butyl Rubber		3 TC GS GS	Rubber	2 200		12 5	175 0	−55 +80	15,000 Peak	
193	1 055" GS Tube TC Braid	Silicone Rubber		3 TC GS GS	Rubber	2 100		12 5	159 0	−55 +80	30,000 Peak	
194	1 055" GS Tube TC Braid	Silicone Rubber		3 TC GS GS	Rubber w Al Armor	1 945		12 5	159 0	−55 +80	30,000 Peak	
195	0 0120" 7/0 004" SCCS	PTFE	0 102	1 SC	PTFE	0 155 max	0 020	95 0	15 4	−55 +250	1,500	Use RG180B
195A	0 0120" 7 0 004" SCCS	PTFE	0 102	1 SC	PTFE	0 155 max	0 020	95 0	15 4	−55 +250	1,500	Use RG180B
196	0 0120" 7/0 004" SCCS	PTFE	0 034	1 SC	PTFE	0 080 max	0 006	50 0	29 4	−55 +250	1,000	Use RG178B
196A	0 0120" 7 0 004" SCCS	PTFE	0 034	1 SC	PTFE	0 080 max	0 006	50 0	29 4	−55 +250	1,000	Use RG178B
197	0 300" BC	PS Helix	0 758	875" OD Al Tube	None	0 875	0 500	50 0	22 0	−55 +80	2 400 Peak	
198	0 1140" BC	PS Helix	0 421	500" OD Al Tube	PE	0 600	0 155	70 0	16 0	−55 +80	1 300 Peak	
199	0 209" BC	PS Helix	0 758	875" OD Al Tube	PE	1 015	0 435	70 0	16 0	−55 +80	2 400 Peak	
200	OD 405" ID 0 301" BC Tube	PS Helix	1 472	1 625" OD Al Tube	PE	1 765	0 900	70 0	16 0	−55 +80	4 600 Peak	
201 208	RECTANGULAR WAVEGUIDE COVERED BY MIL-W-85 See Mil HDBK 216, Para 6 17-6 23											
209	0 189" 19/0 0378" SC	Air-space PTFE	0 500	2 SC	SR and Polyester-VI	0 750 max	0 432	50 0	25 0	−55 +150	3,200	Low loss RG211A
210	0 0253" SCCS	Air-space PTFE	0 146	1 SC	FG Braid-V	0 242	0 040	93 0	13 5	−55 +250	750	High temp Low capacitance
211	0 1900" BC	PTFE	0 620	1 BC	FG Braid-V	0 730	0 641	50 0	29 4	−55 +250	7,000	High Temp High power
211A	0 1900" BC	PTFE	0 620	1 BC	FG Braid-V	0 730	0 641	50 0	29 4	−55 +250	7 000	High temp High power
212	0 0556" SC	PE	0 185	2 SC	PVC-IIA	0 332	0 083	50 0	29 4	−40 +80	3,000	Use to 10,000 MHz
213	0 0888" 7 0 0296" BC	PE	0 285	1 BC	PVC-IIA	0 405	0 099	50 0	30 8	−40 +80	5,000	Use up to 1000 MHz
214	0 0888" 7 0 0296" SC	PE	0 285	2 SC	PVC-IIA	0 425	0 126	50 0	30 8	−40 +80	5,000	Use up to 10,000 MHz
215	0 0888" 7/0 0296 BC	PE	0 285	1 BC	PVC-IIA w Armor	0 475 max	0 121	50 0	30 8	−40 +80	5,000	Amored RG213
216	0 0477" 7/0 0159" BC	PE	0 285	2 BC	PVC-IIA	0 425	0 114	75 0	20 6	−40 +80	5,000	Use up to 1000 MHz

APPENDIX P continued

RG U Type	Inner Conductor	Dielectric Material	DOD (Inch)	Number & Type of Shielding Braids	Jacket Material	O.D. (Inch)	Weight (lbs/ft)	Nominal Imped. (ohms)	Nominal Capac-itance (pf/ft)	Max. Oper. Temp. Range (°C)	Max. Oper. Voltage (Volts RMS)	Comments
217	0 106" BC	PE	0 370	2 BC	PVC-IIA	0 545	0 201	50 0	30 8	40 +80	7 000	Use up to 1000 MHz
218	0 195" BC	PE	0 680	1 BC	PVC-IIA	0 870	0 460	50 0	30 8	40 +80	11 000	Use up to 1000 MHz
219	0 195" BC	PE	0 680	1 BC	PVC-IIA w Armor	0 945 max	0 585	50 0	30 8	-40 +80	11 000	Armored RG218
220	0 260" BC	PE	0 910	1 BC	PVC-IIA	1 120	0 740	50 0	30 8	40 +80	14 000	Use up to 1000 MHz
221	0 260" BC	PE	0 910	1 BC	PVC-IIA w Armor	1 195 max	0 925	50 0	30 8	40 +80	14 000	Armored RG220
222	0 0556" high Resistance wire	PE	0 185	2 SC	PVC-IIA	0 332	0 087	50 0	30 8	40 +80	3.000	High attenuation
223	0 035" SC	PE	0 116	2 SC	PVC-IIA	0 216 max	0 034	50 0	30 8	40 +80	1 900	Usable to 10 000 MHz
224	0 106" BC	PE	0 370	2 BC	PVC-IIA w Armor	0 615 max	0 310	50 0	30 8	40 +80	7.000	Armored RG217
225	0 0936" 7/0 0312" SC	PTFE	0 285	2 SC	FG Braid-V	0 430	0 180	50 0	29 4	-55 +250	5.000	See RG393 for FEP jacket
226	0 1270" 19/0 0254" SC	PTFE	0 370	2 BC	FG Braid-V	0 500	0 445	50 0	29 4	-55 +250	7.000	
227	0 0936" 7/0 0312" SC	PTFE	0 285	2 SC	FG Braid-V w Armor	0 490 max	0 198	50 0	29 4	-55 +250	5.000	Armored RG225
228	0 1900" BC	PTFE	0 620	1 BC	FG Braid-V w Armor	0 795 max	0 682	50 0	29 4	-55 +250	7.000	Armored RG211
228A	0 1900" BC	PTFE	0 620	1 BC	FG Braid-V w Armor	0 795 max	0 682	50 0	29 4	-55 +250	7 000	Armored RG211A
229	0 0960" 7/0 032" SC	PTFE	0 285	1 SC	FG Braid-V w Armor	0 480 max	0 144	50 0	29 4	-55 +250	5.000	Use RG166
230	0 1988" 37 0 0284" TC	Rubber-D	0 455	3 TC. GS. GS	Rubber-IV	0 740 max		25 0	100 0	-40 +80	15 000	
231	OD 0.162 ID 0.112" BC	Foam PE	0 450	0 500" OD Al Tube	None	0 500	0 118	50 0	25 0	-55 +80	5.000 Peak	RG331 for jacketed cable
232	0 300" BC	PE Helix	0 758	0 875" OD Al Tube	PE-IIIA	1 015	0 570	50 0	22 0	-55 +80	2.400 Peak	
233	OD 0 591" BC ID 0 481"	PS Helix	1 472	1 625" OD Al Tube	PE-IIIA	1 765	1 050	50 0	22 0	-55 +80	4.700 Peak	
234	OD-1 1570" BC ID-1 015"	PS Helix	2.775	3 125" OD Al Tube	PE-IIIA	3 295	3 110	50 0	22 0	-55 +80	8.700 Peak	
235	0 0852" 7-0 0284" SC	Taped PTFE	0 255	2 SC	SIL/DAC VI	0 470 Max	0 160	50 0	29 5	-55 +250	5.000	RG115A w VI jacket
236	0 1620" BC	PS Helix	0 421	0 500" OD Al Tube	None	0 500	0 165	50 0	24 0	-55 +80	1.300 Peak	
237	0 1620" BC	PS Helix	0 421	0 500" Al Tube	PE-IIIA	0 600	0 195	50 0	24 0	-55 +80	1.300 Peak	
238	CANCELLED. REPLACE WITH RG197/U											
239	CANCELLED. REPLACE WITH RG232/U											
240	OD 0 591" BC ID 0 481" BC	PS Helix	1 420	1 625 OD Al Tube	None	1 625	0 930	50 0	22.0	-55 +80	4.700 Peak	
241	CANCELLED. REPLACE WITH RG233											
242	OD-1 157" ID-1 0150" BC	PS Helix	2 850	3 125" OD Al Tube	None	3 125	2.700	50 0	22.0	-55 +80	8.700 Peak	
243	CANCELLED. REPLACE WITH RG234											
244	0 102" BC	PS Helix	0 421	0 500" OD Al Tube	None	0 500	0 118	75 0	15.5	-55 +80	1,200 Peak	
245	0 102" BC	PS Helix	0 421	0 500" OD Al Tube	PE-IIIA	0 600	0 148	75 0	15 5	-55 +80	1.200 Peak	

APPENDIX P continued

RG U Type	Inner Conductor	Dielectric Material	DOD (Inch)	Number & Type of Shielding Braids	Jacket Material	O.D. (Inch)	Weight (lbs. ft)	Nominal Imped. (ohms)	Nominal Capac-itance (pf ft)	Max. Oper. Temp. Range (C)	Max. Oper. Voltage (Volts RMS)	Comments
246	0 1880" BC	PS Helix	0 758	0 875" OD Al Tube	None	0 875	0 348	75 0	15 2	-55 +80	2200 eak	
247	0 1880" BC	PS Helix	0 758	0 875" OD Al Tube	PE-IIIA	1 015	0 418	75 0	15 2	-55 +80	2200 Peak	
248	OD-0 3740" BC ID-0 2740"	PS Helix	1 472	1 625" OD Al Tube	None	1 625	0 948	75 0	15 0	-55 +80	4300 Peak	
249	OD-0 374" BC ID-0 2740"	PS Helix	1 472	1 625" OD Al Tube	PE-IIIA	1 765	1 068	75 0	15 0	-55 +80	4300 Peak	
250	0 732" BC 0 632"	PS Helix	2 850	3 125" OD Al Tube	None	3 125	2 395	75 0	15 0	-55 +80	8 500 Peak	
251	0 732" BC 0 632"	PS Helix	2 850	3 125" OD Al Tube	PE-IIIA	3 295	2 805	75 0	15 0	-55 +80	8 500 Peak	
252	0 1670" BC	PE Tubes	0 456	0 530" OD Al Tube	None	0 530	0 175	50 0	24 0	-55 +80	1 000	
253	0 167" BC	PE Tubes	0 456	0 530" OD Al Tube	PE	0 655	0 225	50 0	24 0	-55 +80	1 000	
254	0 3110" BC	PE Tubes	0 833	0 953" OD Al Tube	PE	1 100	0 655	50 0	24 0	-55 +80	1 860	
255	0 3110" BC	PE Tubes	0 833	0 953" OD Al Tube	None	0 953	0 555	50 0	24 0	-55 +80	1 860	
256	OD-0 3110" SC ID-0 2550"	PTFE Tubes	0 833	0 953" OD Al Tube	None	0 953	0 550	50 0	24 0	-55 +80	1 860	
257	OD-0 6060" ID-0 04860" BC	PS Tubes	1 622	1 786" OD Al Tube	None	1 786	1 200	50 0	24 0	-55 +80	3 640	
258	OD-0 606" ID-0 4860" BC Tube	PE Tubes	1 622	1 786" OD Al Tube	PE	1 926	1 380	50 0	24 0	-55 +80	3 640	
259	0 1150" BC Tube	PTFE Tubes	0 318	0 390" OD Al Tube	None	0 390	0 100	50 0	24 0	-55 +80	697	
260	0 1150" BC Tube	PE Tubes	0 318	0 390" OD Al Tube	PE-IIIA	0 450	0 140	50 0	24 0	-55 +80	697	
261-262												Cancelled
263	0 1720" BC	Air-space PTFE	0 421	Al Tube	None	0 500	0 170	50 0	21 5	-40 +250	1 300 Peak	
264	4 cond . 19/0 0142" 2 TC 2 BC	PE	0 176 e/core	2 TC, 2 BC BC over all	PVC-IIA	0 750	0 336	36 8	41 0	-40 +80	2 000	Use RG264C
264A	4 cond . 19/0 0142" 2 TC 2 BC	PE	0 176 e/core	2 TC, 2 BC. BC over all	PUR	0 750	0 327	36 8	41 0	-40 +80	2 000	Use RG264C
264C	4 cond . 0 068" 2 BC. 2 TC	PE	0 186 e/core	2 TC, 2 BC BC overall	PUR	0 765	0 327	40 0	38 4	-40 +80	2 000	Watertight 4 coax
265	0 6770" BC Tube	PE Helix	1 578	Copper clad Mild Steel Tube	PE-IIIA	2 070		50 0	22 3	-40 +80	145KW Peak	
266	0 0113" cond over 0 144" Mag core	PE	0 285	75 spiral wound cond 68 BC & 7 are insulated	PVC-I	0 400	0 120	1530 0	53 0	-40 +80	5 000 DC	Delay line cable 50 ns ft
267	0 3550" BC Tube	PS Helix		Copper clad Mild Steel Corrugated Tubing	PE-IIIA	1 190		50 0	22 2	-40 +80	44 KW Peak	
268	0 1610" BC	PE Helix	0 350	Corrugated BC Tube	None	0 498	0 234	50 0	23 0	-55 +80	10KW Peak	
269	ID-0 2870" OD-0 3580" BC Tube	PE Helix	0 795	Corrugated BC Tube	None	1 005	0 430	50 0	22 2	-55 +80	44KW Peak	
269A	ID-0 2870" OD-0 3580" BC Tube	PE Helix	0 795	Corrugated BC Tube	None	1 005	0 430	50 0	22 2	-55 +80	44KW Peak	
270	ID-0 5880" OD-0 6880" BC Tube	PE Helix	1 578	Corrugated BC Tube	None	1 830	0 875	50 0	22 3	-55 +80	145KW Peak	
270A	ID-0 5880" OD-0 6770" BC Tube	PE Helix	1 578	Corrugated BC Tube	None	1 830	0 875	50 0	22 3	-55 +80	145KW Peak	
271-278	RECTANGULAR WAVEGUIDES COVERED BY MIL-W-85 See Mil HDBK 216, Para 6 17-6 23											
279	0 0250" 19/0 0050" SCCS	Air-space PTFE	0 110	1 SC	FG Braid-V	0 145	0 125	75 0	19 5	-55 +250	1,000	Extra flexible high temp cable

APPENDIX P continued

RG U Type	Inner Conductor	Dielectric Material	DOD (Inch)	Number & Type of Shielding Braids	Jacket Material	O.D. (Inch)	Weight (lbs ft)	Nominal Imped. (ohms)	Nominal Capac-itance (pf ft)	Max. Oper Temp. Range (°C)	Max. Oper. Voltage (Volts RMS)	Comments
280	0 1144" BC	Taped PTFE	0 327	2 SC	FEP-IX	0 468	0 200	50 0	25 4	-55 +200	3 000	Low loss High frequency
281	0 1890" 19 0 0378" SC	Taped PTFE	0 500	2 SC	Sil/DAC-VI	0 750 max	0 400	50 0	25 4	-55 +150	4 000	Low loss High power
282	0 0253" SC	Irradiated PE	0 099	2 SC	FEP	0 200	0 031	54 5	28 2	-40 +150	4 500	
283	0 0585" 19 0 0117" SC	Rubber-D	0 288	2 SC	Sil	0 475	0 145	46 0	50 0	-55 +150	8 000	
284A	0 220" BC	PE Helix	0 795	Corrugated BC Tube	None	1 005	0 410	75 0	15 0	-55 +80	29KW Peak	
285A	0 1140" BC	PTFE Helix	0 795	Corrugated BC Tube	None	1 005	0 430	100 0	13 0	-55 +200	22KW Peak	
286	OD-0 4300" ID-0 3600" BC Tube	PE Helix	1 570	Corrugated BC Tube	None	1 830	0 720	75 0	15 1	-55 +80	100KW Peak	
287	0 1970" BC	PE Helix	1 570	Corrugated BC Tube	None	1 830	0 750	100 0	13 5	-55 +80	73KW Peak	
288	OD-1 3330" ID-1 2221" BC Tube	PE Helix	2 960	3 750" CCS	None	3 750	3 000	50 0	21 6	-40 +80	440KW Peak	
289	OD-0 8200" ID-0 7400" CCS Tube	PE Helix	2 960	3 750" CCS	None	3 750	3 000	75 0	14 7	-40 +80	290KW Peak	
290-291	RECTANGULAR WAVEGUIDE COVERED BY MIL-W-85 See Mil HDBK 216. Para 6 17-6 23											
292	0 4300" BC Tube	PE Helix	1 570	1 8300" Corr BC Tube	PE and flooding comp	2 000	1 040	75 0	15 1	-55 +80	100KW Peak	
293	0 106" BC	PE	0 375	1 SC	PE-IIIA	0 545	0 160	50 0	30 8	-55 +80	7,000	Use RG293A
293A	0 1060" BC	PE	0 370	1 SC	PE-IIIA	0 545	0 160	50 0	30 8	-55 +80	7 000	Watertight RG217
294	2 cond 0 0808" 1 BC 1 TC	PE	0 472	1 TC	PE-IIIA	0 630	0 205	95 0	16 3	-55 +80	3 000	Use RG294A
294A	2 cond 0 0808" 1 BC 1 TC	PE	0 472	1 TC	PE-IIIA	0 630	0 205	95 0	16 3	-55 +80	3 000	Watertight RG130
295	0 1950" BC	PE	0 680	1 BC	PE-IIIA	0 895	0 420	50 0		-55 +80	11 000	Watertight RG218
296	0 2352" 37 0 0336" SC	Silicone Rubber	906	1 SC	Neoprene	1 190		50 0	36 4	-55 +80	13 800	
297	OD0 3550" ID-0 2870" BC Tube	PTFE Helix	0 795	Corr BC Tube	None	1 005		50 0	21 4	-55 +200	44 KW Peak	
298	0 6030" 7/0 0201" CCS	PE	0 115	None	Foam PE	0 650	0 090			-55 +80		Buoyant Per Mil-C-22667
299-300	RECTANGULAR WAVEGUIDE											
301	0 0609" 7 0 0203" HR	PTFE	0 185	1 HR	FEP-IX	0 245	0 056	50 0	29 4	-55 +200	3 000	FEP Jacketed RG126
302	0 0250" SCCS	PTFE	0 146	1 SC	FEP-IX	0 206 max	0 031	75 0	19 5	-55 +200	2 300	FEP Jacketed RG140
303	0 0390" SCCS	PTFE	0 116	1 SC	FEP-IX	0 170	0 030	50 0	29 4	-55 +200	1,900	FEP Jacketed RG141A
304	0 0590" SCCS	PTFE	0 185	2 SC	FEP-IX	0 280	0 088	50 0	29 4	-55 +200	3,000	FEP Jacketed RG143A
305	OD-0 4300" ID-0 3600" BC Tube	FEP	1 570	1 830" BC Tube	PE-IIIA	1 990		75 0	14 4	-55 +80	2,720	
306A	0 1730 "BC	Foam PE	0 801	0 8750" Al Tube	PE-IIIA	1 015	0 545	75 0	16 5	-55 +80	5,700	Per Mil-C-23806
307	0 0290" 19/0 0058 " SC	Foam PE	0 146	2 SC PUR Interlayer	PE-IIIA	0 270	0 070	75 0	16 7	-55 +80	1,000	Triax Use to 100 MHz
307A	0 0290" 19/0 0058" SC	Foam PE	0 146	2 SC PUR Interlayer	PE-IIIA	0 270	0 070	75 0	16 7	-55 +80	1,000	Triax Use to 100 MHz
308-315	BEAD SUPPORTED RIGID LINES See Mil- R-9671											

APPENDIX P continued

RG U Type	Inner Conductor	Dielectric Material	DOD (Inch)	Number & Type of Shielding Braids	Jacket Material	O.D. (Inch)	Weight (lbs/ft)	Nominal Imped. (ohms)	Nominal Capac-itance (pf/ft)	Max. Oper. Temp. Range (°C)	Max. Oper. Voltage (Volts RMS)	Comments
316	0 0201" 7/0 0067" SCCS	PTFE	0 060	1 SC	FEP IX	0 102	0 012	50 0	29 4	−55 +200	1,200	FEP jacketed RG188A
317	2 cond 0 0870" 7/0 0290" BC	FEP	0 446	1 TC	Neoprene	0 710		95 0	15 4	−55 +80	10,000	Water blocked
318	OD-0 3580" BC Tube ID-0 2870"	PE Helix	0 795	1 005" Corr BC Tube	PE IIIA	1 125	0 530	50 0	22 0	−55 +80	44 KW Peak	
319	OD-0 6880" BC Tube ID-0 5880"	PE Helix	1 570	1 830" Corr BC Tube	PE-IIIA	2 000	1 040	50 0	22 0	−55 +80	145 KW Peak	
320	WAVEGUIDE											
321	OD-1 1400" corr BC Tube	PE Helix		2 850" corr BC Tube	None	2 850	1 210	50 0	21 7	−55 +80	320 KW Peak	
322	OD-1 1400" corr BC Tube	PE Helix		2 850" corr BC Tube	PE & flood comp	3 040	1 780	50 0	21 7	−55 +80	320 KW Peak	
323	0 312" BC Tube	Foam PE		0 980" corr BC Tube	PE & flood comp	1 060	0 420	50 0	25 6	−55 +80	1,480	
324	0 312" BC Tube	Foam PE		0 980" corr BC Tube	None	0 980	0 320	50 0	25 6	−55 +80	1,480	
325	0 1000" 19/0 0200" SC Al	PE Spline	0 260	2 SC Strip Braids	PUR	0 350	0 10	50 0	26 3	−55 +80	750	Low loss
326	0 2000" 19/0 0400" SC Al	PE Spline	0 550	2 SC Strip Braids	PUR	0 697	0 24	50 0	26 3	−55 +80	1 700	Low loss
327	0 3200" 19/0 0640" SC Al	PE Spline	0 840	2 SC Strip Braids	PUR	1 010	0 55	50 0	26 3	−55 +80	2,500	Low loss
328	0 4850" TC Braid	Rubber H. J. H	1 065	3 TC GS TC	Neoprene	1 460	1 469	25 0	85 0	−55 +80	20,000	
329	0 0585" 19/0 0117" TC	Rubber H J H	0 380	3 TC GS TC	Neoprene	0 700	0 353	50 0	50 0	−55 +80	15,000	
330	SC	Foam PE		1 SC		0 242		50 0	25 0			
331	0 1620" BC	Foam PE	0 450	0 500" Al Tube	PE-IIIA	0 600	0 187	50 0	25 0	−55 +80	2,500	Jacketed 231 Solid conductor
332	0 2800" BC	Foam PE	0 801	0 8750" Al Tube	None	0 875	0 466	50 0	25 0	−55 +80	4,500	Per Mil-C-23806
333	0 2880" BC	Foam PE	0 801	0 8750" Al Tube	PE-IIIA	1 015	0 548	50 0	25 0	−55 +80	4,500	Jacketed RG332
334	0 0980" BC	Foam PE	0 450	0 500" Al Tube	None	0 500	0 109	75 0	17 0	−55 +80	2,500	Per Mil-C-23806
335	0 0980" BC	Foam PE	0 450	0 500" Al Tube	PE-IIIA	0 625	0 143	75 0	17 0	−55 +80	2,500	Jacketed RG334
336	0 1730" BC	Foam PE	0 801	0 8750" Al Tube	None	0 875	0 315	75 0	17 0	−55 +80	4,000	Per Mil-C-23806
337-359	RECTANGULAR WAVEGUIDES COVERED BY MIL-W-85 See Mil HDBK 216 Para 6 17-6 23											
360	0 2430" BC	Foam PE	0 676	0 7500" Al Tube	PE-IIIA	0 825	0 397	50 0	25 0	−55 +80	4,000	Per Mil-C-23806
361-365	DATA NOT AVAILABLE											
366	0 1600" BC	Foam PE		0 540" Corr BC Tube	PE-IIIA	0 620		50 0	26 6	−55 +80	4,000	
367	Corr BC Tube	PE Helix		5 000" Corr BC Tube	PE-IIIA	5 2000	4 590	50 0	21 7	−55 +80	830 KW Peak	
369	0 1170" BC	PE Tubes	0 318	0 390" Al Tube	PE-IIIA	0 470	0 140	50 0	24 0	−55 +80	700	
370	0 1170" BC	PE Tubes	0 318	0 390" Al Tube	None	0 390	0 100	50 0	24 0	−40 +80	700	
372-373	EXPERIMENTAL BUOYANT COAXIAL TRANSMISSION LINE											
374	0 0285" BC	PE	0 160	None	Foam PE	0 650	0 097			−55 +80		Buoyant antenna

APPENDIX P continued

RG U Type	Inner Conductor	Dielectric Material	DOD (Inch)	Number & Type of Shielding Braids	Jacket Material	O.D. (Inch)	Weight (lbs/ft)	Nominal Imped. (ohms)	Nominal Capac-itance (pf/ft)	Max. Oper. Temp. Range (°C)	Max. Oper. Voltage (Volts RMS)	Comments
375	RECTANGULAR WAVEGUIDE											
376	0 3120" BC Tube	Foam PE		Corr. Al Tube	PE-IIIA	1 060	0 390	50 0	26 0	−55 +80	6,000	
377	0 1650" SC Tube	PTFE Tubes		0 5300" Al Tube	None	0 530	0 170	50 0	24 0	−55 +250	1,000	
378	0 7130" BC Tube	PE Helix		1 830" Corr Al Tube	PE-IIIA	2 000	0 620	50 0	22 1	−55 +80	145KW Peak	
379-381	ELLIPTICAL WAVEGUIDES											
382	RIGID LINE			Al Tube		1 625		50 0				
383	2 cond 0 0403" 2,000 lb Break	PE		None	Foam PE	0 650		100 0		−55 +80		Exper buoyant Twisted pair
384	0 0508" BC	PE		1 Flat BC Braid Waterproofed	Foam PE	0 650		50 0	30 8	−55 +80		Buoy Ant 600 psig
385	0 1530" SC	Semi-solid PTFE	0 425	0 500" Corr Al	Optional	0 660	0 178	50 0	25.0	−55 +250	1,500	Low loss No press req
386	0 0508" CCS	PE		None Non-hosing	Foam PE	0 650				−55 +80		Buoy Ant 400 lb break
387	DATA NOT AVAILABLE											
388	SC	PE		0 444" max SC	PE-IIIA	0 545		50 0	30 8	−55 +80		Watertight See RG14A
389	0 2500" BC Al	PE Spline	0 635	2 SC Strips	PE-IIIA	0 875	0 366	50 0	22 8	−55 +80	2,000	Low Loss Replaces RG189
390	DATA NOT AVAILABLE											
391	0 [illegible]" 7 0 015[illegible]" TC	Cond PE and PE	0 285	1 TC	PVC-IIA	0 405	0 092	72 0		−55 +80	5,000	Low noise cable
392	0 048" 7 0 01[illegible]" TC	Cond PE and PE	0 285	1 TC	PVC-IIA w Armor	0 475	0 114	72 0		−55 +80	5,000	Armored RG391
393	0 0936" 7/0 0312" SC	PTFE	0 285	2 SC	FEP-IX	0 390	0 165	50 0	29 4	−55 +200	5,000	Moisture proof RG225

APPENDIX Q WORKING WITH SEMIRIGID CABLE

Because of the unique construction of semirigid cable, special techniques are needed for working with it. Special precautions are necessary when cutting the cable, stripping the outer conductor, bending the cable, trimming the dielectric, and soldering.

Cutting

A jeweler's saw is adequate in the laboratory; on the production line, a saw having at least 20 teeth per inch and moving at the rate of about 200 inches per second is more efficient. To cut cable with a copper-clad steel inner conductor, a carborundum blade should be used because the wear of a metal saw would be excessive. Unfortunately, cutting with a carborundum blade produces slight burrs on the inner and outer conductor, which must be removed for proper connector or terminal assembly.

The jeweler's saw is still the best tool for the majority of the work done for everyday applications. Under no circumstances should a pair of diagonal pliers be used to cut semirigid cable. This method damages both the outer conductor and the dielectric, and makes much more work for the technician in extra cuts and trimming.

Stripping Outer Conductor

Removal, or stripping, of the outer conductor is the next step in cable end preparation. In principle, the method should avoid cutting. Instead, a score line would be generated, along which the outer conductor is then snapped off. The score and break line must be orthogonal to the long cable axis. Several methods of parting an outer conductor section may be employed. One process involves partial cutting with a jeweler's saw; at least one cable connector manufacturer offers fixtures which allow cutting with a saw at a location compatible with his connector dimensions.

Another approach, predominantly used with cables having relatively large diameters and heavy outer conductor walls, employs a pipe cutter with its cutting tool suitably modified to eliminate excessive penetration into the cable outer conductor. In the case of smaller-diameter and thin-walled cables, the scoring can be done by placing a knife edge against the outer conductor, and then rotating the cable while maintaining suitable pressure on the knife. While this operation may be performed on a workbench top, a fixture providing a flat surface with guides directing the knife and maintaining cable alignment normal to the cutting edge is an improvement, particularly when addition guides allowing scoring at a precise location are incorporated. When the cable is suitably scored, the cable should be grasped with a pair of pliers and snapped all the way around on the score line.

APPENDIX Q continued

After breaking at the score line, the outer conductor may be pulled off with the aid of either pliers or a knife, applied to the outer conductor at an angle of approximately 30 degrees to the long cable axis, and then pulled in the direction of the long cable axis. Occasionally, tolerance buildup (maximum dielectric diameter coinciding with maximum outer conductor wall thickness) may cause a tight fit. In such a case, the remedy is to maintain a steady pull on the outer conductor section to be removed while heating it with a 700°F soldering iron or other source of heat. Of additional help is some arrangement which allows the cable to be held in place without causing cable deformation. In its simplest form, this aid may be a piece of leather which, hand-held against the work-bench, allows an easier and firmer grasp of the cable. It is good practice to never remove more than about one-half inch of outer conductor at a time.

Cable Bending

Forming or bending coaxitube requires, as a minimum, an arrangement consisting of a bending mandrel, and a device which holds one cable end firmly in place while the rest of the cable is being bent. If minimum radius bends are to be made, the bending mandrel must be provided with a groove of diameter equal to the cable diameter, and a forming bar must be employed. Because semirigid cable has resilience, the correct mandrel size required to produce a bend of the desired radius must be determined experimentally. If bends must be applied at a precise location, it is advantageous to construct a fixture with a positioning stop. Forming a cable into complex shapes involving close tolerances, numerous bends, and excursions in more than one plane can be accomplished only with specially designed, rigidly constructed, and, therefore, somewhat expensive fixtures.

Dielectric Trimming

The cable dielectric may be trimmed with a razor blade or on a rotary wire stripper. The cut must be clean and normal to the inner conductor. Excessively deep cuts must be avoided since they would nick the inner conductor and cause failure.

Soldering

Because of the rapid expansion of the cable dielectric upon application of heat, soldering to semirigid cables can become a problem unless proper precautions are observed. These precautions include any measure capable of minimizing the amount of heat fed into the cable, such as careful cleaning of the outer conductor, use of flux-cored 60/40 solder, and use of a soldering iron of suitable rating (about 50

APPENDIX Q continued

watts for .141 inch cable), preferably of the temperature-controlled type. Miniature torches and resistance soldering equipment also make suitable heat source, depending on circumstances. One additional point to aid in soldering of semi-rigid cable is that, following cleaning of the outer conductor, do not touch the cleaned area. Dirt or oil from your hands makes soldering more difficult and requires more heat.

Additional Tips

When using semirigid cable assemblies, complex layouts and extremely close tolerances should be avoided because they contribute to high cost without necessarily improving performance. Most semirigid cables are flexible enough to allow for considerable adjustment in order to achieve the desired configuration; relatively short lengths of cable with copper outer conductors having a diameter of 0.250 inches or more are the exception. Cases have occurred where installation of extremely closely-toleranced assemblies into the "next assembly" was not possible until a switch was made to aluminum-jacketed cable, silverplated to allow for the soldering of connectors.

One point should be emphasized when bending semirigid cable—once the fixture is determined, the cable should be bent very gradually, to ensure a smooth bend with no wrinkles on the inside edge, and with no stress the outside edge (thereby avoiding the possibility of cracking).

Contrary to what many people think, it is possible to straighten a piece of semirigid cable that has been bent, provided that is has not been bent too sharply. Once again, patience is the key. By holding the cable firmly in one hand and straightening a small length at a time, you are able to end up with a piece of cable that is pretty much uneffected by the first bend.

The words to remember when working with semirigid cable are *caution* and *patience*.

APPENDIX R TECHNICAL MATH

Logarithms

Logarithms are used to simplify numerical computations involving multiplications, division, powers, and roots. With logarithms, multiplication is reduced to simple addition, and division is reduced to simple subtraction. Raising to a power is reduced to a single multiplication, and extracting a root is reduced to a single division.

The common logarithm of any number is the power to which 10 must be raised in order to equal that number.

Therefore, since

$$
\begin{aligned}
1000 &= 10^3\\
100 &= 10^2\\
10 &= 10^1\\
1 &= 10^0\\
0.1 &= 10^{-1}\\
0.01 &= 10^{-2}\\
0.001 &= 10^{-3}\\
0.0001 &= 10^{-4}
\end{aligned}
$$

it is true that

$$
\begin{aligned}
\log 1000 &= 3\\
\log 100 &= 2\\
\log 10 &= 1\\
\log 1 &= 0\\
\log 0.1 &= -1\\
\log 0.01 &= -2\\
\log 0.001 &= -3\\
\log 0.0001 &= -4
\end{aligned}
$$

The common system of logarithms has for its base the number 10, and is written $\log_{10}$ or, more commonly, log. The base 10 is always implied unless some other base is indicated specifically. There are formulas, however, which use the natural system of logarithms. This system has for its base the number 2.718, which is represented by the lower-case letter e; it is written either as $\log_e$ or ln.

In observing the following exponential and logarithmic relationships,

$$
\begin{aligned}
100 &= 10^2 & \log 100 &= 2.000\\
15 &= 10^{1.176} & \log 15 &= 1.176\\
10 &= 10^1 & \log 10 &= 1.000\\
7 &= 10^{.845} & \log 7 &= 0.845
\end{aligned}
$$

APPENDIX R continued

1	$= 10^0$	log	1	$= 0.000$
0.1	$= 10^{-1}$	log	0.1	$= -1.000$
0.7	$= 10^{-1.845}$	log	0.7	$= -1.845$
0.015	$= 10^{-2.176}$	log	0.015	$= -2.176$
0.001	$= 10^{-3}$	log	0.001	$= -3.000$

it is seen that only the direct powers of 10 have whole numbers for logarithms, and that the logarithms of all numbers lying between a power of 10 consist of a whole number and a decimal. The whole number is called the *characteristic*; the decimal is the *mantissa*. Since the characteristic serves only to fix the location of the decimal point in the expression indicated by the log, it can be found by inspection and is not included in the log table.

1. The characteristic of any number greater than 1 is always positive and is equal to one less than the number of digits to the left of the decimal.
2. The characteristic of any number less than 1 is always negative and is equal to one plus the number of zeros to the decimal.
3. The characteristic of any number may be determined by expressing the number as a power of 10 and using this power as the characteristic of the logarithm for that number.

Since only the characteristic of a logarithm is ever negative (the mantissa is always positive), it is customary to write a log containing a negative characteristic as follows:

$$\log 0.7 = \bar{1}.845,$$

or by adding +10 to the characteristic and, in order to maintain equality, −10 at the right of the characteristic,

$$\log 0.7 = 9.845 - 10$$

Examples:

150	1.5×10^2	2
15	1.5×10^1	1
1.5	1.5×10^0	0
0.15	1.5×10^{-1}	−1 or 9 − 10
0.015	1.5×10^{-2}	−2 or 8 − 10
0.0015	1.5×10^{-3}	−3 or 7 − 10

Therefore, to find the logarithm of any number:

1. Write the number as a power of 10, and put down the resulting exponent of 10 as the characteristic.

APPENDIX R continued

2. Determine the mantissa from the log tables that follow and write this number as a decimal figure following the characteristic.
3. If the resulting logarithm has a negative characteristic, change this number to the positive form.

Example—Find the logarithm of .00623:

Since $0.00623 = 6.23 \times 10^{-3}$, the characteristic is −3. The mantissa, as shown by the log table, is 7945. The resultant logarithm is $\bar{3}.7945$ or, when written in its positive form, 7.7945 − 10.

To find the log of any number having more than three significant figures (by interpolation):

1. Determine the characteristic.
2. Find the mantissa corresponding to the first three significant figures.
3. Find the next higher mantissa and take the tabular difference.
4. Find the product of the tabular difference and the digit following the first three significant figures of the given number written as a decimal.
5. Add this product to the lesser mantissa.

Example—Find the logarithm of 54.65.

Since $54.65 = 5.465 \times 10^1$, the characteristic is 1.

next higher mantissa =	0.7380
next lower mantissa =	0.7372
tabular difference	0.0008
	× 0.5
product	0.00040
plus lesser mantissa	0.7372
mantissa of 5.465	0.7376
Therefore, log 54.65 =	1.7376

Although a four-place log table is used here, a three place table is usually sufficiently accurate for all practical purposes. Since the mantissa of a logarithm represents only the significant figures of any number, the same mantissa is used for 0.04, 4, 400, *et cetera*; the decimal point is fixed later by the characteristic. Therefore, any number consisting of one or two significant figures may be found in the column marked N, and its mantissa is found on the same line in the column headed by 0. For any number containing three significant figures, locate the first two figures in the N column, and the third figure in the column headed by the corresponding digit. The mantissa is found in this column, on a line even with the first two digits.

APPENDIX R continued

Examples:

$$
\begin{aligned}
\log\ 21 &= 1.3222\\
\log\ 2.1 &= 0.3222\\
\log\ 210 &= 2.3222\\
\log\ 0.0021 &= 7.3222 - 10\\
\log\ 213 &= 2.3284\\
\log\ 0.0213 &= 8.3284 - 10\\
\log\ 3 &= 0.4771\\
\log\ 300 &= 2.4771\\
\log\ 0.003 &= 7.4771 - 10
\end{aligned}
$$

The number corresponding to a given logarithm is called the antilogarithm, and is written "antilog." Example: Since log 692 = 2.8401, antilog 2.8401 = 692.

Finding the antilog of a number is the reverse of finding the logarithm. First locate the mantissa in the log table, then determine its corresponding number. Now, place the decimal as indicated by the characteristic.

Example—To find the antilog of 3.9138, look up 9,138 in the log table. Its corresponding number is 82 or, expressed as a power of 10, 8.2×10^1. A characteristic of three means that 8.2 must be multiplied by 10^3. Therefore, antilog 3.9138 = $8.2 \times 10^3 = 8{,}200$.

Similarly,

$$
\begin{aligned}
&\text{antilog } 5.9138 = 8.2 \times 10^5 = 82{,}000\\
&\text{antilog } 0.9138 = 8.2 \times 10^0 = 8.2\\
&\text{antilog } 7.9138 - 10 = 8.2 \times 10^{-3} = 0.0082\\
&\text{antilog } 9.9138 - 10 = 8.2 \times 10^{-1} = 0.82
\end{aligned}
$$

To find the antilogarithm of a logarithm whose mantissa is not exactly given in the table:

1. Find the tabular difference between the next highest and next lowest mantissas.
2. Divide this value by the difference between the given mantissa and the next lower mantissa.
3. Add the resulting quotient to the significant figures expressed by the next lower mantissa.
4. Place the decimal as indicated by the given characteristic.

APPENDIX R continued

Example—Find the antilog of 1.7376

next higher mantissa =	0.7380
next lower mantissa =	0.7372
tabular difference	0.0008
given mantissa	0.7376
next lower mantissa	0.7372
tabular difference	0.0004
quotient of $\frac{0.0004}{0.0008}$	0.5

The resultant figure is therefore .5 larger than the significant figures expressed by the lesser mantissa 0.7372, or 546. The sequence of figures is thus 546.5.

Therefore, antilog 1.7376 = 54.65

Note—When interpolating as shown above, do not exceed four significant figures in your answer since results interpolated from a four-place table are not accurate past this point.

Logarithms are added or subtracted like arithmetic numbers, provided that they are written with positive characteristics. If the characteristic in the total is greater than nine, and the notation −10, −20, −30, *et cetera*, appears after the mantissa, subtract a multiple of 10 from the positive part, and add the same multiple of 10 to the negative part, in order to make the resultant characteristic less than 10.

Examples

Addition of logarithms:

2.764	6.326 − 10	6.328 − 10
4.304	6.284	7.764 − 10
7.068	12.610 − 10	9.104 − 10
	or	23.196 − 30
	2.610	or
		3.196 − 10

APPENDIX R continued

Subtraction of logarithms:

$$
\begin{array}{ll}
4.107 & = 14.107 - 10 \\
6.986 & \underline{6.986} \\
 & 7.121 - 10
\end{array}
$$

$$
\begin{array}{r}
11.672 - 10 \\
\underline{5.785 - 10} \\
5.887
\end{array}
$$

The relationships of logarithmic operations are expressed by the following formulas:

$$\log(a \times b) = \log a + \log b$$

$$\log\left(\frac{a}{b}\right) = \log a - \log b$$

$$\log (a)^b = b \log a$$

$$\log^b \sqrt{a} = \log a/b$$

Examples

To multiply 1.24 and 246:

$$
\begin{array}{ll}
\log 1.24 & = 0.0934 \\
\log 246 & = \underline{2.3909} \\
\text{total} & \ 2.4843
\end{array}
$$

Antilog 2.4843 = 305, which is as accurate as can be determined with a four-place table. The full answer to this problem is 305.04. To divide 961 by 224:

$$
\begin{array}{ll}
\log 961 & = 2.9827 \\
\log 224 & = \underline{2.3502} \\
\text{difference} & = 0.6325
\end{array}
$$

Antilog 0.6325 = 4.29, which is as accurate as can be determined with a four-place table. The product of 224 and 4.29 is 960.96.

Powers:

Determine 12^2

$$
\begin{array}{lr}
\log 12 = & 1.0792 \\
 & \underline{\times \quad 2} \\
 & 2.1584
\end{array}
$$

APPENDIX R continued

Antilog 2.1584 = 144.

Roots:

Find $\sqrt[3]{343}$

log 343 = 2.5353 ÷ 3 = .8451

antilog 0.8451 = 7.

Logarithms of Negative Numbers:

Because the logarithms of negative numbers are imaginary in character, they cannot be used in computation. However, since the numerical results of multiplying, dividing, *et cetera* are not affected by the signs, you can determine the numerical results by logarithms and later affix the final + or − signs by inspection.

Trigonometry

In any right triangle, if we let

θ = the acute angle formed by the hypotenuse and the base leg
ϕ = the acute angle formed by the hypotenuse and the altitude leg
H = the hypotenuse
A = the side adjacent θ and opposite ϕ
O = the side opposite θ and adjacent ϕ

then

$$\text{sine of } \theta = \sin\theta = \frac{O}{H}$$

$$\text{consine of } \theta = \cos\theta = \frac{A}{H}$$

$$\text{tangent of } \theta = \tan\theta = \frac{O}{A}$$

$$\text{cosecant of } \theta = \csc\theta = \frac{H}{O}$$

$$\text{secant of } \theta = \sec\theta = \frac{H}{A}$$

$$\text{contangent of } \theta = \cot\theta = \frac{A}{O}$$

APPENDIX R continued

also

$$\sin\theta = \cos\phi \quad \csc\theta = \sec\phi$$
$$\cos\theta = \sin\phi \quad \sec\theta = \csc\phi$$
$$\tan\theta = \cot\phi \quad \cot\theta = \tan\phi$$

and

$$\frac{1}{\sin\theta} = \csc\theta \qquad \frac{1}{\csc\theta} = \sin\theta$$
$$\frac{1}{\cos\theta} = \sec\theta \qquad \frac{2}{\sec\theta} = \cos\theta$$
$$\frac{1}{\tan\theta} = \cot\theta \qquad \frac{2}{\cot\theta} = \tan\theta$$

The expression "arc sin" indicates "the angle whose sine is . . ."; likewise, arc tan indicates, "the angle whose tangent is . . .," and so on.

APPENDIX S DECIMAL-TO-METRIC CONVERSION

To Convert	*To Obtain*	*Multiply by*
Feet	Centimeters	3.048×10^{1}
Feet	Meters	3.048×10^{-1}
Inches	Centimeters	2.54
Inches	Meters	2.54×10^{-2}
Yards	Meters	9.144×10^{-1}
Miles (Statute)	Kilometers	1.609
Pounds	Kilograms	4.536×10^{-1}
Pounds	Grams	4.535×10^{2}
Centimeters	Inches	3.937×10^{-2}
Centimeters	Feet	3.281×10^{-1}
Centimeters	Yards	1.094×10^{-2}
Centimeters	Millimeters	10
Meters	Inches	39.37
Meters	Feet	3.281
Meters	Yards	1.094
Meters	Centimeters	100
Kilometers	Miles (Statute)	6.214×10^{-1}
Kilograms	Pounds	2.2046

Bibilography

Adams, Stephan F., *Microwave Theory and Applications,* Prentice-Hall, New York, 1969.

Adams, Stephan F., *Microwave Theory and Applications,* Prentice-Hall, New York, 1969.

Adams, Stephen F., *Microwave Theory and Application,* Prentice-Hall, New Jersey, 1969.

AEI Semiconductor, *Oscillator and Amplifier Devices,* (Issue 1, 4450-50/GUNN), June 1970.

Air Tech Industries, *Tunnel Diodes* (Pub. 137-10M-4/72 NP), April 1972.

Allied Radio Corporation, *Allied Electronics Data Handbook,* Chicago, 1959.

American Radio Relay League, *The ARRL Antenna Book, Eighth Edition,* West Hartford, Connecticut, 1956.

A Practical Guide to Microwave Semiconductors, Microwave Associates, Burlington MA, Reprinted from MSN, Palo Alto, CA.

Avantek, Inc., *High Frequency Transistor Primer* (ATP 1018), Santa Clara, California, July 1971.

Avantek, Inc., *High Frequency Transistor Primer* (ATP 1019), Santa Clara, California, July 1971.

Bechtel, N. G., W. W. Hooper, and D. Mock, *X-Band GaAs FET,* Fairchild Microwave and Optoelectronics, Palo Alto, California, 1972.

Bechtel, N. G., W. W. Hooper, and Phillip C. Howe, "Design and Performance of the GaAs FET," *IEEE Journal of Solid State Circuits, Vol. SC-5, No. 6,* December 1970.

Bechtel, N. G., *Recent Advances in Low-Noise Microwave Transistor Development,* Fairchild Microwave and Optoelectronics, Palo Alto, California.

Bendix Corporation, *RF Connector* (Catalog 1009).

Bering, John J., "HEMT Technology Gains on MM Waves," *Microwaves and RF,* TRW, Inc., Redondo Beach, CA, November, 1985.

Blake, Lamont V., *Antennas,* John Wiley and Sons, New York, 1966.

Borase, Vijay, "Electrical Properties Govern Substrate Effectiveness," *Microwaves and RF,* February, 1983.

Brown, Robert G., Robert A. Sharpe, and William L. Hughes, *Lines, Waves and Antennas,* The Ronald Press Company, New York, 1961.

Carlson, Alan W., and Fred J. Dominick, *The Physics and Applications of Microwave Diodes,* Varian Associates, Beverly, Massachusetts, 1970.

Chen, Daniel and Sui, D. P., "Advanced Solid-State Component Design Looks to Gallium Arsenide," *Microwave Systems Designer's Handbook,* 1984.

Cohn, S., "Problems in Strip Transmission Lines," MTT-3, No. 2, March, 1955.

Communication Transistor Corporation, *Introduction to Transistors,* San Carlos, California.

Cook, Harry F., *Introduction to Microwave FET's* Varian Electron Device Group, Santa Clara, CA, 1982.

Department of the Navy, *Radar Electronic Fundamentals* (NAVSHIPS 900, 016), Bureau of Ships, Washington, DC 1944.

Derewonko, Henri and Delagebeudeuf, Danial, "Thomson Continues Work on 18 to 40 GHz TEGFET's," *Microwave and RF*, Thomson-CSF, November, 1985.

Donegan, Tom, "PTFE Substrates Require Special Care in Fabrication," *Microwaves,* October, 1982.

Ferry, David K., Editor-in-Chief, *Gallium Arsenide Technology,* Howard Sams, Indianapolis, IN, 1985.

Fitchen, Franklin C., *Transistor Circuit Analysis and Design,* D. Van Nostrand Co., Princeton, NJ, 1966.

Fogell, Dr. M., Director, *Modern Microelectronic Circuit Design, I C Applications, Fabrication Technology,* Vol I and II, Research and Education Association, 1981.

Frey, Jeffrey, *Microwave Integrated Circuits,* Artech House, Norwood, MA., 1975

Graf, Rudolf F., *Modern Dictionary of Electronics,* Fifth Edition, Howard Sams, Indianapolis, IN, 1982.

Graf, Rudolf F., *Modern Dictionary of Electronics,* Howard W. Sams, Indianapolis, Indiana, 1972.

Green, Xan B. K., "The Most Popular Connectors Reviewed," *Microwave Journal,* January 1975.

Hamilton, Robert J., and Osbrink, Northe K., *A GaAs FET Primer: Understanding These Vital Devices,* Avantek, Inc. MSN, October, 1982.

Hardeman, Lyman J., "Getting More Out of a Coaxial Connector," *Microwaves,* January 1971.

Hewlett-Packard Company, *Cable Testing with Time Domain Reflectometry* (AN67), Palo Alto, California, 1965.

Hewlett-Packard Company, *Microwave Network (Scalar) Measurement Seminar,* Palo Alto, California, 1969.

Hewlett-Packard Company, *Microwave Power Generation and Amplification Using IMPATT Diodes* (AN935), Palo Alto, California, 1971.

Hewlett-Packard Company, *Microwave Power Measurements* (AN64), Palo Alto, California, 1965.

Hewlett-Packard Company, *Network Analyzer Technical Data,* Palo Alto, California, March 1970.

Hewlett-Packard Company, *Noise Figure Primer* (AN57), Palo Alto, California, 1965.

Hewlett-Packard Company, *Swept-Frequency Technique* (AN65), Palo Alto, California, 1965.

Hewlett-Packard Company, *S-Parameters—Circuit Analysis and Design* (AN95), Palo Alto, California, September 1968.

Hewlett-Packard Company, *Technical Data on HP8755,* Palo Alto, California, August 1972.

Hewlett-Packard Company, *The Hot Carrier Diode: Theory, Design and Application* (AN907), Palo Alto, California, 1967.

Hewlett-Packard Company, *The 423 A Crystal Detector (HP Operating Note),* Palo Alto, California, 1964.

Hewlett-Packard Co., HP8510B Network Analyzer, Technical Data, May 1987.

Hewlett-Packard Co., HP8757A Scalar Network Analyzer, Technical Data, April 1986.

Howe, Harlan, *Stripline Circuit Design,* Artech House, Dedham, Massachusetts, 1974.

Hughes Aircraft Company (Electron Dynamics Division), *TWT's,* Torrance, California.

"Microwave Stripline," *Insulation Magazine* (3M Company), October 1968.

ITT, *Reference Data for Radio Engineers, Fourth Edition,* New York, 1976.

Johnson, Joe, *Circuit Techniques Handbook,* Communications Transistor Corporation, San Carlos, California, March 1973.

Johnson, V. A., *Men of Physics Karl Lark-Horovitz,* Pergammon Press, Oxford, 1969.

Kantor, Robert H., and Peter Pipe, *Personalized Instructions on Klystron Principles,* Varian Associates, 1963.

Kantor, Robert H., and Robert F. Mager, *Personalized Instruction on Klystrons,* Varian Associates (Central Research Laboratory), 1963.

Kennedy, George, *Electronic Communications Systems,* 2nd Edition, McGraw-Hill, New York, 1977.

Lance, A. L., *Introduction to Microwave Theory and Measurements,* McGraw-Hill, New York, 1964.

Laverghetta, Thomas S., *Microwave Materials and Fabrication Techniques,* Artech House, Norwood, MA, 1984.

Laverghetta, Thomas S., *Practical Microwaves,* Howard W. Sams, Indianapolis, IN, 1984.

Laverghetta, Thomas S., *Operating Procedures for the 8410A Network Analyzer,* MR 362 Syracuse University Research Corporation, Syracuse, New York, 1968.

Litton Industries (Electron Tube Division), *Traveling Wave Tubes, Theory and Application,* San Carlos, California, 1971.

Matthaei, George L., Leo Young, and E. M. T. Jones, *Microwave Filters, Impedance-Matching Networks, and Coupling Structures,* McGraw-Hill, New York, 1964.

Microstate Products (Special Microwave Device Operations), *Microstate Microwave Tunnel Diodes* (Bulletin TD-100).

Microwave Semiconductors, Alpha Industries, Woburn, MA.

Miller, E., *Applications of PIN Diodes* (AN922), Hewlett-Packard Company, Palo Alto, California, 1969.

Myers, F. A., "Gunn-Effort Technology," *MSN Magazine* (Plessey Company), April/May 1975.

Nowicki, Thomas E., "Microwave Substrates Present and Future," *Microwave Tech Topics,* Vol. 2, 3M Company, St. Paul, MN.

Olyphant, Murray and Nowicki, Thomas E., "Microwave Substrates Support MIC Technology," Vol. I and II, *Microwaves,* November and December, 1980.

Olyphant, Murray, Demeny, D. D., and Nowicki, Thomas E., *Epsilam 10—A New High Dielectric Constant Conformable Copper-Clad Laminate,* Cu-Tips #6, 3M Company, St. Paul, MN.

Omni-Spectra, Inc., *OSM Connectors and Adapters,* 1974.

Pacific Measurements, Inc., *Swept-Frequency Measurement System with Automemory®* (Technical Data Model 1038), Pacific Measurements, Inc., Palo Alto, California, 1974.

Pacific Measurements, Inc., *Swept-Frequency Return Loss Measurements Using the Model 1038 Measurement System with Automemory®* (AN11), Palo Alto, California, 1974.

Pearce, C. A., *Silicon Chemistry and Applications,* The Chemical Society, London, 1972.

Philco-Ford, *Performance Characteristics of Philco-Ford Silicon Microwave Schottky-Barrier Diodes* (Microelectronic Application Report #300), Spring City, Pennsylvania, July 1968.

PIN Diode Designers' Guide, M/ACOM Silicon Products, Catalog 4013, Burlington, MA, 1983.

Power Hybrid, Inc., *Technical Discussion and Approach,* Torrance, California.

Precision Tube Company, Inc., *Precision Coaxitube* (Catalog 3725M).

Raytheon Company, *Design Considerations for Microwave Tunnel Diode Amplifiers* (Application Note), Waltham, Massachusetts, June 1973.

Raytheon Company, *Theory of Operation of Avalanche Diodes* (Application Note), Waltham, Massachusetts, June 1973.

Riddle, Robert L., and Ristenbatt, Marlin P., *Transistor Physics and Circuits,* Prentice-Hall, Englewood Cliffs, NJ, 1958.

Receiving Diode Handbook, Microwave Associates, Bulletin 4006, Burlington, MA, 1980.

Rose, Robert M., Shephard, Lawrence A., and Wulff, John. "The Structure and Properties of Materials," Volume IV, *Electronic Properties,* John Wiley and Son, New York, 1966.

Saad, Theodore S., ed., *Microwave Engineers' Handbook,* 2 vols., Artech House, Dedham, Massachusetts, 1972.

Saad, Theodore S., *The Microwave Mixer—A Technical Discussion,* Sage Laboratories, Natick, Massachusetts, 1966.

Sage Laboratories, *The Microwave Directional Coupler—An Application Guide,* Natick, Massachusetts, 1968.

Smith, Phillip H., *Electronic Applications of the Smith Chart,* McGraw-Hill, New York, 1969.

SRD Laboratories, *Introduction to Coaxial Magnatrons,* Union, New Jersey.

SRD Laboratories, *Introduction to Pulsed Crossed-Field Amplifiers,* Union, New Jersey.

Startronics, Inc., *Coaxial Transmission Lines* (Design Note #1), Georgetown, Massachusetts.

Swanson, Alan, Herb, John, and Yung, Ming, "First Commercial HEMP Challenges GaAs FET's," *Microwave and RF,* Gould Research Center, San Jose, CA, November, 1985.

Sylvan, T. P., and Eric Gottlieb, *Tunnel Diodes as Amplifiers and Switches,* General Electric Company, Syracuse, New York, 1969.

Tenenholtz, Robert, *The Video Detector—A Technical Discussion,* Sage Laboratories, Natick, Massachusetts, 1968.

Thomas, Harry E., *Handbook for Electronic Engineers and Technicians,* Prentice-Hall, Englewood Cliffs, New Jersey, 1965.

Times Wire and Cable Company, *RF Transmission Line Catalog and Handbook* (Catalog #5L-3), 1970.

Transistor Designer's Guide, Microwave Associates, Bulletin #5210, 1978.

Trans Tech, Inc., *A Discussion of Gyromagnetic Resonances* (Tech Brief #654), Gaithersburg, Maryland.

Trans Tech, Inc., *A Discussion of the Spinning Electron* (Tech Brief #653), Gaithersburg, Maryland, 1967.

Trans Tech, Inc., *A Review of Physical Properties of Ferrites* (Tech Brief #651), Gaithersburg, Maryland, 1967.

Trans Tech, Inc., *Use of Ferrimagnetic Material in Isolators* (Tech Brief #6510), Gaithersburg, Maryland.

Trans Tech, Inc., *Use of Ferrimagnetic Material in Circulators* (Tech Brief #6511), Gaithersburg, Maryland.

Varian Associates, *The RF Characteristics of Coaxial Magnetrons (The Clean-Tube Story),* Beverly, Massachusetts.

Watson, H. A., *Microwave Semiconductor Devices and Their Circuit Applications,* McGraw-Hill, New York, 1969.

Weinschel Engineering, *RF Attenuators and Components,* 1971.

White, Joseph F., *Microwave Semiconductor Engineering,* Van Nostrand Reinhold Co., New York, 1982.

Wiltron Company, *Microwave Measurement Seminar,* Palo Alto, California, 1974.

Wiltron Co., Model 360 Vector Network Analyzer, P/N 11410-00008, Morgan Hills, CA, April 1987.

Wiltron Co., Scalar Network Analyzer Model 561, P/N 11410-00009, Morgan Hills, CA, May 1987.

Young, Victor J., *Understanding Microwaves,* John F. Rider Company, New York, 1965.

3M Company, *Recommended Drilling and Through-Hole Plating Techniques for CuClad® K6098 and 6098 PTFE Glass Laminates,* Minneapolis, Minnesota, May 1971.

Catalogs

Addington Laboratories, *Filters/Multiplexers/Integrated Devices,* 1973.

American Electronic Laboratory, *Antennas and Antenna Systems,* 1973.

Anaren Microwave, *Stripline Wideband Microwave Signal-Processing Components and Subsystems,* 1973.

Atlantic Laminate, 1972.

Bendix Corporation, *RF Connectors* (Catalog 1069)

Duroid, 1970.

E&M Laboratories, *Microwave Devices/Ferrite Components* (Rev. 5172).

Englemann Microwave, *Microwave Components,* 1973.

Hewlett-Packard Company, *Coaxial and Waveguide Measurement Accessories* (#5952-1120D), 1974.

Hewlett-Packard Company, *Master Catalog,* 1975.

Kings, *Catalogs 41-715, 39-7411, 33-691, and 36-714.*

Merrimac Industries, *Catalog M126,* 1973.

Mini-Circuits Laboratories, *Catalog 74-1,* 1974.

Narda Microwave, *Catalog #19,* 1974.

Precision Tube Company, *Precision Coaxitube* (Catalog 3725A).

RYT Industries, *Microwave Ferrite Components,* 1972.

Sage Laboratories, *Catalog of Microwave Components.*

Sanders Microwave, *Product Bulletin M29.*

Texscan Corporation, TEXSCAN *Filter Catalog.*

Times Wire and Cable Company, *RF Transmission Line Catalog and Handbook* (Catalog #TL-3), 1970.

TRG Division Alpha Industries, *Millimeter Microwave Products and Capabilities,* 1974.
UTE Microwave, *Circulators, Isolators, and Filters* (Catalog B/1974), 1974.
Watkins-Johnson, *Microwave Devices* (Catalog #2), 1971.
Weinschel Engineering, *RF Attenuators and Components,* 1971.
Weinschel Engineering, *RF Attenuators and Components,* 1971.
Weinschel Engineering, *Microwave Measuring and Generation Instruments,* 1974.

Manuals

Applications of Microwave GaAs FET's, Application Manual AN82901-1, California Eastern Laboratories, Santa Clara, CA.
G E Transistor Manual, Seventh Edition, The General Electric Co., Syracuse, NY, 1964.
Hewlett-Packard Company, *HP342A Noise Figure Meter,* Palo Alto, California.
RCA Transistor Manual, RCA Semiconductor and Materials Division, Somerville, NJ, 1962.

Index